Signal Transduction and the Gasotransmitters

Signal Transduction and the Gasotransmitters

NO, CO, and H_2S in Biology and Medicine

Edited by

Rui Wang, MD, PhD, FAHA
Department of Physiology,
University of Saskatchewan,
Saskatoon, SK, Canada

Foreword by

Bruce McManus, MD, PhD, FRSC
Canadian Institutes of Health Research,
Vancouver, BC, Canada

Humana Press ✻ Totowa, New Jersey

999 Riverview Drive, Suite 208
Totowa, New Jersey 07512

humanapress.com

This publication is printed on acid-free paper. ∞
ANSI Z39.48-1984 (American National Standards Institute) Permanence of Paper for Printed Library Materials.

Production Editor: Angela Burkey

Cover design by Patricia F. Cleary.

For additional copies, pricing for bulk purchases, and/or information about other Humana titles, contact Humana at the above address or at any of the following numbers: Tel.: 973-256-1699; Fax: 973-256-8341; E-mail: humana@humanapr.com or visit our website at www.humanapress.com

Printed in the United States of America. 10 9 8 7 6 5 4 3 2 1
e-ISBN: 1-59259-806-4

Library of Congress Cataloging-in-Publication Data
Signal transduction and the gasotransmitters : NO, CO, and H2S in biology and medicine / edited by Rui Wang.
p. ; cm.
Includes bibliographical references and index.
ISBN 1-58829-349-1 (alk. paper)
1. Neurotransmitters. 2. Nitric oxide--Physiological effect. 3. Hydrogen sulphide--Physiological effect. 4. Carbon monoxide--Physiological effect. 5. Cellular signal transduction.
[DNLM: 1. Neurotransmitters--physiology. 2. Carbon Monoxide--metabolism. 3. Hydrogen Sulfide--metabolism. 4. Nitric Oxide--metabolism. 5. Signal Transduction--physiology. QV 126 S578 2004] I. Wang, Rui, M.D.
QP364.7.S55 2004
612.8'042--dc22

2003027536

Dedication

To Lily, Jennifer, Jessica, and Clover:

You are my source of inspiration.

***Rui Wang**, MD, PhD, FAHA*

FOREWORD

The current era of biological investigation is among the most transformational in history. The genuine emergence of genomic strategies and the rapid follow-on of proteomic and information technologies have provided scientists with unprecedented opportunities for discovery. The frontiers of knowledge are simply falling back. With this amazing revolution in understanding of the molecular underpinnings of cellular, tissue, and organismic homeostasis, a greater appreciation for the complexity of signals, networks, and linkages has crystallized. Regulation of biological functions rests not only with transcriptional, translational, and post-translational modifications of proteins, but also in the orchestral harmony of ligands and receptors, cell adhesive systems, cytoskeletal organization, ion channel function, membrane dynamics, and a range of transmitters.

Typically, transmitters have been categorized as those participating in neural functions or as humoral amines. In *Signal Transduction and the Gasotransmitters: NO, CO, and H_2S in Biology and Medicine*, Dr. Rui Wang and the sterling group of contributors he has assembled provide a paradigm-shifting assessment of the new category of transmitters, the gasotransmitters. Although nitric oxide was discovered approximately 20 years ago, it has only recently been appreciated that this famous molecule is among a whole group of substances that play critical roles in cell signaling and regulation, arising either environmentally or endogenously. These diverse molecules include, but are not limited to, nitric oxide, carbon monoxide, and hydrogen sulfide. Considering the now-identified roles of these three gasotransmitters in physiology and toxicology, it is understandable that the contributions are accordingly organized in sections corresponding to each. The origin, quantities, and interactions among these transmitters determine their impact on ionic fluxes, the excitability of muscle and nerve, and metabolism. There is an interesting and perhaps not surprising range of availability of any given gasotransmitter that conveys either physiological benefit or toxicological adversity, even when the gases arise endogenously. Individual chapters clearly frame the spectrum of their disease-related and physiological roles.

Like all nascent fields of study, it is often difficult to predict the full magnitude of importance of certain discoveries. Although the discovery of the role of nitric oxide in biological function deserved the Nobel Prize, and nitric oxide is now known to be a pivotal molecule in many organ systems, it is tempting to speculate that knowledge of the role of endogenous gases in a broader scale, especially as it relates to the homeostatic balancing act or that of other species, is barely coming into its own. Dr. Wang is to be congratulated on bringing the subject of gasotransmitters into coherence. I believe that biologists from many fields will welcome the knowledge that is captured here.

Bruce McManus, ***MD, PhD, FRSC***
Director of Institute of Circulatory and Respiratory Health
Canadian Institutes of Health Research
Vancouver, British Columbia, Canada

Preface

The endogenous production and physiological function of many gaseous molecules including, but not limited to, nitric oxide (NO), carbon monoxide (CO), and hydrogen sulfide (H_2S), have been increasingly recognized in recent years. These gaseous molecules, defined as gasotransmitters, share common chemical features and biological action modes, but distinguish themselves from classical neurotransmitters and humoral factors. The concept of gasotransmitters has found its application across a wide spectrum of biological systems. Recent advances in the novel and challenging field of gasotransmitter biology and medicine—encompassing biomedical and clinical issues, health services, and population health studies—are dazzling. Gasotransmitters are important endogenous signaling molecules. Among many cellular and molecular targets of gasotransmitters, membrane ion channels are the key signal transduction link regulated by gasotransmitters. The regulation of ion channels by gasotransmitters can result from the activation of different second messengers or the direct interactions between gasotransmitters and ion channel proteins. The latter is a novel mechanism and has attracted great attention from researchers in every field of biomedical studies.

Many books have been published that focus on neurotransmitters and other classical signal transduction pathways. *Signal Transduction and the Gasotransmitters: NO, CO, and H_2S in Biology and Medicine* reviews the biology and medicine of gasotransmitters with an emphasis on signaling transduction mechanisms in general, and ion channel regulation in particular. Following an account of the historical evolution of the gasotransmitter concept, the endogenous metabolisms of gasotransmitters and their regulation, the comparison of the toxicological profiles and biological actions, and interactions among gasotransmitters in terms of their production and effects are discussed. The physiological roles of NO, CO, and H_2S in the regulation of cardiovascular, neuronal, and gastrointestinal systems, as well as of cell metabolism are reviewed. The interaction of gasotransmitters with K_{Ca} channels, KATP channels, voltage-gated Ca^{2+} channels, voltage-gated Na^+ channels, and cyclic nucleotide-gated ion channels are presented. Included in the array of different mechanisms for the interaction of NO, CO, and H_2S are channel phosphorylation, *S*-nitrosylation, carboxylation, sulfuration, and altered cellular redox status. Guidance and suggestions can be found for exploring and characterizing lesser known gasotransmitters.

Signal Transduction and the Gasotransmitters: NO, CO, and H_2S in Biology and Medicine should serve as a summary and a standard reference source concerning signal transduction mechanisms underlying the physiological functions of gasotransmitters. Clinical scientists and physicians as well as other professional health workers should be excited by the advances in gasotransmitter research described in this book. The authors hope that scientists from both basic biology and health science disciplines find this book useful, interesting, and inspiring.

Rui Wang, MD, PhD, FAHA

Contents

Contributors

GIAN BATTISTA AZZENA, MD • *Institute of Human Physiology, Medical School, Catholic University S. Cuore, Rome, Italy*
PIERRE BONNET, MD • *Laboratoire de Physiopathologie de la Paroi Artérielle (LABPART), Institut Fédératif de Recherche n°120, Faculté de Médecine, Tours, France*
MARIE-CHRISTINE BROILLET, PhD • *Department of Pharmacology and Toxicology, University of Lausanne, Lausanne, Switzerland*
RAY J. CARSON, PhD • *Physiology Section, School of Science and the Environment, Coventry University, Coventry, United Kingdom*
YOUQIN CHENG, MD • *Department of Aged Cardiovascular Internal Medicine, General Hospital of BeijingCommand of PLA, Beijing, People's Republic of China*
NIPAVAN CHIAMVIMONVAT, MD • *Division of Cardiovascular Medicine, Department of Internal Medicine, University of California, Davis, CA*
AUGUSTINE M. K. CHOI, MD • *Division of Pulmonary, Allergy, and Critical Care Medicine, Department of Medicine, University of Pittsburgh Medical Center, Pittsburgh, PA*
MARCELLO D'ASCENZO, PhD • *Institute of Human Physiology, Medical School, Catholic University S. Cuore, Rome, Italy*
HONG DING, PhD • *Microvascular Biology Group, School of Medical Sciences, RMIT University, Bundoora, Victoria, Australia*
ALLAN DOCTOR, MD • *Department of Pediatrics, Pediatric Critical Care, University of Virginia School of Medicine, Charlottesville, VA*
ERIC DUBUIS, PhD • *Department of Physiology, University of Liverpool, Liverpool, United Kingdom*
ANTHIE ELLIS, PhD • *Department of Pharmacology & Therapeutics, Smooth Muscle Research Group, Faculty of Medicine, University of Calgary, Calgary, Alberta, Canada*
BENJAMIN M. GASTON, MD • *Department of Pediatrics, Pediatric Respiratory Medicine, University of Virginia School of Medicine, Charlottesville, VA*
KATHRYN A. GLATTER, MD • *Division of Cardiovascular Medicine, Department of Internal Medicine, University of California, Davis, CA*
CLAUDIO GRASSI, MD, PhD • *Institute of Human Physiology, Medical School, Catholic University S. Cuore, Rome, Italy*
DAVID D. GUTTERMAN, MD • *Department of Medicine, Cardiovascular Center, Medical College of Wisconsin, Milwaukee, WI*
MISATO KASHIBA, PhD • *Department of Biochemistry and Integrative Medical Biology, School of Medicine, Keio University, Shinjuku-ku, Tokyo, Japan*
JIYOONG KIM, MD • *Cardiovascular Division of Medicine, National Cardiovascular Center, Suita, Osaka, Japan*
HIDEO KIMURA, PhD • *Department of Molecular Genetics, National Institute of Neuroscience, Kodaira, Tokyo, Japan*
MASAFUMI KITAKAZE, MD, PhD • *Cardiovascular Division of Medicine, National Cardiovascular Center, Suita, Osaka, Japan*
JAN P. KRAUS, PhD • *Department of Pediatrics, C-233, University of Colorado School of Medicine, Denver, CO*

PREM KUMAR, DPhil • *Department of Physiology, The Medical School, University of Birmingham, Birmingham, United Kingdom*

PATTY J. LEE, MD • *Section of Pulmonary and Critical Care Medicine, Yale University School of Medicine, New Haven, CT*

YANPING LIU, MD, PhD • *Department of Medicine, Cardiovascular Center, Medical College of Wisconsin, Milwaukee, WI*

KENNETH N. MACLEAN, PhD • *Department of Pediatrics, C-233, University of Colorado School of Medicine, Denver, CO*

HELEN MADDOCK, PhD • *Physiology Section, School of Science and the Environment, Coventry University, Coventry, United Kingdom*

RUBINA MIAN, PhD • *Physiology Section, School of Science and the Environment, Coventry University, Coventry, United Kingdom*

PHILIP K. MOORE, PhD • *Department of Pharmacology, Cardiovascular Research Group, National University of Singapore, Singapore*

ELLA S. M. NG, MSc • *Department of Pharmacology & Therapeutics, Smooth Muscle Research Group, Faculty of Medicine, University of Calgary, Calgary, Alberta, Canada*

LEO E. OTTERBEIN, MD • *Division of Pulmonary, Allergy, and Critical Care Medicine, Department of Medicine, University of Pittsburgh Medical Center, Pittsburgh, PA*

SHELDON H. ROTH, PhD • *Division of Toxicology, Departments of Pharmacology & Therapeutics and Anesthesia, Faculty of Medicine, The University of Calgary, Calgary, Alberta, Canada*

STEFAN W. RYTER, PhD • *Division of Pulmonary, Allergy, and Critical Care Medicine, Department of Medicine, University of Pittsburgh Medical Center, Pittsburgh, PA*

SHOJI SANADA, MD, PhD • *Department of Internal Medicine and Therapeutics, Osaka University Graduate School of Medicine, Suita, Japan*

GUNTER SEYFFARTH, PhD • *Division of Biomedical Sciences, School of Applied Science, University of Wolverhampton, Wolverhampton, UK*

CHRIS R. TRIGGLE, PhD • *Department of Pharmacology and Therapeutics, Smooth Muscle Research Group, University of Calgary, Calgary, Alberta, Canada, and Microvascular Biology Group, School of Medical Sciences, RMIT University, Bundoora, Victoria, Australia*

CHRISTOPHE VANDIER, PhD • *Laboratoire de Physiopathologie de la Paroi Artérielle (LABPART), Institut Fédératif de Recherche n°120, Faculté de Médecine, Tours, France*

RUI WANG, MD, PhD, FAHA • *Department of Physiology, University of Saskatchewan, Saskatoon, SK, Canada*

LINGYUN WU, MD, PhD • *Department of Pharmacology, College of Medicine, University of Saskatchewan, Saskatoon, SK, Canada*

ZHAO ZHANG, MD, PhD • *Division of Cardiovascular Medicine, Department of Internal Medicine, University of California, Davis, CA*

I

Gasotransmitters: Past, Present, and Future

1 The Evolution of Gasotransmitter Biology and Medicine

From Atmospheric Toxic Gases to Endogenous Gaseous Signaling Molecules

Rui Wang

CONTENTS

SUMMARY

Overproduction of many atmospheric gases, from natural resources and anthropogenic activities, impose a serious environmental concern with adverse health effects. Among pollutant gases are nitric oxide (NO), carbon monoxide (CO), and hydrogen sulfide (H_2S). Over several decades, studies from numerous laboratories have demonstrated that gases such as NO, CO, and H_2S not only are generated in the human body but also play important physiological roles. These particular gases share many common features in their production and function but carry on their tasks in unique ways, which differ from classic signaling molecules, in the human body Collectively, these endogenous molecules of gases or gaseous signaling molecules compose a family of "gasotransmitters." The regulation of ion channels by gasotransmitters, either directly via chemical modification of ion channel proteins or indirectly via second messengers, exerts significant influence on cellular functions. *S*-nitrosylation, carboxylation, and sulfuration may represent mechanisms of direct interaction of NO, CO, and H_2S with ion channel proteins, respectively.

From: *Signal Transduction and the Gasotransmitters: NO, CO, and H_2S in Biology and Medicine* Edited by: Rui Wang © Humana Press Inc., Totowa, NJ

This chapter summarizes the history and evolution of the concept of the gasotransmitter and outlines the criteria used to identify novel gasotransmitters. Gasotransmitter research is accelerating into the next phase. Many new gasotransmitter candidates are being investigated. Alterations in the metabolism and functions of gasotransmitters under different pathological conditions are being explored, which may shed light on the pathogenesis and management of many diseases. Thus, research on gasotransmitters is certainly as important to clinical practice and community health as it is to basic research, if not more so.

Key Words: Gasotransmitter; nitric oxide; carbon monoxide; hydrogen sulfide; signal transduction.

"Air is a physical substance; it embraces us so intimately that it is hard to say where we leave off and air begins. Inside as well as outside we are minutely designed for the central activity of our existence—drawing the atmosphere into the centre of our being, deep into the moist, delicate membranous labyrinth within our chests, and putting it to use."—David Suzuki, The Sacred Balance

1. INTRODUCTION

Humans tend to treat atmospheric gases, such as oxygen, carbon dioxide (CO_2), nitrogen, carbon monoxide (CO), and hydrogen sulfide (H_2S), like sunshine and water—nature's gifts to us. Accompanying the arrival of the Industrial Revolution, the Third Wave is a high tide of natural gas production from industrial sources. In the public eye, most natural gases are nothing but toxicants, wastes, and pollutants, with oxygen as possibly the only exception. By definition, environmental toxicants are "agents released into the general environment that can produce adverse health effects among large numbers of people" *(1)*. Gas pollutants as environmental toxicants can induce both acute and chronic health hazards at societal as well as individual levels. The health hazards of these toxic gases become magnified in our public life. When this is coupled with public concern about the production of natural gases, it then becomes a health issue impacting both environmental and occupational health.

Scientists have worked with two schools of thought searching for the biological production and the physiological function of natural gases, be it detrimental or beneficial. One ancient frontier is the study of the biological production of gases. Archaea and microbes produce great amounts of gas, not only for their own use, but also for the necessity of life in their environment. Interestingly enough, these studies consistently demonstrate the production of numerous natural gases by microorganisms. For example, many bacterial types, such as Proteus vulgars, produce CO *(2)*. The biological production and utilization of H_2S have been best known for particular bacteria and archaea *(3)*. Human beings sit on top of the genomic life tree. Do we inherit or share any of these abilities from low forms of life to produce gases in our body? Plant life generates oxygen from light, a process of photosynthesis through the use of chlorophyll. Humans are not equipped in this way. However, our bodies do produce CO_2, ammonium, and other gases. The human body is often in this way treated as a pollutant when an analogy is drawn to the automobile or even a restaurant kitchen, which also generates useless gases, toxicants, or other types of harmful byproducts. The records of endogenous production of CO and H_2S in human tissues can be traced back hundreds of years. The human body can generate a myriad of gases with unknown functions—*the truth is still out there.* This body of knowledge, unfortunately, has not been completely used to facilitate the understanding of human physiology.

Scientists working in the second frontier—the physiological function of biological gases, a natural extension of the first frontier—brought about this revolution. In this regard, nitric oxide (NO), a pioneer gas, is doubtless the molecule of a new era. Over the last several decades, studies from thousands of worldwide laboratories have demonstrated that gases such as NO, CO, and H_2S are not only generated in humans but also have important physiological properties.

These gases share many common features in their production and function while carrying out their tasks in unique ways that differ from classic signaling molecules in the human body. Collectively, these endogenous molecules of gases, or gaseous signaling molecules, compose a family of "gasotransmitters," a nomenclature composed of "gas" and "transmitters."

This introductory chapter is devoted to discussing the conceptual transition of biological gases from toxic wastes and pollutants to important physiological gasotransmitters.

2. PRODUCTION AND HEALTH HAZARDS OF ATMOSPHERIC GASES

2.1. Nitric Oxide

Natural causes—lightening, forest fires, and organic decay—lead to the generation of oxides and nitrogen (NOx). Soil microorganisms also produce NOx. NO and N_2O are emitted from anaerobic soils by denitrifiers such as *Pseudomonas* spp. or *Alcaligenes* spp. and from aerobic soil by autotrophic nitrifiers such as *Nitrosomonas europaea (4)*. Motorized vehicles are the major mobile combustion source of NOx production. In 1994, one study showed that in a long, 7.5-km Norwegian road tunnel, with traffic flowing in both directions, the atmospheric NO_2 concentration exceeded the Norwegian air quality limits for road tunnels 17% of the time. When traffic was reduced through the tunnel, the mean NO_2 concentration was significantly lowered *(5)*. Stationary combustion sources of NOx include heat power plants and industrial factories *(6)*. Cigarette smoking generates a considerable amount of NO and NO_2 *(7)*. The biological treatment of nitrogen-rich wastewater with a high concentration of ammonium likewise yields NO and NO_2, although this might not contribute significantly to general environmental pollution with NOx *(8)*.

As the initial product of NOx from a reaction between nitrogen and atmospheric oxygen, NO quickly transforms to NO_2 either through simple oxidation involving molecular oxygen or through a photochemical reaction involving irradiation by sunlight. As a result, health hazards of atmospheric NO must be considered in conjunction with NO_2. Mercer et al. *(9)* found that after adult rats were exposed to 0.5–1 ppm of NO for 9 wk, the fenestration numbers in the alveolar septa of the lung increased more than 30-fold in the control rats, and 3-fold of NO_2 in the exposure group. The number of interstitial cells in the NO group was significantly reduced by 29%. Likewise, a significant reduction in the thickness of interstitial space was observed in the NO-treated rats, but not in the NO_2-treated rats, compared with the control rats *(9)*. Their study demonstrated that a low level of atmospheric NO exposure is more potent than NO_2 in producing interstitial lung damage. It is believed that most NO toxicity is mediated by the interaction of NO with superoxide producing peroxynitrite. This leads to oxidative damage to targeted cells and tissues. Epidemiological data often show controversial results on the adverse health effects of NO_2 *(6)*, partially because of the difficulty in determining the actual atmospheric NO_2 levels to which a specific portion of the population was exposed. Controlled animal and human studies provide evidence that high NO_2 levels weaken pulmonary defense mechanisms and change human airway responsiveness. Lipid peroxidation *(10)* and protein oxidation *(11)* have been described as part of the cellular mechanisms of NO_2-induced health hazards. The most important and consistent conclusion is that exposure to high NO_2 concentrations may exemplify a specific health risk for a subpopulation of people with respiratory diseases, such as asthma and chronic obstructive pulmonary disease.

Over a 1-yr period, Giroux et al. *(12)* examined the correlation of acute myocardial infarction, atmospheric levels of NOx, temperature, and relative humidity. Among 282 patients with acute myocardial infarction, it was determined that the infarction area was reduced when the daily NO level in the atmosphere was higher than 13 $\mu g/m^3$ and the average daily temperature was lower than 13°C.

NO and NO_2 act as phytotoxic agents, damaging plant health as well. The growth of plants becomes poorer and productivity lower when exposed to high NOx levels *(13)*.

2.2. Carbon Monoxide

The toxicology profile of CO has been portrayed for hundreds of years. CO is among the most abundant air pollutants in North America. Because it is colorless, odorless, and noncorrosive, intoxication by CO is hard to detect, which earns CO the reputation of the "silent killer." A report in 1982 by the US Centers for Disease Control revealed that approx 4000 deaths and 10,000 cases of individuals requiring medical attention occur annually because of acute CO intoxication *(14)*.

All types of incomplete combustion of carbon-containing fuels yield CO. Natural processes such as metabolism and production of CO by plants and oceans release CO into the atmosphere. Oxidation of methane and nonmethane hydrocarbons by hydroxyl radicals and ozone, either natural or anthropogenic, is also a significant mode of CO production in the atmosphere. The most notable ways that humans contribute to the production of CO are the operation of internal combustion engines; the fueling of appliances with gas, oil, wood, or coal; and the disposal of solid waste. Cigarette smoking also produces a substantial amount of CO.

Whether an elevated environment of CO levels leads to human intoxication is influenced by the exposure and duration of pulmonary ventilation function, as well as the endogenous buffering capacity (i.e., the level of carbonmonoxy-hemoglobin A [HbCO A]), and the partial pressures of CO and oxygen. Acute ambient CO poisoning occurs as suddenly elevated CO concentration accelerates the binding of CO to normal adult hemoglobin (Hb) (Hb A), forming HbCO A. The formation of HbCO A impairs two functions of Hb. The oxygen storage function of Hb A is significantly reduced because the affinity of CO to Hb A is approx 250 times greater than that of oxygen *(15)*. The affinity of myoglobin to CO is approx 25-fold that of oxygen. The oxygen transportation function of Hb A is also reduced, because the release of oxygen from HbCO A to the recipient tissue becomes more difficult. CO binds to one of the four oxygen-binding sites of Hb A via the formation of a hydrogen bond between CO and the distal histidine residues of Hb A *(16)*. This binding, in turn, increases the affinity of oxygen to HbCO A. With tissue hypoxia being the major toxicological consequence of CO poisoning, the combination of CO with other heme-proteins, such as cytochrome P450, cytochrome-*C* oxidase, catalase, and myoglobin, may also in part account for the toxic effects of CO *(2,17)*. Because of their high demand for oxygen, the brain and heart are the most vulnerable organs, to the CO-induced acute hypoxia. Neurological and myocardial injuries associated with acute CO intoxication can be fatal unless medical treatment is provided immediately. The normal background of the HbCO A level in a healthy nonsmoker is about 0.5–1% *(18)*. Early neurological symptoms such as headaches, dizziness, nausea, vomiting, disorientation, and visual confusion occur when the HbCO A level reaches 10–30%. Depending on the CO exposure level, duration, and treatments, the prognosis in patients with acute CO poisoning varies *(17)*.

Chronic environmental CO exposure may constitute one risk factor for cardiovascular diseases. A retrospective study of 5529 New York City bridge and tunnel officers unmasked the relationship between occupational exposure to CO and mortality from heart disease *(19)*. The CO exposure level of the tunnel officers was much higher than that of the bridge officers. There were 61 deaths from arteriosclerotic heart diseases in tunnel officers, which was higher than the expected 45 deaths based on the New York City population. Once the exposure was eliminated, the high risk of arteriosclerotic heart disease in the tunnel officers dissipated.

There has been a long-lasting debate on whether chronic CO inhalation as intrinsically linked to cigarette smoking acts either alone or with other environmental stressors to induce hypertension *(20,21)*. Increases, decreases, or no change in blood pressure after CO exposure has been reported. What should be remembered is that the adverse health effect of cigarette smoking is not a simple mirror image of CO inhalation. Immediately following cigarette smoking, an acute but transient increase in the smoker's blood pressure occurs, which has been largely ascribed to the nicotine in smoking. This hypertensive effect of nicotine is overcompensated by CO in the end. The blood pressure of these long-term smokers is decreased, or at the very least not increased, without other cardiovascular complications *(22)*. This notion was further supported in an animal study in which borderline hypertensive rats were exposed to chronic CO. This treatment actually led to hypotension, not hypertension, in these animals *(20,22)*. Chronic CO inhalation leads to many diseases, chiefly those linked to hemodynamic responses to CO and hypoxia-adaptive changes *(23)*. Cardiac hypertrophy exemplifies the cardiovascular complications of chronic CO exposure. Continuous exposure in adult male rats to 700 ppm of CO for 27 d *(24)* or 500 ppm CO for 30 d *(25)* induced volume-overload cardiac hypertrophy. Hypertrophy of both the left ventricle (22%) and right ventricle (37%) developed with hematocrit increased nearly 50%. Chronic CO exposure also alters normal development of the cardiovascular and other systems. In one experiment, 1-d-old rat pups were exposed to 500 ppm of CO for 30 d, and cardiac histology analysis was performed at 61 and 110 d of age *(26)*. One notable alteration was the significant increase in small arteries across all heart regions. The diameter of the large arteries in the entire heart region was also greater than that in the control rats. The architectural impact of coronary vessel changes following chronic neonatal CO exposure would be considerable on cardiovascular functions, especially those at different developmental stages and in adulthood.

2.3. Hydrogen Sulfide

The presence of H_2S in our environment is easily recognizable for its peculiar rotten-egg smell *(27,28)*. Atmospheric H_2S has both natural and anthropogenic sources. Volcanic gases, marshes, swamps, sulfur springs, and decaying matter such as from mushrooms all release H_2S into the environment. Emissions from oil and gas refineries, paper mills, and sewer networks also result in odor, health, and corrosion problems. Acute intoxication of H_2S can be lethal *(29)* and is one of the leading causes of sudden death in the workplace *(30)*. At least 5563 cases of intoxication and 29 deaths resulting from H_2S exposure occurred in the United States between 1983 and 1992 *(31)*. Loss of the central respiratory drive is one of the major mechanisms for acute H_2S death *(27,28,32,33)*. The interaction of H_2S with many enzymes and macromolecules, including Hb, myoglobin, and cytochrome oxidase, exerts a profound effect on the vitality of cells *(34–36)*. Disorders of the central nervous, cardiovascular, respiratory, and gastrointestinal systems have been reported with acute H_2S intoxication *(34,37)*.

The health hazard of chronic H_2S exposure has also been observed *(36)*. Bates et al. *(38–40)* carried out a series of studies in the city of Rotorua, New Zealand, which is located over an active geothermal field. Approximately one-quarter of the population had been exposed regularly to high concentrations of H_2S from 143 to 1000 ppb. During 1981–1990, a higher mortality risk for respiratory diseases and a higher morbidity risk for neuronal diseases (both peripheral and central nervous systems) were observed in the Rotorua population compared with the rest of the population of New Zealand *(38,39)*. Another improved survey based on 1993–1996 morbidity data linked adverse health outcomes of Rotorua to other regions within Rotorua with high, medium, or low H_2S exposure levels *(40)*. This recent study again demonstrated an H_2S exposure-response tendency for disorders of the nervous system and sense organs as well as circulatory and respiratory diseases. Furthermore, a retrospective epidemiological study examined 2853 married, adult, nonsmoking women in a petrochemical complex in Beijing, China *(41)*. During their first trimester of pregnancy, about 57% of the surveyed woman had been exposed to petrochemicals. The results showed a significantly increased risk of spontaneous abortion when exposed to H_2S (odds ratio [OR]: 2.3; 95% confidence interval [CI] 1.2–4.4), benzene (OR: 2.5; 95% CI: 1.7–3.7), and gasoline (OR: 1.8; 95% CI: 1.1–2.9).

The average odor threshold for H_2S is about 0.5 ppb *(42)*. A low level of H_2S exposure does not appear to have had any adverse long-term health effect *(42)*. According to the Agency for Toxic Substances and Disease Registry, the acute minimal risk level for H_2S currently is set at 70 ppb, i.e., 24-h daily exposure to 70 ppb of H_2S over a period of 14 d or less *(42)*. An investigation was conducted in a Pennsylvania elementary school that complained of H_2S odors putatively related to the nearby mushroom-composting operations *(43)*. During the spring of 1998, 1-h averages of atmospheric H_2S levels were found to be consistently below 10 ppb at a control school, but between 11 and 59 ppb for 7 d for the outside air, and 5 d for the inside air at the exposed school. During the autumn of 1998, 1-h averages of atmospheric H_2S levels were consistently below 10 ppb at the control school, but between 11 and 129 ppb for 9 d for the outside air, and 7 d for the inside air at the exposed school. The investigators stated: "No consistent association was found between exposure to low levels of hydrogen sulfide and any adverse health effects. It was concluded that the students attending the elementary school near the mushroom-composting operations were not exposed to any significant public health hazard" *(43)*.

More details about the chemical and physical properties and toxicology profile of H_2S are discussed in Chapter 17.

3. PRODUCTION AND PHYSIOLOGICAL EFFECTS OF ENDOGENOUS GASES

Decades of environmental and occupational health studies describe NO, CO, and H_2S as vicious toxicants that exert a detrimental influence only on human health. This conventional thinking has gradually lost ground. First is the evidence that NO is actually endogenously generated with profound biological and physiological effects. The endogenous production of CO, on the other hand, has been known for a long time. The re-evaluation and realization of the physiological importance of CO to the homeostatic control of the human body have been achieved only in the past 10 yr or so *(44)*. Like NO and CO, H_2S at physiologically relevant levels affects structures and functions of the human body at the molecular, cellular, tissue, and system levels.

3.1. Nitric Oxide

Application of nitrate-containing compounds, starting with nitroglycerin, for medicinal purposes can be traced back more than 150 yr. Less than two decades ago, the discovery that a simple gas, NO, was critical for endothelium-dependent vasorelaxation led to a revision of the doctrine about cell signal transduction *(45)*. The enzymatic synthesis of NO from L-arginine occurs in almost every type of cell, catalized by NO synthases. Many endogenous substances modulate the activities of NO synthases. The first discovered was a neurotransmitter, acetylcholine. Decomposition and biotransformation of NO in vivo have also been clearly demonstrated *(46)*. To capitalize on the discovery of endogenous NO, on October 12, 1998, Robert Furchgott, Louis Ignarro, and Ferid Murad were awarded a Nobel Prize in Medicine and Physiology for their discoveries concerning NO as a signaling molecule in the cardiovascular system. Today, the physiological importance of NO has been extended far beyond the cardiovascular system. NO has critical regulatory roles in physiological functions of many different types of cells, tissues, organs, and systems. Abnormal metabolism and/or functions of NO have also been described for pathogenic processes of many diseases. On the incomplete list of diseases involving NO are hypertension, diabetes, ischemia/reperfusion heart damage, cardiac attack, inflammation, stroke, erectile dysfunction, aging, menopause, hyperlipidemias, atherosclerosis, cancer, drug addiction, intestinal motility, memory and learning disorders, neuronal degenerating diseases, septic shock, sunburn, anorexia, tuberculosis, and obesity.

3.2. Carbon Monoxide

In 1898, Saint-Martin and Nicloux gave the first indication of endogenous CO. In 1950, Sjöstrand provided experimental evidence for the endogenous production of CO *(47)*. The biological and physiological function of endogenous CO had been either unknown or ignored for the ensuing half-century. Although lipid peroxidation yields endogenous CO, breakdown of the α-methane bridge of heme is the major route for the endogenous production of CO. Three isoforms of microsomal heme oxygenases (HOs) are involved in the enzymatic CO production in vivo. For more details about endogenous CO production and regulation, refer to Chapter 10.

Endogenous CO plays an important role in long-term potentiation (LTP) as a retrograde messenger in the brain *(48,49)*. This role of CO is similar to that of NO but may be mediated by different mechanisms. One hypothesis is that NO induces LTP by stimulating NMDA receptors, whereas it induces CO by stimulating metabotropic glutamate receptors. The involvement of 5-HT(3) receptors in the induction of ganglionic LTP by CO has also been suggested.

CO released from the vascular wall modulates proliferation and apoptosis of smooth muscle cells as well as endothelial cells. Relaxation of various types of smooth muscles by CO has also been consistently shown. Endogenous cellular levels of CO vary under different pathophysiological conditions, contributing to different disorders. Readers are referred to two recently published books for more detailed descriptions of the different biological effects of CO under physiological and pathophysiological conditions *(50,51)*.

Regarding regulation of heme metabolism, the physiological importance of HO has long been recognized. In addition to the degradation of heme, HO catalyzes the production of CO as well as biliverdine and ferrous iron. However, CO had not been taken into account for its beneficial effects of HO until little more than a decade ago. The breakthrough discovery of NO opened the way to further research on membrane/receptor-independent signaling by gas molecules. In 1991, Marks and colleagues *(52)* hypothesized that CO might be another important endogenous gaseous molecule. This pioneering thinking stirred up the resurgence of CO as a physiological signaling molecule *(44)*.

As CO biology has bloomed in recent years, more and more enthusiasm has been injected into HO biology. Research on CO and HO is now closely interacted and coevolved. This HO/CO field is experiencing phenomenal growth, spurred on by scientists and health workers, from the laboratory bench to the hospital bedside and by trainees from graduate students to postdoctoral fellows.

3.3. Hydrogen Sulfide

A significant amount of H_2S is produced by mammalian cells, and this substance has been measured in both circulatory blood and in isolated tissues and cells *(53)*. Two pyridoxal-5'-phosphate-dependent enzymes, cystathionine β-synthase [CBS] (EC 4.2.1.22) and cystathionine γ-lyase [CSE] (EC 4.4.1.1), are responsible for the majority of the endogenous production of H_2S in mammalian tissues, which use L-cysteine as the main substrate *(53)*. Ammonium and pyruvate are two other end products, in addition to H_2S, of CBS- and/or CSE-catalyzed cysteine metabolism. H_2S is also produced endogenously through the nonenzymatic reduction of elemental sulfur using reducing equivalents obtained from the oxidation of glucose *(53)*.

The elimination of H_2S from the body takes place mainly in the kidney. Mechanisms for biotransformation and scavenging of H_2S in vivo include oxidation in mitochondria, methylation in cytosol, and scavenging by methemoglobin or metallo- or disulfide-containing molecules such as oxidized glutathione. The appendix to this chapter gives detailed descriptions of the metabolism of H_2S *(53)*.

Similar to the story of CO, in which HO captured all of the glories initially, H_2S has lived for a long time in the shadow of H_2S-generating enzymes. These enzymes initially were characterized in the liver and kidney *(54,55)*. The physiological processes modulated by these enzymes were also elucidated in the liver and kidney, but the role played by H_2S was not studied further. Even homocysteine, a precursor of H_2S that is catabolized by the same H_2S-generating enzymes, received more attention from the perspective of atherosclerosis.

Recent studies have contributed significantly to our understanding of the physiological roles of H_2S in the nervous and cardiovascular systems. At physiologically relevant concentrations, H_2S reduced KCl-stimulated releases of the corticotropin-releasing hormone *(56)*. NaHS, a donor of H_2S, induced a concentration-dependent (27–200 μM) hyperpolarization and reduced input resistance of CA1 neurons or dorsal raphe neurons *(34)*. This concentration range is physiologically relevant in the brain *(57)*. Changes in K^+ conductance were identified to be the main ionic basis for these effects of NaHS, and K_{ATP} channels in neurons were speculated as the specific targets.

N-methyl-D-aspartate (NMDA) receptors are another target of H_2S. In the presence of a weak tetanic stimulation, NaHS at 10–130 μM facilitated the induction of hippocampal long-term potentiation in rat hippocampal slices by enhancing the NMDA-induced inward current *(57)*. Activation of the cyclic adenosine monophosphate-dependent protein kinase pathway likely mediates the interaction of H_2S and NMDA receptors *(58)*.

In the cardiovascular system, H_2S has been demonstrated at physiologically relevant concentrations to relax vascular tissues by opening K_{ATP} channels in vascular smooth muscle cells (VSMCs) *(59,60)*. In this case, NO serves as a trigger to increase H_2S production and release *(59)*. Evidence has also been presented for the relaxant effects of NaHS on rabbit isolated ileum, rat vas deferens, and guinea pig isolated ileum at physiologically relevant concentrations *(61)*. Inhibition of the H_2S-generating enzyme CSE caused a slowly developing increase in the contraction of the guinea pig ileum as a result of field stimulation *(61)*.

4. GASOTRANSMITTERS IN EVOLUTION

Table 1 lists organized activities for promoting research on and advancing our understanding of gasotransmitters. A 2-yr span saw the birth of a scientific society, a scientific journal, and the first scientific conference specifically devoted to NO (1996–1998). Since then, NO biology and chemistry have been the subject of many international meetings. Following the first world Internet meeting on cardiovascular effects of CO in 1998, two HO/CO conferences were held in 2000 and 2002 and another HO conference in 2003. **Table 2** lists selective monographs and books on the different types of gasotransmitters. Most of these books are on NO, and two are related to endogenous CO.

While this book was being edited, the *Antioxidants and Redox Signaling* journal published a special forum issue entitled "Gaseous Signal Transducers," discussing the biological roles of NO, CO, and H_2S. Another cheering development was the creation of the first strategic training program for gasotransmitter research in 2003, entitled "Gasotransmitter REsearch And Training" (GREAT). More than 15 researchers from four Canadian universities participated in this 6-yr program, supported by the Canadian Institutes of Health Research. The GREAT program will provide trail-breaking interdisciplinary and transdisciplinary training for local and international students, postdoctoral fellows, and researchers on sabbatical. The training program will be delivered through an array of courses; a trainee exchange program; laboratory, clinical, and community health research; and training-mentoring initiatives. A compulsory component of the GREAT program is a three-credit course, "Gasotransmitter Biology and Medicine." Another course offered through this program is "Career Development Essentials for Gasotransmitter Trainees."

Determination of endogenous levels of NO, CO, and H_2S; identification of the enzymes responsible for the production of these gases; and, most important, elucidation of the physiological functions of these gaseous molecules pave the way for the development of a general concept to envelop all these gases into one family. As can be seen from the aforementioned organized activities, one can only conclude that the era of gasotransmitters is coming and "the medium is the message" (Marshall McLuhan).

5. GASOTRANSMITTERS: DEFINITION OF THE CONCEPT

Vehicles for intercellular communication are either electrical signals via gap junction or chemical substances. The latter category is composed of hormones, autocoids, and transmitters. Hormones are released from endocrine cells into the bloodstream. The concentration of hormones is diluted to a relatively stable level when they reach distant multiple organs and cells. This endocrine mode of action is distinctive from the paracrine action of transmitters, in which transmitters, once released, usually act on adjacent postsynaptic cells. A definition of autocoids is not strictly precise. In general, autocoids (such as prostaglandins, adenosine, and platelet-activating factor) act on the same cells from which they are produced. Similar to the effects of hormones and transmitters,

Table 1
Chronicle of Organized Activities Related to Evolution of Gasotransmitter Biology and Medicine

Event	*Location*	*Year*	*Reference*
Founding of Nitric Oxide Society		1996	http://darwin.apnet.com/no/
Founding of *Journal of Nitric Oxide: Biology and Chemistry*		1997	
First official conference of Nitric Oxide Society: Biochemistry and Molecular Biology of Nitric Oxide[a]	Los Angeles, CA	1998	
Discovery of NO as signaling molecule in cardiovascular system and awarding of Nobel Prize in Medicine and Physiology to Robert Furchgott, Louis Ignarro, and Ferid Murad		1998	www.nobel.se/medicine/laureates/1998/index.html
Invited Symposium of "Carbon Monoxide and Cardiovascular Function"	Internet World Congress '98, INABIS '98	1998	www.mcmaster.ca/inabis98/toc.html
Sixth International Meeting on the Biology of Nitric Oxide	Stockholm, Sweden	1999	www.ki.se/org/nitric-oxide-99/
The 1st International Conference on Heme Oxygenase (HO/CO)	New York, NY	2000	
The 2nd International Conference on the Biology, Chemistry, and Therapeutic Applications of Nitric Oxide	Prague, Czech Republic	2002	
The 2nd International Conference on Heme Oxygenase (HO/CO) and Cellular Stress Response	Catania, Italy	2002	
Initiation of 6-yr GREAT program	University of Saskatchewan; Queen's University; University of Calgary; University of Montreal, Canada	2003	
Conference on HO regulation, functions, and clinical applications	Uppsala, Sweden	2003	

[a] This is the third in a series of conferences on biochemistry and molecular biology of NO.

cognate membrane receptors are still essential for the biological effect of autocoids. Some endocrine hormones, such as melatonin, can also act as autocoids *(62)*. In conventional signal transduction processes, the binding of neurotransmitters, certain endocrine hormones, or autocoids to receptors located on the plasma membrane is the essential triggering event. The ligand-receptor interaction generates intracellular second messengers that relay and direct the extracellular signals to different intracellular destinations, resulting in modulated cellular activity.

A neurotransmitter is a chemical substance that is released from a neuron either by exocytosis or directly from cytoplasm. It binds to specific receptors in the postsynaptic cell membrane and affects the function of postsynaptic cell(s). In some circumstances, neurotransmitters also act on "autoreceptors" located on presynaptic membranes to regulate the progress of synaptic transmission. Since the discovery of acetylcholine release from vagus terminals in frog hearts by Otto Loewi and Henry Dale while studying cholinergic and adrenergic systems in the early 1930s, the neurotransmitter concept has evolved and been constantly redefined.

Generally, a neurotransmitter is gauged against the following criteria:

1. It is synthesized in the neuron.
2. It is present in the presynaptic terminal and is released in amounts sufficient to exert a defined action on the postsynaptic neuron or effector organ.
3. When administered exogenously (as a drug) in reasonable concentrations, it mimics the action of the endogenously released transmitter exactly (for example, it activates the same ion channels or second messengers pathway in the post-synaptic cell).
4. A specific mechanism exists for removing it from its site of action (the synaptic cleft) *(63)*.

Acetylcholine, catecholamines, serotonin, histamine, glutamate, glycine, γ-aminobutyric acid, and adenosine triphosphate or its metabolites are among a handful of the identified low-molecular-weight neurotransmitters.

NO, CO, and H_2S are distinctive from classic neurotransmitters and humoral factors while sharing common characteristics among themselves (**Table 3**). These endogenous gaseous transmitters have been defined as *gasotransmitters*, measured by the following criteria *(53)*:

1. They are small molecules of gas.
2. They are freely permeable to membranes. As such, their effects do not rely on the cognate membrane receptors. They can have endocrine, paracrine, and autocrine effects. In their endocrine mode of action, for example, gasotransmitters can enter the bloodstream, be carried to remote targets by scavengers and released there, and modulate functions of remote target cells.
3. They are endogenously and enzymatically generated and their production is regulated.
4. They have well-defined and specific functions at physiologically relevant concentrations. Thus, manipulating the endogenous levels of these gases evokes specific physiological changes.
5. Their functions can be mimicked by their exogenously applied counterparts.
6. Their cellular effects may or may not be mediated by second messengers but should have specific cellular and molecular targets.

The gasotransmitter family may consist of many as-yet-unknown endogenous gaseous molecules, such as NH_3 and acetaldehyde. It is also worth noting that the effects of gasotransmitters may not always be beneficial. Under certain circumstances or in specific cellular environments, some gasotransmitters may inhibit physiological cellular function.

Table 2
Selective Monographs and Books on Different Gasotransmitters[a]

Authors/editors	Title and publisher	Year
Topics on NO		
S. Moncada, M. A. Marletta, J. B. Hibbs Jr., E. A. Higgs	*The Biology of Nitric Oxide: Part 1—Physiological and Clinical Aspects.* California Princeton Fulfillment	1992
S. Moncada, E. A. Higgs, J. B. Hibbs, M. A. Marletta	*The Biology of Nitric Oxide: Part 2—Enzymology, Biochemistry and Immunology.* California Princeton Fulfillment	1993
M. A. Titheradge	*Nitric Oxide Protocols.* Humana	1993
Peter Jenner, Steven R. Vincent	*Nitric Oxide in the Nervous System. Academic*	1995
N. Allon, S. Shapira, B. A. Weissman	*Biochemical, Pharmacological, and Clinical Aspects of Nitric Oxide.* Kluwer Academic/Plenum	1995
M. P. Fink, D. Payen	*Role of Nitric Oxide and Sepsis and ARDS.* Springer-Verlag	1995
S. Moncada, S. Gross, A. E. Higgs, *J. Stamler*	*Biology of Nitric Oxide: Proceedings of the Fourth International Meeting on the Biology of Nitric Oxide Held at Amelia Island, Florida, on September 17–21, 1995.* Portland Press	*1996*
E. Kenneth Weir, Stephen L. Archer, John T. Reeves	*Nitric Oxide and Radicals in the Pulmonary Vasculature.* Blackwell Publishing, Futura Division	1996
Martin Feelisch, Jonathan S. Stamler	*Methods in Nitric Oxide Research.* John Wiley & Sons	1996
Jack Lancaster Jr.	*Nitric Oxide Principles and Actions.* Academic	1996
Mahin Maines, Michael Conn	*Nitric Oxide Synthase: Characterization and Functional Analysis.* Academic	1996
Helmut Sies, John Abelson, Melvin Simon, Enrique Cadenas, Lester Packer	*Nitric Oxide, Part A–Part D (Methods in Enzymology).* Academic	1996–2002
Jack R. Lancaster, J. F. Parkinson	*Nitric Oxide, Cytochromes P450, and Sexual Steroid Hormones.* Springer-Verlag	1997
Jeffrey Burnstock, Jill Lincoln, Charles H. Hoyle	*Nitric Oxide in Health and Disease.* Cambridge University Press	1997
Michael S. Goligorsky, Steven S. Gross	*Nitric Oxide and the Kidney—Physiology and Pathophysiology.* Kluwer Academic	1997
Yann A. Henry, Annie Guissani, Béatrice Ducastel	*Nitric Oxide Research from Chemistry to Biology: EPR Spectroscopy of Nitrosylated Compounds.* Landes	1997
S. Moncada, G. Nisticò, G. Bagetta, E. A. Higgs	*Nitric Oxide and the Cell: Proliferation, Differentiation, and Death.* Princeton University Press	1998
S. Moncada, R. Busse, E. A. Higgs	*The Biology of Nitric Oxide: Physiological and Clinical Aspects.* California Princeton Fulfillment	1998

R. Ranney Mize, Ted M. Dawson, Valina L. Dawson, Michael J. Friedlander	*Nitric Oxide in Brain Development, Plasticity, and Disease.* Elsevier Health Sciences	1998
Stanislaw Lukiewicz, Jay L. Zweier	*Nitric Oxide in Transplant Rejection and Anti-tumor Defense.* Kluwer Academic	1998
Robert T. Mathie, Tudor M. Griffith	*Haemodynamic Effects of Nitric Oxide.* World Scientific Publishing	1998
Mika V. J. Hukkanen, Julia M. Polak, Sean P. F. Hughes	*Nitric Oxide in Bone and Joint Disease.* Cambridge University Press	1998
Robert T. Mathie, Tudor M. Griffith	*The Haemodynamic Effects of Nitric Oxide.* Imperial College Press	1999
M. Belvisi, J. Mitchell	*Nitric Oxide in Pulmonary Processes: Role in Physiology and Pathophysiology of Lung Disease.* Birkhäuser Verlag AG	1999
Debra L. Laskin, Jeffrey D. Laskin	*Cellular and Molecular Biology of Nitric Oxide.* Marcel-Dekker	1999
Ferric C. Fang	*Nitric Oxide and Infection.* Kluwer Academic/Plenum	1999
Julio A. Panza, Richard O. Cannon	*Endothelium, Nitric Oxide, and Atherosclerosis.* Futura Publishing	1999
Joseph Loscalzo, Joseph A. Vita	*Nitric Oxide and the Cardiovascular System.* Humana	2000
Louis J. Ignarro	*Nitric Oxide Biology and Pathobiology.* Harcourt	2000
Stanley Kaslner	*Nitric Oxide and Free Radicals in Peripheral Neurotransmission.* Springer-Verlag	2000
P. Kadowitz	*Nitric Oxide and the Regulation of the Peripheral Circulation.* Birkhauser Boston	2000
B. Mayer	*Nitric Oxide.* Springer	2000
R. J. Gryglewski and P. Minuz	*Nitric Oxide—Basic Research and Clinical Applications.* IOS Press	2001
Daniela Salvemini, Timothy R. Billiar, Yoram Vodovotz	*Nitric Oxide and Inflammation.* Birkhäuser	2001
Chuang C. Chiueh, Jau-Shyong Hong, Seng Kee Leong	*Nitric Oxide: Novel Actions, Deleterious Effects, and Clinical Potential.* New York Academy of Sciences	2002
A. Tomasi, T. Özben, V. P. Skulachev	*Free Radicals, Nitric Oxide and Inflammation.* IOS Press	2003
Topics on Carbon Monoxide		
Mahin D. Maines	*Heme Oxygenase: Clinical Applications and Functions.* CRC Press	1992
Rui Wang	*CO and Cardiovascular Functions.* CPC Press	2001
Nader G. Abraham, James Alam, Karl Nath, Jawed Alam	*Heme Oxygenase in Biology and Medicine.* Plenum	2002

[a] Not including books and monographs on atmospheric gases or toxicology and environmental concerns.

Table 3
Comparison of the Action Modes of Neurotransmitters and Gasotransmitters

	Release	*Re-uptake*	*Removal mechanism*	*Revert direction*	*Membrane receptors*
Neurotransmitters	Exocytotic vesicle	Yes	Enzyme dependent	Pre- to postsynaptic membrane (one direction)	Necessary
Gasotransmitters	Cytoplasm release	No	Nonenzymatic: oxidation, scavenging, methylation, etc.	Bidirectional	Not necessary

In this scenario, these gasotransmitters under physiological conditions would be maintained at low levels, thereby ensuring homeostasis of specific organs or cells. Significant differences among gasotransmitters exist regarding their mechanisms of production and function. For example, the effects of NO and H_2S, but not CO, may involve the production of free radicals. The biological outcome of the activation of NO synthase can be easily explained by its end product of NO, but upregulation of HO may alter cellular functions via its end products of iron and biliverdin, other than CO, in some cases. Differences in chemical and physical properties, cellular production levels, signal transduction pathways involved, and so on for various gasotransmitters are discussed in detail in other chapters of this book.

The birth of the gasotransmitter concept—with NO, CO, and H_2S its current stars—is just the first exploratory step on an unknown path. Much more work remains to be done. The physiological roles and importance of CO and H_2S still need to be vigorously tested. More gasotransmitters may be discovered and identified in the future. The interaction among gasotransmitters should be investigated. Physiological levels, both circulatory and cellular, of gasotransmitters as well as molecular switches to turn on or off the production of gasotransmitters should be determined. Pathological actions of gasotransmitters should be assessed. Before all these concerns and challenges can be addressed, answered, and articulated, the jury is still out on the case of gasotransmitters *(64)*.

The concept of gasotransmitters initially was framed in a *FASEB Journal* article by Wang *(53)*, which is provided as an appendix at the end of this chapter.

6. GASOTRANSMITTERS AND ION CHANNELS

Gasotransmitters are freely permeable to biological membranes and very likely interact with ion channels in the plasma membrane and intracellular organelle membranes. The interaction of gasotransmitters with ion channels is the focus of discussion in this book for the following considerations: First, direct modification of ion channels by gasotransmitters, independent of conventional second messengers, has been demonstrated in many cases for NO, CO, and H_2S. Key discoveries in this regard are summarized in **Table 4**. This direct modulation of ion channels by gasotransmitters represents a novel class of signal transduction mechanism. Conventional dogma argues that membrane ion channels can only be modified by endogenous substances via membrane receptor–related second-messenger systems. Second, the structure and function of ion

Table 4
Key Discoveries on Direct Interaction of Gasotransmitters with Ion Channels

Authors	*Discovery*	*Year*	*References*
Bolotina VM, Najibi S, Palacino JJ, Pagano PJ, Cohen RA	NO opens K_{Ca} channels in VSMCs by a direct interaction with cysteine residue of K_{Ca} channels.	1994	Nature 368:850–853.
Wang R, Wu L	CO opens K_{Ca} channels in VSMCs by a direct interaction with cysteine residue of K_{Ca} channels.	1997	J Biol Chem 272:8222–8226.
Shin JH, Chung S, Park EJ, Uhm DY, Suh CK	NO directly activates neuronal K_{Ca} channels reconstituted into planar lipid bilayer.	1997	FEBS Lett 415:299–302.
Ahern GP, Hsu SF, Jackson MB	NO directly opens neurohypophysial K_{Ca} channels.	1999	J Physiol 520(Pt 1):165–176
Hammarstrom AK, Gage PW	NO increases persistent Na^+ channel current in rat hippocampal neurons through an oxidizing action directly on the channel protein.	1999	J Physiol 520 (Pt 2):451–461
Liu H, Mount DB, Nasjletti A, Wang W	CO directly activates an apical 70pS K^+ channel of the rat thick ascending limb.	1999	J Clin Invest 103:963–970
Broillet MC	NO activates the olfactory cyclic nucleotide gated channel by acting on a single intracellular cysteine residue.	2000	J Biol Chem 275:15,135–15,141
Kaide JI, Zhang F, Wei Y, Jiang H, Yu C, Wang WH, Balazy M, Abraham NG, Nasjletti A	CO directly activates a tetraethylammonium-sensitive K^+ channel in VSMCs.	2001	J Clin Invest 107:1163–1171
Zhao W, Zhang J, Lu Y, Wang R.	H_2S opens K_{ATP} channels in VSMCs by a direct interaction.	2001	EMBO J 20:6008–6016
Wu L, Cao K, Lu Y, Wang R.	NO acts on β-subunit, but CO on α-subunit, of K_{Ca} channels in VSMCs to open these channels.	2002	J Clin Invest 110:691–700
Jaggar JH, Leffler CW, Cheranov SY, Tcheranova DES, Cheng X	CO increases the activity of K_{Ca} channels in VSMCs by shifting the Ca^{2+} sensitivity, suggesting a priming mechanism.	2002	Circ Res 91:610–617
Liu Y, Terata K, Chai Q, Li H, Kleinman LH, Gutterman DD	NO inhibits K_{Ca} channels in VSMCs by a direct interaction mediated by the intermediate reactive oxygen species.	2002	Circ Res 91:1070–1076
Renganathan M, Cummins TR, Waxman SG	NO inhibits neuronal Na^+ channels by a direct interaction with sulfhydryl groups.	2002	J Neurophysiol 87:761–775

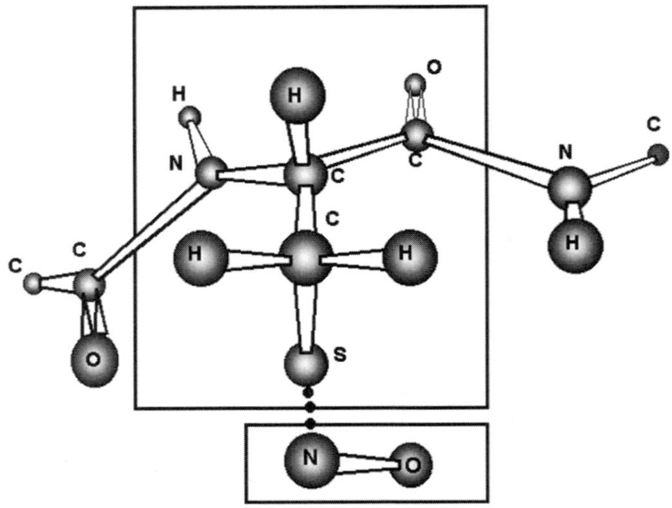

Fig. 1. Modification of ion channel proteins by NO through a *S*-nitrosylation mechanism.

channels on cell membranes affect general as well as many specific cellular functions. Third, by conducting specific ions, ion channels themselves serve as important signal transduction links. Fourth, the complexity of ion channel families is directly coupled to diverse biological functions. Finally, the modulation and mobilization of classic second messengers by gasotransmitters has been the topic of numerous peer-reviewed articles, monographs, and books. The direct interaction of gasotransmitters with membrane ion channels has not previously been systematically described and encapsulated.

Three specific modes of direct interaction of gasotransmitters with membrane ion channels are discussed in this book. NO covalently modifies free cysteine residues in proteins via *S*-nitrosylation *(65)*. The *S*-nitrosylation of ion channel proteins by NO would directly change the function of these channels (**Fig. 1**). This mechanism is specifically discussed in Part II of this book. Direct interaction of gasotransmitters with ion channel proteins also applies in the case of CO. Many reported effects of CO on K^+ channels are not regulated by known second messengers. Chemical modification of histidine residues of K^+ channel proteins by CO via the formation of hydrogen bond (**Fig. 2**), a process of carboxylation, has been indicated *(66–68)*. Chapters 12 and 13 of this book give detailed descriptions of the ion channel carboxylation. Direct modulation of K_{ATP} channels by H_2S is recently reported, which is not mediated by cyclic guanosine *S*′-monophosphate or other known second messengers *(59)*. A chemical interaction of H_2S with sulfhydryl groups of ion channel proteins is, as such, hypothesized. The formation of adduct of HS^- with free sulfhydryl group, a sulfuration mechanism, and the breakdown of disulfide bonds, a reducing mechanism, by H_2S are alternative molecular mechanisms, which are further discussed in Chapter 21.

Notwithstanding the focus on the direct interaction of gasotransmitters on ion channel proteins, this book also gives a balanced view to include the effect of gasotransmitters on ion channels mediated by different second messengers.

7. PERSPECTIVES ON GASOTRANSMITTER RESEARCH

Gasotransmitters are recent discoveries emanating from both the laboratory and clinical research ends of the health research spectrum. There are already more than 97,000

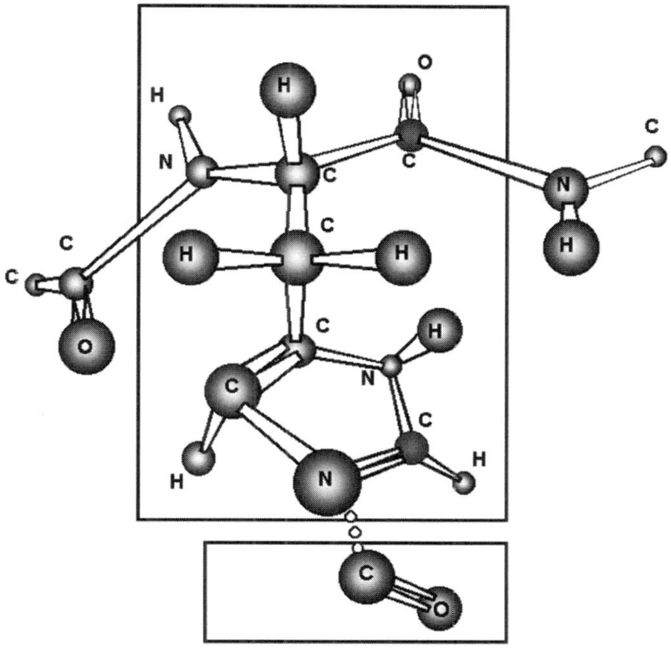

Fig. 2. Modification of ion channel proteins by CO through a carboxylation mechanism.

articles incorporating the terms NO, CO, or H_2S. Numerous laboratories worldwide are studying these gasotransmitters. It seems probable, if not certain, that new members of the gasotransmitter family will come to light in a few years. Chapter 22 discusses many other candidates for potential gasotransmitters.

7.1. Growth of Gasotransmitter Research

By March 2004, Medline searchers found approx 60,000 articles incorporating the term *nitric oxide*, with some 17,000 using the term *carbon monoxide* and more than 21,000 using the term *hydrogen sulfide*. Not only is the information base in this area exploding, but some researchers expect the roster of proven gasotransmitters to grow dramatically in the future, citing such biomolecules as formaldehyde (CH_2O), ethylene (CH_2CH_2), and ammonia (NH_3) as potential new members in this class. As new gasotransmitter molecules appear on the scene, membership in the gasotransmitter family will be enlarged and updated by incorporating these substances. This new and challenging field of gasotransmitter medicine encompasses biomedical, clinical, health services, and population health studies.

7.2. Link Between Gasotransmitters and Human Diseases

Numerous human diseases are linked to abnormal metabolism and functions of gasotransmitters. This knowledge will significantly affect the pathogenesis, diagnosis, therapeutics, and prevention strategies for gasotransmitter-related diseases. Therefore, gasotransmitter research is as important for clinical researchers and practitioners as it is for basic researchers—if not more so. **Table 5** summarizes Medline search results reflecting links between gasotransmitters and circulatory and respiratory diseases.

Table 5
Relevant Medline-Indexed Disease-Related Publications
(Dated to March 15, 2004)

	NO	*CO*	*H_2S*
Stroke	821	106	12
Hypertension	5065	214	73
Transplantation	1321	153	115
Atherosclerosis	1840	171	110
Ischemia and reperfusion	1919	66	8
Heart failure	1087	105	14
Asthma	825	221	29
Chronic obstructive pulmonary disease	98	222	5

As research progresses, more implications of the role gasotransmitter molecules play in human health are emerging. Elucidation of the roles of NO, CO, and H_2S in the mechanisms of specific human diseases will enable future discovery, development, and clinical use of innovative therapeutic interventions *(69)*. For instance, Medline search reveals about 2000 publications on hyperhomocysteinemia *(70,71)*. This clinical problem is now known to be related to the metabolism of homocysteine, an endogenous precursor of H_2S. A better understanding of the metabolism of H_2S will greatly illuminate clinical practice for many hyperhomocysteinemia-related cardiovascular diseases.

Another example is the application of inhaled gasotransmitters to treat human diseases, including the understanding of the technology to administer gasotransmitters, their action mechanisms, and their indication for the treatment of different pathologies. These gasotransmitters can be administered alone or in combination. Human clinical trials have been conducted to determine the role of inhaled NO in the treatment of severe acute respiratory distress syndrome in adults and in the treatment of pulmonary hypertension during surgery *(72)*. Inhaled NO decreases the pulmonary inflammation induced by the extracorporeal circulation in swine *(73)*. The inhaled NO also has major extrapulmonary effects particularly on renal function, preventing the detrimental renal effects of cyclooxygenase inhibitors.

The application of gene therapy to protect the heart from ischemia/reperfusion damage by the overexpression of HO has proven to be effective *(74)*. Intramyocardial delivery of the human HO-1 gene by the adenoassociated virus protected the heart from reperfusion injury. An upregulated HO/CO system inhibited cardiac anaphylaxis *(75)* and lowered blood pressure in young spontaneously hypertensive rats *(76)*. As the link between HO/CO and heart transplants becomes increasingly known *(77,78)*, it will be important for clinicians to learn about the analysis and regulation of HO/CO level to perform successful heart transplantation to protect the heart from ischemia. In this case, gasotransmitter research at the laboratory level will elucidate the mechanisms for HO/CO protection. Clinical research will bring a new paradigm into organ and tissue transplant technology. Community and population health research will provide specific populations with abnormal CO metabolism more knowledge regarding their cardiac health, stress management, and prevention of cardiovascular diseases.

7.3. Triad Frame of Transdisciplinary Gasotransmitter Research

The clinical relevance of gasotransmitter research has been delineated. Linkages between gasotransmitter research and community health can be viewed from at least two

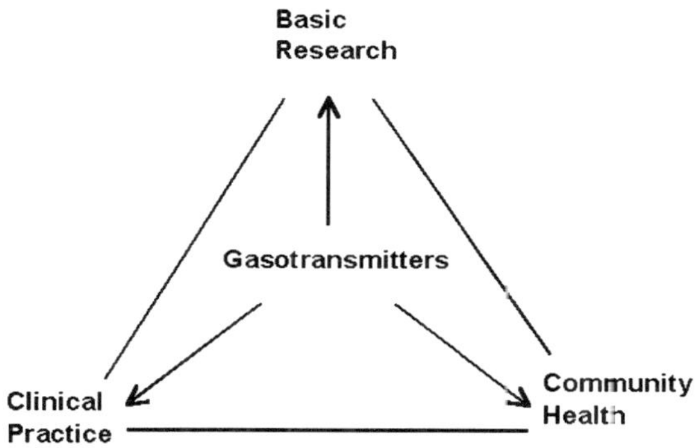

Fig. 3. Bridging and branching of gasotransmitter research.

angles (**Fig. 3**): (a) several community health issues are related to diseases caused by abnormal metabolism and functions of gasotransmitter molecules; and (b) the long-recognized health hazards of these gases, at higher ambient and pollution levels, pose environmental and community health questions, which may have relevance to or share characteristics with their endogenous levels. High concentrations of CO, NO, NO_2 *(5)*, and H_2S *(34,79)* are especially hazardous for people working or living in specific environments and communities. Levels of NO and NO_2 are constantly increasing in the urban community atmosphere, particularly in areas close to automobile traffic and airports *(13)*. Exposure to high concentrations of these gases may present a specific health concern for the population's health *(5)*. Gasotransmitter research will arm community health researchers and workers with a better knowledge of and expertise in the metabolism of these gases at toxic levels in our bodies, their specific cellular targets, toxicological mechanisms, and specific detoxification maneuvers.

7.4. Future Directions

More organized activities for promoting research on gasotransmitters are expected. Capacity building by recruiting new researchers into the field of gasotransmitters and training more highly qualified personnel is becoming a priority in research agendas throughout the globe.

ACKNOWLEDGMENTS

I wish to thank Dr. K. Cao for preparing figures for this work. This work was supported by Canadian Institutes of Health Research and Natural Sciences and Engineering Research Council of Canada.

APPENDIX

Two's company, three's a crowd: can H_2S be the third endogenous gaseous transmitter?

RUI WANG[1]

Department of Physiology, University of Saskatchewan, Saskatoon, SK, Canada S7N 5E5

ABSTRACT **Bearing the public image of a deadly "gas of rotten eggs," hydrogen sulfide (H_2S) can be generated in many types of mammalian cells. Functionally, H_2S has been implicated in the induction of hippocampal long-term potentiation, brain development, and blood pressure regulation. By acting specifically on K_{ATP} channels, H_2S can hyperpolarize cell membranes, relax smooth muscle cells, or decrease neuronal excitability. The endogenous metabolism and physiological functions of H_2S position this gas well in the novel family of endogenous gaseous transmitters, termed "gasotransmitters." It is hypothesized that H_2S is the third endogenous signaling gasotransmitter, besides nitric oxide and carbon monoxide. This positioning of H_2S will open an exciting field—H_2S physiology—encompassing realization of the interaction of H_2S and other gasotransmitters, sulfurating modification of proteins, and the functional role of H_2S in multiple systems. It may shed light on the pathogenesis of many diseases related to the abnormal metabolism of H_2S.—Wang, R. Two's company, three's a crowd: can H_2S be the third endogenous gaseous transmitter? *FASEB J.* 16, 1792–1798 (2002)**

Key Words: carbon monoxide · cardiovascular system · gasotransmitter · neuron · nitric oxide

THE CELLULAR SIGNALING process is usually initiated by the binding of neurotransmitters or humoral factors to receptors located on the plasma membrane. The ligand–receptor interaction generates intracellular second messengers that relay and direct the extracellular signals to different intracellular destinations, resulting in modulated cellular activity. The discovery of nitric oxide (NO) elucidates more than just the nature of the endothelium-derived relaxing factor (1). It presents a membrane receptor-independent signaling mechanism, emphasizing the necessity to modify the conventional doctrine about cellular signal transduction. The subsequent resurgence of carbon monoxide (CO) as another important endogenous signaling gas is embraced by researchers in almost every field of life sciences (2). To distinguish NO and CO from the classical neurotransmitters and humoral factors while acknowledging the common nature of these two gases, an effort has been made to classify these endogenous gaseous transmitters against several criteria (**Table 1**). I would recommend designating these gaseous transmitters as gasotransmitters. NO and CO are the first two identified gasotransmitters. In this hypothesis study, arguments are made to entitle hydrogen sulfide (H_2S) as the third gasotransmitter. Important implications of this identification are explained.

Physical and chemical properties of H_2S

H_2S is a colorless gas with a strong odor of rotten eggs. The detectable level of this gas by the human nose is at a concentration 400-fold lower than the toxic level. Oxidation of H_2S yields elemental sulfur, sulfur oxide (SO_2), and sulfates such as sulfuric acid. H_2S can be hydrolyzed to hydrosulfide and sulfide ions in the following sequential reactions: $H_2S \Leftrightarrow H^+ + HS^- \Leftrightarrow 2H^+ + S^{2-}$. Even in an aqueous solution, about one-third of H_2S remains undissociated at pH 7.4. H_2S is permeable to plasma membranes as its solubility in lipophilic solvents is ~ fivefold greater than in water.

Endogenous generation and metabolism of H_2S

The biological production and utilization of H_2S have been best known for certain bacteria and archae (3). A sobering fact is that mammalian cells also produce H_2S. The H_2S concentration of rat serum is ~ 46 μM (4). Aside from circulating H_2S, a significant amount of H_2S is produced in various tissues. For instance, the physiological concentration of H_2S in brain tissue has been reported to be 50–160 μM (5, 6). Recent studies have shown that vascular tissues generate measurable amounts of H_2S (4, 5).

Two pyridoxal-5′-phosphate-dependent enzymes—cystathionine β-synthase or CBS (EC 4.2.1.22) and cystathionine γ-lyase or CSE (EC 4.4.1.1)—are responsible for the majority of the endogenous production of H_2S in mammalian tissues that use L-cysteine as the main substrate (7–9). In some tissues CBS and CSE are both needed for generation of H_2S, whereas in others one enzyme suffices (**Fig. 1**). Thus, it comes as no surprise that the expression of CBS and/or CSE is tissue specific. The expression of CBS (5, 10) and CSE (11–14) has been identified in many human and other mammalian cells, including those from liver, kidney,

[1] Correspondence: Department of Physiology, University of Saskatchewan, 107 Wiggins Road, Saskatoon, SK, Canada, S7N 5E5. E-mail: wangrui@duke.usask.ca

0892-6638/02/0016-1792 © FASEB

TABLE 1. *Classification of gasotransmitters (gaseous transmitters)*

(1) They are small molecules of gas, like nitric oxide (NO) and carbon monoxide (CO).
(2) They are freely permeable to membrane. As such, their effects will not rely on cognate membrane receptors.
(3) They are endogenously and enzymatically generated and their generation is regulated.
(4) They have well-defined specific functions at physiologically relevant concentrations. For instance, NO and CO both participate in vasorelaxation and synaptic transmission in the central nervous system.
(5) Their cellular effects may or may not be mediated by second messengers but should have specific cellular and molecular targets. For instance, NO and CO activate K_{Ca} channels in plasma membrane either directly or mediated by the cGMP pathway.

brain, skin fibroblasts, and blood lymphocytes. As the end product of CBS- and CSE-catalyzed cysteine metabolism, H_2S exerts a negative feedback effect on the activity of these enzymes. Elevated H_2S level inhibited CSE activity (15) and the rate of gluconeogenesis from cysteine (16). Another less important endogenous source of H_2S is the nonenzymatic reduction of elemental sulfur to H_2S using reducing equivalents obtained from the oxidation of glucose (17) (**Fig. 2**). All essential components of this nonenzymatic pathway are present in vivo, including the supply of reducible sulfur. The presence of millimolar concentration of sulfur in blood circulation has been reported in humans (18) or mice (19).

H_2S in vivo is metabolized by oxidation in mitochondria or by methylation in cytosol (Fig. 1). H_2S can be scavenged by methemoglobin (20) or metallo- or disulfide-containing molecules such as oxidized glutathione (21). H_2S is excreted mainly by the kidney as free or conjugated sulfate (20). The interaction of hemoglobin and H_2S calls for special attention. Hemoglobin may be the common "sink" for CO in forming scarlet carboxyhemoglobin (22), for NO in forming nitrosyl hemoglobin, and for H_2S in forming green sulfhemoglobin (23). If this sink is filled with one gas, the binding of other gases would be affected and their individual availability to act on targeted cells would be altered. A

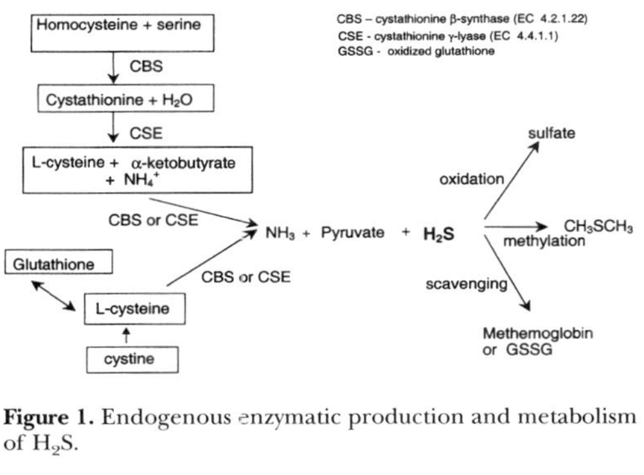

Figure 1. Endogenous enzymatic production and metabolism of H_2S.

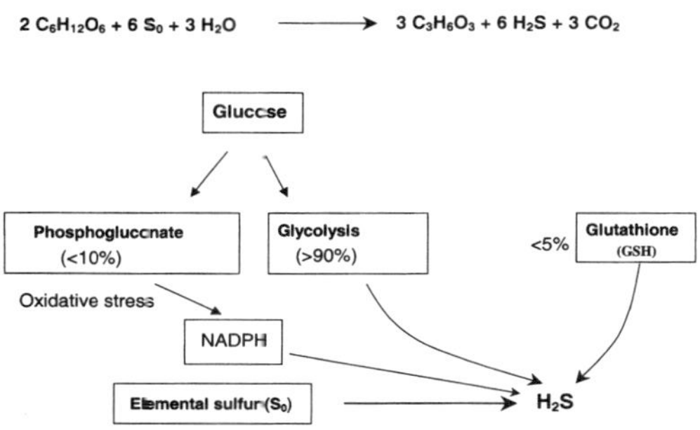

Figure 2. Endogenous nonenzymatic production of H_2S.

case in point is the observation that after pretreatment of human erythrocytes with CO to saturate the hemoglobin sink, the accumulated amount of endogenous H_2S was significantly enhanced (17).

Physiological effects of H_2S and the underlying mechanisms

The physiological functions of endogenous H_2S may be multifaceted. In liver and kidney, activities of the H_2S-generating enzymes have been studied in great detail (8, 9, 24, 25). To be succinct, a discussion of this study focuses on the physiological role of H_2S in nervous and cardiovascular systems.

Physiological effects of H_2S on the nervous system

The first and most important evidence for the physiological role of H_2S was obtained in 1989 when endogenous sulfide levels in rat brain tissues (1.6 μg/g) (26) and in normal human postmortem brainstem (0.7 μg/g) were reported (26, 27). Endogenous sulfide level in mice brain (28) was similar to that of rats, but threefold lower than that of bovine cerebral cortex (29). The study by Awata et al. in 1995 (30) provided the enzymatic mechanisms for this endogenous H_2S in rat brain, in which activities of CBS and CSE in six different brain regions were detected though the activity of CBS was > 30-fold greater than that of CSE. Brain activities of CBS and CSE gradually increased after birth and reached adult level at 2–4 wk. The transcriptional expression of CBS in rat brain (hippocampus, cerebellum, cerebral cortex, and brainstem) was later confirmed using Northern blot analysis but no CSE mRNA was detected (6). The reduced H_2S production after the inhibition of CBS further pinpointed CBS to be the major endogenous enzyme for H_2S production in brain (6).

The functional role of H_2S at physiologically relevant concentrations in brain was gradually uncovered in early 1990s. Chronic exposure of neonatal rats to H_2S altered the release of neurotransmitters in brain with increased serotonin and norepinephrine levels in rat cerebellum and frontal cortex (31, 32). Application of NaHS, which generates H_2S once in solution, to rat hypothalamic explants in vitro did not affect the basal secretion of corticotropin-releasing hormone (CRH), but consistently reduced KCl-stimulated CRH release from the explants (33). This effect of exogenous H_2S was consistent with the observation that the intramuscular application of S-adenosyl-L-methionine, an endogenous precursor of H_2S, to conscious rats reduced the hypothermia-induced increase in serum level of corticosterone (33).

Voltage-dependent and TTX-sensitive Na^+ channels may be targeted by H_2S in neurons. In cultured neuroblastoma cells, NaHS or taurine alone did not alter Na^+ channel currents. After pretreatment of these cells with NaHS, taurine dramatically inhibited Na^+ channels in a reversible fashion (34). This effect of NaHS was mimicked by disulfide-reducing agents dithiothreitol and β-mercaptoethanol. A reduction of disulfide bonds between Na^+ channel subunits by H_2S was thus suggested. Since taurine is an inhibitory neurotransmitter and a short exposure to NaHS (<2 min) resulted in a twofold increase in taurine levels in brainstem (35), the interaction between NaHS and taurine suggests that certain neuronal effects of H_2S could be mediated by the alteration in taurine levels. However, the physiological importance of this study is limited since the concentration used for NaHS (10 mM) was far outside the physiological range.

NaHS induced a concentration-dependent (27–200 μM) hyperpolarization and reduced input resistance of CA1 neurons or dorsal raphe neurons (36). This concentration range is physiologically relevant in the brain (6). Changes in K^+ conductance were identified to be the main ionic basis for these effects, since the presence of extracellular barium or intracellular cesium abolished the NaHS-induced membrane hyperpolarization. NaHS-induced neuronal hyperpolarization was blocked by a high concentration of TEA (50 mM) but not by a low concentration of TEA (10 mM) or 4-aminopyridine (1 mM). Thus, the involvement of either calcium-activated K^+ channels or voltage-dependent K^+ channels in NaHS effect was not supported. Activation of ATP-sensitive K^+ (K_{ATP}) channels by NaHS was proposed in these experiments as the consequence of ATP depletion due to the inhibition by sulfide of the oxidative phosphorylation (36). This hypothesis was not without ambiguity, since in the same experiments manipulation of intracellular ATP concentrations did not affect the NaHS-induced membrane hyperpolarization and no K_{ATP} channel currents were directly examined. Electrophysiological measurement of K^+ channel currents in neurons with tight control of intracellular ATP levels in the presence of NaHS/H_2S would help clarify the interaction of H_2S and neuronal K_{ATP} channels.

In addition to K_{ATP} channels, NMDA receptors may be the target of H_2S. In the presence of a weak tetanic stimulation, NaHS at 10–130 μM facilitated the induction of hippocampal long-term potentiation (LTP) in rat hippocampal slices by enhancing the NMDA-induced inward current (6). Interaction of H_2S and NMDA receptors was possibly mediated by the activation of a cAMP-dependent protein kinase pathway. NaHS (1–100 μM) increased cAMP production in primarily cultured rat cerebral and cerebellar neurons or in selected rat brain neuronal and glial cell lines (37). By enhancing the production of cAMP, NaHS increased the sensitivity to NMDA stimulation of NMDA receptors expressed in oocytes (37).

Physiological effects of H_2S on the cardiovascular system

It has been a conventional view that H_2S interferes with cardiovascular function as a result of the secondary anoxia rather than a direct action of the gas on cardiac myocytes or vascular smooth muscle cells (SMCs) (36). However, this doctrine has started to become shaky in light of two aspects of development. The location of the H_2S-generating enzymes as well as the detection of endogenous levels of H_2S in cardiovascular system provides the endogenous sources of H_2S. In-depth study of the whole animal and at tissue and cellular levels defines the functional role of H_2S in the cardiovascular system.

Chen et al. (38) found no activity or expression of CBS in human atrium and ventricle tissues. The activity and/or expression of CBS were also lacking in human internal mammary arteries, saphenous veins, coronary arteries, or aortic arteries (38, 39). Thus, CBS does not appear to play a major role in generating H_2S in cardiovascular tissues under physiological conditions. On the other hand, expression of CSE and the endogenous production of H_2S have been shown in rat portal vein and thoracic aorta (5). In rat mesenteric artery and other vascular tissues, CSE is the only H_2S-generating enzyme that has been identified, cloned, and sequenced (4). mRNA of this enzyme was expressed solely in vascular SMCs as detected by RT-PCR and in situ hybridization (4). No transcript of CSE was found in the endothelium layers of intact vascular tissues or cultured endothelial cells (4). Expression levels of CSE mRNA varied in different types of vascular tissues, with an intensity rank of pulmonary artery > aorta > tail artery > mesenteric artery (4). Endogenous production of H_2S depends on the types of vascular tissues. For instance, the homogenates of thoracic aortas yielded more H_2S than that of portal vein of rats (5).

The physiological function of H_2S in the cardiovascular system has been studied recently. An intravenous bolus injection of H_2S transiently decreased blood pressure of rats by 12–30 mmHg, an effect mimicked by pinacidil (a K_{ATP} channel opener) and antagonized by glibenclamide (a K_{ATP} channel blocker) (4). At the tissue level, H_2S at physiologically relevant concentrations (IC_{50}, 125 μM) induced in vitro relaxation of aorta and portal vein of rats (4, 5). Whether this vasorelaxant effect was due to a direct action of H_2S on vascular SMCs has been questioned. Zhao et al. (4) showed that the H_2S-induced relaxation of rat aortic tissues was due mainly to a direct interaction of H_2S and SMCs, based on the failure of denervation of vascular tissues in vitro to alter H_2S effects and on the observation that H_2S still significantly relaxed vascular tissues after endothelium removal. Zhao et al. (4) showed that a small portion of the H_2S-induced vasorelaxation was attenuated by either removal of the endothelium or the application of L-NAME (an inhibitor of NO synthase) in the presence of the endothelium. This endothelium-dependent effect of H_2S could be explained by the release of endothelium-derived vasorelaxant factors in response to H_2S stimulation. The presence of an intact endothelium might serve as a buffer to retain H_2S in the blood vessel wall so that its vasorelaxant effect can be potentiated and prolonged. Another interesting observation was that the coapplication of apamin and charybdotoxin, a protocol to block the effect of endothelium-derived hyperpolarizing factor (EDHF) (40), to the endothelium-intact rat aortic tissues reduced the vasorelaxant effect of H_2S. It seems that H_2S might release EDHF from vascular endothelium. It should be borne in mind that endothelium dependency of the vascular effects of H_2S has been controversial. One study concluded that the vasorelaxant effect of H_2S was independent of endothelium, even though no experimental data were shown to support this conclusion (5).

Mechanisms for the direct effect of H_2S on vascular SMCs have been explored. Unlike NO or CO, H_2S relaxed vascular tissues independent of the activation of cGMP pathway. Whereas the vasorelaxation induced by NO was virtually abolished by ODQ, a specific inhibitor of soluble guanylyl cyclase, the H_2S-induced vasorelaxation was not inhibited by ODQ (4). In fact, ODQ even potentiated the vasorelaxant effect of H_2S. The synergistic actions of H_2S and ODQ cannot be fully understood yet. Hypothetically, the interaction between ODQ and H_2S may have generated vasorelaxant free radicals, which further relaxed vascular tissues.

The most recent significant advance in our understanding of the vascular effects of H_2S was the identification of K_{ATP} channels in vascular SMCs as the target protein of H_2S. When isolated rat aortic tissues were precontracted with 20 or 100 mM KCl, the maximum vascular relaxation induced by H_2S was ~ 90% or 19%, respectively (4). This difference in relaxation potency of H_2S represents the portion of relaxation possibly mediated by potassium conductance. Furthermore, H_2S-induced relaxation of the aortic tissues precontracted with phenylephrine was mimicked by a K_{ATP} channel opener pinacidil but concentration-dependently inhibited by glibenclamide. Results from these tissue contractility studies were substantiated in isolated single SMCs. K_{ATP} channel currents in rat aortic SMCs were significantly and reversibly increased by either H_2S or pinacidil. A direct action of H_2S on K_{ATP} channel proteins, rather than the interfered ATP metabolism by H_2S, was proposed based on three lines of evidence. First, intracellular ATP concentration in these studies was clamped at a fixed level (e.g., 0.5 mM) by dialyzing cells with the pipette solution. Second, the effect of H_2S on K_{ATP} channels was quickly reversed on washing out H_2S from the bath solution. Third, intentionally varying ATP concentrations inside the cell (from 0.2 to 3 mM) did not change the excitatory effect of H_2S on K_{ATP} channels. Together, these results demonstrate that H_2S is an important endogenous vasoactive factor and is the first identified gaseous opener of K_{ATP} channels in vascular SMCs.

Physiological vs. toxicological effects of H_2S

The toxicity of H_2S has been known for ~ 300 years. The major lethal consequence of H_2S intoxication is the loss of central respiratory drive due to biochemical lesions of the respiratory centers of the brainstem (41). For a complete toxicological profile of H_2S, readers are redirected to two excellent reviews by Beauchamp et al. (20) and Reiffenstein et al. (36). Note first that the endogenously generated H_2S under physiological conditions is hardly accumulated or toxic to cells due to the balanced cellular metabolism of the gas (Fig. 1). In the presence of > 30 μM HS^-, no apparent disturbance in oxidative phosphorylation could be observed likely due to the rapid oxidation of H_2S in mitochondria (42, 43). Second, the line between toxicological and physiological effects of H_2S is very thin. The reported toxic level of H_2S is < twofold greater than its endogenous level in rat brain tissues (26). Intoxication of mice with NaSH only elevated the sulfide concentration from the endogenous level by 57%, 18%, and 64% in brain, liver, and kidney, respectively (28). It is thus reasoned that the dose-response relationship of H_2S at the physiological concentration range must be very steep before the physiological effect of H_2S sharply transforms into a highly toxic effect (4). Moreover, mammalian cells must possess a delicate regulatory mechanism to control the endogenous H_2S level within the physiological range.

Interaction of H_2S with other gasotransmitters

Given that H_2S, NO, and CO can all be gasotransmitters, they are not redundant (**Table 2**). For example, H_2S, NO, and CO facilitate the induction of hippocampal LTP. This effect of H_2S depends on the activation of NMDA receptors (6) whereas that of NO and CO does not. NO can act as a reactive oxygen species by impairing the reduced glutathione/oxidized glutathione balance and/or by inhibiting enzymes and ion channels through S-nitrosylation processes. H_2S may also be involved in the reduction of thiols, whereas CO is not directly involved in redox reactions. Gasotransmitters may interact with each other. As discussed above, competition for the common hemoglobin sink by one gasotransmitter would potentiate or unmask the biological effect of other gasotransmitters.

Published data have shown that the endogenous production of H_2S from rat aortic tissues is enhanced by NO donor treatment (4). The NO donor also enhances the expression level of CSE in cultured vascular SMCs. Similar to the release of NO by acetyl-

TABLE 2. *Metabolism and function of gasotransmitters*[a]

	H_2S	CO	NO
Main substrates	L-cysteine	Heme	L-arginine
Generating enzymes	CBS, CSE	Heme oxygenases	NO synthases
Inducer	NO	Free radicals	Acetylcholin, endotoxin
Scavenger	Hemoglobin	Hemoglobin	Hemoglobin
Inhibitor	D,L-propargylglycerine	Zinc-PPIX	L-NAME
Protein targets	K_{ATP} channel, cAMP (?)	cGMP, K_{Ca} channel	cGMP, K_{Ca} channel
Amino acid targets	?	Histidine	Cysteine
Half-life in solution	Minutes	Minutes	Seconds
Production tissue source	SMC, not in EC	EC < SMC	EC > SMC

[a] Only examples, not a complete list, are given. SMC, smooth muscle cell; EC, endothelial cell; zinc-PPIX, zinc protoporphyrin-IX; L-NAME, N^G-nitro-L-arginine methyl ester.

choline, release of H_2S by NO adds a line of essential evidence for the physiological role of H_2S.

Finally, the integrated vascular effect of H_2S and NO may not be a simple algebraic summation of their individual actions. Hosoki et al. (5) observed that the vasorelaxant effect of sodium nitroprusside (SNP), a NO donor, was enhanced by incubating rat aortic tissues with 30 μM NaHS. On the contrary, pretreating aortic tissues in another study with 60 μM H_2S inhibited the vasorelaxant effect of SNP. This discrepancy may be partially explained by the experimental conditions of these studies, including differences in tissue preparations and tension development before the application of H_2S. The putative interactions of NO and H_2S are hypothetically presented in **Fig. 3**.

CONCLUDING REMARKS AND PERSPECTIVES

In keeping with the criteria listed in Table 1, H_2S might be classified as the third gasotransmitter besides NO

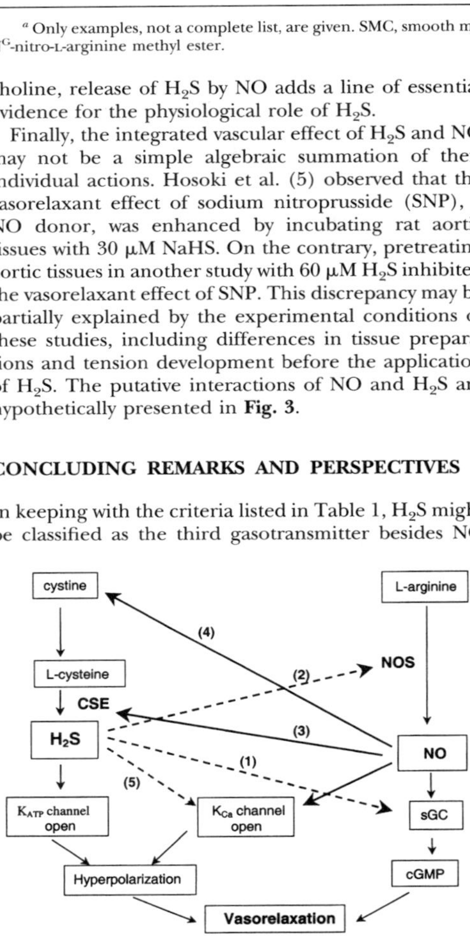

Figure 3. Hypothesized scheme of the interaction of H_2S and NO in vascular tissues. The solid lines indicate the stimulatory inputs and the dashed lines, inhibitory inputs. (*1*) H_2S may decrease the sensitivity of the cGMP pathway to NO (27). (*2*) H_2S may reduce the expression level of NO synthase (NOS). (*3*) NO may increase the expression of CSE. (*4*) NO may increase the cellular uptake of cystine. (*5*) H_2S may modify K_{Ca} channels to decrease their sensitivity to NO.

and CO. This gas is endogenously generated and manifests significant effects at physiologically relevant concentrations. The effect of H_2S on K_{ATP} channels may represent an important endogenous mechanism in vascular SMCs, neurons, and other excitable cells to couple cellular metabolism to excitability. By demonstrating the role of NO as an inducer or as a molecular switch for endogenous H_2S production, we can begin to understand how the interaction between H_2S and NO provides an integrated regulation of vascular tone. These advances in H_2S research may revolutionize many conventional doctrines. For example, hyperhomocystinemia is a disease with a deficient expression of CBS. The role of a low level of endogenous H_2S in the pathogenesis of this disease has been largely overlooked or simply neglected (13), yet it may be an important cause of atherosclerosis and thrombotic complications associated with hyperhomocystinemia. We still have a long way to go before a complete understanding of cellular metabolism and functions of H_2S is achieved. The following future studies of H_2S physiology serve only as examples.

1) Molecular mechanisms of the interaction of H_2S and K_{ATP} channels should be further investigated. As expression of different K_{ATP} channel subunits is tissue-type specific, whether H_2S stimulates K_{ATP} channels in other tissues (e.g., lungs, kidney, pancreas) as it does in vascular SMCs and neurons may be a key to the differential effects of H_2S on different tissues. Direct evidence, including single channel recording on heterologously expressed K_{ATP} channels in the presence of H_2S, should be collected. H_2S may interact with membrane and/or cytosol proteins to form reactive and unstable persulfides (44). These persulfides may take different forms, including protein-SSH, thiotaurine, thiocysteine, thiocystine, or mercaptopyruvate (45). The persulfide-related sulfuration and structural changes of the targeted proteins are recognized mechanisms for the biological effects of sulfide donors. This mechanism may underlie the interaction of H_2S and K_{ATP} channel proteins.

2) H_2S may alter cellular redox status. H_2S in an aqueous solution is a weak reducing agent. Vasorelaxation induced by H_2S was not mimicked by the disulfide bond-reducing agents (5) but the H_2S-induced modulation of Na^+ channels in neurons was (34). This controversy supports, rather than denounces, the importance of the reducing capability of H_2S. Quite likely, manifestation of the reducing effect of H_2S depends on the tissue-specific targets and the tissue-specific redox environment. Does H_2S have an oxidative potential? This is unsettled given the reported yield of free radicals from H_2S. In the presence of peroxidase and H_2O_2, H_2S produced thiyl free radicals (SH$^\bullet$ and S$^\bullet$) (46). More vigorous studies are needed to investigate the physiological effects of H_2S in the presence of different antioxidants, especially the scavengers for thiyl free radicals.

3) The endogenous inhibitors and stimulators for H_2S production should be explored. Since CBS is a heme-containing protein (10) and heme-containing proteins are common targets of NO and CO, the activity of CBS might be under the influence of both CO and NO (47). CSE activity is increased by L-cysteine (48), but this substance is not stable and may have neurotoxicity. Steroid hormones are putative modulators of CBS functions; one such example is the testosterone-induced increase in the activity of CBS (49). The expression of CBS is also inducible. Although no CBS protein could be detected in freshly isolated human aortic tissues, primarily cultured human aortic SMCs within five passages exhibited clear CBS activity and protein expression (58). This may imply a regulatory role of endogenous H_2S in the proliferation of vascular SMCs, which are normally quiescent.

4) Pharmacological or genomic manipulation of H_2S production is an underdeveloped area with great potential. Enhancement of CBS activity by S-adenosyl-methionine (6, 9, 50) may find novel applications in dealing with some brain disorders. However, S-adenosyl-methionine may have other effects unrelated to the endogenous generation of H_2S due to its methyl donor role. Specific activators of CSE, which is uniquely expressed in vascular tissues, are not available at present, but these agents can be important tools in the regulation of abnormal cardiovascular functions related to the altered endogenous H_2S metabolism. Most if not all of the currently available inhibitors for different types of the H_2S-generating enzymes are not membrane permeable, which significantly impedes their applications under physiological conditions. A heterozygous deficiency of CBS mice has been established (51). The transgenic animal model with CSE deletion will be needed to establish the contribution of this enzyme to endogenous H_2S levels in vascular tissues.

5) Investigations should begin to look into the pathological role of endogenous H_2S. Deficiency in CBS expression causes hyperhomocystinemia, which leads to premature peripheral and cerebral occlusive arterial disease (52). The pathogenic role of low levels of H_2S in this disease has not been explored. Similarly, homocystinuria is an autosomal recessively inherited disorder (53) that may be closely related to the low endogenous production of H_2S. On the other hand, Down syndrome with elevated CBS expression, low plasma homocysteine, and significantly increased thiosulfate urinary excretion (54) may couple to abnormally high H_2S levels. These observations have led to the hypothesis that the accumulation of H_2S in the brain could cause the metabolic intoxication (55). Sudden infant death syndrome may be related to the abnormally higher taurine levels induced by H_2S (34). The development of vascular diseases after heart transplantation is accompanied by increased total plasma homocysteine concentrations (56). In this case and other vasculopathy circumstances, a potentially lower endogenous level of H_2S may be an important pathogenic factor.

Now that the role of H_2S has been identified as sharing metabolic mechanisms and cellular effects similar to NO and CO, it is the time to call the family of gasotransmitters to 'please stand up.' It is expected that the gasotransmitter family will be expanded to include other yet undefined endogenous gaseous molecules. FJ

The author thanks Dr. J. Thornhill for reading through this study, and thanks to the Natural Sciences and Engineering Research Council of Canada for supporting this project.

REFERENCES

1. Furchgott, R. F., and Zawadski, J. V. (1980) The obligatory role of endothelial cells in the relaxation of arterial smooth muscle by acetylcholine. *Nature (London)* **228,** 373–376
2. Cao, K., and Wang, R. (2001) Carbon monoxide, vascular contractility, and K^+ channels. In *CO and Cardiovascular Functions.* (Wang, R., ed) pp. 83–109, CRC Press, Boca Raton, LA
3. Pace, N. R. (1997) A molecular view of microbial diversity and the biosphere. *Science* **276,** 734–740
4. Zhao, W. Zhang, J., Lu, Y., and Wang, R. (2001) The vasorelaxant effect of H_2S as a novel endogenous gaseous K_{ATP} channel opener. *EMBO J.* **20,** 6008–6016
5. Hosoki, R., Matsiki, N., and Kimura, H. (1997) The possible role of hydrogen sulfide as an endogenous smooth muscle relaxant in synergy with nitric oxide. *Biochem. Biophysic. Res. Commun.* **237,** 527–531
6. Abe, K., and Kimura, H. (1996) The possible role of hydrogen sulfide as an endogenous neuromodulator. *J. Neurosci.* **16,** 1066–1071
7. Bukovska, G., Kery, V., and Kraus, J. P. (1994) Expression of human cystathionine beta-synthase in *Escherichia coli*: purification and characterization. *Protein Exp. Purif.* **5,** 442–448
8. Erickson P. F., Maxwell, I. H., Su, L. J., Baumann, M., and Glode, L. M. (1990 Sequence of cDNA for rat cystathionine γ-lyase and comparison of deduced amino acid sequence with related *Escherichia coli* enzymes. *Biochem. J.* **269,** 335–340
9. Stipanuk, M. H., and Beck, P. W. (1982) Characterization of the enzymic capacity for cysteine desulphhydration in liver and kidney of the rat. *Biochem. J.* **206,** 267–277
10. Meier, M., Janosik, M., Kery, V., Kraus, J. P., and Burkhard, P. (2001) Structure of human cystathionine beta-synthase: a unique pyridoxal 5′-phosphate-dependent heme protein. *EMBO J.* **20,** 3910–3916
11. Levonen, A. L., Lapatto, R., Saksela, M., and Raivio, K. O. (2000) Human cystathionine gamma-lyase: developmental and in vitro expression of two isoforms. *Biochem. J.* **347,** 291–295
12. Lu, Y., O'Dowd, B. F., Orrego, H., and Isreal, Y. (1992) Cloning and nucleotide sequence of human liver cDNA encoding for cystathionine gamma-lyase. *Biochem. Biophys. Res. Commun.* **189,** 749–758
13. Yap, S., Naughten, E. R., Wilcken, B., Wilcken, D. E., and Boers, G. H. (2000) Vascular complications of severe hyperhomocysteinemia in patients with homocystinuria due to cystathionine beta-synthase deficiency: effects of homocysteine-lowering therapy. *Semin. Thromb. Hemost.* **26,** 335–340
14. van der Molen, E. F., Hiipakka, M. J., van Lith-Zanders, H., Boers, G. H., van den Heuvel, L. P., Monnens, L. A., and Blom, H. J. (1997) Homocysteine metabolism in endothelial cells of a

patient homozygous for cystathionine beta-synthase (CS) deficiency. *Thromb. Haemost.* **78,** 827–833
15. Kredich, N. M., Foote, L. J., and Keenen, B. S. (1973) The stoichiometry and kinetics of the inducible cysteine desulfhydrase from *Salmonella typhimurium. J. Biol. Chem.* **248,** 6187–6197
16. Simpson, R. C., and Freedland, R. A. (1976) Factors affecting the rate of gluconeogenesis from L-cysteine in the perfused rat liver. *J. Nutr.* **106,** 1272–1278
17. Searcy, D. G., and Lee, S. H. (1998) Sulfur reduction by human erythrocytes. *J. Exp. Zool.* **282,** 310–322
18. Westely, A. M., and Westley, J. (1991) Biological sulfane sulfur. *Biochemistry* **195,** 63–67
19. Buzaleh, A. M., Vazquez, E. S., and del Carmen, Batlle. (1990) Cyanide intoxication—III. On the analogous and different effects provoked by non-lethal and lethal challenged doses. *Gen. Pharmacol.* **21,** 27–32
20. Beauchamp, R. O., Bus, J. S., Popp, J. A., Boreiko, C. J., and Andjelkhovich, D. A. (1984) A critical review of the literature on hydrogen sulfide toxicity. *CRC Crit. Rev. Toxicol.* **13,** 25–97
21. Smith, R. P., and Abbanat, R. A. (1966) Protective effect of oxidized glutathione on acute sulfide poisoning. *Toxicol. Appl. Pharmacol.* **9,** 209–217
22. Wang, R. (1998) Resurgence of carbon monoxide: an endogenous gaseous vasorelaxing factor. *Can. J. Physiol. Pharmacol.* **76,** 1–15
23. Arp, A. J., Childress, J. J., and Vetter, R. D. (1987) The sulphide-binding protein in the blood of the vestimentiferan tubeworm, *Riftia pachyptila*, is the extracellular hemoglobin. *J. Exp. Biol.* **128,** 139–158
24. Awata, S., Nakayama, K., Suzuki, I., and Kodama, H. (1989) Effect of cysteine on the inactivation of cystathionine gamma-lyase by D,L-propargylglycine. *Acta Med. Okayama* **A3,** 329–335
25. Swaroop, M., Bradley, K., Ohura, T., Tahara, T., Roper, M. D., Rosenberg, L. E., and Kraus, J. P. (1992) Rat cystathionine β-synthase. *J. Biol. Chem.* **267,** 11455–11461
26. Warenycia, M. W., Goodwin, L. R., Benishin, C. G., Reiffenstein, R. J., Francom, D. M., Taylor, J. D., and Dieken, F. P. (1989) Acute hydrogen sulfide poisoning: demonstration of selective uptake of sulfide by the brainstem by measurement of brain sulfide levels. *Biochem. Pharmacol.* **38,** 973–981
27. Goodwin, L. R., Francom, D., Dieken, F. P., Taylor, J. D., Warenycia, M. W., Reiffenstein, R. J., and Dowling, G. (1989) Determination of sulfide in brain tissue by gas dialysis/ion chromatography: postmortem studies and two case reports. *J. Analyt. Toxicol.* **13,** 105–109
28. Mitchell, T. W., Savage, J. C., and Gould, D. H. (1993) High-performance liquid chromatography detection of sulfide in tissues from sulfide-treated mice. *J. Appl. Toxicol.* **13,** 389–394
29. Savage, J. C., and Gould, D. H. (1990) Determination of sulfide in brain tissue and rumen fluid by ion interaction, reversed-phase high-performance liquid chromatography. *J. Chromatogr.* **526,** 540–545
30. Awata, S., Nakayama, K., Suzuki, I., Sugahara, K., and Kodama, H. (1995) Changes in cystathionine gamma-lyase in various regions of rat brain during development. *Biochem. Mol. Biol. Int.* **35,** 1331–1338
31. Skrajny, B., Hannah, R. S., and Roth, S. N. (1992) Low concentrations of hydrogen sulfide alter monoamine levels in the developing rat central nervous system. *Can. J. Physiol. Pharmacol.* **70,** 1515–1518
32. Roth, S. H., Skrajny, B., and Reiffenstein, R. J. (1995) Alteration of the morphology and neurochemistry of the developing mammalian nervous system by hydrogen sulfide. *Clin. Exp. Pharmacol. Physiol.* **22,** 379–380
33. Dello Russo, C., Tringali, G., Ragazzoni, E., Maggiano, N., Menini, E., Vairano, M., Preziosi, P., and Navarra, P. (2000) Evidence that hydrogen sulphide can modulate hypothalamo-pituitary-adrenal axis function: in vitro and in vivo studies in the rat. *J Neuroendocrinol.* **12,** 225–233
34. Warenycia, M. W., Steele, J. A., Karpinski, E., and Reiffenstein, R. J. (1989) Hydrogen sulfide in combination with taurine or cysteic acid reversibly abolishes sodium currents in neuroblastoma cells. *Neurotoxicology* **10,** 191–199
35. Kombian, S. B., Warenycia, M. W., Mele, F., and Reiffenstein, R. J. (1988) Effects of acute intoxication with hydrogen sulfide on central amino acid transmitters systems. *Neurotoxicology* **9,** 587–596
36. Reiffenstein, R. J., Hulbert, W. C., and Roth, S. H. (1992) Toxicology of hydrogen sulfide. *Annu. Rev. Pharmacol. Toxicol.* **32,** 109–134
37. Kimura, H. (2000) Hydrogen sulfide induces cyclic AMP and modulates the NMDA receptor. *Biochem. Biophys. Res. Commun.* **267,** 129–133
38. Chen, P., Poddar, R., Tipa, E. V., Dibello, P. M., Moravec, C. D., Robinson, K., Green, R., Kruger, W. D., Garrow, T. A., and Jacobsen, D. W. (1999) Homocysteine metabolism in cardiovascular cells and tissues: implications for hyperhomocysteinemia and cardiovascular disease. *Adv. Enzyme Regul.* **39,** 93–109
39. Bao, L., Vlcek, C., Paces, V., and Kraus, J. P. (1998) Identification and tissue distribution of human cystathionine beta-synthase mRNA isoforms. *Arch. Biochem. Biophys.* **350,** 95–103
40. Doughty, J. M., Plane, F., and Langton, P. D. (1999) Charybdotoxin and apamin block EDHF in rat mesenteric artery if selectively applied to the endothelium. *Am. J. Physiol.* **276,** H1107–H1112
41. Guidotti, T. L. (1996) Hydrogen sulfide. *Occup. Med.* **46,** 367–371
42. Bartholomew, T. C., Powell, G. M., Dodgson, K. S., and Curtis, C. G. (1980) Oxidation of sodium sulphide by rat liver, lungs and kidney. *Biochem. Pharmacol.* **29,** 2431–2437
43. Nicholls, P., and Kim, J. K. (1981) Oxidation of sulphide by cytochrome aa3. *Biochim. Biophys. Acta* **637,** 312–320
44. Valentine, W. N., Toohey, J. I., Paglia, D. E., Nakatani, M., and Brockway, R. A. (1987) Modification of erythrocytes enzyme activities by persulfides and methanethiol: possible regulatory role. *Proc. Natl. Acad. Sci. USA* **84,** 1394–1398
45. Wood, J. L. (1982) Biochemical functions of persulfides. *Adv. Exp. Med. Biol.* **148,** 327–342
46. Nicholls, P. (1961) The formation and properties of sulphmyoglobin and sulphcatalase. *Biochem. J.* **81,** 374–383
47. Bruno, S., Schiaretti, F., Burkhard, P., Kraus, J. P., Janosik, M., Mozzarelli, A. (2001) Functional properties of the active core of human cystathionine beta-synthase crystals. *J. Biol. Chem.* **276,** 16–19
48. Kredich, N. M., Keenan, B. S., and Foote, L. J. (1972) The purification and subunit structure of cysteine desulfhydrase from *Salmonella typhimurium. J. Biol. Chem.* **244,** 7157–7162
49. Manteuffeil-Cymborowska, M., Chmurzynska, W., and Grzelakowska-Sztabert, B. (1992) Tissue-specific effects of testosterone on S-adenosylmethionine formation and utilization in the mouse. *Biochim. Biophys. Acta* **1116,** 166–172
50. Finkelstein, J. D., Kyle, W. E., Martin, J. J., and Pick, A. M. (1975) Activation of cystathionine synthase by adenosylmethionine and adenosylmethionine. *Biochem. Biophys. Res. Commun.* **66,** 81–87
51. Eberhardt, R. T., Forgione, M. A., Cap, A., Leopold, J. A., Rudd, M. A., Trolliet, M., Heydrick, S., Stark, R., Klings, E. S., Moldovan, N. I., Yaghoubi, M., Goldschmidt-Clermont, P. J., Farber, H. W., Cohen, R., and Loscalzo, J. (2000) Endothelial dysfunction in a murine model of mild hyperhomocyst(e)inemia. *J. Clin. Invest.* **106,** 483–491
52. Boers, G. H., Smals, A. G. H., Trijbels, F. J. M., Fowler, B., Bakkeren, J. A. J. M., Schoonderwaldt, H. C., Kleijer, W. J., and Kloppenborg, P. W. C. (1985) Heterozygosity for homocystinuria in premature peripheral and cerebral occlusive arterial disease. *New Engl. J. Med.* **313,** 709–715
53. Mudd, S. H., Levy, H. L., and Skovby, F. (1995) In *The Metabolic Basis of Inherited Disease* (Scriver, C. R., Beaudet, A. L., Sly, W. S., and Valle, D., eds) 7th Ed, Vol. 1, pp. 1279–1327, McGraw-Hill, New York
54. Chadefaux, B., Ceballos, I., Hamet, M., Coude, M., Poissonnier, M., Kamoun, P., and Allard, D. (1988) Is absence of atheroma in Down syndrome due to decreased homocysteine levels? *Lancet* **2,** 741
55. Kamoun, P. (2001) Mental retardation in Down syndrome: a hydrogen sulfide hypothesis. *Med. Hypotheses* **57,** 389–392
56. Berger, P. B., Jones, J. D., Olson, L. J., Edwards, B. S., Frantz, R. P., Rodeheffer, R. J., Kottke, B. A., Daly, R. C., and McGregor, C. G. A. (1995) Increase in total plasma homocysteine concentration after cardiac transplantation. *Mayo Clin. Proc.* **70,** 125–131

Received for publication April 18, 2002.
Accepted for publication July 17, 2002.

REFERENCES

1. Lippmann M. Introduction and background. In: Lippmann M, ed. Environmental Toxicants: Human Exposure and Their Health Effects. John Wiley: New York, 2000.
2. Jain KK. Carbon Monoxide Poisoning. Warren H. Green: St. Louis, MO, 1990.
3. Pace NR. A molecular view of microbial diversity and the biosphere. Science 1997;276:734–740.
4. Anderson IC, Poth M, Homstead J, et al. A comparison of NO and N_2O production by the autotrophic nitrifier Nitrosomonas europaea and the heterotrophic nitrifier Alcaligenes faecalis. Appl Environ Microbiol 1993;59:3525–3533.
5. Indrehus O, Vassbotn P. CO and NO_2 pollution in a long two-way traffic road tunnel: investigation of NO_2/NOx ratio and modelling of NO_2 concentration. J Environ Monit 2001;3:221–225.
6. Schlesinger RB. Nitrogen oxides. In: Lippmann M, ed. Environmental Toxicants—Human Exposure and Their Health Effects. John Wiley: New York, 2000.
7. Vleeming W, Rambali B, Opperhuizen A. The role of nitric oxide in cigarette smoking and nicotine addiction. Nicotine Tob Res 2002;4:341–348.
8. Stuven R, Bock E. Nitrification and denitrification as a source for NO and NO_2 production in high-strength wastewater. Water Res 2001;35:1905–1914.
9. Mercer RR, Costa DL, Crapo JD. Effects of prolonged exposure to low doses of nitric oxide or nitrogen dioxide on the alveolar septa of the adult rat lung. Lab Invest 1995;73:20–28.
10. Sagai M, Ichinose T, Kubota K. Studies on the biochemical effects of nitrogen dioxide. IV. Relation between the change of lipid peroxidation and the antioxidative protective system in rat lungs upon life span exposure to low levels of NO_2. Toxicol Appl Pharmacol 1984;73:444–456.
11. Prutz WA, Monig H, Butler J, et al. Reactions of nitrogen dioxide in aqueous model systems: oxidation of tyrosine units in peptides and proteins. Arch Biochem Biophys 1985;243:125–134.
12. Giroux M, Ruidavets JB, Ferrieres J. Atmospheric NO, temperature and ischaemic heart disease: study in Toulouse and its conurbation. Sci Total Environ 2000;246 293, 294.
13. Rowland A, Murray AJ, Wellburn AR. Oxides of nitrogen and their impact upon vegetation. Rev Environ Health 1985;5:295–342.
14. U.S. Centers for Diseases Control. Carbon monoxide intoxication—a preventable environmental health hazard. Morb Mortal Weekly Rep 1982;31:529–531.
15. Roughton FJW. The equilibrium of carbon monoxide with human hemoglobin in whole blood. In: Biological Effects of Carbon Monoxide, Proceedings of a Conference. Ann NY Acad Sci 1970; 174:177–188.
16. Lukin JA, Simplaceanu V, Zou M, et al. NMR reveals hydrogen bonds between oxygen and distal histidines in oxyhemoglobin. Proc Natl Acad Sci USA 2000;97:10,354–10,358.
17. World Health Organization. Environmental Health Criteria 213. Carbon Monoxide. 2nd ed. World Health Organization: Geneva, 1999.
18. Stewart RD, Baretta ED, Platte LR, et al. Carboxyhemoglobin levels in American blood donors. JAMA 1974;229:1187–1195.
19. Stern FB, Halperin WE, Hornung RW, et al. Heart disease mortality among bridge and tunnel officers exposed to carbon monoxide. Am J Epidemiol 1988;128:1276–1288.
20. Penney DG, Howley JW. Is there a connection between carbon monoxide exposure and hypertension? Environ Health Perspect 1991;95:191–198.
21. Krupski WC. The peripheral vascular consequences of smoking. Ann Vasc Surg 1991;5:291–304.
22. Penney DG, Skikun RM. Hypertension is not exacerbated by chronic carbon monoxide exposure, with or without added salt, in the borderline hypertensive rat. Arch Toxicol Suppl 1991;14:118–123.
23. Penney DG. A review: hemodynamic response to carbon monoxide. Environ Health Perspect 1988;77:121–130.
24. Penney DG, Bugaisky LB. Non-coordinate expression of collagen mRNAs during carbon monoxide-induced cardiac hypertrophy. Mol Cell Biochem 1992;109:37–41.
25. Penney DG, Giraldo AA, Van Egmond EM. Coronary vessel alterations following chronic carbon monoxide exposure in the adult rat. J Appl Toxicol 1994;14:47–54.
26. Penney DG, Giraldo AA, Van Egmond EM. Chronic carbon monoxide exposure in young rats alters coronary vessel growth. J Toxicol Environ Health 1993;39:207–222.
27. Smith RP, Gosselin RE. Hydrogen sulfide poisoning. J Occup Med 1979;21:93–97.
28. Winder CV, Winder HO. The seat of action of sulfide on pulmonary ventilation. Am J Physiol 1933;105:337–352.
29. Carlos V. Lozano. Gas leak near Ventura kills 3 oil workers (hydrogen sulfide gas leak near Vintage Petroleum's plant). Los Angeles Times. August 11, 1994:113,A1, col 5 (21 col in).

30. U.S. Environmental Protection Agency (EPA). Health assessment document for hydrogen sulfide. Environmental Criteria and Assessment Office, Office of Health and Environmental Assessment, Office of Research and Development. Research Triangle Park, NC: EPA, 1992;EPA/600/8–86/026F.
31. Snyder JW, Safir EF, Summerville GP, et al. Occupational fatality and persistent neurological sequela after mass exposure to hydrogen sulfide. Am J Emerg Med 1995;13:199–203.
32. Evans CL. The toxicity of hydrogen sulfide and other sulfides. J Exp Physiol 1967;52:231–248.
33. Finklea JF. Criteria for a recommended standard: occupational exposure to hydrogen sulfide. DHEW (NIOSH). 1997:77–158.
34. Reiffenstein RJ, Hulbert WC, Roth SH. Toxicology of hydrogen sulfide. Annu Rev Pharmacol Toxicol 1992;32:109–134.
35. Guidotti TL. Occupational exposure to hydrogen sulfide in the, sour gas industry: some unresolved issues. Int Arch Occup Environ Health 1994;66:153–160.
36. Legator MS, Singleton CR, Morris DL, et al. Health effects from chronic low-level exposure to hydrogen sulfide. Arch Environ Health 2001;56:123–131.
37. Beauchamp RO Jr., Bus JS, Popp JA, et al. A critical review of the literature on hydrogen sulfide toxicity. CRC Crit Rev Toxicol 1984;13:25–97.
38. Bates MN, Garrett N, Graham B, et al. Air pollution and mortality in the Rotorua geothermal area. Aust N Z J Public Health 1997;21:581–586.
39. Bates MN, Garrett N, Graham B, et al. Cancer incidence, morbidity, and geothermal pollution in Rotorua, New Zealand. Int J Epidemiol 1998;27:10–14.
40. Bates MN, Garrett N, Shoemack P. Investigation of health effects of hydrogen sulfide from a geothermal source. Arch Environ Health 2002;57:405–411.
41. Xu X, Cho SI, Sammel M, et al. Association of petrochemical exposure with spontaneous abortion. Occup Environ Med 1998;55:31–36.
42. U.S. Department of Health and Human Services. Toxicological Profile for Hydrogen Sulfide. Agency for Toxic Substances and Disease Registry: Atlanta, GA, 1999.
43. Logue JN, Ramaswamy K, Hersh JH. Investigation of illness associated with exposure to hydrogen sulfide among Pennsylvania school students. J Environ Health 200;63:9–13.
44. Wang R. Resurgence of carbon monoxide: an endogenous gaseous vasorelaxing factor. Can J Physiol Pharmacol 1998;76:1–15.
45. Ignarro LJ. Nitric oxide: a unique endogenous signaling molecule in vascular biology. Biosci Rep 1999;19:51–71.
46. Bian K, Murad F. Nitric oxide (NO)—biogeneration, regulation, and relevance to human diseases. Front Biosci 2003;8:D264–D278.
47. Sjöstrand T. Endogenous formation of carbon monoxide. Acta Physiol Scand 1950;22:137–141.
48. Stevens CF, Wang Y. Reversal of long-term potentiation by inhibitors of haem oxygenase. Nature 1993;364 :147–149.
49. Zhuo M, Small SA, Kandel ER, et al. Nitric oxide and carbon monoxide produce activity-dependent long-term synaptic enhancement in hippocampus. Science 1993;260:1946–1950.
50. Wang R. CO and cardiovascular functions. CPC Press: Boca Raton, FL, 2001.
51. Abraham NG, Alam J, Nath K, et al. Heme Oxygenase in Biology and Medicine. Plenum Press, New York, 2002.
52. Marks GS, Brien JF, Nakatsu K, et al. Does carbon monoxide have a physiological function? Trends Pharmacol Sci 1991;12:185–188.
53. Wang R. Two's company, three's a crowd—can H_2S be the third endogenous gaseous transmitter? FASEB J 2002;16:1792–1798.
54. Stipanuk MH, Beck PW. Characterization of the enzymic capacity for cysteine desulphhydration in liver and kidney of the rat. Biochem J 1982;206:267–277.
55. Awata S, Nakayama K, Suzuki I, et al. Effect of cysteine on the inactivation of cystathionine gamma-lyase by D,L-propargylglycine. Acta Med Okayama 1989;43:329–335.
56. Dello Russo C, Tringali G, Ragazzoni E, et al. Evidence that hydrogen sulphide can modulate hypothalamo-pituitary-adrenal axis function: in vitro and in vivo studies in the rat. J Neuroendocrinol 2000;12:225–233.
57. Abe K, Kimura H. The possible role of hydrogen sulfide as an endogenous neuromodulator. J Neurosci 1996;16:1066–1071.
58. Kimura H. Hydrogen sulfide induces cyclic AMP and modulates the NMDA receptor. Biochem Biophys Res Commun 2000;267:129–133.

59. Zhao W, Zhang J, Lu Y, et al. H_2S is an endogenous K_{ATP} channel opener in vascular smooth muscle cells. EMBO J 2001;20:6008–6016.
60. Zhao W, Wang R. H_2S-induced vasorelaxation and underlying cellular and molecular mechanisms. Am J Physiol 2002;283:H474–H480.
61. Teague B, Asiedu S, Moore PK. The smooth muscle relaxant effect of hydrogen sulphide in vitro: evidence for a physiological role to control intestinal contractility. Br J Pharmacol 2002;137:139–145.
62. Tan DX, Manchester LC, Hardeland R, et al. Melatonin: a hormone, a tissue factor, an autocoid, a paracoid, and an antioxidant vitamin. J Pineal Res 2003;34:75–78.
63. Kandel ER, Schwartz JH, Jessell TM. Principles of Neural Science. 4th ed. McGraw-Hill: New York, 2000.
64. Cary SP, Marletta MA. The case of CO signaling: why the jury is still out. J Clin Invest 2001;107:1071–1073.
65. Stamler JS, Simon DI, Osborne JA, et al. S-Nitrosylation of proteins with nitric oxide: synthesis and characterization of biologically active compounds. Proc Natl Acad Sci USA 1992;89:444–448.
66. Wang R, Wu L. The chemical modification of K_{Ca} channels by carbon monoxide in vascular smooth muscle cells. J Biol Chem 1997;272:8222–8226.
67. Wang R, Wu L, Wang ZZ. The direct effect of carbon monoxide on K_{Ca} channels in vascular smooth muscle cells. Pflügers Arch 1997;434:285–291.
68. Wu L, Cao K, Lu Y, et al. Different mechanisms underlying the stimulation of K_{Ca} channels by nitric oxide and carbon monoxide. J Clin Invest 2002;110:691–700.
69. Cooke JP. The 1998 Nobel Prize in medicine: clinical implications for 1999 and beyond. Vasc Med 1999;4:57–60.
70. Brattstrom LE, Hardebo JE, Hultberg BL. Moderate homocysteinemia—a possible risk factor for arteriosclerotic cerebrovascular disease. Stroke 1984;15:1012–1016.
71. Chen P, Poddar R, Tipa EV, et al. Homocysteine metabolism in cardiovascular cells and tissues: implications for hyperhomocysteinemia and cardiovascular disease. Adv Enzyme Regul 1999;39:93–109.
72. Jiang ZY, Costachescu T, Derouin M, et al. Treatment of pulmonary hypertension during surgery with nitric oxide and vasodilators. Can J Anaesth 2000;47:552–555.
73. Hubert MB, Salazkin I, Desjardins J, et al. Cardiopulmonary bypass surgery in swine: a research model. JEANS 2004;in press.
74. Melo LG, Agrawal R, Zhang L, et al. Gene therapy strategy for long-term myocardial protection using adeno-associated virus-mediated delivery of heme oxygenase gene. Circulation 2002;105:602–607.
75. Ndisang JF, Wang R, Vannacci A, et al. Haeme oxygenase-1 and cardiac anaphylaxis. Br J Pharmacol 2001;134:1689–1696.
76. Ndisang JF, Wang R. Mechanisms underlying selective regulation of blood pressure by heme oxygenase-1 in hypertension. Hypertension 2002;40:315–321.
77. Katori M, Busuttil RW, Kupiec-Weglinski JW. Heme oxygenase-1 system in organ transplantation. Transplantation 2002;74:905–912.
78. Chauveau C, Bouchet D, Roussel JC, et al. Gene transfer of heme oxygenase-1 and carbon monoxide delivery inhibit chronic rejection. Am J Transplant 2002;2:581–592.
79. Partlo LA, Sainsbury RS, Roth SH. Effects of repeated hydrogen sulphide (H_2S) exposure on learning and memory in the adult rat. Neurotoxicology 2001;22:177–189.

2 Interactions Between Gasotransmitters

Ray J. Carson, Gunter Seyffarth, Rubina Mian, and Helen Maddock

CONTENTS

SUMMARY

It is well established that nitric oxide (NO), carbon monoxide (CO), and hydrogen sulfide (H_2S) have signaling roles in the body. There are important similarities among them in their actions and generation, but there are also intriguing differences. The mechanism of action of H_2S still has not been fully elucidated. It is becoming increasingly clear that there are important interactions among the gasotransmitters. There is clear evidence of links between the NO- and CO-generating systems. So far, this is most apparent in the control of the cardiovascular system, and knowledge of the function of NO has led to new therapeutic interventions. There is also a suggestion of synergy between NO and H_2S that is not yet fully understood. Interactions between CO and H_2S have not yet been explored, and more research is required in this area. Interactions in the immune system also require more research, and increased understanding of this area could lead to novel therapies.

Key Words: Nitric oxide; carbon monoxide; hydrogen sulfide; gasotransmitters; interactions; signal transduction.

1. INTRODUCTION

Research in the field of gas signaling molecules has increased exponentially since Palmer et al. *(1)* published their seminal article. It is now well established that nitric oxide

From: *Signal Transduction and the Gasotransmitters: NO, CO, and H_2S in Biology and Medicine* Edited by: Rui Wang © Humana Press Inc., Totowa, NJ

Table 1
Comparison of Metabolism and Function of Gasotransmitters[a]

	H_2S	*CO*	*NO*
Main substrates	L-cysteine	Heme	L-arginine
Generating enzymes	CBS, CSE	Heme oxygenases	NO synthases
Inducer	NO	Free radicals	Acetylcholin, endotoxin
Scavenger	Hemoglobin	Hemoglobin	Hemoglobin
Inhibitor	D,L-Propargylglycerine	Zinc-PPIX	L-NAME
Protein targets	K_{ATP} channel cAMP (?)	cGMP, K_{Ca} channel	cGMP, K_{Ca} channel
Amino acid targets	?	Histidine	Cysteine
Half life in solution	Minutes	Minutes	Seconds
Production tissue source	SMC, not in EC	EC < SMC	EC > SMC

[a] Only examples, not a complete list, are given. SMC, smooth muscle cell; EC, endothelial cell; zinc-PPIX, zinc protoporphyrin-IX; L-NAME, N^G-nitro-L-arginine methyl ester.
(Reproduced from ref. *2*.)

(NO), carbon monoxide (CO), and hydrogen sulfide (H_2S) have signaling roles in the body. There are some similarities in their structure, properties, and actions. For example, because of their structural similarity to molecular oxygen (O_2), they all bind to heme groups in key protein molecules. It is not surprising, then, that they all relax smooth muscle and that there are interactions among them. Some of the similarities and differences among NO, CO, and H_2S are summarized in **Table 1** *(2)*. It is important to state that this chapter covers interactions among NO, CO, and H_2S at physiological levels and that the remit does not include toxicological effects at higher concentrations.

2. COMPARISON OF CELLULAR EFFECTS OF NO, CO, AND H_2S

It is well established that NO is a neurotransmitter in the central and peripheral nervous systems, a smooth muscle relaxant, and an inflammatory mediator. The main signaling target for NO is the enzyme guanylate cyclase, which converts guanosine *S'*-triphosphate to cyclic guanosine *S'*-monophosphate (cGMP). High levels of intracellular cGMP are known to cause relaxation of smooth muscle. Because NO is also a free radical (•NO) and can react with oxygen to produce peroxynitrite ($ONOO^-$), it also has an important role as an inflammatory mediator. Similarly to NO, CO has been described as a gaseous muscle relaxant and a neuronal messenger *(3)*. Like NO, exogenously administered CO relaxes isolated blood vessels and inhibits platelet aggregation, presumably by increasing intracellular cGMP levels *(4)*. Alternatively, CO may dilate blood vessels by interference with a cytochrome P450-based constrictor mechanism as described by Coceani *(5)*. His team demonstrated CO-induced vascular relaxation that remained unchanged after treatment with methylene blue, a guanylate cyclase inhibitor, which suggests that guanylate cyclase did not have a role in the relaxation. He found the involvement of a cytochrome P450 hemoprotein, which limited the effect of the vasoconstrictor endothelin *(5)*.

The CO produced in the body originates mainly from the breakdown of hemoglobin *(6)*. This degradation is catalyzed by the group of enzymes known as heme oxygenases (HOs), which oxidize the α-methine bridge of the heme porphyrin structure and thereby yield CO and biliverdin. HOs resemble mixed function oxidases as they require the reducing cofactor NADPH and oxygen for the oxidation of their substrate *(7)*. Two isoforms of HO have been found, an inducible form (HO-1) and a constitutive form (HO-2). HO-1 is induced by oxidative stress and is abundant in spleen and liver tissue, where it decomposes heme-containing proteins. HO-2 activity is mainly found in the brain and testes *(3,4)*. All HOs are inhibited by zinc- and tin-containing porphyrin analogs *(3,4,7)*.

The mechanisms of action of H_2S have not yet been fully elucidated. There is evidence that some of the effects of H_2S result from an increase in intracellular cyclic adenosine monophosphate (cAMP) and activation of the protein kinase A pathway *(8)*. Increased levels of cAMP are known to relax smooth muscle. H_2S has been shown to enhance *N*-methyl-D-aspartate receptor-mediated responses in neurons and neuronal cell lines *(9)*, which appears to be a specific action of H_2S compared to NO and CO. Some published studies have used *S*-nitroso-L-cysteine *(10,11)* or other sulfur-containing agents as NO donors; however, it is not clear whether these can also act as H_2S donors and perhaps cause relaxation of smooth muscle via this mechanism.

Large-conductance K_{Ca} channels are a common target for NO and CO; however, the exact mechanism of interaction at the molecular level is not known. These gasotransmitters excite K_{Ca} channels leading to opening, which increases K^+ conductance, causing hyperpolarization in smooth muscle cells and thus relaxation. New evidence has suggested that the interaction between NO and CO with K_{Ca} channels may be different. The effects of CO on K_{Ca} channels have been shown to be mediated via interactions with histidine residues and the α-subunit of the channel protein *(12)*. By contrast, NO modifies sulfydryl groups and interacts with the β-subunit *(12)*.

CO is much more stable than NO, and, therefore, its effects may be longer lasting and it could act at a distance from its site of production. The metabolism of sulfur-containing compounds in cells is highly complex and H_2S is probably rapidly metabolized after being produced. Alternatively, various groups, such as heme groups, bind H_2S, so its effects may be truncated by being taken up.

One difference between NO and the other gas signaling molecules is that its redox state varies and this changes its biological effects *(13)*. Different NO donors release different redox state forms of NO in biological systems; thus, (+)*S*-nitroso-*N*-acetylpenicillamine releases the free radical $^{\bullet}NO$, 3-morpholino-sydnonime forms NO and superoxide, and sodium nitroprusside generates the nitrosonium ion NO^+ *(4)*. A further complication is that exogenously applied NO can be converted from one redox form to another depending on the local conditions *(13)*. Similarly, it has been found that inducible nitric oxide synthase (iNOS) generates different redox forms of NO depending on the intracellular conditions *(14)*. This makes it difficult to determine exactly which form of NO is responsible for which actions, and perhaps this is not always fully taken into account.

3. INTERACTIONS BETWEEN THE CO AND NO SYSTEMS

Because there are some similarities between the production and effects of NO and CO, interactions between the two systems might be expected. In the vasculature, exogenously administered CO, like NO, relaxes blood vessels by increasing intracellular cGMP levels in vascular smooth muscle cells (VSMCs). Unlike HO-2, HO-1 is found in VSMCs, and NO selectively induces HO-1 gene expression and CO release in these cells *(4)*. Recently, it has been shown that NO triggers the release of free heme from heme proteins, and unbound heme is known to induce HO-1 expression. Unfortunately, the exact mechanism of this upregulation is not entirely clear. The ability of NO to induce CO production in

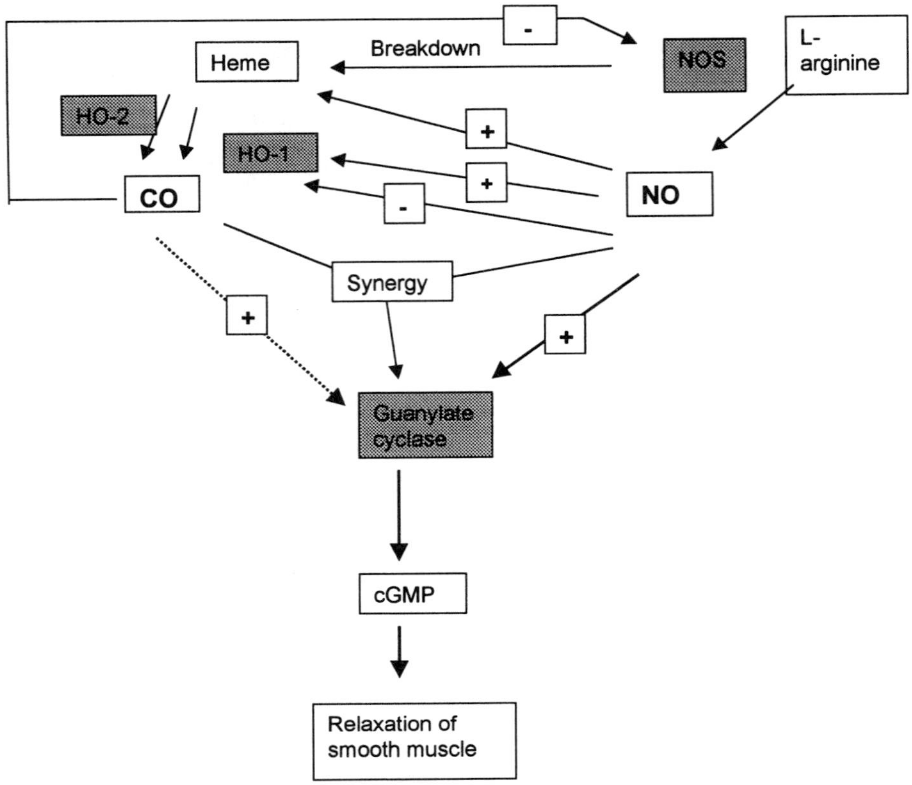

Fig. 1. Summary of interactions between NO and CO systems.

VSMCs may provide another mechanism by which NO activates guanylate cyclase and regulates vascular tone. Interestingly, CO directly inhibits iNOS activity by binding to the heme moiety of the enzyme. Thus, CO might act as a cytoprotector by limiting excessive NO synthesis, such as because of oxidative stress *(4)*.

Ingi et al. *(3)* showed that CO produced via HO-2 stimulates guanylate cyclase, such as in olfactory neurons, which do not show NOS activity. The findings of this study on a possible CO–NO interaction in cerebellar cells are controversial. HO-2 activity seems to peak in immature cerebellar cells, whereas NO output increases as these cells mature. CO obviously does not affect the NO–cGMP system in immature cells; however, Infi et al. *(3)* report that HO activity suppresses cGMP levels in later culture development, when NO is present. Furthermore, they found that CO inhibits purified NOS in vitro. They conclude that rather than CO interfering with the NO–cGMP system at the stage of NO synthesis, it interferes at the level of NO activation of guanylate cyclase. Here, CO may function as a partial agonist or inhibitory modulator of the enzyme, whereas NO acts as a full agonist of guanylate cyclase. Presumably, CO may bind to the enzyme and thereby induce conformational changes that may affect NO-mediated activation *(3)*. Interactions between the CO and NO systems are illustrated in **Fig. 1**.

3.1. Interactions of the CO- and NO-Generating Systems

There is increasing evidence of a link between the regulation of HO activity and NO production, but the purpose of this link is not fully clear. NO is a highly reactive free radical as well as a signaling molecule, and, therefore, its production must be controlled in cells. It seems that the HO system is one way in which this control is exerted *(15)*. NOS is a heme-containing protein, and binding of CO to the heme groups could inactivate the enzyme. The neuronal isoform of NOS, nNOS, has been shown to bind CO *(16)*. The heme groups in NOS could also act as a substrate for HO, thus decreasing the production of NO *(15)*.

NO has been shown to inhibit as well as stimulate HO enzyme. Incubation with the NO donors L-arginine and sodium nitroprusside has been found to both reduce HO activity *(17)* and to increase HO activity *(18)*. Maines *(15)* has proposed an explanation for this apparent paradox: NO as a free radical could inactivate HO by attacking cysteine residues in the protein, but it can also induce HO-1 expression, and by displacing O_2 from heme groups, NO could inhibit HO activity. In addition, iron is known to be involved in gene expression and iron metabolism can be influenced by NO *(15)*.

NO donors were found to selectively increase mRNA and protein expression for HO-1 in rat VSMCs in culture and to increase the production of CO, although a nonspecific bioassay for CO was used *(4)*. The mechanism by which NO induces the expression of HO-1 is not yet clear, although it does not seem to involve the cGMP signaling pathway *(4)*. It may involve the liberation of free heme, which is known to induce HO-1 expression *(4,15)*.

Hemin, a CO donor, has been found to potentiate L-arginine-stimulated insulin secretion from mice islet cells, suggesting a link between the CO and NO systems *(19)*. It has also been demonstrated that HO-1 expression and activity can be increased by both hemin and sodium nitroprusside in a rat skeletal muscle cell line *(20)*; however, the mechanism involved has not yet been fully elucidated.

In a study using cerebellar granule cell cultures, Ingi et al. *(3)* found that exogenously applied CO blocked an increase in cGMP mediated by NO. They also showed that inhibitors of endogenous CO production potentiated the increase in cGMP mediated by NO. By comparison, an inhibitor of endogenous NO production, N^{ω}-nitro-L-arginine, significantly inhibited dilation caused by CO in porcine pial arterioles in vivo, and the addition of an NO donor, sodium nitroprusside, restored vasodilation to CO *(21)*. This suggests that NO is essential for the vasodilatory effect of CO. Maines *(15)* has previously summarized interactions between NO and CO (**Fig. 2**).

4. INTERACTIONS BETWEEN H_2S AND NO

There is some published evidence to suggest that there is some synergy between H_2S and NO in the relaxation of vascular smooth muscle. Indeed, the relaxant effect of H_2S in vascular smooth muscle may be at least partially dependent on NO. In some elegant experiments, Zhao et al. *(22)* showed that N^{ω}-Nitro-L-arginine methyl ester (L-NAME), an inhibitor of endogenous NO production, significantly shifted the H_2S dose-response relaxation curve to the right, decreasing the potency of H_2S, in rat aortic rings (**Fig. 3**). Similar effects were obtained by removing the endothelium from the aortic rings. These findings suggest that H_2S stimulates the endogenous production of NO and this is at least partially responsible for the vasorelaxant effect of H_2S. These investigators also showed that the addition of a specific inhibitor of soluble guanylate cyclase (sGC), 1 H-[1,2,4]oxadiazolo[4,3,-*a*]quinoxalin-1-one, significantly increased the relaxant effect of H_2S, showing that it could not be working directly via stimulating guanylate cyclase. However, Teague et al. *(23)* found that L-NAME had no significant effect on relaxation of guinea pig ileum by H_2S. Interestingly, they also found that a combination of NaHS (an H_2S donor) and sodium nitroprusside (an NO donor) produced a significantly greater

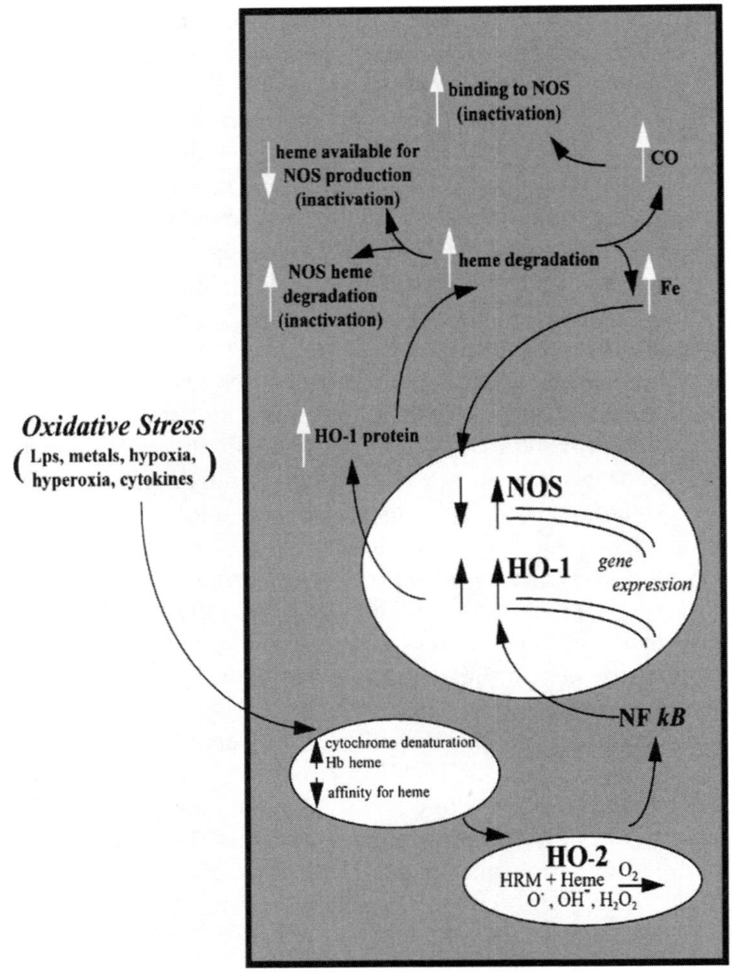

Fig. 2. Schematic presentation of regulatory interactions between HO and NOS systems proposed by Maines (Reprinted with permission from the *Annual Review of Pharmacology and Toxicology,* Volume 37 ©1997 by Annual Reviews www.annualreviews.org.)

inhibition of twitch response of guinea pig ileum than either agent alone, suggesting some synergy between the two agents. In our laboratory, we have shown that H_2S relaxes pregnant rat uterus in vitro *(24)*. However, methylene blue (an inhibitor of guanylate cyclase) did not significantly affect this tocolytic action (unpublished data), suggesting that H_2S does not cause relaxation by the activation of guanylate cyclase and increased production of cGMP in this tissue. The tocolytic action of H_2S was also not inhibited by glibenclamide (a K_{ATP} channel blocker) or tetraethylammonium (a nonspecific K^+ channel blocker) (unpublished data).

It is becoming increasingly clear that H_2S exerts its effects via different mechanisms in different tissues. In vascular smooth muscle, there seems to be involvement of NO production, because of the neighboring endothelium. However, in other smooth muscle NO production does not seem to be involved, yet there is still synergy between H_2S and NO. Hosoki et al. *(25)* were the first to suggest synergy between H_2S and NO in relaxing

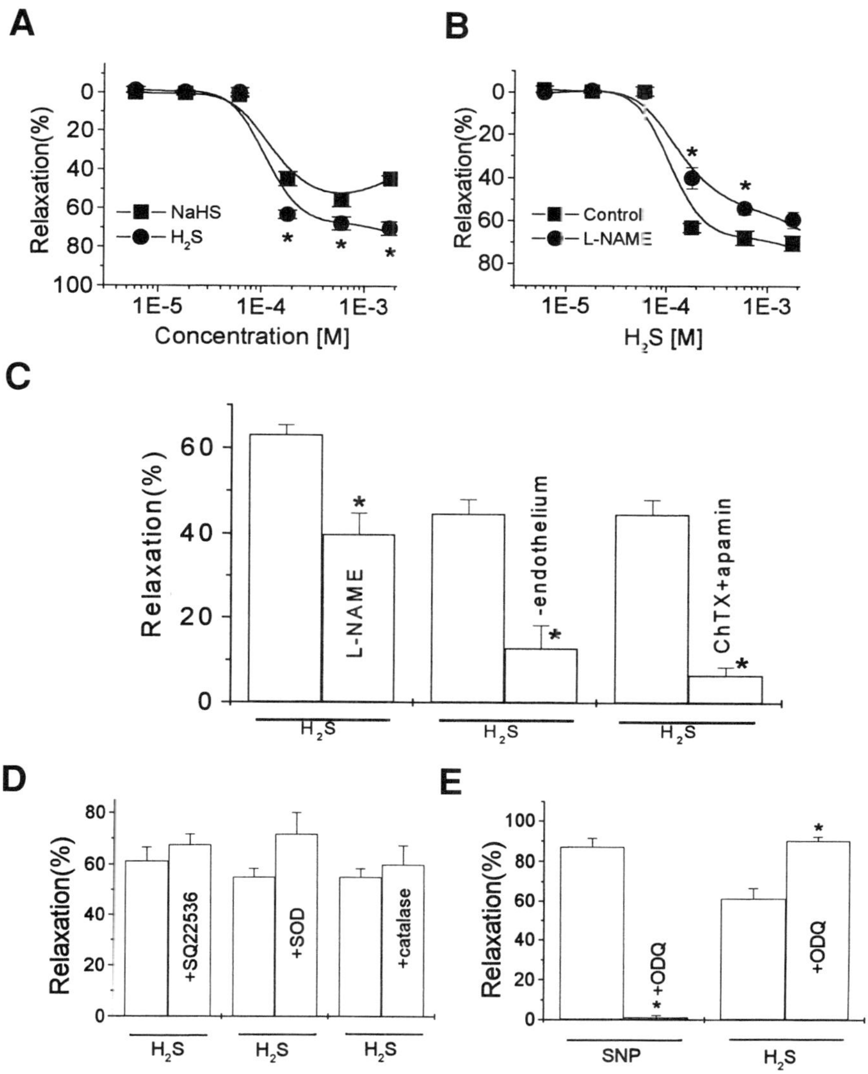

Fig. 3. Dose-response curves to H_2S and NaHS and underlying mechanisms. **(A)** Relaxation of phenylephrine-precontracted tissues by H_2S in form of either standard NaHS solution (■) or H_2S gas-saturated solution (●); **(B)** inhibitory effect of L-NAME (100 μ*M*, 20 min) (●) on H_2S-induced relaxation (control) (■); **(C)** effects of H_2S (180 μ*M*) on endothelium-free or endothelium-intact aortic tissues pretreated with L-NAME or charybdotoxin (ChTX)/apamin; **(D)** the relaxant effect of H_2S was not affected by pretreating the tissues with SQ22536, SOD, or catalase, respectively; **(E)** effect of 1H-[1,2,4]oxadiazolo[4,3,-a]quinoxalin-1-one (ODQ) treatment (10 μ*M* for 10 min) on relaxant effects of SNP (0.1 μ*M*) or H_2S (600 μ*M*). $^*p < 0.05$ compared to control. (Reproduced from ref. *22*.)

vascular smooth muscle. They demonstrated a leftward shift in the dose-response curve for relaxation of rat thoracic aorta by NaHS in the presence of two different NO donors, sodium nitroprusside and morpholinosydnonimine (**Fig. 4**). They reported that a low concentration of H_2S enhanced the smooth muscle relaxant effect of NO by up to 13-fold. However, Zhao and Wang *(26)* found that low doses of NaHS shifted the dose–response

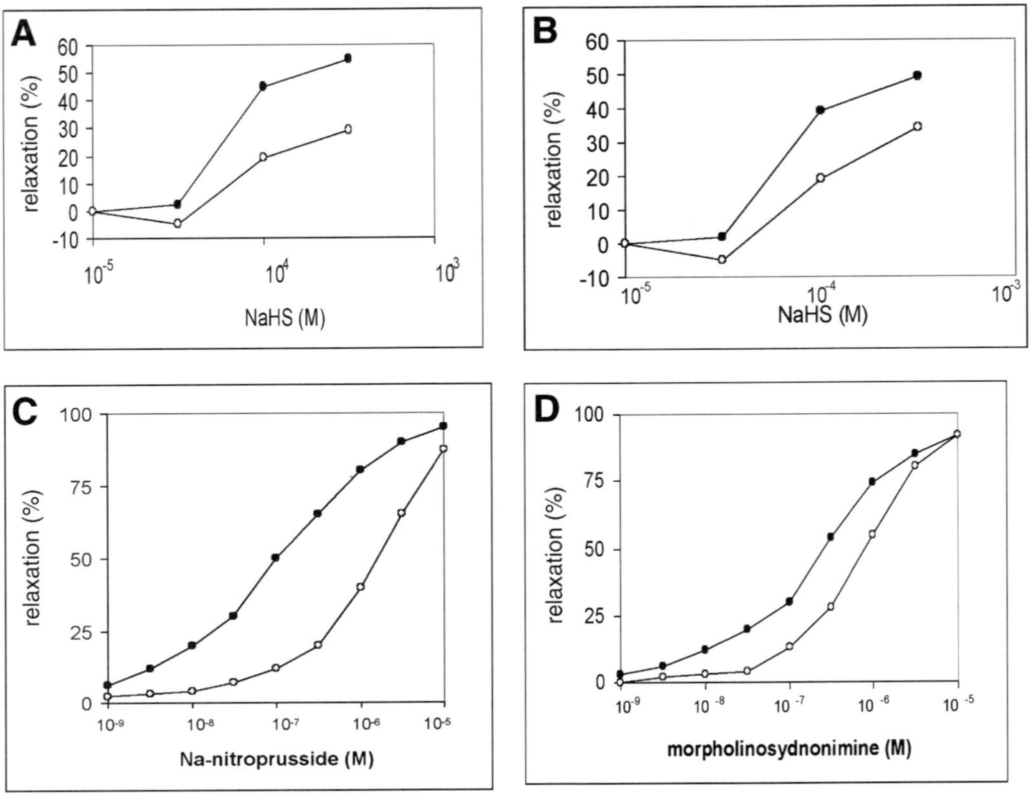

Fig. 4. Dose–response curves for relaxation of rat thoracic aorta helical strips in vitro because of NaHS and NO donors. **(A,B)** Potentiation of relaxant effects of NaHS by NO donors: **(A)** 10 nM sodium nitroprusside and **(B)** 30 nM morpholinosydnonimine; (○) control, (●) with NO donor. **(C,D)** Potentiation of relaxant effects of sodium nitroprusside **(C)** and morpholinosydnonimine **(D)** by NaHS, (○) control, (●) with 30 μ*M* NaHS. (Reprinted from *Biochem. Biophys. Res. Commun.*, 237, Hosoki et al., The possible role of hydrogen sulfide as an endogenous smooth muscle relaxant in synergy with nitric oxide, 527–531, copyright 1997, with permission from Elsevier.)

relaxation curve for sodium nitroprusside to the right in rat aortic rings, suggesting that H_2S inhibited the vasorelaxant effect of NO (**Fig. 5**). The contradiction of these results is not easy to explain. Different preparations, helical strips of Wistar rat aorta were used vs Sprague-Dawley rat aortic rings and different methods of precontraction were used, 1 μ*M* norepinephrine vs 0.3 μ*M* phenylephrine. If H_2S does decrease the production of cGMP, then this could explain how it decreases the response to NO, which works via this pathway. By contrast, if H_2S causes relaxation via a totally independent pathway to NO, then it could augment the relaxant effect of NO in an additive manner. It is not yet clear which mechanism is correct.

Li et al. *(27)* demonstrated that L-cysteine, an H_2S donor, inhibited NO-induced relaxation of rabbit aortic rings. L-Cysteine inhibited an increase in cGMP induced by NO, and superoxide dismutase (SOD) decreased the inhibitory effect of L-cysteine. These investigators concluded that the inhibitory effect of L-cysteine on NO was partly because of superoxide generation by the autooxidation of L-cysteine and partly via a direct interaction of SH groups with NO *(27)*.

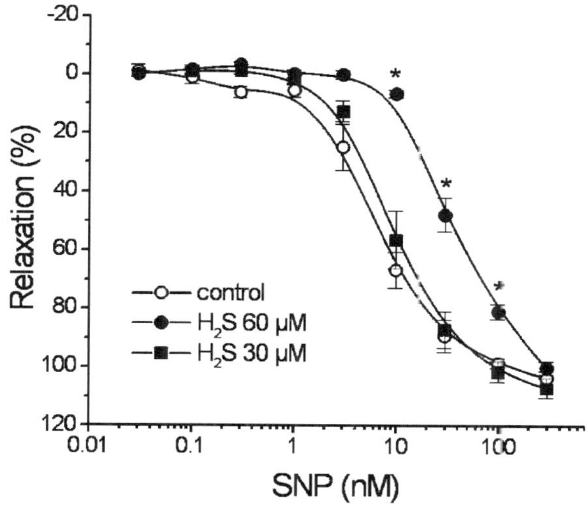

Fig. 5. Dose-response of precontracted rat aortic tissues to sodium nitroprusside. The tissues were pretreated with either 30 or 60 μM H_2S. * $p < 0.05$ vs control. (Reproduced from ref. *26*.)

It has been found that the endogenous production of H_2S by homogenized rat vascular tissue was increased by sodium nitroprusside in a dose-dependent manner *(22)*, suggesting a direct stimulatory effect of NO on the enzymes that produce H_2S, cystathionine β-synthase (CBS) and cystathionine γ-lyase (CSE). The mechanism of action could be downstream of cGMP, involving the stimulation of cGMP-dependent kinases, which could phosphorylate and activate the enzymes or could involve a direct effect on the enzyme protein, perhaps via nitrosylation. It was also shown that incubating rat VSMCs in culture with a dose range of sodium nitroprusside for 6 h significantly increased mRNA levels for CSE *(22)*; however, the mechanism of action of NO here is not yet clear, but it could involve nuclear factor-κB (NF-κB).

H_2S is produced endogenously from L-cysteine, but there is some debate as to whether L-cysteine is transported into cells as it is or as L-cystine, which consists of two molecules of L-cysteine joined by a disulfide bond *(24,28)*. Zerangue and Kavanaugh *(29)* have reported that L-cysteine was transported into cells by the neuronal EAAT3 excitatory amino acid transporter, which is known to be expressed in tissues other than neurons. Li et al. *(30)* showed that a NO donor increased cystine uptake into bovine vascular endothelial cells in a dose-dependent manner. This increase was found to require both RNA and protein synthesis and appears to be because of the induction of expression of a cystine transport system *(30)*. In theory, then, NO could increase the production of H_2S cells by increasing the availability of substrate.

Further interactions among NO, CO, and H_2S may become apparent because it has been reported that the activity of human CBS may be regulated by heme-mediated redox-linked mechanisms *(31)*.

Care must be taken in interpreting and comparing the results of different studies because there are differences in preparations and methods, such as the use of precontracted or spontaneously contracting smooth muscle preparations.

Wang *(2)* has previously summarized interactions between NO and H_2S (**Fig. 6**). Known interactions between the NO and H_2S systems are illustrated in **Fig. 7**.

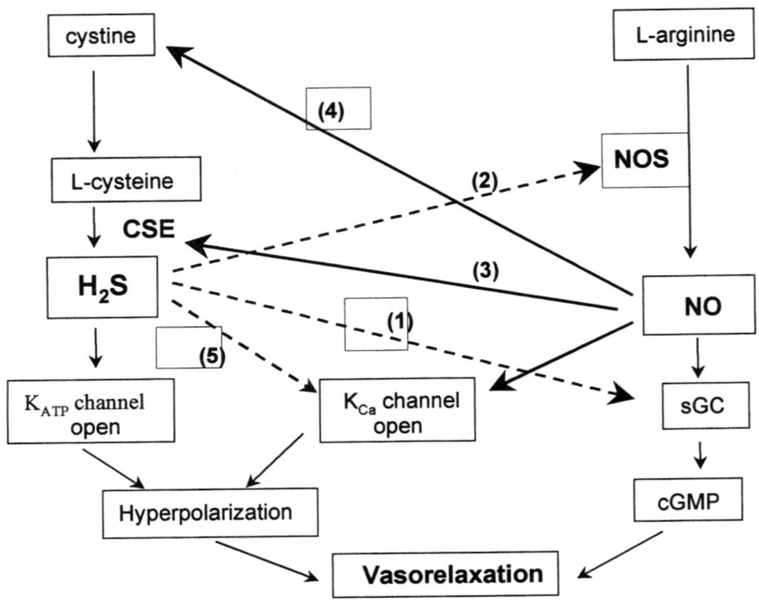

Fig. 6. Hypothesized scheme of interaction of H_2S and NO in vascular tissues as proposed by Wang *(2)*. The solid lines indicate the stimulatory inputs and the dashed lines, inhibitory inputs. (1) H_2S may decrease the sensitivity of the cGMP pathway to NO. (2) H_2S may reduce the expression level of NOS. (3) NO may increase the expression of CSE. (4) NO may increase the cellular uptake of cystine. (5) H_2S may modify K_{Ca} channels to decrease their sensitivity to NO. (Reproduced from ref. *2*.)

At the time of this writing, there are no published reports of interactions between the CO and H_2S systems; however, further research is required because interactions are likely.

5. INVOLVEMENT OF FREE RADICALS

The role of NO as a free radical itself and a generator of other free radicals has been well documented. A role for CO as protection against free radicals, in contrast to NO, is becoming increasingly apparent. Increased production of CO via HO is thought to be involved in the protection of tissue against oxidative stress *(32)*. Indeed, HO may be an endogenous protection mechanism against free radicals in acute inflammation *(33)*. Upregulation of HO-1 and consequent overproduction of intracellular bilirubin are associated with protection against $ONOO^-$-mediated apoptosis *(34)*, suppression of oxidant-induced microvascular leukocyte adhesion *(35)*, and amelioration of postischemic myocardial function *(36)*. After treatment with hemin, there is resultant HO-1 expression and bilirubin production, and it has been discovered that cells display high resistance to oxidant damage only when actively producing the bile pigment, strongly implicating the HO-1 pathway in cytoprotection against oxidative stress *(37,38)*. Cell injury caused by oxidative stress appears to contribute extensively to the pathogenesis of vascular disease, and HO-1 is widely considered to be valuable in the restoration of vascular function under conditions of increased generation of reactive oxygen species (ROS). It has been recently hypothesized that the HO-1 system may act in a similar fashion to counteract the excessive production of NO and reactive nitrogen species (RNS) *(39)*. The ability of HO-1 to

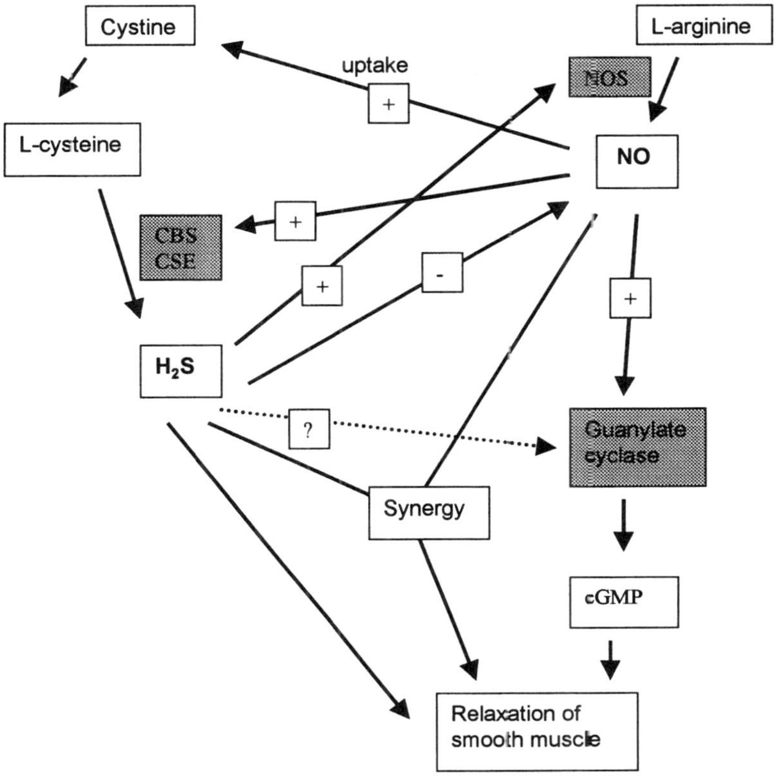

Fig. 7. Summary of interactions between NO and H_2S systems.

be highly induced in eukaryotes in response to NO and NO-related species makes this stress protein a likely contender to participate in NO detoxification *(40)*.

It is now known that the antioxidant protein HO can somehow detect NO and act successfully as a key player in cytoprotection against insults from ROS and RNS. For example, NO and NO-related species induce HO-1 expression and increase HO activity in aortic vascular cells *(4,18,39,41,42)*. Furthermore, cells pretreated with various NO-releasing agents acquire increased resistance to H_2O_2-mediated cytotoxicity at the time HO is maximally activated *(18)*. In addition, bilirubin, one of the end products of heme degradation by HO, has been shown to protect against the cytotoxic effects caused by the strong oxidants H_2O_2 and $ONOO^-$ *(18,34,37)*. Given that further investigations have revealed that NO-mediated activation of the HO-1 pathway is a stress response that can be extended to various mammalian systems *(39)*, several important issues on the possible signal transduction mechanism(s) that leads to HO-1 induction by NO and RNS remain to be carefully examined *(40)*.

The physiological significance and the mode of regulation of the HO-1 system by NO have not yet been fully elucidated *(39,40)*. In endothelial and smooth muscle cells, isolated aortic tissue, and cardiac myocytes, certain NO-releasing agents have been shown to induce HO-1 and augment HO activity *(18,34,38,42)*, suggesting that the effect mediated by the NO group is independent of its redox state. It has been concluded that reaction of NO with $O_2^{\bullet-}$ and the extent of the conversion of NO to NO^+ or NO^- by intracellular components could be critical to determine the modulation of HO-1 gene expression *(40)*.

Although H_2S is not a free radical, like •NO, in aqueous solution it is in fact a reducing agent and should be protective against oxygen free radicals. However, it has been shown that under certain conditions, in the presence of peroxidase and hydrogen peroxide, H_2S produced thiyl free radicals, SH• and S• *(43)*. Perhaps H_2S could be produced locally in relatively high concentrations to have a cytotoxic role, as is the case for NO. Because H_2S is cytotoxic, we propose that it could also be an inflammatory mediator and be involved in host defense, although there is currently no published evidence to support this. The fact that H_2S can affect xenobiotic metabolic enzymes *(44)* in the liver and affect the generation of ROS supports the notion that the immune system could be modulated. This may be particularly true in times of stress when there is an increase in the production of ROS *(45,46)*.

6. GAS SIGNALING MOLECULES IN THE CARDIOVASCULAR SYSTEM

6.1. Action of H_2S Within the Cardiovascular System

To date, the cardiovascular effects of both endogenous and exogenous H_2S have not been elucidated fully. The opinion previously has been that H_2S interfered with the cardiovascular function as a result of secondary anoxia rather than a direct action on cardiac myocytes or VSMCs *(47)*. However, more recent evidence has revealed that H_2S may have more of an endogenous physiological role to play within the cardiovascular system.

CBS, one of the enzymes responsible for generating H_2S, has been shown to have no activity or expression in human cardiovascular-related tissues *(48,49)*. On the other hand, CSE expression and endogenous production of H_2S have been shown in rat portal vein and thoracic aorta *(25)*. The expression of H_2S-generating enzyme has been identified in VSMCs, but not in the endothelium *(22)*.

Recently, the physiological function of H_2S in the cardiovascular system has been studied. When H_2S is injected intravenously, a transient decrease in blood pressure is observed in rats, which is antagonized by glibenclamide (a K_{ATP} channel blocker). This concurs with preliminary results from our laboratory showing that H_2S causes a dose-dependent decrease in left ventricular developed pressure (LVDP) and heart rate while increasing coronary flow rate in an isolated rat Langendorff rat heart model (unpublished data). Preliminary results would also indicate a role for K_{ATP} channel, because glibenclamide blocks both H_2S-induced changes in LVDP and heart rate.

6.2. Interactions Between CO and NO in the Cardiovascular System

6.2.1. CO and the Cardiovascular System

First, we discuss briefly the actions of CO and NO on the cardiovascular system, and then the interaction between the two gas signaling molecules.

HO is the rate-limiting step in heme degradation; it catalyzes the oxidation of the α-*meso* carbon of the protoporphyrin ring leading to the formation of CO, free iron, and biliverdin *(50)*. Three isoforms of HO are known: HO-1, HO-2, and HO-3; for the purpose of this section, we concentrate on HO-1 (also termed hsp32), a stress-inducible enzyme. HO-1 is induced in response to oxidant stress, has been shown itself to be cytoprotective, and plays an important role in the regulation of cardiovascular function *(51)*.

The dangers of CO within the cardiovascular system have been well defined—at high concentrations, CO is unquestionably lethal. However, recent studies (for a review, *see* ref. *52*) consistently support the emerging idea that CO at low concentrations exerts distinctly different effects on physiological and cellular functions, with this revelation motivating the need to re-evaluate its role.

CO, which is the byproduct of HO activity, has recently been attributed to being an important modulator of many physiological processes. It particularly plays a role in the homeostatic control of cardiovascular function (for reviews, *see* refs. *39* and *53*). The possible beneficial effects of HO-1 induction in stress are mainly attributed to the vasoactive CO. When the heart is stressed, there is an impressive increase in HO-1 mRNA expression in the heart (for a review, *see* ref. *15*). In the stressed heart, HO-1 protein is expressed in particularly high levels in the atrioventricular node and in the myocytes *(15)*. Experimental evidence suggests that in the blood vessels and the heart, CO, generated by HO activity, is a regulator of cGMP production (this is discussed in greater detail later in this chapter). It is highly relevant to the pathophysiology of the cardiovascular system and is integral to the heart's response to oxidative stress *(15)*.

Endogenously released CO is known to cause vasodilation and have antiproliferative actions. CO also has indirect actions producing vasoconstrictors and vascular growth factors, such as ET-1 and platelet-derived growth factors, which may be involved in combating chronic hypoxic stress *(54)*. Motterlini et al. *(55)* have recently shown that an increased CO production by HO-1 in vascular tissue contributes to the suppression of acute hypertensive responses under stress conditions in vivo. Sammut et al. *(42)* have also reported that HO-1-derived CO significantly suppresses phenylephrine-mediated contraction of isolated aortic rings.

It has been demonstrated in an in vivo vascular injury model of xenotransplantation that CO not only can confer protection as effectively as HO-1 but can also confer cytoprotection in the absence of HO-1 *(56,57)*.

Reports have also demonstrated that the HO-1/CO pathway is markedly upregulated by hypoxia in VSMC, cardiomyocytes, and heart tissue *(54,58,59)*. It has been suggested that aortic vasoconstriction following chronic hypoxia in rats involves the induction of endothelial HO-1 and the enhanced production of CO *(60)*.

Recent observations suggest that CO may impart potent antiinflammatory and antiapoptotic effects via the mitogen kinase pathway in macrophages and endothelial cells, respectively *(52,61)*. It has been shown that the cytoprotection via CO requires the activation of NF-κB transcription factor and is dependent on p38 kinase activity *(61)*. It is well documented that activation of p38 kinase and other mitogen kinase pathways within the cardiovascular system can transduce signals to provide downstream cytoprotection against cellular stresses such as myocardial ischemia reperfusion injury. Therefore, future studies are required to unify the possible importance of myocardial CO production during myocardial ischemeia and its influence on the activity of several cytoprotective transcription factors and kinases.

In summary, HO-1 has been shown to have antiinflammatory, antiapoptotic, and antiproliferative effects, and it is now known to have salutary effects in diseases as diverse as atherosclerosis and sepsis. The mechanism by which HO-1 confers its protective effect is still poorly understood, although recently direct links have been postulated concerning stress-induced production of CO *(62)*.

6.2.2. NO and the Cardiovascular System

NO plays an important role in the homeostatic regulation of the cardiovascular system *(40,63)*. NO is produced by vascular endothelium and smooth muscle, cardiac muscle, and many other cell types *(64,65)*. NO serves many important functions in the cardiovascular system, including vasodilation, inhibition of vasoconstrictor influences (e.g., inhibits angiotensin II and sympathetic vasoconstriction), inhibition of leukocyte adhesion to vascular endothelium (anti-inflammatory), antiproliferative (e.g., inhibits smooth muscle hyperplasia following vascular injury), as well as inhibition of platelet adhesion to the vascular endothelium (antithrombotic) and scavenging superoxide anion (anti-inflammatory) *(66,67)*.

The mechanism of many of these actions of NO involves the formation of cGMP. The antiplatelet aggregatory effects of NO are also related to the increase in cGMP. When NO production is impaired, as occurs when the vascular endothelium becomes damaged or dysfunctional, the following can result: vasoconstriction (e.g., coronary vasospasm, elevated systemic vascular resistance, hypertension); platelet aggregation and adhesion leading to thrombosis, vascular stenosis, or restenosis, as occurs following balloon angioplasty and stent placement; increased inflammation and tissue damage mediated by ROS such as superoxide anion and hydroxyl radical.

There is considerable evidence that cardiovascular-related diseases/conditions such as hypertension, dyslipidemia, diabetes, heart failure, atherosclerosis, cigarette smoking, aging, and vascular injury are associated with endothelial dysfunction and reduced NO production and/or bioavailability.

NO is necessary for normal cardiac physiology, but it is potentially toxic in excess concentrations. The role that NO plays in apoptosis is not known because NO has been shown to exert both proapoptotic and antiapoptotic effects in the myocardium *(68,69)*. NO also spontaneously interacts with molecular oxygen and reactive oxygen metabolites to yield potentially injurious oxidizing and nitrosating agents *(70)*, as discussed elsewhere.

6.3. Coexpression of HOs and NOSs in the Cardiovascular System

HO and NOS are the enzymes responsible for generating CO and NO, respectively. They have intriguing similarities in their isoforms, requirements for activity, and regulation *(71)*. For example, both the CO- and NO-generating systems have constitutive (HO-2, HO-3, endothelial NOS [eNOS], and NOS) and inducible (HO-1 and iNOS) isoforms. Production of CO and NO arises from different substrates (heme for HO and L-arginine for NOS); however, both enzymes require molecular oxygen and the reducing agent NADPH for activity. The differences are that NO synthesis requires additional cofactors (tetrahydrobiopterin, flavin adenine dinucleotide, and flavin mononucleotide) and that the constitutive isoforms of NOS are calcium/calmodulin dependent *(63)*.

Zakhary et al. *(72)* have reported marked similarities in the localization of HO and NOS in endothelial cells and adventitial nerves of blood vessels, suggesting a possible coordinated function for CO and NO. Indeed, in vitro studies show that under certain pathophysiological conditions, such as hypoxia, downregulation of constitutive eNOS is concurrent with transient increases in inducible HO-1 protein, indicating a potential compensatory regulation between the two systems *(54)*.

It is well established that NO donors can activate HO-1 gene expression and activity in various tissues *(40,71)*, although extensive studies of the cardiovascular system have not concurred this finding.

This section concentrates on literature concerning coexpression of HOs and NOSs solely within the cardiovascular system. It appears that HO and NOS require molecular oxygen for activation, although modulation of these enzymes by hypoxia remains unclear.

Experiments conducted using the isolated heart model confirmed the vasoactive properties of CO-releasing molecules within the cardiovascular system. The metal carbonyl markedly attenuated an L-NAME-mediated (NO inhibitor) increase in coronary perfusion pressure. Hearts expressing high HO-1 in the vasculature following treatment of animals with hemin also displayed reduced contractility when challenged with L-NAME, and inhibition of HO activity abolished the effect; this confirms the important role of endogenously produced CO in vascular control *(42,55)*. Thus, augmented HO-1-derived CO can profoundly modulate cardiac vessel functions, and this effect can be mimicked by exogenously applied CO-releasing molecules. Motterlini et al. *(55)* found that induction of the HO-1 system by hemin pretreatment considerably suppressed the increase in mean arterial pressure elicited by intravenous administration of L-NAME, a finding consistent with a previous report by this group; as observed for the isolated aortic ring and heart preparations, SnPPIX (HO inhibitor) restored the vasoconstrictor responses to L-NAME.

In the carotid body, basal levels of CO and NO act together to suppress sensory discharge. However, during acute hypoxia decreased synthesis of CO and NO has been implicated in contributing to the augmentation of sensory discharge *(73)*. Other studies have shown that, in contrast to the apparent decrease in synthesis of CO and NO in the carotid body, hypoxic conditions induce the gene expression of both HO-1 and iNOS, although the mechanisms involved remain unclear *(59,74)*. This apparent discrepancy is most likely attributable to the difference in regulation between constitutive and inducible isoforms. Interestingly, CO and NO themselves have been shown to suppress the hypoxic induction of vascular endothelial growth factor *(75)* and to inhibit hypoxia-inducible factor-1 (HIF-1) DNA-binding activity by abrogating hypoxia-induced accumulation of HIF-1a protein *(76)*.

Maulik et al. *(77)* demonstrated that in isolated working rat hearts made ischemeic for 30 min followed by 30 min of reperfusion NO activates HO, which further stimulates the production of cGMP presumably by CO signaling. This study revealed that NO not only potentiates cGMP-mediated intracellular signaling but it also functions as a retrograde messenger for CO signaling in heart.

Studies performed on smooth muscle cells revealed that increases in HO-1 transcript by the NO donor, spermine NONOate, is associated with enhanced activator protein-1 (AP-1) DNA-binding activity *(41)*. By contrast, recent work using HeLa cells suggests that mitogen-activated protein kinase (MAPK) extracellular signal-related kinase (ERK) and p38 pathways, but not the AP-1 transcription factor, are involved in NO-mediated induction of HO-1 *(78)*; in this case, the mechanism of activation would be unrelated to cyclic GMP and also appears to be independent of redox signaling events.

6.4. Interaction Between NO and CO Signaling Pathways and Atherosclerosis

Increased expression of the stress response protein HO-1 in human atherosclerotic lesions *(79)* and vascular endothelial and smooth muscle cells exposed to oxidized low-density lipoprotein (LDL) *(80,81)* may serve a multipurpose role, via metabolism to the antioxidant biliverdin and the vasodilator CO *(15,82)*. These adaptive responses may contribute to the maintenance of vascular tone and patency in atherosclerotic vessels and compensate for diminished NO generation and activity *(83,84)*.

Endothelium-derived CO or NO diffuses to subjacent smooth muscle cells where activation of SGC results in elevated intracellular cGMP levels, leading to smooth muscle relaxation *(85)*. As shown in **Fig. 8** and pointed out by Siow et al. *(81)*, CO and NO can also be generated in smooth muscle cells in response to atherogenic stimuli. The metabolic functional links between CO and NO suggest that vasodilator actions of

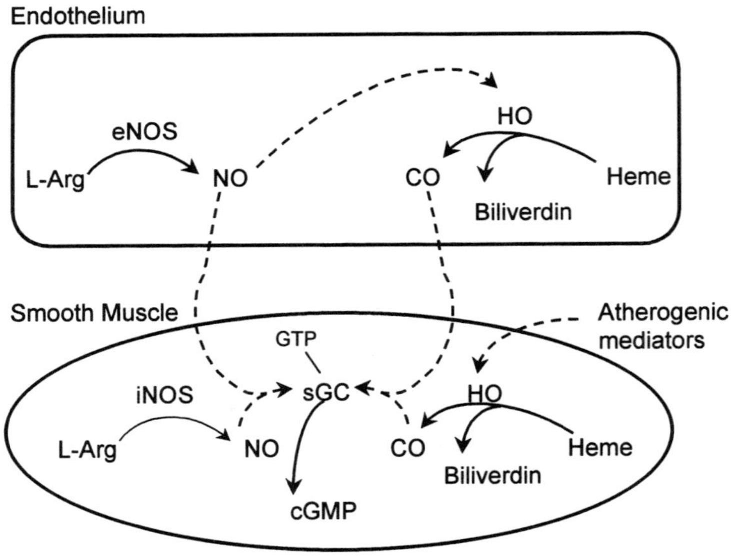

Fig. 8. Importance of the HO-CO and L-arginine–NO signaling pathways in vascular endothelial and smooth muscle cells in atherogenesis. HOs metabolize heme to generate the antioxidant biliverdin and CO, which, like NO, stimulates sGC, resulting in increased intracellular cGMP levels. Atherogenic and proinflammatory mediators such as oxidized LDL and cytokines decrease the expression and activity of eNOS while inducing HO-1 and iNOS in smooth muscle cells. Diminished production or activity of NO by the endothelium in atherogenesis could be compensated for by induction of HO-1. Increased cGMP levels in VSMCs would sustain blood flow, whereas catabolism of heme and generation of biliverdin would attenuate cellular oxidative stress in atherogenesis. (Adapted from ref. *81*.)

CO may become important in atherogenesis, where endothelium-derived NO production is inhibited.

As mentioned earlier, there is accumulating evidence demonstrating that NO donors and endogenously generated NO induce expression of HO-1 in vascular endothelial and smooth muscle cells *(4,34,39,41)*. This provides an endogenous adaptive defense mechanism against the oxidative stress associated with sustained production of NO *(18,86)*. It has been shown that the heme moiety of NOS and sGC can serve as alternative substrates for HO; their activity may, under certain conditions, be downregulated. In addition, CO is able to bind to the heme moiety of NOS and thereby inhibit L-arginine turnover and NO production *(15)*. NOS and NADPH–cytochrome P450 reductase are extremely similar, and, therefore, electron transfer from NOS to HO can also occur, fuelling heme catabolism by HO *(15)*. By reducing intracellular heme levels in vascular cells, HOs may limit *de novo* synthesis of iNOS, whereas the iron generated by heme catabolism would further limit synthesis of iNOS through inhibition of nuclear transcription *(87)*. It has been shown that iNOS is expressed and has been detected in animal and human atherosclerotic lesions and contributes to the formation of $ONOO^-$ *(88–90)*. Induction of HO-1 in atherogenesis in response to $ONOO^-$ *(34)* may attenuate vascular injury.

In the setting of atherosclerosis, impairment of the vascular NO signaling pathway could tip the balance in favor of HO as a salvage mechanism required to maintain vascular tone and function *(81)*.

7. NO, CO, H_2S, AND THE IMMUNE SYSTEM

7.1. NO and the Immune System

The notion that NO has an important role to play in the functioning of the immune system has taken time to be accepted, which is interesting because in evolutionary terms, more than 500 million years ago, the horseshoe crab was using the NO pathway to prevent blood coagulation. It is now well established that NO produced via iNOS is an important inflammatory mediator in the body; thus, NO has a proinflammatory role.

7.1.1. Immune Cells

NO is produced by macrophages and will eradicate many parasites and bacteria that are otherwise difficult to kill. Studies performed on mice infected with *Leishmania major*, a pathogenic protozoan, demonstrated that host defense against this infection depends on the macrophages releasing NO *(91)*. Studies have clearly demonstrated that immunological activation of mouse macrophages induces the activity of NOS, producing NO *(92)*. Much of the antimicrobial activity of mouse macrophages against some fungal, helminthic, protozoal, and bacterial pathogens has been attributed to alterations in the activity of NO. Production of large amounts of NO by activated macrophages contributes to their ability to suppress lymphocyte proliferation. However, no compelling evidence yet exists however that NO synthesis can occur directly in lymphocytes. However, cytokines secreted by activated lymphocytes can certainly regulate NO synthesis by macrophages. Constitutive NOS is activated in neutrophils in response to inflammatory stimuli, and NO has diverse, often biphasic, effects on neutrophil functions *(93)*.

NO acts as an inter- and intracellular messenger molecule, coordinating cross talk between immune cells and endothelial cells; thus, NO plays a vital role in the inflammatory processes. This involves NO derived from constitutive NOS, which appears important in the early stages of an inflammatory response through to high-output production of NO by iNOS, which is fundamental to chronic inflammatory disease *(94)*. Constitutively produced NO released by endothelial cells has been shown to act as an endogenous agent that inhibits the rolling and adhesion of leukocytes in the microcirculation, and the importance of iNOS in modulating leukocyte recruitment can vary according to the type of inflammatory response *(95)*. The significance of NO in this capacity has been reported as almost an incidental observation. The rolling and adhesion of leukocytes within the microcirculation is a significant step in determining endothelial-leukocyte crosstalk *(96)* and could well play a role in subsequent leukocyte activation and formation of edema.

7.1.2. Therapeutic Potential

Increased knowledge of the role of gasotransmitters in host defense may lead to novel therapeutic interventions. Researchers from the University of North Carolina led by Schoenfisch have developed a gel that can release NO when in contact with biological fluids, such as blood, resulting in a technique that mimics the body's own NO-producing capabilities. Medical implants such as catheters and artificial organs have been coated with such gel-based materials. This targeted release of NO extends the normal, extremely short duration half-life of NO from a few seconds to minutes. The overall effect is that of selectively mimicking phagocytosis, the process by which immune cells release a host of antibacterial agents including NO. The NO released also serves the dual function of reducing bacterial adhesion *(97)*. The genes coding for eNOS, nNOS, and iNOS enzymes are on chromosomes 12, 7, and 17, respectively, and will no doubt offer scope for future therapeutic targeting.

7.2. CO and the Immune System

CO mediates potent anti-inflammatory effects and has been shown to suppress the inflammatory response *(32)*. Both in vivo and in vitro, CO at low concentrations differentially and selectively inhibited the expression of the lipopolysaccharide-induced proinflammatory cytokines tumor necrosis factor-β, interleukin (IL)-1β, and macrophage inflammatory protein-1α and increased the lipopolysaccharide-induced expression of the anti-inflammatory cytokine IL-10, involving the MAPK pathway *(98)*. All of these cytokines have significant modulating effects on immune cells. There have been few studies on the effect of CO on the actual functioning of immune cells, primarily because CO has been previously associated with poisoning. Studies examining the effects of CO on the immune system have focused largely on the effects of CO toxicity, with a paucity of research considering its potential as a signaling molecule.

CO poisoning has been reported to temporarily inhibit B2 integrin adherence molecules on leukocytes *(99)*. This has the potential of modifying leukocyte-endothelial cell interactions and could be of tremendous benefit in modifying stress-induced leukocyte activation, which has been in part attributed to B2 integrins *(45)*.

The fact that CO can upgrade the production of free radicals and modulate leukocyte adherence has yet to persuade researchers that it has remarkable therapeutic potential.

7.3. H_2S and the Immune System

H_2S dissociates into free sulfide in the circulation and sulfide binds to many macromolecules, among them cytochrome oxidase *(100)*. Exposure to toxic levels of H_2S resulted in inhibition of complement activity along with the bacteriocidal activity of blood serum *(101)*, both indirect indicators that immune cell function has been modified.

7.4. A New Look at Phagocytic Killing: Therapeutic Possibilities

Phagocytic leukocytes play a pivotal role in the innate immune response against bacteria, fungi, foreign particles, and stress-induced immunosuppression *(45,46,102)*. On the surface of phagocytic leukocytes is NADPH oxidase, a multi-subunit enzyme that can assemble and "shoot" pathogens with highly ROS. These NADPH oxidases are highly controlled and thus prevented from blasting highly reactive superoxide anions into healthy tissues. Recently, it has been shown that once "superoxide shooting" commences, the leukocyte initiates a highly coordinated sequence of events that includes fusion and release of several types of granules and activation of antimicrobial enzymes *(103)*. Therefore, the role of ROS is not just that of a reactive oxygen free radical but may be a signal for subsequent alteration of electrons, movement of ions, and ultimately release of granular contents *(103)*. Thus, an alteration of pH, undoubtedly possible by any one of the intracellular gas signaling molecules, in particular H_2S and CO, could result in selective leukocyte activation. Clearly this offers a novel therapeutic approach to modulating leukocyte activation (**Fig. 9**).

8. CONCLUSION

There are some similarities among NO, CO, and H_2S, in terms of their production and effects, but also some important differences. There is now clear evidence of interactions among NO, CO, and H_2S. In particular, the negative interactions among NO, CO, and their generating systems allows a degree of control in the cardiovascular system, via

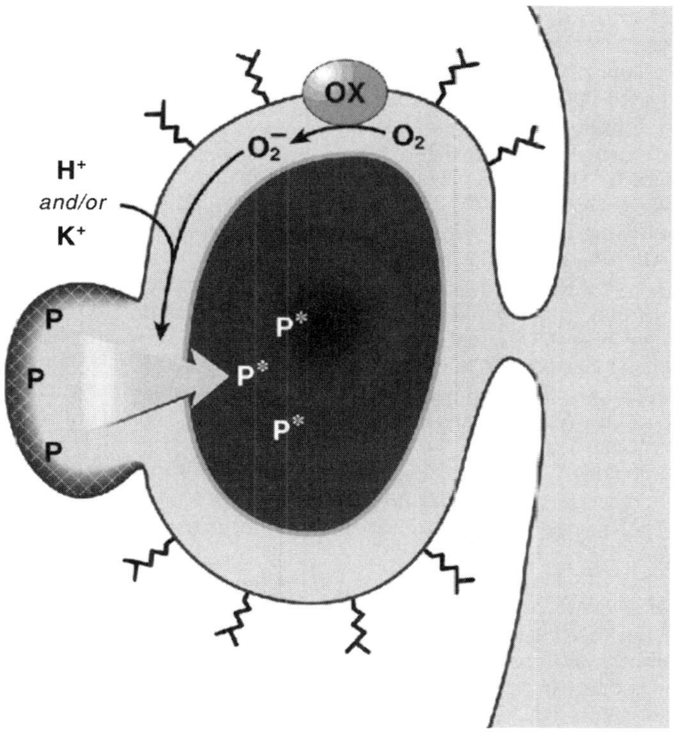

Fig. 9. Fresh look at phagocytic killing. The single electron reduction of molecular oxygen to form superoxide anion by the phagocyte NADPH oxidase (OX), stimulated by bacterial uptake, results in the transfer of electrons into the enclosed phagocytic vesicle. Dismutation of the superoxide generates OH^-, and the accumulating negative charge must be compensated by the influx of H^+ and/or K^+. The hypertonicity resulting from K^+ transport promotes the release of inactive cationic granule proteases (P) bound to an anionic sulfated proteoglycan matrix (crosshatching). The released and active proteases (P^*) encounter the bacterium under optimal pH conditions within the phagocytic vesicle and degrade it. Cytoskeletal elements associated with the phagocytic vesicle (wavy lines) indirectly affect the killing process by modulating vesicular volume. The pH and movement of ions may well be affected by gas signaling molecules. (From ref. *103* with permission.)

opposing effects. Initial reports of synergy between NO and H_2S in their actions are interesting but require further investigation. Interactions among the three gasotransmitters in the immune system and free-radical production require further research but could open up the possibility of novel therapies.

REFERENCES

1. Palmer RM, Ferrige AG, Moncada S. Nitric oxide release accounts for the biological activity of endothelium-derived relaxing factor. Nature 1987;327:524–526.
2. Wang R. Two's company, three's a crowd: can H_2S be the third endogenous gaseous transmitter? FASEB J 2002;16:1792–1798.
3. Ingi T, Cheng J, Ronnett GV. Carbon monoxide: an endogenous modulator of the nitric oxide-cyclic GMP signalling system. Neuron 1996;6:835–842.
4. Durante W, Kroll MH, Christodoulides N, et al. Nitric oxide induces heme oxygenase-1 gene expression and carbon monoxide production in vascular smooth muscle cells. Circ Res 1997;80:557–564.
5. Coceani F. Carbon monoxide and dilation of blood vessels. Science 1993; 260:739.

6. Linderholm H, Lundström P. Endogenous carbon monoxide production and blood loss at delivery. Acta Obstet Gynaecol Scand 1969;48:362–370.
7. Frydman RB, Tomaro ML, et al. Specificity of heme oxygenase: a study with synthetic hemins. Biochemistry 1981;20:5177–5182.
8. Kimura H. Hydrogen sulfide induces cyclic AMP and modulates the NMDA receptor. Biochem Biophys Res Commun 2000;267:129–133.
9. Abe K, Kimura H. The possible role of hydrogen sulphide as an endogenous neuromodulator. J Neurosci 1996;16:1066–1071.
10. Bradley KK, Buxton IL, et al. Nitric oxide relaxes human myometrium by a cGMP-independent mechanism. Am J Physiol 1998;275:C1668–C1673.
11. Buxton ILO, Kaiser RA, Malmquist NA, et al. NO-induced relaxation of labouring and non-labouring human myometrium is not mediated by cyclic GMP. Br J Pharmacol 2001;134:206–214.
12. Wu L, Cao K, Lu Y, et al. Different mechanisms underlying the stimulation of K_{Ca} channels by nitric oxide and carbon monoxide. J Clin Invest 2002;110:691–700.
13. Feelisch M. The biochemical pathways of nitric oxide formation from nitrovasodilators: appropriate choice of exogenous NO donors and aspects of preparation and handling of aqueous NO solutions. J Cardiovasc Pharmacol 1991;17:S25–S33.
14. Kim YM, Bergonia HA, Muller C, et al. Loss and degradation of enzyme-bound heme induced by cellular nitric oxide synthesis. J Biol Chem 1995;270:5710–5713.
15. Maines MD. The heme oxygenase system: a regulator of second messenger gases. Annu Rev Pharmacol Toxicol 1997;37:517–554.
16. McMillan K, Bredt DS, Hirsch DJ, et al. Cloned, expressed rat cerebellar NOS containing stoichiometric amounts of heme which binds CO. Proc Natl Acad Sci USA 1992;89:11,141–11,145.
17. Willis D, Tomlinson A, Frederick R, et al. Modulation of heme oxygenase activity in rat brain and spleen by inhibitors and donors of nitric oxide. Biochem Biophys Res Commun 1995;214:1152–1156.
18. Motterlini R, Foresti R, Intaglietta M, et al. NO-mediated activation of heme oxygenase: endogenous cytoprotection against oxidative stress to endothelium. Am J Physiol 1996;270:H107–H114.
19. Henningsson R, Alm P, Ekstrom P, et al. Heme oxygenase and carbon monoxide: regulatory roles in islet hormone release. A biochemical, immunohistochemical, and confocal microscopic study. Diabetes 1999;48:66–76.
20. Vesely MJJ, Exon DJ, et al. Heme oxygenase-1 induction in skeletal muscle cells: hemin and sodium nitroprusside are regulators in vitro. Am J Physiol Cell Physiol 1998;275:C1087–C1094.
21. Leffler CW, Nasjletti A, Johnson RA, et al. Contributions of prostacyclin and nitric oxide to carbon monoxide-induced cerebrovascular dilation in piglets. Am J Physiol Heart Circ Physiol 2001;280: H1490–H1495.
22. Zhao W, Zhang J, Lu Y, et al. The vasorelaxant effect of H_2S as a novel endogenous gaseous K_{ATP} channel opener. EMBO J 2001;20:6008–6016.
23. Teague B, Asiedu S, Moore PK. The smooth muscle relaxant effect of hydrogen sulphide in vitro: evidence for a physiological role to control intestinal contractility. Br J Pharmacol 2002;137: 139–145.
24. Sidhu R, Singh M, Samir G, et al. L-Cysteine and sodium hydrosulphide inhibit spontaneous contractility in isolated pregnant rat uterine strips in vitro. Pharmacol Toxicol 2001;88:198–203.
25. Hosoki R, Matsuki N, Kimura H. The possible role of hydrogen sulfide as an endogenous smooth muscle relaxant in synergy with nitric oxide. Biochem Biophys Res Commun 1997;237:527–531.
26. Zhao W, Wang R. H_2S-induced vasorelaxation and underlying cellular and molecular mechanisms. Am J Physiol Heart Circ Physiol 2002;283:H474–H480.
27. Li J, Liu XJ, Furchgott RF. Blockade of nitric oxide-induced relaxation of rabbit aorta by cysteine and homocysteine. Chung Kuo Yao Li Hsueh Pao 1997;18:11–20.
28. Deves R, Boyd CAR. Transporters for cationic amino acids in animal cells: discovery, structure and function. Physiol Rev 1998;78:487–545.
29. Zerangue N, Kavanaugh MP. Interaction of L-cysteine with a human excitatory amino acid transporter. J Physiol 1996;493:419–423.
30. Li H, Marshall ZM, Whorton AR. Stimulation of cystine uptake by nitric oxide: regulation of endothelial cell glutathione levels. Am J Physiol 1999;276:C803–C811.
31. Taoka S, Ohja S, Shan X, et al. Evidence for heme-mediated redox regulation of human cystathionine beta-synthase activity. J Biol Chem 1998;273:25,179–25,184.
32. Willis D, Moore AR, Frederick R, et al. Heme oxygenase: a novel target for the modulation of the inflammatory response. Nat Med 1996;2:87–90.

33. Willis D. Expression and modulatory effects of heme oxygenase in acute inflammation in the rat. Inflamm Res 1995;44:S218–S220.
34. Foresti R, Sarathchandra P, Clark JE, et al. Peroxynitrite induces haem oxygenase-1 in vascular endothelial cells: a link to apoptosis. Biochem J 1999;339:729–736.
35. Hayashi S, Takamiya R, Yamaguchi T, et al. Induction of heme oxygenase-1 suppresses venular leukocyte adhesion elicited by oxidative stress: role of bilirubin generated by the enzyme. Circ Res 1999;85:663–671.
36. Clark JE, Foresti R, Sarathchandra P, et al. Heme oxygenase-1-derived bilirubin ameliorates post-ischemic myocardial dysfunction. Am J Physiol Heart Circ Physiol 2000;278:H643–H651.
37. Clark JE, Foresti R, Green CJ, et al. Dynamics of haem oxygenase-1 expression and bilirubin production in cellular protection against oxidative stress. Biochem J 2000:348:615–619.
38. Foresti R, Goatly H, Green CJ, et al. Role of heme oxygenase-1 in hypoxia-reoxygenation: requirement of substrate heme to promote cardioprotection. Am J Physiol Heart Circ Physiol 2001;281: H1976–H1984.
39. Foresti R, Motterlini R. The heme oxygenase pathway and its interaction with nitric oxide in the control of cellular homeo-stasis. Free Radic Res 1999;31:459–75.
40. Motterlini R, Green CJ, Foresti R. Regulation of heme oxygenase-1 by redox signals involving nitric oxide. Antioxid Redox Signal 2002;4:615–624.
41. Hartsfield CL, Alam J, Cook JL, et al. Regulation of heme oxygenase-1 gene expression in vascular smooth muscle cells by nitric oxide. Am J Physiol 1997;273:L980–L988.
42. Sammut IA, Foresti R, Clark JE, et al. Carbon monoxide is a major contributor to the regulation of vascular tone in aortas expressing high levels of haeme oxygenase-1. Br J Pharmacol 1998;125:1437–1444.
43. Nicholls P. The formation and properties of sulphmyoglobin and sulphcatalase. Biochem J 1961;81:374–383.
44. Boev VM, Nikonorov AA, Perepelkin SV, et al. Effects of hydrogen sulfide containing gas condensate on the hepatic microsomal monooxygenase system. Gigiena Sanitariia 1997;5:5–6.
45. Mian R, Shelton-Rayner G, Harkin B, et al. Observing a fictitious stressful event: haematological changes, including circulating leukocyte activation. Stress 2003;6:41–47.
46. McLaren GW, Macdonald DW, Georgiou C, et al. Leukocyte coping capacity: a novel technique for measuring the stress response in vertebrates. Exp Physiol 2003;88:541–546.
47. Reiffenstein RJ, Hulbert WC, Roth SH. Toxicology of hydrogen sulfide. Annu Rev Pharmacol Toxicol 1992;32:109–134.
48. Chen P, Poddar R, Tipa EV, et al. Homocysteine metabolism in cardiovascular cells and tissues: implications for hyperhomocysteinemia and cardiovascular disease. Adv Enzyme Regul 1999;39:93–109.
49. Bao L, Vlek C, Paces V, et al. Identification and tissue distribution of human cystathionine beta-synthase mRNA isoforms. Arch Biochem Biophys 1998;350:95–103.
50. Tenhunen R, Marver HS, Schmid R. Microsomal heme oxygenase: characterization of the enzyme. J Biol Chem 1969;244:6388–6394.
51. Chen YH, Yet SF, Perella MA. Role of heme oxygenase-1 in the regulation of blood pressure and cardiac function. Exp Biol Med 2003;228:447–453.
52. Otterbein LE. Choi AMK. Heme oxygenase: colors of defense against cellular stress. Am J Physiol Lung Cell Mol 2000;279:L1029–L1037.
53. Baranano DE, Snyder SH. Neural roles for heme oxygenase: contrasts to nitric oxide synthase. Proc Natl Acad Sci USA 2001;98:10,996–11,002.
54. Morita T, Kourembanas S. Endothelial cell expression of vasoconstrictors and growth factors is regulated by smooth muscle cell–derived carbon monoxide. J Clin Invest 1995;96:2676–2682.
55. Motterlini R, Gonzales A, Foresti R, et al. Heme oxygenase-1-derived carbon monoxide contributes to the suppression of acute hypertensive responses in vivo. Circ Res 1998;83:568–577.
56. Soares MP, Lin Y, Anrather J, et al. Expression of heme oxygenase-1 can determine cardiac xenograft survival. Nat Med 1998;4:1073–1077.
57. Sato K, Balla J, Otterbein L, et al. Carbon monoxide generated by heme oxygenase-1 suppresses the rejection of mouse-to-rat cardiac transplants. J Immunol 2001;166:4185–4194.
58. Borger DR, Essig DA. Induction of HSP 32 gene in hypoxic cardiomyocytes is attenuated by treatment with N-acetyl cysteine. Am J Physiol 1998;274:H965–H973.
59. Lee PJ, Jiang BH, Chin BY, et al. Hypoxia-inducible factor-1 mediates transcriptional activation of the heme oxygenase-1 gene in response to hypoxia. J Biol Chem 1997;272:5375–5381.
60. Caudill TK, Resta TC, Kanagy NL, et al. Role of endothelial carbon monoxide in attenuated vasoreactivity following chronic hypoxia. Am J Physiol 1998;275:R1025–R1030.

61. Brouard S, Otterbein LE, Anrather J, et al. Carbon monoxide generated by heme oxygenase-1 suppresses endothelial cell apoptosis. J Exp Med 2001;92:1015–1026.
62. Morse D, Choi AM. Heme oxygenase-1:the "emerging molecule" has arrived. Am J Respir Cell Mol Biol 2002;27:8–16.
63. Moncada S, Palmer RMJ, Higgs EA. Nitric oxide: physiology, pathophysiology, and pharmacology. Pharmacol Rev 1991;43:109–142.
64. Curtis MJ, Pabla R. Nitric oxide supplementation or synthesis block—which is the better approach to treatment of heart disease? Trends Pharmacol Sci 1997;18:239–-244.
65. Palacin M, Estevez R, Bertran J, et al. Molecular biology of mammalian plasma membrane amino acid transporters. Physiolog Rev 1998;78:969–1054.
66. Hare JM, Comerford ML. Role of nitric oxide in the regulation of myocardial function. Prog Lipid Res 1995;38:155–166.
67. Xie YW, Wolin MS. Role of nitric oxide and its interaction with superoxide in the suppression of cardiac muscle mitochondrial respiration: involvement in response to hypoxia/reoxygenation. Circulation 1996;94:2582586.
68. Gao F, Gao E, Yue TL, et al. Nitric oxide mediates the antiapoptotic effect of insulin in myocardial ischaemia-reperfusion: the roles of PI3–kinase, Akt, and endothelial nitric oxide synthase phosphorylation. Circulation 2002;105:1497–1502.
69. Taimor G, Hofstaetter B, Piper HM. Apoptosis induction by nitric oxide in adult cardiomyocytes via cGMP-signaling and its impairment after simulated ischaemia. Cardiovasc Res 2000;45:588–594.
70. Turko IV, Murad F. Protein nitration in cardiovascular diseases. Pharmacol Rev 2002;54:619–634.
71. Hartsfield CL. Cross talk between carbon monoxide and nitric oxide. Antioxid Redox Signal 2002;4:301–307.
72. Zakhary R, Gaine SP, Dinerman JL, et al. Heme oxygenase 2:endothelial and neuronal localization and role in endothelium-dependent relaxation. Proc Natl Acad Sci USA 1996;93:795–798.
73. Prabhakar NR. NO and CO as second messengers in oxygen sensing in the carotid body. Respir Physiol 1999;115:161–168.
74. Jung F, Palmer LA, Zhou N, et al. Hypoxic regulation of inducible nitric oxide synthase via hypoxia inducible factor-1 in cardiac myocytes. Circ Res 2000;86:319–325.
75. Liu Y, Christou H, Morita T, et al. Carbon monoxide and nitric oxide suppress the hypoxic induction of vascular endothelial growth factor gene via the 5' enhancer. J Biol Chem 1998;273:15,257–15,262.
76. Huang LE, Willmore WG, Gu J, et al. Inhibition of hypoxia-inducible factor 1 activation by carbon monoxide and nitric oxide: implications for oxygen sensing and signaling. J Biol Chem 1999;274:9038–9044.
77. Maulik N, Engelman DT, Watanabe M, et al. Nitric oxide/carbon monoxide: a molecular switch for myocardial preservation during ischemia. Circulation 1996;94:398–406.
78. Chen K, Maines MD. Nitric oxide induces heme oxygenase-1 via mitogen-activated protein kinases ERK and p38. Cell Mol Biol 2000;46:609–617.
79. Wang LJ, Lee TS, Lee FY, et al. Expression of heme oxygenase-1 in atherosclerotic lesions. Am J Pathol 1998;152:711–720.
80. Ishikawa K, Navab M, Leitinger N, et al. Induction of heme oxygenase-1 inhibits the monocyte transmigration induced by mildly oxidized LDL. J Clin Invest 1997;100:1209–1216.
81. Siow RC, Sato H, Mann GE. Heme oxygenase-carbon monoxide signalling pathway in atherosclerosis: anti-atherogenic actions of bilirubin and carbon monoxide? Cardiovasc Res 1999;41:385–394.
82. Marks G, Brien J, Nakatsu K, et al. Does carbon monoxide have a physiological function? Trends Pharmacol Sci 1991;12:185–188.
83. Cox DA, Cohen ML. Effects of oxidized low-density lipoprotein on vascular contraction and relaxation: clinical and pharmacological implications in atherosclerosis. Pharmacol Rev 1996;48:3–19.
84. Jay MT, Chirico S, Siow RCM, et al. Modulation of vascular tone: effects on L-arginine transport and oxide synthesis. Exp Physiol 1997;82:349–360.
85. Schmidt HHHW, Lohmann SM, Walter U. The nitric oxide cGMP signal transduction system: regulation and mechanism of action. Biochim Biophys Acta 1993;1178:153–175.
86. Yet SF, Pellaconi A, Patterson C, et al. Induction of heme oxygenase-1 expression in vascular smooth muscle cells: a link to endotoxic shock. J Biol Chem 1997;272:4295–4301.
87. Weiss G, Wernerfelmayer G, Werner ER, et al. Iron regulates nitric oxide synthase activity by controlling nuclear transcription. J Exp Med 1994;180:969–976.
88. Darley-Usmar V, White R. Disruption of vascular signalling by the reaction of nitric oxide with superoxide: implications for cardiovascular disease. Exp Physiol 1997;82:305–316.

89. Buttery LD, Springall DR, Chester AH, et al. Inducible nitric oxide synthase is present within human atherosclerotic lesions and promotes the formation and activity of peroxynitrite. Lab Invest 1996;75:77–85.
90. Luoma JS, Yla-Herttuala S. Expression of inducible nitric oxide synthase in macrophages and smooth muscle cells in various types of human atherosclerotic lesions. Virchows Arch 1999;434:561–568.
91. Brandonisio O, Panaro MA, Sisto M, et al. Interactions between Leishmania parasites and host cells. Parasitologia 2000;42:183–190.
92. Nathan CF, Hibbs JB Jr. Role of nitric oxide synthesis in macrophage antimicrobial activity. Curr Opin Immunol 1991;3:65–70.
93. Moilanen E, Vapaatalo H. Nitric oxide in inflammation and immune response. Ann Med 1995;27:359–367.
94. Hobbs AJ, Higgs A, Moncada S. Inhibition of nitric oxide synthase as a potential therapeutic target. Annu Rev Pharmacol Toxicol 1999;39:191–220.
95. Hickey MJ. Role of inducible nitric oxide synthase in the regulation of leucocyte recruitment. Clin Sci 2001;100:1–12.
96. Mian R, Marshall JM. Effect of acute systemic hypoxia on vascular permeability and leucocyte adherence in the anaesthetised rat. Cardiovasc Res 1993;27:1531–1537.
97. Nablo BJ, Chen TY, Schoenfisch MH. Sol-gel derived nitric-oxide releasing materials that reduce bacterial adhesion. J Am Chem Soc 2001;123:9712, 9713.
98. Otterbein LE, Bach FH, Alam J, et al. Carbon monoxide mediates anti-inflammatory effects via the p38 mitogen activated protein kinase pathway. Nat Med 2000;6:422–428.
99. Thom SR, Ohnishi ST, Ischiropoulos H. Nitric oxide released by platelets inhibits neutrophil B2 integrin function following acute carbon monoxide poisoning. Toxicol Appl Pharmacol 1994;128:105–110.
100. Guidotti TL. Hydrogen sulphide. Occup Med 1996;46:367–371.
101. Setko NP, Zheludeva GN, Mazaeva SG. Reaction of the immune system of the body to combined effects of chemical agents. Gigiena Sanitariia 1989;3:24–26.
102. Ellard DR, Castle PC, Mian R. The effect of a short term mental stressor on leukocyte activation. Int J Psychophysiol 2001;41:93–100.
103. Bokoch GM. Microbial killing: hold the bleach and pass the salt. Nat Immunol 2002;3:340–342.

II

The Emergence of the First Gasotransmitter: Nitric Oxide

3 Nitric Oxide

Synthesis and Metabolism, Tissue Stores, and the Relationship of Endothelium-Derived Nitric Oxide to Endothelium-Dependent Hyperpolarization

Chris R. Triggle, Hong Ding, Ella S. M. Ng, and Anthie Ellis

CONTENTS

SUMMARY

Nitric oxide (NO) is a gas that was first shown to be synthesized by endothelial cells and macrophages but was subsequently shown to be synthesized by most if not all cell types, including neuronal tissue. NO plays important functions as a signaling substance in mammalian and nonmammalian species. Despite the simple nature of the molecule, the chemistry of NO and its adjuncts have proved to be quite complex and its cellular actions are now known to extend beyond a role as a short-lived cell-signaling substance. This chapter discusses the physiochemical characteristics of NO and nitrosothiols; measurement of NO; metabolic pathways for NO; synthesis of NO; and the important question of whether tissues can store NO, possibly as a nitrosothiol, in a stable form. In addition, we discuss the relationship of NO to another important, but still putative, mediator termed *endothelium-derived hyperpolarizing factor*.

Key Words: Nitric oxide; nitroxyl anion; nitrosonium cation; nitrosothiol; peroxynitrite; superoxide ion; nitric oxide synthase; tissue stores; endothelium-dependent relaxation; endothelium-dependent hyperpolarizing factor.

From: *Signal Transduction and the Gasotransmitters: NO, CO, and H_2S in Biology and Medicine* Edited by: Rui Wang © Humana Press Inc., Totowa, NJ

1. INTRODUCTION

In 1980, Furchgott and Zawadzki *(1)* described a putative endothelium-derived relaxing factor (EDRF) and revolutionized researchers' thinking of the importance of the endothelium in the regulation of cardiovascular function. Ultimately, EDRF was shown to be nitric oxide (NO) *(2)* and is now recognized as a key, likely ubiquitous, signaling molecule in many tissues from many species including the invertebrates *(3)*.

2. CHEMISTRY OF NO

NO is a colorless gas that in the absence of oxygen dissolves in water up to concentrations in the low millimolar range (approx 1.9 m*M* at 25°C) *(4)*.

2.1. Reactions of NO

On exposure to oxygen (O_2), NO becomes unstable and generates various reactive nitrogen oxide species, such as nitrogen dioxide (NO_2) and dinitrogen trioxide (N_2O_3), which are harmful to biological tissues *(5)*. In aqueous solution, NO undergoes auto-oxidation to form NO_2, a reaction similar to that in the gas phase *(6)*. This reaction follows third-order kinetics, which involves two molecules of NO and one of O_2 (**reaction 1**):

$$2NO + O_2 \rightarrow 2NO_2 \tag{1}$$

NO_2 can further react with NO to form N_2O_3 (**reaction 2**) or dimerize to N_2O_4 (**reaction 3**):

$$NO_2 + NO \leftrightarrow N_2O_3 \tag{2}$$

$$2NO_2 \leftrightarrow N_2O_4 \tag{3}$$

The intermediate, N_2O_3, formed in aqueous solution undergoes hydrolysis to form nitrite (NO_2^-) (**reaction 4**), whereas N_2O_4 hydrolyzes to equimolar amounts of NO_2^- and nitrate (NO_3^-) (**reaction 5**) *(4,7)*.

$$N_2O_3 + H_2O \rightarrow 2\ NO_2^- + 2H+ \tag{4}$$

$$N_2O_4 + H_2O \rightarrow 2H+ + NO_2^- + NO_3^- \tag{5}$$

Reactions 2 and **4** are approx 10-fold faster than **reaction 5**. Therefore, in the absence of other NO scavengers, the major breakdown product of NO in aqueous solutions is NO_2^-, and the formation of NO_3^- is comparatively low *(4)*.

The kinetics of auto-oxidation of NO in aqueous medium are dependent on the concentration of NO *(6)*. Interestingly, the half-life of NO is inversely proportional to its concentration ($t_{1/2} = 1/k[NO][O_2]$). This means that the half-life of NO is much longer when it becomes more dilute *(8)*. In biological systems, NO will diffuse from its site of origin and thus decrease in concentration with distance. As the NO concentration dilutes, the lifetime increases, which allows NO to react with other biological molecules such as guanylate cyclase, oxyhemoglobin, and plasma proteins *(9)*. Generally, the rate of NO metabolism is dependent on its own concentration, its diffusibility, and the surrounding concentration of other bioreactants.

2.1.1. Half-Life of NO

The half-life of NO reported in different studies ranged from seconds to minutes *(10–12)* or even hours *(13)*. These variations may be because of the different methods for NO detection and to the diversity of the experimental conditions applied *(14)*. For instance, Kharitonov et al. *(13)* reported that the first half-life (i.e., the time required for halving the NO concentration) of NO was 2 h at physiologically relevant concentrations of NO (100 n*M*), whereas Ford et al. *(6)* demonstrated that the half-life of NO was approx 100–500 s when studied at more physiological levels of oxygen. Note that the reaction of NO with O_2 within the membranes has been shown to be approx 300 times more rapid than in the aqueous solution *(15)*. Apparently, the hydrophobic compartments in tissues are important sites for the auto-oxidation of NO and for the formation of NO-derived reactive species.

2.1.2. Superoxide and Peroxynitrite

Because NO contains an unpaired electron and is paramagnetic, it reacts rapidly with other radical species, such as the one-electron reduction product of oxygen, superoxide anion (O_2^-). The reaction between NO and O_2^- has been shown to occur at near-diffusion controlled rates with a rate constant of $6.7 \times 10^9\ M^{-1}\ s^{-1}$ to produce peroxynitrite ($ONOO^-$) (**reaction 6**) *(8,16)*.

$$NO + O_2^- \rightarrow ONOO^- \qquad k = 6.7 \times 10^9\ M^{-1}\ s^{-1} \tag{6}$$

In biological systems, the formation of $ONOO^-$ is dependent on the relative amount of NO and O_2^- produced, as well as the reaction of these radicals with other biological components *(9)*. For instance, superoxide dismutase (SOD) has a reaction rate similar constant to that of the NO/O_2^- reaction; therefore, SOD will compete effectively with NO for O_2^- *(9)*. Under normal circumstances, the estimated cellular concentrations of O_2^- and NO are 1 p*M* and 0.1–1 μ*M*, respectively *(17)*. Because O_2^- concentration is significantly less than NO, the amount of $ONOO^-$ formed is likely to be controlled by O_2^- production. O_2^- is formed in the mitochondria; endoplasmic reticulum; and membranes of various cells, including activated macrophages and endothelial cells *(8)*. Sources of O_2^- in the vascular wall include NADPH oxidase *(18,19)*, xanthine oxidase *(20)*, cytochrome P-450 (CYP) *(21)*, and cyclooxygenase *(22)*. In addition, NO synthase (NOS) itself may generate O_2^- anions under certain conditions and in a cofactor-dependent manner *(23)*.

$ONOO^-$ is a powerful oxidant *(24)*. It inhibits mitochondrial respiration *(25)*, initiates lipid peroxidation *(9)*, and causes DNA damage *(26)*. $ONOO^-$ decays rapidly once protonated to produce peroxynitrous acid (ONOOH). The decomposition of ONOOH results in the formation of hydroxyl ($HO^\bullet$) and NO_2 radicals ($NO_2^\bullet$), both of which are tissue-damaging agents *(27)*. However, in the absence of adequate substrate, ONOOH rearranges to form nitrate (NO_3^-), which can be considered a detoxification pathway (**reaction 7**) *(24)*:

$$O_2^- + NO \rightarrow ONOO^- + H^+ \leftrightarrow ONOOH \rightarrow HO^\bullet + NO_2^\bullet \rightarrow NO_3^- + H^+ \tag{7}$$

The detrimental effects caused by $ONOO^-$ are associated with the involvement of NO in inflammatory processes, in which NO is produced excessively by inducible NOS (iNOS), or during oxidative stress characterized by many cardiovascular diseases, in which O_2^- is generated. Nevertheless, several investigators have reported the unexpected observation that $ONOO^-$ could, indeed, contribute to physiological processes. For example, $ONOO^-$ has been shown to induce vasodilation *(28,29)*, stimulate guanylate cyclase *(30–32)*, prevent platelet aggregation *(33)*, and inhibit leukocyte–endothelial cell interactions *(34)*. The most likely explanation for these beneficial effects appears to be the reaction occurring between $ONOO^-$ and thiols to form *S*-nitrosothiols (RSNOs), which are thiol substances with a bound NO moiety *(35–37)*; however, the mechanism of this reaction has not been fully elucidated.

2.1.3. Reactions of NO With Heme Proteins

One of the most important pathways for NO metabolism is the reaction of NO with five-coordinate heme proteins (iron-containing proteins) to form nitrosyl complexes *(9)*. The most notable example is the interaction between NO and the heme moiety of soluble guanylate cyclase (sGC), which results in the formation of cyclic guanosine 5'-monophosphate (cGMP), a key intracellular second messenger that mediates numerous regulatory functions *(38,39)*. Because NO is freely diffusible, it reacts rapidly with oxyhemoglobin ($Hb[Fe^{2+}]O_2$) (k approx $10^7\ M^{-1}\ s^{-1}$) to produce methemoglobin ($HbFe^{3+}$) and nitrate (NO_3^-) (**reaction 8**). This reaction has been considered as the major route for the destruction of NO in vivo *(7,40,41)*. However, a small proportion of NO may interact with deoxyhemoglobin ($Hb[Fe^{2+}]$) to form nitrosylhemoglobin ($Hb[Fe^{2+}]NO$) (**reaction 9**) or with the cysteine residue at position 93 on the β-globin chains to form *S*-nitrosohemoglobin (SNO-Hb) (**reaction 10**) *(7,42–44)*. The fraction of each is dependent on the ratio of oxygenated and deoxygenated hemoglobin (Hb) within the erythrocytes *(7,8)*.

$$\underset{\text{oxyhemoglobin}}{Hb[Fe^{2+}]O_2} + NO \rightarrow \underset{\text{methemoglobin}}{HbFe^{3+}} + NO_3^- \tag{8}$$

$$\underset{\text{deoxyhemoglobin}}{Hb[Fe^{2+}]} + NO \rightarrow \underset{\text{nitrosylhemoglobin}}{Hb[Fe^{2+}]NO} \tag{9}$$

$$\underset{S\text{-nitrosohemoglobin}}{\text{Cysteine-}\beta 93} + NO \rightarrow \text{SNO-Hb} \tag{10}$$

The concentration of circulating Hb is 10–20 m*M* in heme, whereas the local concentration of NO produced by vascular endothelial cells is at nanomolar levels *(45,46)*. Accordingly, Hb should scavenge NO produced by the endothelium and thus inhibit its role as a vasodilator. Nevertheless, the relaxation properties of NO are unequivocal *(2,47)*, indicating that NO is not entirely inactivated by Hb *(44,48)*. Therefore, specific mechanisms must exist to reduce the NO consumption by Hb. Stamler and colleagues proposed that SNO-Hb may function as a vasodilator *(42,43)*. In fact, Jia and colleagues *(43)* showed that the concentrations of SNO-Hb were higher in the arterial blood than the venous blood, suggesting that this species might act as an NO donor in the systemic circulation. Moreover, it has been suggested that SNO-Hb elicits vasodilation via transnitrosation reactions with low-molecular-weight thiols, such as glutathione, or with a high-molecular-weight thiol (specifically AE1, the anion exchanger in the erythrocyte membrane) *(46,49)*. In addition, the release of NO occurs at sites of low-oxygen tension where Hb undergoes transition to the T-conformational state (i.e., "tense" or low oxygen affinity of Hb). This system facilitates efficient delivery of oxygen and NO to tissues, and thus dilating vessels, to ensure an adequate supply of oxygenated blood *(42–44)*. Although it is tempting to suggest that SNO-Hb is accountable for the regulation of vascular tone and blood flow, there is controversy regarding the mechanisms responsible for the formation of SNO-Hb, and the release of NO from this species *(50,51)* remains to be investigated.

2.1.4. NO, Thiols, and Nitrosothiols

NO does not react directly with thiols (RSH) *(52)*. In fact, the direct reaction between NO and RSH is a very slow oxidation that yields thiol disulfide (RSSR) and nitroxyl anion (NO^-, **reaction 11**) *(53,54)*:

$$2RSH + NO \rightarrow NO^- + RSSR + 2H^+ \quad (11)$$

As mentioned earlier, the auto-oxidation of NO in an aqueous environment leads to the formation of N_2O_3, which is a potent nitrosating agent *(9,17,55)* that can nitrosate nucleophiles such as thiols. Nitrosation is an electrophilic reaction involving attack by NO^+ or a carrier of NO^+, such as N_2O_3, at a nucleophilic center *(56)*. For example, N_2O_3 may interact with amines to produce nitrosamines *(9,55)* or may undergo nitrosation with molecules containing thiol groups to form RSNOs (**reaction 12**) *(57–59)*. The nitrosation of thiols is a more physiologically relevant reaction pathway, because nitrosation of amines is usually associated with pathophysiological conditions *(60)*. Indeed, RSNOs have been shown to possess EDRF-like vasorelaxant effects *(61,62)*, and antiplatelet properties *(59,63–65)*.

$$RSH + N_2O_3 \rightarrow RSNO + H^+ + NO_2^- \quad (12)$$

Despite the well-characterized interactions between the biological thiols and the derivative oxides of nitrogen (N_2O_3), the exact mechanisms by which RSNOs are formed in vivo remain uncertain. The formation of RSNO depends largely on whether NO reacts with O_2 in vivo. Because of the presence of other competitive reactions in biological systems *(57,66)*, many investigators consider the auto-oxidation of NO to be too slow to lead to meaningful nitrosation reactions *(57,67,68)*. Nevertheless, recent studies have supported the view that the formation of RSNOs is oxygen dependent and N_2O_3 is an important intermediate for nitrosation *(66,69)*. In fact, Nedospasov et al. *(69)* have demonstrated that N_2O_3 forms inside protein-hydrophobic cores and thus causes nitrosation within the protein interior.

In the laboratory, RSNO can be easily synthesized from the reaction between thiols and nitrous acid (HNO_2) (**reactions 13** and **14**) *(70)*. This method of synthesis does not occur at physiological pH and is only suitable for low-molecular-weight thiols, such as glutathione and cysteine. To synthesize protein RSNO, the most common method used is the spontaneous transfer of the nitroso group from a low-molecular-weight RSNO (i.e., *S*-nitrosocysteine) to the protein thiol *(15)*.

$$HNO_2 + H^+ \rightarrow NO^+ + H_2O \quad (13)$$
$$RSH + NO^+ \rightarrow RSNO + H^+ \quad (14)$$

Thiols also react with NO^-. However, the product of this reaction does not possess significant relaxant activity. As a result, thiols can be considered as NO^- scavengers *(71)*. For example, the vasorelaxant activity of the NO^- donor sodium trioxodinitrate ($Na_2N_2O_3$; Angeli's salt) is significantly reduced in the presence of high concentrations of thiols such as L-cysteine *(72,73)*. This occurs through a two-stage reaction that results in the formation of hydroxylamine (NH_2OH) and disulfide cysteine (**reactions 15** and **16**) *(74)*. Although NH_2OH may induce relaxations *(75)*, its activity is dependent on catalase, whose activity can be reduced by thiols *(76)*.

$$H^+ + NO^- + RSH \rightarrow RSNHOH \quad (15)$$
$$RSNHOH + RSH \rightarrow RSSR + NH_2OH \quad (16)$$

2.2. NO Free Radical, Nitroxyl, and Nitrosonium Ions

The conventional view of NO-mediated communication is that NOS produces NO, which, as a free radical, readily diffuses across cell membranes and acts on paracrine targets to elicit its intended effect. Although this may be true for the most part, there are indications that the actions of NO may also be carried out by alternative chemical forms of NO, which could potentially be derived from sources other than NOSs *(77)*.

NO is a paramagnetic or free-radical species, because it carries a single, unpaired electron in its outer shell (in total, 11 electrons). The positioning of the unpaired electron in the antibonding orbital renders NO more likely to accept or lose other electrons in this orbital and in the process able to form different oxides or redox forms of NO. These include the free-radical form of NO ($NO^•$), nitrosonium cation (NO^+), nitroxyl anion (HNO/NO^-), and hydroxylamine (NH_2OH). Although it is often thought that it is $NO^•$ alone that carries out its physiological functions, the physicochemical properties of the other redox forms of NO also warrant consideration regarding their possible roles.

2.2.1. Nitrosonium/Nitrosyl Cation

The direct oxidation of $NO^•$ into nitrosonium cations (NO^+) does not appear to occur as readily as previously thought, because the reduction potential for the $NO^•/NO^+$ pair is 1.21 V *(78)*; however, there is substantial evidence indicating the involvement of this species in nitrosation reactions *(58)*. The existence of free NO^+ in the cellular environment is fleeting because it will immediately nitrosate thiols or reduce metal centers to form RSNOs or nitrosyl-metal complexes, respectively *(71)*. It is the NO^+ species that reacts with cysteine residues in proteins, that transmit the NO signal during transnitrosation reactions *(15)*. Studies comparing the bioactivity of a putative NO^+ donor, nitrosonium tetrafluoroborate, against other redox forms of NO found that it was almost 100-fold less potent as a relaxant compared to authentic $NO^•$ or Angeli's salt *(79)* and had negligible effects on guanylate cyclase activity *(80)*. Hughes *(71)* subsequently argued that using this substance with the expectation that NO^+ cations would be generated was misguided because on its addition to aqueous solutions, dissociated NO^+ cations would immediately be hydrolyzed into nitrous acid, thus accounting for its very low potency. Overall, the physiological relevance of the NO^+ species would mostly pertain to *S*-nitrosation or transnitrosation reactions.

2.2.2. Nitroxyl Anion

The one-electron reduction product of $NO^•$, HNO/NO^-, has been proposed by some researchers to mediate the physiological functions of NO and to be the primary product formed from NOS *(79,81,82)*. HNO/NO^- can also be generated by the reaction between $NO^•$ and thiols *(83)*, during the decomposition of RSNOs *(84)*, by the catalase- or hydrogen peroxide (H_2O_2)-mediated oxidation of *N*-hydroxy-L-arginine (an intermediate of NO synthesis *[85,86]*), the reaction between NO and Hb *(44)*; and during the reaction between thiols and RSNOs *(87)*. It is unlikely, though, that $NO^•$ is directly reduced into NO^-, because the reduction potential of $NO^•$ has been deemed thermodynamically unviable under physiological conditions *(88)*. This figure has also led to revised estimates of the pK_a for NO^- to about 11.6, indicating that at physiological pH the species is protonated (HNO), uncharged, and therefore cell permeable. HNO readily dimerizes to form the nonrelaxant terminal product nitrous oxide (N_2O); therefore, the half-life of HNO is likely to be shortened at high concentrations.

Under experimental conditions, the most often used approach to generate HNO/NO^- is to use Angeli's salt, which dissociates into HNO/NO^- and NO_2^- anions in solution. HNO/NO^-, being 100 times more potent a relaxant than NO_2^-, almost entirely accounts for the relaxant activity of Angeli's salt *(89)*. It was earlier speculated that the cellular effects of Angeli's salt arose because HNO/NO^- was being immediately oxidized into $NO^•$ *(80,90)*. However, arguing against this theory, recent studies have shown that HNO/NO^- produces cellular effects that are distinct from or even oppose those of $NO^•$ *(91–94)*. The chemical basis for this difference may lie in the redox-dependent effect exerted on heme-containing proteins by HNO/NO^-. HNO/NO^- preferentially binds with ferric (Fe III) heme groups and reduces them to the ferrous (Fe II) state, in which a stable ferrous-nitrosyl complex can form. $NO^•$, on the other hand, does not require reductive modification and directly nitrosates ferrous groups to form a complex *(95)*. Therefore, proteins with a functional heme group in the ferric state would be expected to be targets of HNO/NO^-; these include methemoglobin, catalases, and cytochrome-*c*. The conventional receptor for $NO^•$, guanylate cyclase, has a ferrous heme group, which interacts with $NO^•$ to form a ferrous-nitrosyl complex that induces conformational changes to the enzyme and exposes the catalytic site to guanosine *S*′-triphosphate (GTP) *(96)*. Being a ferrous heme, HNO/NO^- would be less likely to mediate relaxation by a direct action on guanylate cyclase, contrasting the findings of studies that have shown 1H-[1,2,4]oxadiazolo[4,3,-a]quinoxalin-1-one (ODQ) sensitivity to relaxations by Angeli's salt *(79,97,98)*. The difference in effects between HNO/NO^- and $NO^•$ might also be related to the contrasting influence thiols have on the bioactivity of HNO/NO^- compared to $NO^•$ as described earlier (*see* **Subheading 2.1.4.** on thiols). HNO/NO^- is also coupled to a redox reaction that oxidizes NADPH; therefore, its presence could potentially alter many NADPH-dependent processes *(99)*.

2.2.3. Hydroxylamine

A two-electron oxidation of ammonia (NH_3) generates NH_2OH, which itself can be subsequently oxidized into HNO/NO^-. Smooth muscle relaxations to NH_2OH have been demonstrated on numerous occasions *(75,100–102)* and are thought to occur following its conversion into NO by endogenous catalases *(103)*. However, this view does not fit with the reported lack of effect of the catalase inhibitor 3-amino-1,2,4-triazole on relaxations to NH_2OH in rat duodenum *(104)*. NH_2OH can be generated during the synthesis of NO, which can be derived from the hydroxylated intermediate of L-arginine, *N*-hydroxy-L-arginine *(101)*, and as mentioned earlier, is an intermediate product of the reaction between HNO/NO^- and thiols. Feelisch et al. *(75)* ruled out the likelihood that NH_2OH could be the actual mediator of endothelium-dependent relaxations because it displayed lower sensitivity to oxyhemoglobin and had a longer half-life than the acetylcholine (ACh)-stimulated release of NO in rabbit aorta.

3. *S*-NITROSOTHIOLS

The binding of a NO moiety onto low-molecular-weight thiols, or thiols incorporated into proteins (e.g., nitrosoglutathione and nitrosoalbumin), is a process critical to the bioactivity of NO. The formation of RSNOs has implications for the stability, transport, and storage of NO, and for the intracellular signaling processes performed by NO. Although the earlier assertion that RSNOs accounted for the activity of endothelium-derived NO and/or the nitrergic transmitter *(61,105–107)* now appears unlikely, there is evidence of the existence of preformed tissue stores of RSNOs *(108,109)*. Given that the half-life of NO is comparably shorter than that of RSNOs *(64,75)*, the ability to form tissue stores of RSNOs would provide a stable reservoir of NO that would not be dependent on the immediate synthesis of NO from NOSs and that could release NO in a controlled manner to act on more distal targets. Although it has been suggested that the biological effects of RSNOs could be stereospecific and, therefore, potentially receptor mediated *(110)*, the general consensus is that the bioactivity of RSNOs results from the release of NO from RSNOs. The decomposition of RSNOs to liberate the NO moiety can be induced by metal ions (most likely Cu[I]) *(111)*; changes in oxygen tension *(112)*; light illumination *(109)*; diethyldithiocarbamate *(113)*; ascorbate *(114)*; superoxide anions *(115)*; and during the reaction between RSNOs and other thiols, which releases nitroxyl anions *(84)*. Transnitrosation reactions in which the NO moiety (as NO^+) is transferred from RSNO onto an acceptor thiol group assist in the transmission of the NO signal. A series of transnitrosation reactions originating from a donor RSNO onto thiol groups incorporated in enzymes or other proteins *(15)* occur until ensuing alterations to protein structure elicit changes to cellular function. For example, ion channel function can be modulated by S-nitrosation, such as *N*-methyl-D-aspartate (NMDA) receptor-coupled Ca^{2+} channels *(116)*, TRP3 channels *(117)*, Na^+ channels *(118)*, and ryanodine receptors *(119)*. Many enzymes are also regulated by *S*-nitrosation including ornithine decarboxylase *(120)*, GAPDH *(121)*, and methionine adenosyltransferase *(122)*.

In addition, although there is some dispute on this issue, it was hypothesized that nitrosation of proteins might follow consensus motif targeting, analogously to protein phosphorylation, which preferentially targets serine or threonine residues. Stamler et al. *(123)* and others suggested that nitrosation might favorably target proteins with the peptide sequence XY-Cys-Z, in which X denotes either a glycine, serine, threonine, tyrosine, or glutamate; Y denotes a lysine, arginine, histidine, asparagine, or glutamate; and Z is an asparagine or glutamate residue. However, Ascenzi et al. *(124)* and Butler et al. *(60)* concluded that the likelihood of proteins undergoing nitrosation was more likely dictated by cellular environment rather than peptide sequence.

4. DINITROSYL IRON COMPLEXES

Dinitrosyl iron complexes (DNICs) are stable paramagnetic substances that form when NO binds with metalloproteins that contain iron-sulfur cluster centers *(125)*. They produce distinctive electron paramagnetic resonance spectra, which can be quantified by electron paramagnetic resonance spectroscopy *(126,127)*. In a manner analogous to the transnitrosation reactions of RSNOs, DNICs are able to transfer their NO moiety to other metalloproteins, causing conformational changes that result in relaxation or other cellular effects *(128)*. Release of the NO moiety can be induced by thiols, which causes its displacement as free NO or allows it to transfer onto other metalloproteins *(125)*. DNICs display stable and potent relaxant activity of vascular smooth muscle *(75,126)*, can be formed in vivo *(128)*, can be stored in both endothelial cell and smooth muscle cells (SMCs) *(129)*, and have been suggested to account for the activity of endogenous NO *(125,130)*. Although this appears unlikely *(75)*, the thiol *N*-acetyl-L-cysteine was demonstrated to form a DNIC following pretreatment of rat aorta with lipopolysaccharide to induce NO production by iNOS *(129)*. This detection of a DNIC following induction of iNOS highlighted the possibility that a stable store of NO could arise following any number of inflammatory processes.

5. MEASUREMENT OF NO AND RSNOs

5.1. NO Electrodes

The gaseous and somewhat labile nature of NO, as well as the low physiological output by endothelial cells and neurons (nanomolar), has made the ability to directly detect it from tissues a difficult task. Indirect approaches (e.g., nitrite measurements) had been the conventional way to measure the endogenous production of NO. An innovative approach for real-time or in vivo situations emerged following the development of amperometric electrodes that purported to selectively detect NO in solution and tissues, and with a greater sensitivity than other methods *(131)*. NO selective electrodes are widely available from many suppliers or they can be manufactured in the laboratory *(131,132)*. Briefly, NO electrodes consist of a carbon fiber tip connected to electrode wiring, which is first coated with a porphyrinic heme substance (usually a nickel-porphyrin complex). The porphyrin surface provides a reactive surface for $NO^{\bullet}$ to oxidize and generate a current that is displayed as an amperometric signal on a data acquisition system. To ensure selectivity for $NO^{\bullet}$, the electrode tip is also coated with a membrane (Nafion) that is permeable only to electroneutral $NO^{\bullet}$, thus discriminating against other oxides of NO. More accurately, the Nafion membrane is a fluorocarbon polymer substance with sulfonate side chains, which carry a negative charge; thus, a Nafion membrane would exclude anionic substances (e.g., nitrite) but permit the passage of gases (i.e., $NO^{\bullet}$) and some cations *(133)*. However, because NO is such a strong reductant of porphyrinic heme, any signal would be assumed to be entirely mediated by $NO^{\bullet}$. Electrodes are calibrated with known amounts of NO using NO donors or solutions of NO before test measurements are taken. Most commercially available NO electrode probes have a physiological detection limit, being able to detect levels of NO in the nanomolar range *(131,134–138)*.

Despite this, there are a few disadvantages to using this system. For example, because NO has a short half-life, quantifying the amount of NO produced by a cell or tissue may be greatly dependent on the response time of the particular electrode, in which case, if it responds too slowly, then the sum amount of NO being produced may not be entirely captured or registered by the detector system. Furthermore, any changes in temperature and fluctuations in flow within blood vessels will give rise to artifactual signals. However, one way to circumvent this is to ensure that the instrument is calibrated in the same medium or environment in which real-time measurements are to be taken. For example, calibration of the electrode should be carried out in the same tissue in which NO production is to be measured, by spiking the tissue with known amounts of a NO donor. Only if a reasonably linear curve is obtained would it be suitable to proceed and use the electrode for real-time measurements. This is also important because more often the tissue medium will contain substances (e.g., radical species, transition metal cations) that can interfere with the bioactivity of NO, and this will give erroneous measures of NO production. Calibrating the device under such conditions may provide a means to extrapolate a physiological figure amount of NO.

5.2. NO Fluorophores

Another innovative approach for detecting and measuring NO has emerged following the development of NO-reactive fluorophores, which yield a fluorescent signal when they come into contact with NO in the presence of O_2 and, therefore, provide a unique method for visualizing the generation of NO. These fluorophores are essentially diamino derivatives of the existing fluorophore, fluoroscein (e.g., 4,5-diaminofluorescein, 4-amino-5-methylamino-2'7'-fluorescein, diaminorhodamine), which react with NO in the presence of O_2 to form the highly fluorescent diamino-derivative triazole, and are detected by spectrofluorometry or fluoromicroscopy *(139)*. Most diamino fluorophores are already cell permeable or have an ester group incorporated into their structures to assist in cell permeability. Once these substances have entered a cell, cytosolic esterases hydrolyze the fluorophores, rendering them membrane impermeable and thus trapped within the cell. At this point, the fluorophores only yield a very faint signal—only when NO is produced does the level of fluorescence dramatically increase to signify the synthesis or presence of NO. These useful tools were first described by Nakatsubo et al. *(140)* who loaded bovine aortic endothelial cells with a diaminofluorescein derivative and subsequently observed a NO-dependent fluorescent signal. They also successfully used these fluorophores to visualize the neuronal production of NO in rat brain slices *(141)*. According to Kojima et al. *(139)*, the limit of detection for NO is <10 n*M*, making such fluorophores very useful and highly sensitive tools for the real-time measurement of NO.

However, some problems have been encountered with the use of fluorophores regarding their specificity. The presence of divalent cations, variable exposure to light, changes in pH, and pro-oxidant substances all have been reported to enhance the fluorescent signal produced from these fluorophores *(142–144)*. Furthermore, antioxidant substances and catecholamines were reported to have a quenching effect on signals produced by these fluorophores *(145)*; thus, for example, the use of antioxidant agents to examine changes in the bioactivity of NO would yield confounding results. However, one study reported that to avoid getting nonspecific fluorescence with the diamino fluorophores, cells or tissues should be loaded with a 100-fold less concentration than is usually employed (0.1 vs 10 μ*M*) to get more reliable fluorescent signals *(146)*. Overall, NO-reactive fluorophores can be useful and highly sensitive experimental tools to visualize the production of NO *in situ* provided that the proper controls and optimization protocols are performed to be certain of their specificity.

5.3. Measurement of Nitrosothiols

Over the years, the findings on biological roles of RSNOs have greatly stimulated interest in the examination of biochemical properties of these compounds *(57,84)*, as well as in the development of analytical techniques for their reliable detection and quantification in biological samples. RSNOs have been detected in vivo by many investigators *(59,64,147–151)*. Examples of RSNOs include low-molecular-weight *S*-nitrosocysteine (CysNO), *S*-nitrosoglutathione (GSNO), and high-molecular-weight *S*-nitrosoalbumin (SNO-Alb). Among these RSNOs, SNO-Alb tends to be more stable than low-molecular-weight molecules and is the principal molecule formed *(64)*. The normal endogenous levels of RSNOs have been reported to be as high as 7 μ*M* *(64)* and as low as 28 n*M* *(66)*, with numerous values falling within this range *(151–153)*. The variations in concentration may reflect the differences in species, as well as the analytical techniques used.

RSNOs in biological samples have been measured using a wide spectrum of different techniques. High-performance liquid chromatography (HPLC) with ultraviolet (UV) absorbance detection at 333 nm is frequently employed for the measurement of low-molecular-weight RSNO compounds *(154)*. Nevertheless, the majority of other methods are based on the measurement of NO released by chemicals from the *S*-nitroso group of the *S*-nitroso compound *(155)* followed by chemiluminescent detection of released NO *(156)*. Moreover, gas chromatography-mass spectrometry (GC-MS) has been shown to be suitable for the analysis of RSNOs *(157,158)*. However, GC-MS is an indirect method that is based on the detection of nitrite from the conversion of *S*-nitroso group after treatment with mercuric ions. Recently, electrospray ionization-mass spectrometry (ESI-MS) *(159)* in combination with HPLC *(160)* has been described for the detection of RSNOs.

Table 1
Absorption Maxima and Molar Extinction Coefficients (ε) for Some Common RSNOs

RSNO	λ_{max1} *(nm)*	ε $(M^{-1}\ cm^{-1})$	λ_{max2} *(nm)*	ε $(M^{-1}\ cm^{-1})$
GSNO	335	586	544	17.2
CysNO	335	503	544	14.9
SNO-Alb	335	3869	545	47

This technique allows discrimination among different RSNO species and seems to be the most promising technology in RSNO research. Herein, we summarize the most common methods used for the detection of RSNOs with the emphasis on GSNO, which is predominantly found in blood and in tissues such as the lungs *(59,161)*.

5.3.1. Spectrophotometry

RSNOs exhibit UV and visible absorption maxima in the regions of 320–360 and 540–600 nm. Molar absorptivities of these species vary at different wavelengths (**Table 1**) *(59,162)*. In the past, RSNOs were measured directly by this conventional spectroscopic technique. However, because of the low molar absorptivity of the *S*-nitroso group of these compounds, this method only permits the detection of RSNOs at submillimolar or millimolar concentrations using the absorption band at 335 nm *(162)*. Moreover, lack of specificity is another limitation of this method because there is no direct distinction between each individual nitrosothiol present in the biological matrices.

5.3.2. Saville Reaction

The Saville assay *(163)*, originally used for the detection of thiols, has been modified for the measurement of RSNO levels in biological fluids. This method uses mercuric chloride ($HgCl_2$) to disrupt the S–NO bond of RSNOs, followed by the addition of sulfanilamide and *N*-1-naphthylethylenediamine dihydrochloride to form a colored azo compound. The concentration of the azo compound can be determined by measuring the absorbance at 540 nm (ε approx 50,000 M^{-1} cm^{-1}). In fact, the second step used after mercury ion-mediated bond cleavage is identical to the classic Griess assay for the quantification of nitrite *(162)*.

Although the modified Saville reaction is simple, there are two major limitations to this method of detection: sensitivity and interference from nitrate contamination. First, this technique has the same limitation as the Griess assay, including sensitivity limited to the micromolar range, which is insufficient for many biological applications *(164)*. Second, nitrite contamination in all reagents may interfere with RSNO detection. In addition, the procedure can be hindered by free thiols and the background concentration of nitrite present in biological samples, which may lead to artifactual formation of RSNOs *(66)*. Interference prevents reliable information for the quantification of RSNOs.

5.3.3. Direct Detection: HPLC

Concentrations of RSNOs can be measured directly by employing separation techniques such as HPLC. HPLC offers several advantages. It allows distinctive separation of RSNO from a complex mixture and provides the capability of detecting the compound of interest directly. Moreover, this method requires only minimal sample preparation; it does not need elaborate clean-up procedures when working with biological media. In general, the two major components of HPLC, the mobile phase and the stationary phase, facilitate the separation. The extent or degree of separation is mostly determined by the choice of mobile phase and stationary phase, which is dependent on the chemical and physical properties of the compounds. When a mixture is dissolved in the mobile phase, the solution passes through the stationary phase (chromatographic column), where separation occurs. The separation process is governed by the distribution of substances between two phases that is based on their different migration rates through the column. Compounds that are weakly held by the stationary phase travel faster through the column. The compound that emerges from the column is detected by UV-visible absorbance, fluorescence, or electrochemical detector *(165)*. Generally, compounds are identified and quantified by comparison with authentic standards.

5.3.3.1. HPLC With UV-Visible Detection. The most challenging task in measuring the concentrations of RSNOs by HPLC is the development of specific and interference-free methods. HPLC analysis of RSNOs has been dependent on the use of mobile phases at low pH ranging from 2.0 to 4.0 *(31,61,154,156)*. Nevertheless, mobile phases with acidic pH have been associated with problems regarding artifactual formation of RSNOs in thiol and nitrite-rich matrices. Because RSNOs occur in biological fluids and tissues at very low concentrations *(155)*, artifactual formation of *S*-nitroso compounds from nitrite and thiols, even of very low extent, may reach or exceed endogenous levels, leading to significant misinterpretations. To eliminate the problems of interference, several investigators have developed artifactual-free HPLC assays. For example, Pfeiffer et al. *(166)* have shown that 20 m*M* sodium phosphate buffer at neutral pH (7.4) consisting 5% of methanol is capable of separating GSNO through a C_{18} reversed-phase column. This technique allows the detection of GSNO at 338 nm in the hundreds to thousands of nanomolar concentration range. Similarly, Tsikas et al. *(167)* have employed mobile phases at pH 7.0. Specifically, an anion-pairing agent, tetrabutylammoniumhydrogen sulfate, is added in the mobile phase that enables the analysis of GSNO at neutral pH. This method prevents the reaction between thiols glutathione (GSH) and nitrite and thus provides artifactual-free analysis of GSNO.

An alternative way to overcome the problem of artifactual formation of RSNO compounds is to eliminate excess thiols in biological samples prior to HPLC analysis. This can be accomplished by adding *N*-ethylmaleimide (NEM) to samples during sample preparation. In principle, NEM alkylates thiol groups and minimizes the artifactual formation of RSNOs *(66)*, as well as transnitrosation reactions *(114)*. In fact, Ng et al. *(168)* developed an HPLC assay for GSNO using a mobile phase that contained 50 m*M* phosphate buffer (pH 5.5) in acetonitrile (99:1). GSNO was separated on a Hypersil BDS C_{18} column (125 × 4 mm, 5 μm; Agilent, Mississauga, ON, Canada). In addition, this group has examined the effect of NEM on the removal of thiols in biological samples. This was carried out by adding 100 μ*M* GSH to plasma samples that contained different concentrations of spiked GSNO followed by the treatment of NEM. Ng et al. *(168)* demonstrated that a standard curve of GSNO in plasma in the presence of NEM was identical to that in phosphate-buffered saline (PBS) in the absence of the same reagent, suggesting that NEM was capable of removing high concentrations of GSH in plasma, because the slopes of both standard curves were comparable (**Fig. 1**). **Figure 1** illustrates that GSH did not interfere with the HPLC assay after the treatment of NEM. Conceivably, the addition of NEM to biological matrices prior to HPLC analysis eliminates the problems of interference.

5.3.3.2. HPLC With Electrochemical Detection. GSNO and CysNO can be measured by HPLC with electrochemical detection *(156)*. These molecules are separated by

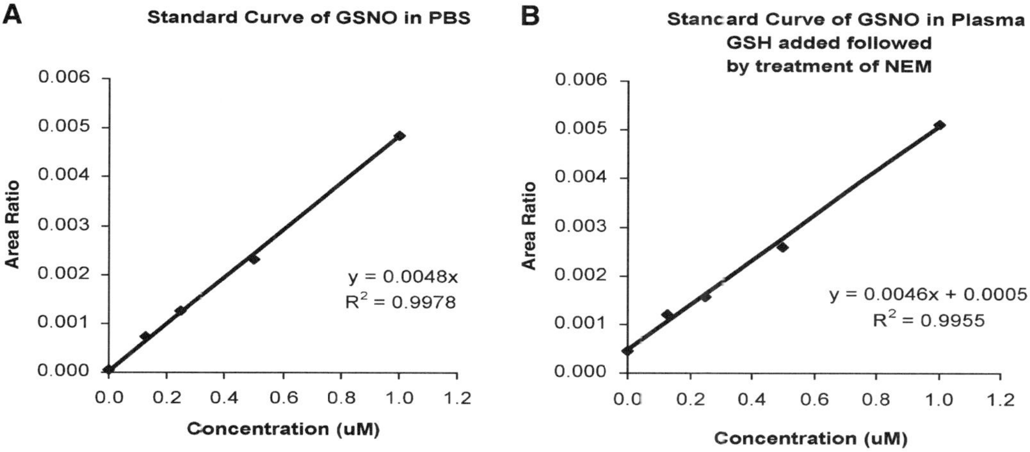

Fig. 1. Standard curves of GSNO in **(A)** PBS and **(B)** plasma at concentrations 0, 0.125, 0.25, 0.5, and 1.0 μ*M*. The *y*-intercept of the standard curve in PBS is zero because this buffer contained no GSNO at 0 μ*M*, whereas that in plasma is shifted upward because of the background concentrations of GSNO in the matrix. In **(B)**, 100 μ*M* glutathione (GSH) was added to plasma followed by treatment with NEM. The slope of GSNO in PBS resembles that in plasma ($y = 0.00482 \cdot x$ vs $y = 0.00461 \cdot x$). The similarity of the slopes suggests that the detection of GSNO in plasma was not different from that in PBS. Apparently, NEM was able to remove GSH in the samples to prevent artifactual formation from the interaction between thiol and nitrite. NEM also increased the stability of GSNO, which would otherwise decompose when combined with other thiol groups in biological samples.

HPLC using a C_{18} reversed-phase column coupled to an electrochemical detector with dual working electrodes that consist of mercury-gold amalgam (Au|Hg electrode) and a reference electrode (Ag|AgCl). The working electrodes are configured in series with the reducing electrode set at –0.15 V and the oxidizing electrode set at +0.15 V. RSNOs are detected by both electrodes, first as RSNO at the electrode during a reduction step that generates a thiol, which, in turn, is detected at the oxidizing electrode. The half reactions that occur at the reducing and oxidizing electrode are as follows *(156)*:

Reducing electrode:
$$2RSNO + 2H^+ + 2e^- \rightarrow 2RSH + 2NO$$

Oxidizing electrode:
$$2RSH + Hg \rightarrow Hg(SR)_2 + 2H+ + 2e^-$$

Under these configurations, RSNOs can be separated and detected in the nanomolar concentration range *(149)*, and this method is usually one to two orders of magnitude more sensitive than the UV detection method *(162)*.

5.3.4. Indirect Detection

5.3.4.1. Chemiluminescence-Based Assay. An alternative approach for the determination of RSNOs is based on an indirect method that involves the decomposition of the S-NO bond to NO followed by its subsequent detection by chemiluminescence *(64,156)*. Essentially, a vacuum draws the NO gas released from RSNOs into a reaction chamber where NO is oxidized with ozone to form an excited state of nitrogen dioxide (NO_2^*) species. The excited NO_2^* radical rapidly decays back to its ground state and emits an electromagnetic radiation, which is detected by a sensitive photomultiplier tube *(155,162)*.

$$NO + O_3 \rightarrow NO_2^* + O_2$$
$$NO_2^* \rightarrow NO_2 + h\nu$$

Samouilov and Zweier *(164)* developed a chemiluminescence-based method for the measurement of RSNOs. They employed a mixture of quinone and hydroquinone in alkaline solution (pH > 10.0) that selectively cleaves the S–NO bond of GSNO at 60°C. These investigators found that the release of NO from GSNO occurred rapidly within 1 min, and the quantitative detection was possible down to 10-n*M* levels in either plasma or buffer. Similarly, Marley et al. *(66)* have developed a highly sensitive and reproducible chemiluminescence-based assay for the measurement of total plasma RSNOs in healthy humans. Cleavage of the S–NO bond is achieved by the reaction with a mixture of copper (I), iodine, and iodide, and the NO released is then detected by its chemiluminescent reaction with ozone. This method allows specific and effective cleavage of RSNOs. Additionally, this group has demonstrated that the removal of endogenous nitrite by sulfanilamide and the stabilization of RSNOs by NEM and EDTA allow reliable quantification of low nanomolar concentrations of endogenous RSNOs in complex biological matrices. In fact, these two steps represent a major methodological advancement in the quantitative analysis of low concentrations of RSNOs, because this is a common problem with most if not all other assays. Interestingly, using the chemiluminescence assay just described, Rassaf et al. *(169)* demonstrated that plasma nitroso species increased during infusion of aqueous NO solutions to human arteries, and vasodilation was observed in the peripheral vasculature.

Early chemiluminescence-based techniques for the quantification of RSNOs were performed mostly in plasma, and only a few methods have been validated. Considerably less attention has been paid to the analysis of these molecules in cells or tissues, although Feelisch et al. *(170)* and Rodriguez et al. *(109)* were able to apply the chemiluminescence technique for the trace-level detection of *S*-nitroso species in complex biological matrices, such as tissues. In fact, Rodriguez et al. *(109)* demonstrated that rat aortic tissues contained 40 n*M* *S*-nitroso compounds.

5.3.4.2. Fluorometric Detection. Park and Kostka *(171)* described a fluorometric technique for the analysis of GSNO, CysNO, and SNO-Alb in biological fluids. The methodology is a variation of the Saville reaction in combination with fluorescence technique, which is based on the detection of a fluorescent compound, 1-[H]-naphthotriazole (NTA), formed in the reaction between acidified 2,3-diaminonaphthalene (DAN) and nitrous acid (protonated NO_2^- ion) released from RSNOs by treatment with $HgCl_2$ *(171,172)* (**Fig. 2**). Consequently, the intensity of the fluorescent signal produced by NTA is measured. Fluorometric readings are taken at excitation and emission wavelengths of 380 and 450 nm, respectively *(171)*. This assay provides a rapid and sensitive determination of RSNOs that can be applied to the analysis of complex reaction mixtures without the need for excessive sample preparation.

Recently, Ng et al. *(168)* applied a similar fluorometric technique for the detection of high-molecular-weight nitrosothiol compound (SNO-Alb) in plasma, except that a few pretreatment steps were performed prior to the addition of DAN for (a) alkylation of thiols (with NEM), (b) elimination of nitrite (with sulfanilamide), and (c) discrimination between high and low-molecular-weight RSNOs. Briefly, plasma samples were diluted with an equal volume of water. NEM (final concentration of 5 m*M*) was immediately added followed by 50 μL of 0.5% sulfanilamide (incubated at room temperature for 10 min).

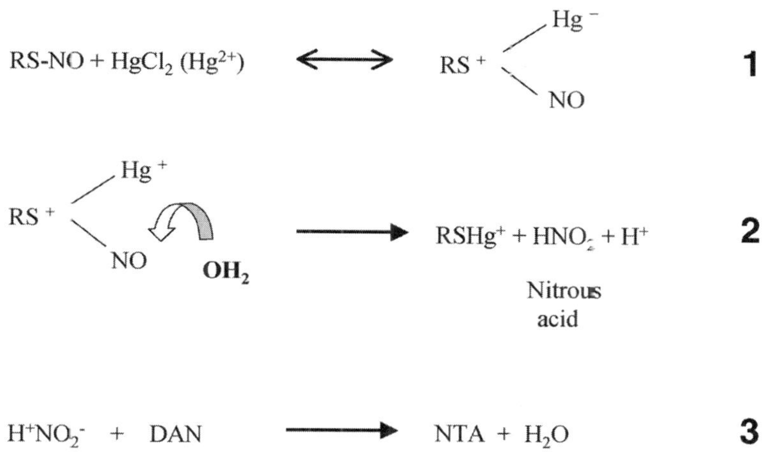

Fig. 2. Simplified scheme for detection of RSNO using DAN fluorescence assay. Mercury (II) ion (Hg^{2+}) catalyzes the decomposition of RSNO by breaking the S-NO bond, which results in the release of nitrous acid (HNO_2) (**Eqs. 1** and **2**). The protonated NO_2^- ion (nitrous acid) is then added to DAN to form a fluorescent compound, NTA, for fluorescence detection (**Eq. 3**).

Desalting was carried out using Bio-Rad Econo-Pac® 10 DG columns (Bio-Rad, Hercules, CA) to recover the high-molecular-weight fraction from plasma samples and to remove nitrite/nitrate and the low-molecular-weight RSNOs. Collected fractions were incubated with $HgCl_2$ (final concentration of 200 μ*M*) at room temperature for 10 min. Acidified DAN (final concentration of 5 μg/mL) was then added and incubated with samples at 37°C for 15 min in the dark. To neutralize the acidic solution, samples were treated with K_2HPO_4 followed by the addition of 5-sulfosalicylic acid for precipitation of proteins. The precipitate was removed by centrifuging at 1500*g* for 10 min, and the supernatant was treated with 50 μL of 2.8 *M* NaOH. The alkalinization of samples with NaOH maximized the fluorescent signal *(173)*. Fluorometric readings were taken using an Aminco SPF-500 Spectrofluorometer (SLM Instruments, Rochester, NY) with excitation and emission wavelengths of 380 and 450 nm, respectively. The detection limit of this method for plasma SNO-Alb was 50 n*M*. This technique should be useful for the measurement of high-molecular-weight RSNOs and allows artifact-free quantification of SNO-Alb in plasma.

5.3.4.3. GC-MS Detection. A major drawback of chemiluminescence and fluorometric detection of RSNOs is that they do not use internal standards for reliable quantification. By contrast, GC-MS allows specific, highly sensitive, and accurate quantification of analytes because stable isotope-labeled internal standards are feasible with this methodology. Over the past few years, GC-MS methods have been developed for the determination of GSNO, CysNO *(174)*, and SNO-Alb *(158)* in biological fluids using ^{15}N-labeled analogs ($GS^{15}NO$, $Cys^{15}NO$, $S^{15}NO$-Alb) as internal standards. The principle of these methods is based on specific cleavage of the *S*-nitroso group by Hg^{2+} to nitrite. Essentially, the *S*-nitroso groups of GSNO, CysNO, and SNO-Alb and their ^{15}N-labeled analogs are converted to nitrite and [^{15}N]nitrite. Nitrite and [^{15}N]nitrite are further derivatized by pentafluorobenzyl (PFB) bromide to the corresponding PFB derivatives. The products of the derivatives are analyzed by capillary GC, ionized by negative-ion

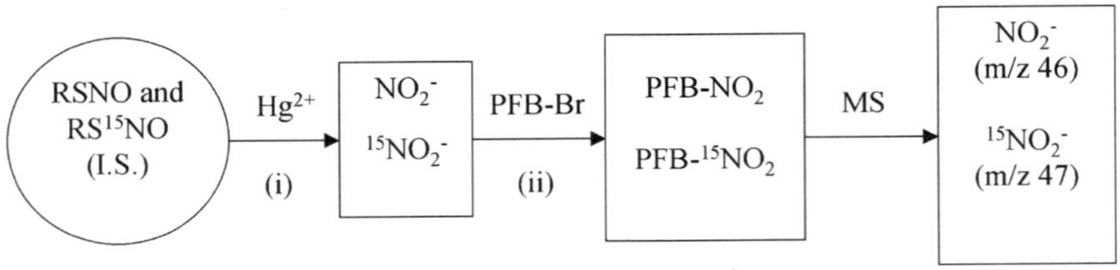

Fig. 3. Simplified scheme of GC-MS methods for the detection of **(i)** RSNOs using isotope-labeled internal standards. RSNOs and internal standards $RS^{15}NO$ are converted by $HgCl_2$ (Hg^{2+}) to nitrite (NO_2^-) and [^{15}N]nitrite ($^{15}NO_2^-$). **(ii)** Nitrite and [^{15}N]nitrite are then derivatized by PFB bromide to the corresponding PFB derivatives. The resultant derivatives are detected by MS based on their m/z.

chemical ionization, followed by MS analysis according to their mass-to-charge ratios (m/z) (m/z 46 for nitrite and m/z 47 for [^{15}N]nitrite) and detection by an electron multiplier *(158,174)*. A simplified scheme of the GC-MS methods for the analysis of RSNOs is depicted in **Fig. 3**. Note that GC-MS detection generally is not possible for labile and nonvolatile compounds, such as GSNO and CysNO. However, these compounds can be separated by HPLC and converted to nitrite prior to derivatization for MS analysis *(160)*. However, GC-MS has been described as an exquisitely sensitive technique, and relatively large and expensive instrumentation is required that is not commonly available in most laboratories.

5.3.4.4. Electrospray Ionization-Mass Spectrometry. The majority of the methods used for the detection of RSNOs are based on the measurement of NO or nitrite released by chemicals from the *S*-nitroso group of the *S*-nitroso compounds *(155)*. In general, these methods lack specificity. Although direct measurement of RSNOs by HPLC is plausible, insufficient sensitivity is problematic. Recently, ESI-MS has been developed to become a powerful technique for the analysis of low- and high-molecular-weight RSNO compounds. ESI-MS is a very accurate method of determining molecular mass *(175)*, and conditions can be varied such that the protein of interest is subjected to gentle perturbations during analysis in an effort to preserve labile structures *(159)*. Moreover, ESI-MS is a highly sensitive technique that offers the potential advantage of identifying the specific RSNO involved in different biological matrices. In addition to specificity, an advantage of this technique is that it can detect RSNOs in very small amounts of sample and can determine the stoichiometry of substitution.

Indeed, Mirza et al. *(159)* performed the pioneering work for the identification of RSNO formation using ESI-MS. Additionally, Tsikas et al. *(160)* applied ESI-MS in combination with HPLC for the detection of low-molecular-weight RSNOs. In brief, low-molecular-weight RSNOs were separated by the HPLC system. The extracts from the HPLC eluates were diluted with 2 vol % of formic acid in acetonitrile (1:1 [v/v]), and the sample extracts were analyzed on a triple-quadrupole mass spectrometer equipped with the ion spray ionization source. Mass spectra were generated in the positive ion mode. The analyte was introduced into the mass spectrometer using a Harvard syringe pump through a 50-μm-inner-diameter fused silica capillary directly into the ion spray at 5 μL/min. An orifice voltage of 40–50 V was used. The capillary interface tube temperature was kept at 55°C. Tandem quadrupole mass spectrometry (MS-MS) was performed under identical ionization conditions. ESI-MS analysis of extracts of HPLC eluates was performed by subjecting the parent ion $[M+H]^+$ at m/z 337 for GSNO to collision-activated dissociation using argon as the collision gas.

The inherent accuracy and high sensitivity of MS would help researchers to better understand the formation, metabolism, and physiological roles of RSNOs in biological systems. ESI-MS in combination with HPLC seems to be an essential technique for the characterization of RSNOs in complex biological matrices. Nevertheless, future work is needed for the development of such a method.

6. SYNTHESIS OF NO

NO is involved in many diverse biological processes. This variety in actions may be partly explained by the different isoforms of enzymes that synthesize NO and their site of expression, as well as the various cellular mechanisms by which NO elicits its actions.

6.1. Endothelial Cell-Derived NO

The endothelium plays a key role in the short- and long-term regulation of the cardiovascular system and is the source of many factors that influence blood flow, blood coagulation, and angiogenesis. Collectively, the endothelium is a major "organ" in the body and, in an adult, the total endothelium mass is about 500 g; mainly in the pulmonary circulation. A key substance produced by the endothelium is NO. It is an important multitasking substance in the regulation of cardiovascular function:

1. NO activates sGC by binding to the heme moiety of sGC and increases cellular GMP and, hence, activates g-kinase and is the key mediator of endothelium-dependent vasodilatation in conduit and elastic arteries.
2. NO possesses antiplatelet aggregatory.
3. NO has pro- (and anti-) angiogenic activity.
4. NO is anti-inflammatory.
5. NO decreases leukocyte adhesion.

6.2. Synthesis of NO by NOSs

NO is synthesized in mammalian cells by a family of three NOSs: endothelial NOS (eNOS), neuronal NOS (nNOS), and iNOS. eNOS (also known as NOS III) is constitutively produced by endothelial cells and encoded by genes on chromosome 7 *(176)*. eNOS is located uniquely within the caveolae of the plasma membrane *(177)*. NOS catalyzes an NADPH- and oxygen-dependent five-electron oxidation of L-arginine to generate L-citrulline and NO, via formation of N^{ω}-hydroxy-L-arginine as an intermediate. All NOSs are homodimers, with each subunit composed of two functional domains. The carboxy-terminal domain contains a binding site for NADPH, flavin adenine dinucleotide (FAD), and flavin mononucleotide (FMN) and bears close homology to the CYP reductases; thus, this region of the enzyme is called the *reductase domain*. The reductase domain also contains a calcium-dependent binding site for calmodulin, which reversibly binds to the constitutive NOS isoforms *(178)*. The amino-terminal half of the each NOS isoform is an oxygenase domain that binds heme, L-arginine, and the cofactor tetrahydrobiopterin (BH_4). The NOS oxygenase domains contain a cysteine thiolate, which coordinates to the heme iron as an axial ligand. In the presence of bound calmodulin, the heme iron can accept NADPH-derived electrons from the reductase flavins. When reduced, the ferrous heme can bind and then proceed with the oxygenation of the substrate. BH_4 is an essential cofactor for the proper flow of electrons to oxidize L-arginine *(179)*, and NOS is the only heme-containing enzyme known to require a pterin cofactor for activity. Of significant interest, therefore, is that a reduced availability of BH_4 may

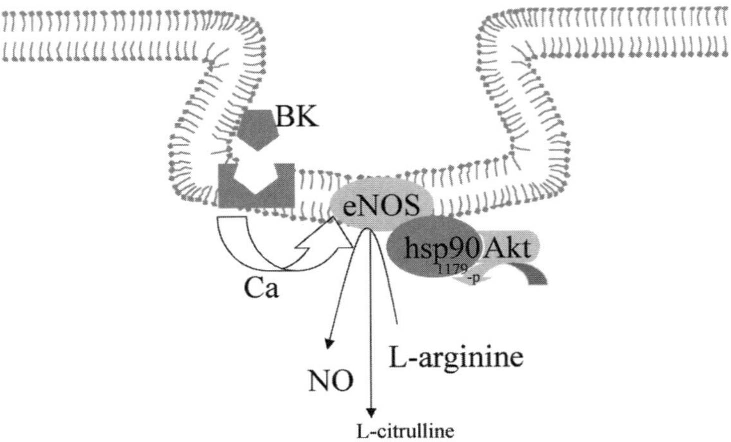

Fig. 4. Basal activity of eNOS is maintained by myristoylated and palmitoylated membrane-associated eNOS associated with caveolin-1. Less active forms of eNOS are also noted following protein kinase C- and mitogen-activated protein kinase-mediated phosphorylation. Mechanical (shear stress) and/or chemical activation (bradykinin [BK]) of the endothelial cell leads to a rise in intracellular calcium and subsequent calcium-activated calmodulin and hsp90-facilitated activation of eNOS. An activated dimeric eNOS facilitates the transfer of five electrons and the production of NO from a guanadino group of the amino acid L-arginine, with L-citrulline as the "byproduct." The critical phosphorylation of serines 1177/1179 of eNOS by Akt and protein kinase A also enhances eNOS activity.

be a key factor in the etiology of the endothelial dysfunction associated with cardiovascular disease. In addition to calmodulin, eNOS requires several chaperone proteins including heat shock protein 90 (hsp90) and caveolin *(177)*. eNOS can also be activated by direct phosphorylation by serine/threonine protein kinase Akt (protein kinase B) on serine 1179 *(180)*; key components of these regulatory pathways are depicted in **Fig. 4**.

It has also been suggested that shear stress might activate tyrosine kinase *(181,182)*, and, thus, conceivably this process could lead to the activation of eNOS by way of Akt-mediated phosphorylation of the enzyme *(180)*. Furthermore, NOS may be negatively regulated by NO *(183)*. The activity of eNOS can also be modified in disease states, and in diabetes hyperglycemia inhibits eNOS as a result of a posttranslational modification at the protein kinase Akt regulatory site (serine 1177) *(184)*.

Production of endothelium-derived NO (also known as EDRF) occurs following a rise in intracellular concentrations of calcium, usually stimulated by the activation of G-protein-coupled receptors or shear stress. Activation of G-protein-coupled receptors will activate a phospholipase C-dependent mobilization of intracellular stores of calcium or activate calcium channels to allow an extracellular influx *(185)*. Shear stress is thought to activate eNOS through either calcium-dependent or -independent pathways. Shear stress-mediated activation of store-operated calcium channels stimulates an extracellular calcium flux that results in the activation of various calcium-dependent pathways including NO synthesis. This mode of calcium entry is likely to be the physiological stimulus for eNOS activation and might also explain why store-operated calcium channels are found to be proximal to caveolae that usually house eNOS *(186)*.

Endothelial cells, in addition to NO, also produce superoxide anions, although there is some uncertainty as to whether this is a physiological or pathophysiological occurrence. Nevertheless, endothelial cells are endowed with antioxidant substances and/or enzymes to potentially remove superoxide anions. Endothelial cells express various forms of superoxide dismutase (SOD), and these may be bound extracellularly on cell-surface proteoglycans *(187)* or they may exist intracellularly *(188)*. The reaction catalyzed by SOD transforms superoxide anions into H_2O_2, which itself has been reported to possess vasorelaxant properties *(189,190)*; thus, it remains to be seen whether SOD is providing antioxidant protection for NO against superoxide or is promoting the production of vasorelaxant H_2O_2. Some researchers have proposed that H_2O_2 is an endothelium-derived relaxing substance and, potentially, the elusive endothelium-derived hyperpolarizing factor (EDHF) *(62,191,192)*.

6.3. Neuronal NO

nNOS, the neuronal isoform of NOS (also known as NOS I), synthesizes NO in nerves of the central nervous system (CNS) and autonomic nerves in the periphery. The peripheral nerve-derived NO, also termed the *nitrergic transmitter (193)*, is released from nonadrenergic, noncholinergic (NANC) nerves that innervate visceral smooth muscle. Release of the nitrergic transmitter mediates relaxation of smooth muscle and is involved in regulating gastrointestinal (GI) motility, sphincter function, erectile function, and bronchodilation. In the CNS, NO is noted for its involvement in long-term potentiation and modulation of NMDA receptors *(194)*. Also, it is worthwhile to note that in some instances vascular smooth muscle itself may also express nNOS *(195,196)*, although a function for the product of this NOS has not yet been elucidated. It is presumed that the NO signal from nerves is transmitted by the permeation of NO across cell membranes, which acts on target proteins to elicit its cellular effects. However, because of the presumed gaseous nature of the transmitter, it is unclear how this type of transmission operates.

Neurotransmission requires a rise in calcium levels within the nerve terminal to trigger exocytotic release of transmitter prepackaged in vesicles. Although it has been confirmed that nitrergic transmission is dependent on the activation of ω-conotoxin-sensitive N-type calcium channels *(197,198)*, it is not clear whether this dependence on calcium entry is for exocytotic release of the transmitter or for its synthesis. Findings from one study suggest that the latter option is more likely because the guinea pig intestinal tissue botilinum toxin, which inhibits the docking of vesicles at the synaptic membrane during exocytosis, only inhibited noradrenergic and cholinergic transmission, whereas nitrergic responses persisted *(199)*, suggesting that release of the nitrergic transmitter from vesicles does not occur. Furthermore, prolonged periods of field stimulation applied to rabbit anococcygeus muscles did not diminish nitrergic relaxations, suggesting that production of the transmitter by nNOS occurs on a demand basis *(200)*. In most cases, NANC nerves release more than one type of transmitter from the same nerve terminal, as is the case with some inhibitory NANC nerves in the GI tract that release both NO and vasoactive intestinal peptide (VIP) *(201,202)*. This process may be analogous to the production of other relaxant mediators in addition to NO by endothelial cells of some vessels, or the dual production of superoxide and NO by macrophages and neutrophils. Therefore, this may reflect a common requirement for NO-mediated processes in which a backup mechanism is essential to preserve or augment the actions of NO in the event of disturbances in the NO pathway.

6.4. NO From iNOS

iNOS, the inducible isoform of NOS (also known as NOS 2), produces NO at a rate relative to its rate of translation, in which it generates NO as soon as the protein is assembled and packaged into macrophagic lysosomes. Although the expression of iNOS has been best characterized in macrophages and neutrophils, many other cell types, including vascular smooth muscle cells *(203)*, are capable of expressing iNOS under appropriate conditions. Expression of iNOS can be induced by inflammatory mediators or antigenic proteins, and sequences for various cytokine-responsive elements have been located in the promoter region of the iNOS gene *(204)*. The NO product of this isoform of NOS initiates host-mediated cytotoxicity, and although the prospect that NO may fulfill a cytotoxic role seems to contrast its involvement in physiological processes, this difference can be largely explained by the different yields of NO by each isoform (iNOS: nano- to micromoles; eNOS/nNOS: pico- to nanomoles *[38]*). Furthermore, cytotoxicity against pathogens may be enacted by $ONOO^-$ derived from the reaction between NO and NADPH oxidase-derived superoxide anions *(205)*.

6.5. Non-NOS Sources of NO

Nitrite itself possesses negligible relaxant activity *(85)*; however, under certain conditions it can be reduced back into NO. A point in case is the fact that earlier studies investigating the biological properties of NO relied on using nitrite prepared in acidic solutions as a source of NO *(206)*—it now appears that this process could also hold some biological relevance. For example, the detection of NO in the expired air of human subjects was traced back to the amount of nitrite anions in the stomach, which were being converted into NO in the acidic gastric environment *(207,208)*. From a chemical standpoint, NO is produced from nitrite at low pH when the conjugated acid of nitrite reacts with another nitrite anion to produce N_2O_3, which then liberates NO (or NO^+) *(209,210)*. Impaired circulation during ischemia also may provide suitable conditions in which NO could be generated from nitrite. For example, rat hearts that were subjected to 30 min of ischemia displayed a distinctive electron paramagnetic resonance signal that was indicative of the presence of NO *(210)*. Furthermore, this signal was only partially attenuated by inhibitors of NOS, leading the investigators to speculate that although the origin of the nitrite was likely to have been from NOS at some point in time, its reduction back into NO did not depend directly on NOS activity.

Among the number of non-NOS enzymatic sources of NO are cytochrome oxidase, CYP, catalase, and XO. XO was reported to produce NO by reducing nitrite in an NADH-dependent manner irrespective of changes in oxygen tension *(211)*. Because under normoxic conditions XO generates superoxide as well as NO, this mechanism may be more important in maintaining levels of NO under conditions that do not usually favor the production of NO by NOS (i.e., hypoxia), rather than representing a viable or physiological alternative source of NO. Furthermore, under low oxygen tensions, superoxide production by XO is minimized, and, as such, any nitrite that is reduced into NO by XO will not be as readily consumed by superoxide to form $ONOO^-$.

SMCs are capable of storing NO in the form of RSNOs. Illuminating tissues with UV light produces relaxation mediated by the release of NO from a photosensitive store of NO *(212,213)*, which is thought to be specifically through the photolytic cleavage of the sulfur-nitrogen bond of RSNOs. The photorelaxation response is largely endothelium independent and NOS-inhibitor insensitive *(212,214,215)*, and studies in eNOS-deficient mice suggest that eNOS is not a requirement for photorelaxation *(215)*, indicating that a non-NOS source of NO may be responsible for this photosensitive store of NO.

7. CELLULAR BASIS OF ACTION OF NO

7.1. Guanylate Cyclase-Dependent Effects

The main cellular target of NO in smooth muscle is the enzyme sGC, which catalyzes the conversion of GTP into the second-messenger molecule cGMP. sGC exists as a heterodimer composed of subunits $\alpha 1$ and $\beta 1$ *(216)*. The N-terminal of the $\beta 1$-subunit contains a histidine residue, which functions as an axial ligand for the enzyme heme group. The catalytic domain is located toward the C-terminal region of the subunits *(217)*. Activation of the enzyme arises when NO interacts with the heme, dissociating it from the histidine residue and in effect exposing its catalytic site to GTP *(96)*. cGMP relaxes SMCs through numerous mechanisms, including the activation of protein kinases and by targeting ion channels. The functional consequence of this is that cellular events take place to reduce intracellular concentrations of calcium and signal relaxation.

7.2. Potassium Channel Activation

The opening of potassium channels allows K^+ ions to exit from SMCs and cause membrane hyperpolarization. Changes in membrane potential then lead to the closure of voltage-sensitive calcium channels, preventing an increase in intracellular concentrations of calcium to cause relaxation. This therefore presents an alternative, cGMP-independent mechanism for NO-mediated relaxation. However, cGMP is able to modulate, either directly or by activating kinases, the activity of numerous cellular proteins including potassium channels; thus, the influence of NO on potassium channel activity may be secondary to the activation of sGC *(218,219)*.

On the other hand, some studies indicate that NO may directly affect potassium channels. In 1994, Bolotina et al. *(220)* were the first to show that relaxations to exogenous or endothelium-derived NO, which could not be fully inhibited by inhibition of guanylate cyclase, were sensitive to inhibitors of large-conductance, calcium-dependent potassium channels. The investigators surmised that NO elicited this effect by interacting with channel protein thiol groups because the thiol-depleting agent NEM prevented NO-induced activation of potassium channels. Accordingly, it is likely, then, that NO nitrosates the thiols of the channel protein and these confer the activation of the potassium channel. NO also was reported to directly activate potassium channels in rat mesenteric myocytes *(221)*. Patch-clamp studies in guinea pig taenia caeci showed that NO donor compounds increased the open probability of large-conductance, calcium-dependent potassium channels and did so in an ODQ-insensitive manner *(222)*. However, a study in cultured endothelial cells could not identify a direct stimulatory effect by nitrosocysteine on large-conductance, calcium-dependent potassium channel conductance and reasoned that, because there was only a modest change in intracellular calcium following exposure to the NO donor, this was not sufficient to activate large conductance, calcium-dependent potassium channels *(223)*. However, inside-out patches from HEK293 cells transfected with large-conductance, calcium-dependent potassium channels did display increased conductance when the NO donor (±) *S*-nitroso-*N*-acetylpenicillamine was applied to the patches by an action on channel sulfhydryl groups, suggesting that, in this case, channel activation does not depend on the presence of a cGMP-generating system *(224)*.

8. NO AND CARDIOVASCULAR DISEASE

As predicted from the list of actions of NO, a reduction in the bioavailability of NO (defined as endothelial dysfunction resulting from a reduced vasodilatory response to ACh) can result in cardiovascular dysfunction and increase morbidity and mortality.

Despite the universally recognized importance of NO in both the short- and long-term regulation of the cardiovascular system, it is now recognized that the synthesis and release of vasoactive factors other than NO may also be altered in disease states and contribute to both the physiological and pathophysiological regulation of endothelial function; this remains a largely unexplored field. Endothelium dysfunction, thought to result from a reduction in the bioavailability of NO, is considered the major risk factor for cardiovascular complications of types 1 and 2 diabetes. Impaired endothelium-dependent vasodilatation is induced mainly by a decreased synthesis of the endothelium-derived NO and/or an increase in the production of reactive oxygen species (ROS) such as superoxide. Administration of BH_4, an important cofactor for NOS, has been demonstrated to enhance NO production in prehypertensive rats, restore endothelium-dependent vasodilatation in coronary arteries following reperfusion injury in aortae from streptozotocin-induced diabetic rats and in patients with hypercholesterolemia. BH_4 supplementation improves endothelium-dependent relaxation in healthy individuals, in patients with type 2 diabetes, and in smokers.

These findings from different animal models as well as in clinical trials led to the hypothesis that BH_4, or a precursor thereof, could be a new and an effective therapeutic approach for the improvement of endothelium function in pathophysiological conditions. We have studied endothelium function in spontaneously diabetic (db/db) mice, a model of type 2 diabetes, as well as human vascular tissue harvested for coronary artery bypass grafting. Endothelium-dependent relaxation responses to acetylcholine have been shown to be reduced in vessels from both humans and mice but were enhanced by acute incubation with BH_4 *(224a,224b)*. These data suggest that a deficiency in the availability of BH_4 plays an important role in the vascular dysfunction associated with type 2 diabetes and leads to the decreased bioavailability of NO. In addition, changes in the contribution of EDHF occur in vascular tissue from the db/db mice, suggesting a compensatory increase in EDHF production. Further studies are needed to elucidate the nature of EDHF(s) and whether EDHF serves a physiological or pathophysiological function (or both) in blood vessels.

9. ENDOTHELIUM-DERIVED HYPERPOLARIZING FACTOR

EDHF is a term used to describe a still unidentified endothelium-derived factor that mediates vascular relaxation via the hyperpolarization, presumably via the activation of potassium channels, of vascular smooth muscle. EDHF has also been referred to as the "third pathway" that, in addition to NO and prostaglandin I_2 (PGI_2), mediates endothelium-dependent vascular relaxation and is thought to be particularly important because of a likely greater role for EDHF than NO in the resistance vessels *(225,226)*. Both NO and prostacyclin (PGI_2) can hyperpolarize vascular smooth muscle via the activation of K-channels, but, following the definition that EDHF is the non-NO and non-PGI_2 mediator of endothelium-dependent relaxation, neither NO nor PGI_2 are considered in this review as candidate molecules.

Over the past 20 yr, several investigators have vigorously reviewed the nature of EDHF, and we thus focus on a few key issues. The reader is referred to refs. *225–228*.

Several chemical entities have been proposed to be candidate molecules for EDHF, but no one molecule or cellular process can be identified with certainty as the EDHF. Multiple EDHFs may exist because both tissue and species differences are evident; the prospect of multiple EDHFs with selectivity for different vascular beds is, of course, an exciting prospect for pharmaceutical development *(229)*. The following substances have been proposed as candidate molecules for EDHF.

9.1. Putative EDHFs

1. An arachidonic acid product and most likely an epoxyeicosatrienoic acid (EET).
2. An endogenous cannabinoid, anandamide, which is also an arachidonic acid product.
3. Adenosine.
4. Carbon monoxide.
5. An isoprostane.
6. L-Citrulline.
7. H_2O_2.
8. A small increase, 1–5 m*M*, in extracellular potassium.
9. Vasoactive peptides such as calcitonin-gene related peptide, VIP, and ghrelin.

In addition, myoendothelial gap junctions, which provide low-resistance electrical coupling and also allow the passage of low-molecular-weight (<1000 Daltons) water-soluble molecules, may be of particular importance in the resistance vessels and thus can be included in this list of possible pathways accounting for EDHF-10 myoendothelial gap junctions.

To be an EDHF, the molecule must meet the following criteria:

1. Mimic the effects of an endothelium-dependent vasodilator (i.e., ACh or bradykinin) on vascular smooth muscle and, notably, be shown to hyperpolarize vascular smooth muscle.
2. Be shown to be synthesized in and released from endothelial cells.
3. Have an action on smooth muscle that is modified by appropriate pharmacological intervention.

In most instances, an EDHF-mediated event is inhibited by the combination of two potassium channel inhibitors: apamin (a small-conductance, calcium-activated potassium channel [SK_{Ca}] blocker) and charybdotoxin (ChTx) (an intermediate [IK_{Ca}] and a large-conductance [BK_{Ca}] calcium-activated K-channel, as well as K_V 1.2 and 1.3 channels inhibitor) *(230–232)*. Of interest is that iberiotoxin (IbTx), a selective inhibitor of BK_{Ca} channels, does not substitute for ChTx, and this has led to the view that it is the IK_{Ca} inhibitory action of ChTx that is important for the inhibitory action on EDHF. Support for an endothelial site of action of apamin and ChTx is strong, and a recent study with the selective IK inhibitor 2-(2-chlorophenyl)-2,2-diphenyl acetonitrile (TRAM-39) demonstrated that TRAM-39 had no significant effect on EDHF-mediated relaxation in rat mesenteric arteries, but EDHF-mediated vasorelaxation and hyperpolarization were abolished by a combination of TRAM-39 and apamin *(233)*. Generalizations are, of course, dangerous, and in the guinea pig cerebral circulation, ChTx alone completely inhibited the EDHF response and an IbTx-sensitive process was also apparent *(234)*.

Of the putative EDHFs listed, the most recent attention has been placed on those discussed in the following sections (in alphabetical sequence).

9.1.1. Epoxyeicosatrienoic Acids

EETs are CYP-derived epoxides, and several CYPs are expressed in endothelial cells and may thus represent *EDHF synthases*. The most convincing data that an EET, specifically 11,12 EET, can serve as an EDHF comes from the work of Fisslthaler et al. *(235)*, who combined bioassay, microelectrode, and molecular techniques to demonstrate that 11,12 EET had the anticipated properties expected for an EDHF in porcine coronary arteries. Specifically, the induction of CYP 2C8/34 in native porcine coronary artery endothelial cells by β-naphthoflavone enhanced the formation of 11,12 EET (as determined by HPLC) and bradykinin-mediated vasorelaxation and membrane hyperpolarization. In addition, the transfection of coronary arteries with CYP 2C8/34 antisense oligonucleotides resulted in decreased CYP 2C8/34 and attenuated EDHF-mediated vasorelaxation and hyperpolarization. Furthermore, the vascular effects of bradykinin were inhibited by the combination of apamin and ChTx. These data provide a strong argument that 11,12 is an EDHF in the coronary circulation and add to a large body of literature that supports the hypothesis that EETs may serve as EDHFs, particularly in the coronary circulation. Nonetheless, Fisslthaler et al.'s *(235)* study can be criticized for not definitively demonstrating that the increased production of EETs by β-naphthoflavone was of endothelial cell origin and, furthermore, that EETs are thought to hyperpolarize smooth muscle via activation of the IbTx-sensitive BK_{Ca} channels, whereas EDHF-mediated relaxation and hyperpolarization is generally an apamin/ChTx-sensitive process.

9.1.2. Gap Junctions

Gap junctions couple endothelial cells to endothelial cells and SMCs to SMCs and also provide myoendothelial coupling. Gap junctions, via intercellular channels, provide a pathway for water-soluble molecules of up to 1000 Daltons and, thus, cyclic adenosine monophosphate, cGMP, inositol trisphospate, as well as Ca^{2+} ions can, in principal, move from endothelial cell to vascular SMC, or vice versa. Thus, an "EDHF" could be synthesized in an endothelial cell and, rather than be released, can be transferred to SMCs via the gap junction channels. Connexins are the principal proteins that make up a gap junction, each connexin molecule has four transmembrane domains, six connexin subunits (molecules) form a connexon, and the gap junction itself is formed when two connexons (one from each cell) *dock* to connect the two connexon hemichannels. Four connexins are expressed in vascular tissue: 37, 40, 43, and 45. Because gap junctions are formed from 12 connexins coming together and each connexon can be made up from different connexin molecules, there is considerable potential for heterogeneity in the properties of the hemichannels, and this could explain the heterogeneous nature of EDHF. Myoendothelial gap junctions are more abundant in resistance vessels than in conduit arteries, and this may also explain the predominance of endothelium-dependent hyperpolarization in the resistance arteries. This suggestion is nicely supported by ultrastructural data from Sandow and Hill *(236)* who demonstrated a greater abundance of myoendothelial gap junctions in the smaller distal vs larger proximal vessels of the rat mesenteric vascular bed. All in all, the concept of gap junctions provides an attractive explanation for EDHF. However, a difficulty in studying EDHF has been the lack of good pharmacological tools. Compounds such as heptanol in particular, but also the lipophilic saponins obtained from the licorice root *Glycyrrhizia glabra*, notably 18α- and 18β-glycyrrhetinic and the water-soluble derivative carbenoxolone, although effective uncouplers have significant nonspecific effects *(237)*. However, a major advance has been made by Griffith and his group in Cardiff, Wales, UK. They have developed inhibitors that are based on the amino acid sequence of a portion of an extracellular loop of certain connexins. Applying these peptides (e.g., Gap27 is an 11 amino acid peptide of sequence SRPTEKTIFII) seems to demonstrate specificity *(238)* as well as block cell-to-cell coupling *(239)*. Despite these advances in designing specific gap junction inhibitors, it is still not possible to conclude that such inhibitors (i.e, Gap27) are selectively blocking myoendothelial cell communication and not, in addition, inhibiting endothelial to endothelial cell or vascular smooth muscle to vascular SMC communication *(240)*. More recent studies by Griffith's group *(241)* have used a combination of inhibitory peptides directed toward connexins 37, 40, and 43, and their data indicate that more than one connexin subtype can be involved in facilitating endothelium-dependent relaxation.

Elegant studies have also been performed by Segal's group at Yale and support the importance of the endothelial cell layer as a pathway for the EDHF signal *(242,243)*. Thus, in the preparation of retractor muscle feed artery in the hamster, electrical signals can be transmitted directly to the smooth muscle layer and mediate the endothelium-dependent vasodilatory response to ACh. Interestingly Segal's group also has reported what appears to be an important role for a CYP product as the mediator of EDHF in the preparation of hamster cheek pouch *(244)*, and these vessels are of a size similar to that of the retractor muscle feed arteries. Again, these findings indicate the heterogeneous nature of the EDHF signal and support the view of the authors of this chapter that no one "molecule" or "cellular process" can explain EDHF.

9.1.3. Hydrogen Peroxide

H_2O_2 elicits vascular smooth muscle hyperpolarization and vasorelaxation and has been proposed as an EDHF largely based on the studies by Matoba and colleagues *(191,245,246)*, and the evidence for this hypothesis has been reviewed *(247)*. In support of this hypothesis is the considerable documentation that H_2O_2 mediates the opening of calcium-activated potassium channels and that H_2O_2 can be produced by endothelial cells in response to receptor activation. Therefore, an attractive argument can be made that H_2O_2 mediates vasorelaxation attributed to EDHF. Matoba et al. *(191)* demonstrated that ACh produced EDHF-mediated relaxations that were almost abolished by catalase in mesenteric arteries obtained from eNOS-deficient mice as well as from the wild-type control mice strains. They compared the EDHF-mediated relaxations with those elicited by H_2O_2, and in both cases the relaxations were sensitive to high potassium and the K_{Ca} blocker tetrabutylammonium, although only the EDHF-mediated relaxation was inhibited by the combination of apamin and ChTx. In inhibiting EDHF, apamin and ChTx are thought to target potassium channels expressed on endothelial cells *(227,231)*, and thus, their lack of effect on relaxation mediated by exogenously applied H_2O_2 would be consistent with the actions of an exogenous putative EDHF. However, sensitivity of the EDHF-mediated relaxations to tetrabutylammonium would be unexpected because others have shown that in the rat small mesenteric artery the combination of apamin and ChTx is required *(231)*

Also of interest is that catalase inhibition of EDHF-mediated relaxations from wild-type mice was only observed when an NOS inhibitor was present and catalase in combination with indomethacin alone did not affect relaxations, whereas in eNOS-deficient mice, catalase did inhibit relaxations even in the absence of an NOS inhibitor. The data from eNOS-deficient mice indicate that NO modulates the activity of ROS-generating enzymes, such that ROSs are produced only when the synthesis of NO is reduced or inhibited. Thus, NO donors attenuate EDHF-mediated vasodilatation in rabbit carotid (although this has since been identified as being entirely mediated by NO) *(248)*, and porcine coronary arteries *(249)*. A link with the generation of EETs can also be shown because a putative EDHF synthase, CYP2C9, also produces ROS in coronary arteries *(21)*.

That H_2O_2 is an EDHF is controversial because the cellular effects of H_2O_2 have previously been associated with pathophysiological events *(250)*, high concentrations of H_2O_2 are required to initiate relaxation *(251)*, and others have been unable to confirm the association between H_2O_2 and an EDHF-mediated event *(189,252–254)*; Ellis and Triggle *(255)* have recently reviewed these controversies.

9.1.4. Potassium as an EDHF

The concept that potassium ions, K^+, could serve as an EDHF is attractively simple and supported by considerable direct and indirect data. The concept is that an endothelium-dependent vasodilator, such as ACh or bradykinin, activates endothelial cells and results in an increase in endothelial cell calcium that leads to the activation of apamin-sensitive SK_{Ca} and ChTx-sensitive IK_{Ca} channels, resulting in the efflux of endothelial cell-derived K^+ into the myoendothelial space. Edwards et al. *(231)* initially advanced this hypothesis an elegant 1998 article that demonstrated in both rat hepatic and small mesenteric arteries membrane potential measurements from both endothelial and smooth muscle cells as well as myograph data. These data included the observation that the combination of apamin and ChTx was required to inhibit ACh-mediated hyperpolarization of the endothelial cells and vascular smooth muscle but not the effects of exogenously applied K^+. A small increase in K^+ was also recorded using a K-sensitive electrode in the myoendothelial space of the hepatic vessel following ACh application. The effects of K^+ were inhibited by the combination of barium (an inward-rectifying K-channel [K_{ir}] inhibitor) and ouabain (a Na^+,K^+-ATPase inhibitor).

There is a considerable body of literature indicating that, primarily in small vessels, where K_{ir} predominate, small increases in K^+ (low concentration of 2–10 m*M*) relax blood vessels and hyperpolarize the smooth muscle *(256,257)*. Small increases in extracellular K^+, as would occur in exercising skeletal muscle, would also explain or contribute to the blood flow changes that are seen in reactive hyperemia. Of interest is whether the beneficial effects of dietary potassium supplementation to hypertensives may involve an endothelium-dependent action. Endothelium-dependent relaxation is impaired in cardiovascular disease, including essential hypertension, and intrabrachial potassium infusion has been reported to improve endothelium-dependent increases in forearm blood flow in patients with essential hypertension but not in normotensive patients *(258)*. It is unlikely that K^+ is the EDHF in all vessels, and considerable debate has also centered on this hypothesis, but nonetheless, K^+ remains an attractive candidate for EDHF.

9.2. What Is EDHF?

The question "What is EDHF?" cannot be answered with certainty at this time. Most likely no lone mediator or cellular event will explain EDHF-mediated vasodilatation, and, therefore, this makes further studies of EDHF very important. If multiple EDHFs and cellular processes do exist, then EDHF is an attractive target for vessel-specific modulation by mimics or inhibitors *(229)*. Furthermore, because endothelial dysfunction, as defined as a reduced vasodilator response to ACh, is seen at an early stage in cardiovascular disease, including diabetes, the study of the role of EDHF in such disease states becomes very important *(259)*. In addition, in preeclampsia the EDHF system appears to be defective or absent, and because endothelial dysfunction is thought to lead to the cardiovascular complications of preeclampsia, this is another important area to study with respect to restoring EDHF function *(130)*. Finally, EDHF has been shown to be an important mediator of endothelium-dependent relaxation in human penile resistance vessels and may prove to be a new target for the treatment of erectile dysfunction *(260)*.

REFERENCES

1. Furchgott RF, Zawadzki JV. The obligatory role of endothelial cells in the relaxation of arterial smooth muscle by acetylcholine. Nature 1980;288(5789):373–376.
2. Palmer RM, Ferrige AG, Moncada S. Nitric oxide release accounts for the biological activity of endothelium-derived relaxing factor. Nature 1987;327(6122):524–526.
3. Martinez A. Nitric oxide synthase in invertebrates. Histochem J 1995;27(10):770–776.
4. Feelisch M. The biochemical pathways of nitric oxide formation from nitrovasodilators: appropriate choice of exogenous NO donors and aspects of preparation and handling of aqueous NO solutions. J Cardiovasc Pharmacol 1991;17:S25–S33.
5. Schwartz SE, White WH. Kinetics of reactive dissolutions of nitrogen oxides into aqueous solutions. In: Trace Atmospheric Constituents: Properties, Transformation and Fates. (Schwartz SE, ed.) John Wiley: New York, 1983, pp. 1–17.
6. Ford PC, Wink DA, Stanbury DM. Autoxidation kinetics of aqueous nitric oxide. FEBS Lett 1993; 326(1–3):1–3.
7. Kelm M., Yoshida K. Metabolic fate of nitric oxide and related N-oxides. In: Feelisch MSJ, ed. Methods in Nitric Oxide Research. John Wiley: West Sussex, 1996, pp. 47–58.
8. Kelm M. Nitric oxide metabolism and breakdown. Biochim Biophys Acta 1999;1411(2–3):273–289.
9. Wink DA, Mitchell JB. Chemical biology of nitric oxide: insights into regulatory, cytotoxic, and cytoprotective mechanisms of nitric oxide. Free Radic Biol Med 1998:25(4–5):434–456.
10. Grube R., Kelm M, Motz W, et al. Measurement of nitric oxide in biological fluids: implication for in vivo studies. In: The Biology of Nitric Oxide. (Moncada S, ed.) Portland Press: London, 1994, pp. 201–204.
11. Kelm M, Schrader J. Control of coronary vascular tone by nitric oxide. Circ Res 1990;66(6):1561–1575.
12. Borland C, Demoncheaux E, Constable E. A chemiluminescent method for studying the reactions of nitric oxide with biologically important molecules in aqueous liquids. In: The Biology of Nitric Oxide. (Feelisch M, Busse R, Higgs EA, eds.) Portland Press: London, 1994, pp. 212–215.
13. Kharitonov VG, Sundquist AR, Sharma VS. Kinetics of nitric oxide autoxidation in aqueous solution. J Biol Chem 1994;269(8):5881–5883.
14. Kelm M, Feelisch M, Spahr R, et al. Quantitative and kinetic characterization of nitric oxide and EDRF released from cultured endothelial cells. Biochem Biophys Res Commun 1988;154(1):236–244.
15. Liu Z, Rudd MA, Freedman JE, et al. *S*-Transnitrosation reactions are involved in the metabolic fate and biological actions of nitric oxide. J Pharmacol Exp Ther 1998;284(2):526–534.
16. Huie RE, Padmaja S. The reaction of NO with superoxide. Free Radic Res Commun 1993;18(4):195–199.
17. Wink DA, Miranda KM, Espey MG, et al. Balancing nitric oxide with oxidative and nitrosative stress. In: Mayer B, ed. Springer-Verlag: New York, 2000, pp. 7–24.
18. Griendling KK, Sorescu D, Ushio-Fukai M. NAD(P)H oxidase: role in cardiovascular biology and disease. Circ Res 2000;86(5):494–501.
19. Babior BM. NADPH oxidase: an update. Blood 1999;93(5):1464–1476.
20. Suzuki H, DeLano FA, Parks DA, et al. Xanthine oxidase activity associated with arterial blood pressure in spontaneously hypertensive rats. Proc Natl Acad Sci USA 1998;95(8):4754–4759.
21. Fleming I, Michaelis UR, Bredenkotter D, et al. Endothelium-derived hyperpolarizing factor synthase (Cytochrome P450 2C9) is a functionally significant source of reactive oxygen species in coronary arteries. Circ Res 2001;88(1):44–51.
22. Cosentino F, Sill JC, Katusic ZS. Role of superoxide anions in the mediation of endothelium-dependent contractions. Hypertension 1994;23(2):229–235.
23. Xia Y, Tsai AL, Berka V, et al. Superoxide generation from endothelial nitric-oxide synthase: a Ca^{2+}/calmodulin-dependent and tetrahydrobiopterin regulatory process. J Biol Chem 1998;273(40):25804–25808.
24. Koppenol WH, Moreno JJ, Pryor WA, et al. Peroxynitrite, a cloaked oxidant formed by nitric oxide and superoxide. Chem Res Toxicol 1992;5(6):834–842.
25. Stadler J, Billiar TR, Curran RD, et al. Effect of exogenous and endogenous nitric oxide on mitochondrial respiration of rat hepatocytes. Am J Physiol 1991;260(5 pt 1):C910–C916.
26. Nguyen T, Brunson D, Crespi CL, et al. DNA damage and mutation in human cells exposed to nitric oxide in vitro. Proc Natl Acad Sci USA 1992;89(7):3030–3034.
27. Beckman JS, Beckman TW, Chen J, et al. Apparent hydroxyl radical production by peroxynitrite: implications for endothelial injury from nitric oxide and superoxide. Proc Natl Acad Sci USA 1990;87(4):1620–1624.

28. Liu S, Beckman JS, Ku DD. Peroxynitrite, a product of superoxide and nitric oxide, produces coronary vasorelaxation in dogs. J Pharmacol Exp Ther 1994;268(3):1114–1121.
29. Wu M, Pritchard KA Jr. Kaminski PM, et al. Involvement of nitric oxide and nitrosothiols in relaxation of pulmonary arteries to peroxynitrite. Am J Physiol 1994;266(5 pt 2):H2108–H2113.
30. Mayer B, Schrammel A, Klatt P, et al. Peroxynitrite-induced accumulation of cyclic GMP in endothelial cells and stimulation of purified soluble guanylyl cyclase: dependence on glutathione and possible role of S-nitrosation. J Biol Chem 1995;270(29):17355–17360.
31. Mayer B, Pfeiffer S, Schrammel A, et al. A new pathway of nitric oxide/cyclic GMP signaling involving S-nitrosoglutathione. J Biol Chem 1998;273(6):3264–3270.
32. Tarpey MM, Beckman JS, Ischiropoulos H, et al. Peroxynitrite stimulates vascular smooth muscle cell cyclic GMP synthesis. FEBS Lett 1995;364(3):314–318.
33. Moro MA, Darley-Usmar VM, Goodwin DA, et al. Paradoxical fate and biological action of peroxynitrite on human platelets. Proc Natl Acad Sci USA 1994;91(14):6702–6706.
34. Lefer DJ, Scalia R, Campbell B, et al. Peroxynitrite inhibits leukocyte-endothelial cell interactions and protects against ischemia-reperfusion injury in rats. J Clin Invest 1997;99(4):684–691.
35. Nakamura M, Thourani VH, Ronson RS, et al. Glutathione reverses endothelial damage from peroxynitrite, the byproduct of nitric oxide degradation, in crystalloid cardioplegia. Circulation 2000;102(19 Suppl 3):III332–III338.
36. Cheung PY, Wang W, Schulz R. Glutathione protects against myocardial ischemia-reperfusion injury by detoxifying peroxynitrite. J Mol Cell Cardiol 2000;32(9):1669–1678.
37. Ma XL, Gao F, Lopez BL, et al. Peroxynitrite, a two-edged sword in post-ischemic myocardial injury-dichotomy of action in crystal. J Pharmacol Exp Ther 2000;292(3):912–920.
38. Moncada S, Palmer RM, Higgs EA. Nitric oxide: physiology, pathophysiology, and pharmacology. Pharmacol Rev 1991;43(2):109–142.
39. Ignarro LJ. Biosynthesis and metabolism of endothelium-derived nitric oxide. Annu Rev Pharmacol Toxicol 1990;30:535–560.
40. Beckman JS. The physiological and pathological chemistry of nitric oxide. In: Lancaster J Jr., ed. Nitric Oxide: Principles and Actions. Academic: San Diego, 1996, pp. 1–71.
41. Wennmalm A, Benthin G, Petersson AS. Dependence of the metabolism of nitric oxide (NO) in healthy human whole blood on the oxygenation of its red cell haemoglobin. Br J Pharmacol 1992;106(3):507, 508.
42. Stamler JS, Jia L, Eu JP, et al. Blood flow regulation by S-nitrosohemoglobin in the physiological oxygen gradient. Science 1997;276(5321):2034–2037.
43. Jia L, Bonaventura C, Bonaventura J, et al. S-nitrosohaemoglobin: a dynamic activity of blood involved in vascular control. Nature 1996;380(6571):221–226.
44. Gow AJ, Stamler JS. Reactions between nitric oxide and haemoglobin under physiological conditions. Nature 1998;391(6663):169–173.
45. Lancaster JR Jr. Simulation of the diffusion and reaction of endogenously produced nitric oxide. Proc Natl Acad Sci USA 1994;91(17):8137–8141.
46. Hobbs AJ, Gladwin MT, Patel RP, et al. Haemoglobin: NO transporter, NO inactivator or NOne of the above? Trends Pharmacol Sci 2002;23(9):406–411.
47. Ignarro LJ, Buga GM, Wood KS, et al. Endothelium-derived relaxing factor produced and released from artery and vein is nitric oxide. Proc Natl Acad Sci USA 1987;84(24):9265–9269.
48. Gow AJ, Luchsinger BP, Pawloski JR, et al. The oxyhemoglobin reaction of nitric oxide. Proc Natl Acad Sci USA 1999;96(16):9027–9032.
49. Pawloski JR, Hess DT, Stamler JS. Export by red blood cells of nitric oxide bioactivity. Nature 2001;409(6820):622–626.
50. Patel RP, Hogg N, Spencer NY, et al. Biochemical characterization of human S-nitrosohemoglobin: effects on oxygen binding and transnitrosation. J Biol Chem 1999;274(22):15487–15492.
51. Wolzt M, MacAllister RJ, Davis D, et al. Biochemical characterization of S-nitrosohemoglobin: mechanisms underlying synthesis, NO release, and biological activity. J Biol Chem 1999;274(41): 28983–28990.
52. Clancy RM, Miyazaki Y, Cannon PJ. Use of thionitrobenzoic acid to characterize the stability of nitric oxide in aqueous solutions and in porcine aortic endothelial cell suspensions. Anal Biochem 1990;191(1):138–143.
53. Hogg N, Singh RJ, Kalyanaraman B. The role of glutathione in the transport and catabolism of nitric oxide. FEBS Lett 1996;382(3):223–228.
54. Pryor WA, Church DF, Govindan CK, et al. Oxidation of thiols by nitric oxide and nitrogen dioxide synthetic utility and toxicological implications. J Org Chem 1982;47:159–161.

55. Tamir S, Tannenbaum SR. The role of nitric oxide (NO) in the carcinogenic process. Biochim Biophys Acta 1996;1288(2):F31–F36.
56. Bonner FT, Stedman G. The chemistry of nitric oxide and redox-related species. In: Feelisch MSJ, ed. Methods in Nitric Oxide Research. John Wiley: West Sussex, 1996, pp 3–18.
57. Kharitonov VG, Sundquist AR, Sharma VS. Kinetics of nitrosation of thiols by nitric oxide in the presence of oxygen. J Biol Chem 1995;270(47):28158–28164.
58. Stamler JS, Singel DJ, Loscalzo J. Biochemistry of nitric oxide and its redox-activated forms. Science 1992;258(5090):1898–1902.
59. Stamler JS, Simon DI, Osborne JA, et al. S-nitrosylation of proteins with nitric oxide: synthesis and characterization of biologically active compounds. Proc Natl Acad Sci USA 1992;89(1):444–448.
60. Butler AR, Flitney FW, Williams DL. NO, nitrosonium ions, nitroxide ions, nitrosothiols and iron-nitrosyls in biology: a chemist's perspective. Trends Pharmacol Sci 1995;16(1):18–22.
61. Myers PR, Minor RL Jr, Guerra R Jr, et al. Vasorelaxant properties of the endothelium-derived relaxing factor more closely resemble S-nitrosocysteine than nitric oxide. Nature 1990;345(6271):161–163.
62. Rubanyi GM, Vanhoutte PM. Oxygen-derived free radicals, endothelium, and responsiveness of vascular smooth muscle. Am J Physiol 1986;250(5 pt 2):H815–H821.
63. Keaney JF Jr, Simon DI, Stamler JS, et al. NO forms an adduct with serum albumin that has endothelium-derived relaxing factor-like properties. J Clin Invest 1993;91(4):1582–1589.
64. Stamler JS, Jaraki O, Osborne J, et al. Nitric oxide circulates in mammalian plasma primarily as an S-nitroso adduct of serum albumin. Proc Natl Acad Sci USA 1992;89(16):7674–7677.
65. Radomski MW, Rees DD, Dutra A, et al. S-nitroso-glutathione inhibits platelet activation in vitro and in vivo. Br J Pharmacol 1992;107(3):745–749.
66. Marley R, Feelisch M, Holt S, et al. A chemiluminescense-based assay for S-nitrosoalbumin and other plasma S-nitrosothiols. Free Radic Res 2000;32(1):1–9.
67. Gaston B. Nitric oxide and thiol groups. Biochim Biophys Acta 1999;1411(2–3):323–333.
68. Beckman JS, Koppenol WH. Nitric oxide, superoxide, and peroxynitrite: the good, the bad, and ugly. Am J Physiol 1996;271(5 pt 1):C1424–C1437.
69. Nedospasov A, Rafikov R, Beda N, et al. An autocatalytic mechanism of protein nitrosylation. Proc Natl Acad Sci USA 2000;97(25):13,543–13,548.
70. Hogg N. Biological chemistry and clinical potential of S-nitrosothiols. Free Radic Biol Med 2000;28:1478–1486.
71. Hughes MN. Relationships between nitric oxide, nitroxyl ion, nitrosonium cation and peroxynitrite. Biochim Biophys Acta 1999;1411(2–3):263–272.
72. Pino RZ, Feelisch M. Bioassay discrimination between nitric oxide (·NO) and nitroxyl (NO^-) using L-cysteine. Biochem Biophys Res Commun 1994;201(1):54–62.
73. Zamora R, Grzesiok A, Weber H, et al. Oxidative release of nitric oxide accounts for guanylyl cyclase stimulating, vasodilator and anti-platelet activity of Piloty's acid: a comparison with Angeli's salt. Biochem J 1995;312 (pt 2):333–339.
74. Doyle MP, Mahapatro SN, Broene RD, et al. Oxidation and reduction of hemoproteins by trioxodinitrate(II): the role of nitrosyl hydride and nitrite. J Am Chem Soc 1988;110:593–599.
75. Feelisch M, TePoel PM, Zamora R, et al. Understanding the controversy over the identity of EDRF. Nature 1994;368(6466):62–65.
76. Khan AA, Schuler MM, Coppock RW. Inhibitory effects of various sulfur compounds on the activity of bovine erythrocyte enzymes. J Toxicol Environ Health 1987;22(4):481–490.
77. Andrews KL, Triggle CR, Ellis A. NO and the vasculature: where does it come from and what does it do? Heart Fail Rev 2002;7(4):423–445.
78. Koppenol WH. Thermodynamics of reactions involving nitrogen-oxygen compounds. Methods Enzymol 1996;268:7–12.
79. Li CG, Karagiannis J, Rand MJ. Comparison of the redox forms of nitrogen monoxide with the nitrergic transmitter in the rat anococcygeus muscle. Br J Pharmacol 1999;127(4):826–834.
80. Dierks EA, Burstyn JN. Nitric oxide (NO), the only nitrogen monoxide redox form capable of activating soluble guanylyl cyclase. Biochem Pharmacol 1996;51(12):1593–1600.
81. Schmidt HH, Hofmann H, Schindler U, et al. No NO from NO synthase. Proc Natl Acad Sci USA 1996;93(25):14,492–14,497.
82. Adak S, Wang Q, Stuehr DJ. Arginine conversion to nitroxide by tetrahydrobiopterin-free neuronal nitric-oxide synthase: implications for mechanism. J Biol Chem 2000;275(43):33,554–33,561.
83. Hogg N. The biochemistry and physiology of S-nitrosothiols. Annu Rev Pharmacol Toxicol 2002;42:585–600.

84. Arnelle DR, Stamler JS. NO+, NO, and NO– donation by S-nitrosothiols: implications for regulation of physiological functions by S-nitrosylation and acceleration of disulfide formation. Arch Biochem Biophys 1995;318(2):279–285.
85. Fukuto JM, Chiang K, Hszieh R, et al. The pharmacological activity of nitroxyl: a potent vasodilator with activity similar to nitric oxide and/or endothelium-derived relaxing factor. J Pharmacol Exp Ther 1992;263(2):546–551.
86. Pufahl RA, Wishnok JS, Marletta MA. Hydrogen peroxide-supported oxidation of N^G-hydroxy-L-arginine by nitric oxide synthase. Biochemistry 1995;34(6):1930–1941.
87. Wong PS, Hyun J, Fukuto JM, et al. Reaction between S-nitrosothiols and thiols: generation of nitroxyl (HNO) and subsequent chemistry. Biochemistry 1998;37(16):5362–5371.
88. Bartberger MD, Liu W, Ford E, et al. The reduction potential of nitric oxide (NO) and its importance to NO biochemistry. Proc Natl Acad Sci USA 2002;99(17):10958–10963.
89. Fukuto JM, Wallace GC, Hszieh R, et al. Chemical oxidation of N-hydroxyguanidine compounds: release of nitric oxide, nitroxyl and possible relationship to the mechanism of biological nitric oxide generation. Biochem Pharmacol 1992;43(3):607–613.
90. Murphy ME, Sies H. Reversible conversion of nitroxyl anion to nitric oxide by superoxide dismutase. Proc Natl Acad Sci USA 1991;88(23):10,860–10,864.
91. Kim WK, Choi YB, Rayudu PV, et al. Attenuation of NMDA receptor activity and neurotoxicity by nitroxyl anion, NO^-. Neuron 1999;24(2):461–469.
92. Ma XL, Gao F, Liu GL, et al. Opposite effects of nitric oxide and nitroxyl on postischemic myocardial injury. Proc Natl Acad Sci USA 1999;96(25):14,617–14,622.
93. Paolocci N, Saavedra WF, et al. Nitroxyl anion exerts redox-sensitive positive cardiac inotropy in vivo by calcitonin gene-related peptide signaling. Proc Natl Acad Sci USA 2001;98(18):10,463–10,468.
94. Paolocci N, Katori T, Champion HC, et al. Positive inotropic and lusitropic effects of HNO/NO^- in failing hearts: independence from beta-adrenergic signaling. Proc Natl Acad Sci USA 2003;100(9):5537–5542.
95. Miranda KM, Nims RW, Thomas DD, et al. Comparison of the reactivity of nitric oxide and nitroxyl with heme proteins: a chemical discussion of the differential biological effects of these redox related products of NOS. J Inorg Biochem 2003;93(1–2):52–60.
96. Hobbs AJ. Soluble guanylate cyclase: the forgotten sibling. Trends Pharmacol Sci 1997;18:484–491.
97. Ellis A, Li CG, Rand MJ. Differential actions of L-cysteine on responses to nitric oxide, nitroxyl anions and EDRF in the rat aorta. Br J Pharmacol 2000;129:315–322.
98. Wanstall JC, Jeffery TK, Gambino A, et al. Vascular smooth muscle relaxations mediated by nitric oxide donors: a comparison with acetylcholine, nitric oxide and nitroxyl ion. Br J Pharmacol 2001;134:463–472.
99. Reif A, Zecca L, Riederer P, et al. Nitroxyl oxidizes NADPH in a superoxide dismutase inhibitable manner. Free Radic Biol Med 2001;30(7):803–808.
100. Thomas G, Ramwell PW. Vascular relaxation mediated by hydroxylamines and oximes: their conversion to nitrites and mechanism of endothelium dependent vascular relaxation. Biochem Biophys Res Commun 1989;164(2):889–893.
101. DeMaster EG, Raij L, Archer SL, et al. Hydroxylamine is a vasorelaxant and a possible intermediate in the oxidative conversion of L-arginine to nitric oxide. Biochem Biophys Res Commun 1989;163(1): 527–533.
102. Huang Y. Hydroxylamine-induced relaxation inhibited by K^+ channel blockers in rat aortic rings. Eur J Pharmacol 1998;349(1):53–60.
103. Craven PA, DeRubertis FR, Pratt DW. Electron spin resonance study of the role of NO catalase in the activation of guanylate cyclase by NaN3 and NH_2OH: modulation of enzyme responses by heme proteins and their nitrosyl derivatives. J Biol Chem 1979;254(17):8213–8222.
104. Correia NA, Oliveira RB, Ballejo G. Pharmacological profile of nitrergic nerve-, nitric oxide-, nitrosoglutath. Life Sci 2000;68(6):709–717.
105. Bates JN, Harrison DG, Myers PR, et al. EDRF: nitrosylated compound or authentic nitric oxide. Basic Res Cardiol 1991;86(Suppl 2):17–26.
106. Thornbury KD, Ward SM, Dalziel HH, et al. Nitric oxide and nitrosocysteine mimic nonadrenergic, noncholinergic hyperpolarization in canine proximal colon. Am J Physiol 1991;261(3 pt 1): G553–G557.
107. Liu X, Gillespie JS, Martin W. Non-adrenergic, non-cholinergic relaxation of the bovine retractor penis muscle: role of S-nitrosothiols. Br J Pharmacol 1994;111(4):1287–1295.
108. Muller B, Kleschyov AL, Alencar JL, et al. Nitric oxide transport and storage in the cardiovascular system. Ann NY Acad Sci 2002;962:131–139.

109. Rodriguez J, Maloney RE, Rassaf T, et al. Chemical nature of nitric oxide storage forms in rat vascular tissue. Proc Natl Acad Sci USA 2003;100(1):336–341.
110. Davisson RL, Travis MD, Bates JN, et al. Stereoselective actions of *S*-nitrosocysteine in central nervous system of conscious rats. Am J Physiol 1997;272(5 pt 2):H2361–H2368.
111. Al Sa'doni HH, Megson IL, Bisland S, et al. Neocuproine, a selective Cu(I) chelator, and the relaxation of rat vascular smooth muscle by S-nitrosothiols. Br J Pharmacol 1997;121(6):1047–1050.
112. McMahon TJ, Moon RE, Luschinger BP, et al. Nitric oxide in the human respiratory cycle. Nat Med 2002;8(7):711–717.
113. Arnelle DR, Day BJ, Stamler JS. Diethyl dithiocarbamate-induced decomposition of S-nitrosothiols. Nitric Oxide 1997;1(1):56–64.
114. Scorza G, Pietraforte D, Minetti M. Role of ascorbate and protein thiols in the release of nitric oxide from S-nitroso-albumin and S-nitroso-glutathione in human plasma. Free Radic Biol Med 1997;22(4):633–642.
115. Aleryani S, Milo E, Rose Y, et al. Superoxide-mediated decomposition of biological S-nitrosothiols. J Biol Chem 1998;273(11):6041–6045.
116. Choi YB, Tenneti L, Le DA, et al. Molecular basis of NMDA receptor-coupled ion channel modulation by S-nitrosylation. Nat Neurosci 2000;3(1):15–21.
117. Favre CJ, Ufret-Vincenty CA, Stone MR, et al. Ca^{2+} pool emptying stimulates Ca^{2+} entry activated by S-nitrosylation. J Biol Chem 1998;273(47):30,855–30,858.
118. Renganathan M, Cummins TR, Waxman SG. Nitric oxide blocks fast, slow, and persistent Na+ channels in C-type DRG neurons by S-nitrosylation. J Neurophysiol 2002;87(2):761–775.
119. Sun J, Xu L, Eu JP, et al. Nitric oxide, NOC-12, and *S*-nitrosoglutathione modulate the skeletal muscle calcium release channel/ryanodine receptor by different mechanisms: an allosteric function for O2 in S-nitrosylation of the channel. J Biol Chem 2003;278(10):8184–8189.
120. Bauer PM, Buga GM, Fukuto JM, et al. Nitric oxide inhibits ornithine decarboxylase via S-nitrosylation of cysteine 360 in the active site of the enzyme. J Biol Chem 2001;276(37):34,458–34,464.
121. Padgett CM, Whorton AR. S-nitrosoglutathione reversibly inhibits GAPDH by S-nitrosylation. Am J Physiol 1995;269(3 pt 1):C739–C749.
122. Perez-Mato I, Castro C, Ruiz FA, et al. Methionine adenosyltransferase S-nitrosylation is regulated by the basic and acidic amino acids surrounding the target thiol. J Biol Chem 1999;274(24): 17,075–17,079.
123. Stamler JS, Toone EJ, Lipton SA, et al. (S)NO signals: translocation, regulation, and a consensus motif. Neuron 1997;18(5):691–696.
124. Ascenzi P, Colasanti M, Persichini T, et al. Re-evaluation of amino acid sequence and structural consensus rules for cysteine–nitric oxide reactivity. Biol Chem 2000;381(7):623–627.
125. Vanin AF. Endothelium-derived relaxing factor is a nitrosyl iron complex with thiol ligands. FEBS Lett 1991;289(1):1–3.
126. Vedernikov YP, Mordvintcev PI, Malenkova IV, et al. Similarity between the vasorelaxing activity of dinitrosyl iron cysteine complexes and endothelium-derived relaxing factor. Eur J Pharmacol 1992;211(3):313–317.
127. Vanin AF, Mordvintcev PI, Hauschildt S, et al. The relationship between L-arginine-dependent nitric oxide synthesis, nitrite release and dinitrosyl-iron complex formation by activated macrophages. Biochim Biophys Acta 1993;1177(1):37–42.
128. Ueno T, Suzuki Y, Fujii S, et al. In vivo nitric oxide transfer of a physiological NO carrier, dinitrosyl dithiolato iron complex, to target complex. Biochem Pharmacol 2002;63(3):485–493.
129. Muller B, Kleschyov AL, Stoclet JC. Evidence for N-acetylcysteine-sensitive nitric oxide storage as dinitrosyl-iron complexes in lipopolysaccharide-treated rat aorta. Br J Pharmacol 1996;119(6): 1281–1285.
130. Gillham JC, Kenny LC, Baker PN. An overview of endothelium-derived hyperpolarising factor (EDHF) in normal and compromised pregnancies. Eur J Obstet Gynecol Reprod Biol 2003;109(1):2–7.
131. Malinski T, Taha Z. Nitric oxide release from a single cell measured in situ by a porphyrinic-based microsensor. Nature 1992;358(6388):676–678.
132. Ledo A, Barbosa RM, Frade J, et al. Nitric oxide monitoring in hippocampal brain slices using electrochemical methods. Methods Enzymol 2002;359:111–125.
133. Allen BW, Coury LA Jr, Piantadosi CA. Electrochemical detection of physiological nitric oxide: materials and methods. Methods Enzymol 2002;359:125–134.
134. Freedman JE, Loscalzo J, Barnard MR, et al. Nitric oxide released from activated platelets inhibits platelet recruitment. J Clin Invest 1997;100(2):350–356.

135. Malinski T, Mesaros S, Patton SR, et al. Direct measurement of nitric oxide in the cardiovascular system. Physiol Res 1996;45(4):279–284.
136. Ellis A, Lu H, Li CG, et al. Effects of agents that inactivate free radical NO (NO*) on nitroxyl anion-mediated relaxations, and on the detection of NO* released from the nitroxyl anion donor Angeli's salt. Br J Pharmacol 2001;134(3):521–528.
137. Guo JP, Murohara T, Buerke M, et al. Direct measurement of nitric oxide release from vascular endothelial cells. J Appl Physiol 1996;81(2):774–779.
138. Liu ZG, Ge ZD, He GW. Difference in endothelium-derived hyperpolarizing factor-mediated hyperpolarization and nitric oxide release between human internal mammary artery and saphenous vein. Circulation 2000;102(19 suppl 3):III296–III301.
139. Kojima H, Nakatsubo N, Kikuchi K, et al. Detection and imaging of nitric oxide with novel fluorescent indicators: diaminofluoresceins. Anal Chem 1998;70(13):2446–2453.
140. Nakatsubo N, Kojima H, Kikuchi K, et al. Direct evidence of nitric oxide production from bovine aortic endothelial cells using new fluorescence indicators: diaminofluoresceins. FEBS Lett 1998;427(2): 263–266.
141. Kojima H, Nakatsubo N, Kikuchi K, et al. Direct evidence of NO production in rat hippocampus and cortex using a new fluorescent indicator: DAF-2 DA. Neuroreport 1998;9(15):3345–3348.
142. Broillet M, Randin O, Chatton J. Photoactivation and calcium sensitivity of the fluorescent NO indicator 4,5–diaminofluorescein (DAF-2): implications for cellular NO imaging. FEBS Lett 2001;491(3): 227–232.
143. Espey MG, Miranda KM, Thomas DD, et al. Distinction between nitrosating mechanisms within human cells and aqueous solution. J Biol Chem 2001;276(32):30,085–30,091.
144. Jourd'heuil D. Increased nitric oxide-dependent nitrosylation of 4,5-diaminofluorescein by oxidants: implications for the measurement of intracellular nitric oxide. Free Radic Biol Med 2002;33(5):676–684.
145. Nagata N, Momose K, Ishida Y. Inhibitory effects of catecholamines and anti-oxidants on the fluorescence reaction of 4,5-diaminofluorescein, DAF-2, a novel indicator of nitric oxide. J Biochem (Tokyo) 1999;125(4):658–661.
146. Leikert JF, Rathel TR, Muller C, et al. Reliable in vitro measurement of nitric oxide released from endothelial cells using low concentrations of the fluorescent probe 4,5-diaminofluorescein. FEBS Lett 2001;506(2):131–134.
147. Marzinzig M, Nussler AK, Stadler J, et al. Improved methods to measure end products of nitric oxide in biological fluids: nitrite, nitrate, and S-nitrosothiols. Nitric Oxide 1997;1(2):177–189.
148. MacMicking J, Xie QW, Nathan C. Nitric oxide and macrophage function. Annu Rev Immunol 1997;15:323–350.
149. Scharfstein JS, Keaney JF Jr, Slivka A, et al. In vivo transfer of nitric oxide between a plasma protein-bound reservoir and low molecular weight thiols. J Clin Invest 1994;94(4):1432–1439.
150. Minamiyama Y, Takemura S, Inoue M. Effect of thiol status on nitric oxide metabolism in the circulation. Arch Biochem Biophys 1997;341(1):186–192.
151. Goldman RK, Vlessis AA, Trunkey DD. Nitrosothiol quantification in human plasma. Anal Biochem 1998;259(1):98–103.
152. Jourd'heuil D, Hallen K, Feelisch M, et al. Dynamic state of S-nitrosothiols in human plasma and whole blood. Free Radic Biol Med 2000;28(3):409–417.
153. Rossi R, Giustarini D, Milzani A, et al. Physiological levels of S-nitrosothiols in human plasma. Circ Res 2001;89(12):E47.
154. Tsikas D, Gutzki FM, Rossa S, et al. Measurement of nitrite and nitrate in biological fluids by gas chromatography-mass spectrometry and by the Griess assay: problems with the Griess assay—solutions by gas chromatography-mass spectrometry. Anal Biochem 1997;244(2):208–220.
155. Butler AR, Rhodes P. Chemistry, analysis, and biological roles of S-nitrosothiols. Anal Biochem 1997;249(1):1–9.
156. Welch GN, Upchurch GR Jr, Loscalzo J. S-Nitrosothiol detection. Methods Enzymol 1996;268:293–298.
157. Tsikas D, Sandmann J, Holzberg D, et al. Determination of S-nitrosoglutathione in human and rat plasma by high-performance liquid chromatography with fluorescence and ultraviolet absorbance detection after precolumn derivatization with o-phthalaldehyde. Anal Biochem 1999;273(1):32–40.
158. Tsikas D, Sandmann J, Gutzki FM, et al. Measurement of S-nitrosoalbumin by gas chromatography-mass spectrometry. II. Quantitative determination of S-nitrosoalbumin in human plasma using S-[^{15}N]nitrosoalbumin as internal standard. J Chromatogr 1999;726:13–24.
159. Mirza UA, Chait BT, Lander HM. Monitoring reactions of nitric oxide with peptides and proteins by electrospray ionization-mass spectrometry. J Biol Chem 1995;270(29):17,185–17,188.

160. Tsikas D, Raida M, Sandmann J, et al. Electrospray ionization mass spectrometry of low-molecular-mass S-nitroso compounds and their thiols. J Chromatogr B Biomed Sci Appl 2000;742(1):99–108.
161. Gaston B, Reilly J, Drazen JM, et al. Endogenous nitrogen oxides and bronchodilator S-nitrosothiols in human airways. Proc Natl Acad Sci USA 1993;90(23):10 957–10,961.
162. Stamler JS, Feelisch M. Preparation and detection of S-nitrosothiols. In: Methods in Nitric Oxide Research. (Feelisch M, Stamler JS, ed.) John Wiley: West Sussex, 1996, pp. 522–539.
163. Saville B. A scheme for the colorimetric determination of microgram amounts of thiols. Analyst 1958;(83):670–672.
164. Samouilov A, Zweier JL. Development of chemiluminescence-based methods for specific quantitation of nitrosylated thiols. Anal Biochem 1998;258(2):322–330.
165. Krstulovic AM, Brown PR. Basic theory and terminology. In: Reversed-phase High-Performance Liquid Chromatography: Theory, Practice and Biomedical applications. (Krstulovic AM, Brown PR, ed.) John Wiley: New York, 1982, pp. 4–32.
166. Pfeiffer S, Schrammel A, Schmidt K, et al. Electrochemical determination of S-nitrosothiols with a Clark-type nitric oxide electrode. Anal Biochem 1998;258(1):68–73.
167. Tsikas D, Denker K, Frolich JC. Artifactual-free analysis of S-nitrosoglutathione and S-nitroglutathione by neutral-pH, anion-pairing, high-performance liquid chromatrography: study on peroxynitrite-mediated S-nitration of glutathione to S-nitroglutathione under physiological conditions. J Chromatogr A 2001;915:107–116.
168. Ng ESM, Jourd'heuil D, McCord JM, et al. S-nitroso-albumin formed from inhaled NO is a physiologic nitric oxide delivery system during ischemia/reperfusion. Circ Res 2004;94(4):559–565.
169. Rassaf T, Preik M, Kleinbongard P, et al. Evidence for in vivo transport of bioactive nitric oxide in human plasma. J Clin Invest 2002;109(9):1241–1248.
170. Feelisch M, Rassaf T, Mnaimneh S, et al. Concomitant S-, N-, and heme-nitros(yl)ation in biological tissues and fluids: implications for the fate of NO in vivo. FASEB J 2002;16(13):1775–1785.
171. Park JK, Kostka P. Fluorometric detection of biological S-nitrosothiols. Anal Biochem 1997;249(1): 61–66.
172. Wang K, Zhang W, Xian M, et al. New chemical and biological aspects of S-nitrosothiols. Curr Med Chem 2000;7(8):821–834.
173. Misko TP, Schilling RJ, Salvemini D, et al. A fluorometric assay for the measurement of nitrite in biological samples. Anal Biochem 1993;214(1):11–16.
174. Tsikas D, Sandmann J, Rossa S, et al. Gas chromatographic-mass spectrometric detection of S-nitrosocysteine and S-nitrosoglutathione. Anal Biochem 1999;272:117–122.
175. Chait BT, Kent SB. Weighing naked proteins: practical, high-accuracy mass measurement of peptides and proteins. Science 1992;257(5078):1885–1894.
176. Michel T, Feron O. Nitric oxide synthases: which, where, how, and why? J Clin Invest 1997;100(9): 2146–2152.
177. Fulton D, Gratton JP, Sessa WC. Post-translational control of endothelial nitric oxide synthase: why isn't calcium/calmodulin enough? J Pharmacol Exp Ther 2001;299(3):818–824.
178. Mayer B, Werner ER. In search of a function for tetrahydrobiopterin in the biosynthesis of nitric oxide. Naunyn Schmiedebergs Arch Pharmacol 1995;351(5):453–463.
179. Werner-Felmayer G, Golderer G, Werner ER. Tetrahydrobiopterin biosynthesis, utilization and pharmacological effects. Curr Drug Metab 2002;3(2):159–173.
180. Fulton D, Gratton JP, McCabe TJ, et al. Regulation of endothelium-derived nitric oxide production by the protein kinase Akt. Nature 1999;399(6736):597–601.
181. Ishida Y, Hashimoto M, Fukushima S, et al. A nitric oxide-sensitive electrode: requirement of lower oxygen concentration for detecting nitric oxide from the tissue. J Pharmacol Toxicol Methods 1996;35(1):19–24.
182. Ayajiki K, Kindermann M, Hecker M, et al. Intracellular pH and tyrosine phosphorylation but not calcium determine shear stress-induced nitric oxide production in native endothelial cells. Circ Res 1996;78(5):750–758.
183. Griscavage JM, Fukuto JM, Komori Y, et al. Nitric oxide inhibits neuronal nitric oxide synthase by interacting with the heme prosthetic group: role of tetrahydrobiopterin in modulating the inhibitory action of nitric oxide. J Biol Chem 1994;(269):21,644–21,649.
184. Du XL, Edelstein D, Dimmeler S, et al. Hyperglycemia inhibits endothelial nitric oxide synthase activity by posttranslational modification at the Akt site. J Clin Invest 2001;108(9):1341–1348.
185. Flavahan NA, Vanhoutte PM. Endothelial cell signaling and endothelial dysfunction. Am J Hypertens 1995;8(5 pt 2):28S-41S.

186. Lin S, Fagan KA, Li KX, et al. Sustained endothelial nitric-oxide synthase activation requires capacitative Ca^{2+} entry. J Biol Chem 2000;275(24):17,979–17,985.
187. Abrahamsson T, Brandt U, Marklund SL, et al. Vascular bound recombinant extracellular superoxide dismutase type C protects against the detrimental effects of superoxide radicals on endothelium-dependent arterial relaxation. Circ Res 1992;70(2):264–271.
188. Woodman CR, Muller JM, Rush JW, et al. Flow regulation of ecNOS and Cu/Zn SOD mRNA expression in porcine coronary arterioles. Am J Physiol 1999;276(3 pt 2):H1058–H1063.
189. Beny JL, der Weid PY. Hydrogen peroxide: an endogenous smooth muscle cell hyperpolarizing factor. Biochem Biophys Res Commun 1991;176(1):378–384.
190. Iida Y, Katusic ZS. Mechanisms of cerebral arterial relaxations to hydrogen peroxide. Stroke 2000;31(9):2224–2230.
191. Matoba T, Shimokawa H, Nakashima M, et al. Hydrogen peroxide is an endothelium-derived hyperpolarizing factor in mice. J Clin Invest 2000;106(12):1521–1530.
192. Miura H, Bosnjak JJ, Ning G, et al. Role for hydrogen peroxide in flow-induced dilation of human coronary arterioles. Circ Res 2003;92(2):E31–E40.
193. Rand MJ, Li CG. Nitric oxide as a neurotransmitter in peripheral nerves: nature of transmitter and mechanism of transmission. Annu Rev Physiol 1995;57:659–682.
194. Luo D, Vincent SR. NMDA-dependent nitric oxide release in the hippocampus in vivo: interactions with noradrenaline. Neuropharmacology 1994;33(11):1345–1350.
195. Loesch A, Burnstock G. Ultrastructural localization of nitric oxide synthase and endothelin in coronary and pulmonary arteries of newborn rats. Cell Tissue Res 1995;279(3):475–483.
196. Boulanger CM, Heymes C, Benessiano J, et al. Neuronal nitric oxide synthase is expressed in rat vascular smooth muscle cells: activation by angiotensin II in hypertension. Circ Res 1998;83(12): 1271–1278.
197. De Luca A, Li CG, Rand MJ, et al. Effects of omega-conotoxin GVIA on autonomic neuroeffector transmission in various tissues. Br J Pharmacol 1990;101(2):437–447.
198. Zygmunt PM, Zygmunt PK, Hogestatt ED, et al. Effects of omega-conotoxin on adrenergic, cholinergic and NANC neurotransmission in the rabbit urethra and detrusor. Br J Pharmacol 1993;110(4): 1285–1290.
199. Olgart C, Gustafsson LE, Wiklund NP. Evidence for nonvesicular nitric oxide release evoked by nerve activation. Eur J Neurosci 2000;12(4):1303–1309.
200. Kasakov L, Cellek S, Moncada S. Characterization of nitrergic neurotransmission during short and long-term electrical stimulation of the rabbit anococcygeus muscle. Br J Pharmacol 1995;115(7): 1149–1154.
201. Costa M, Furness JB, Pompolo S, et al. Projections and chemical coding of neurons with immunoreactivity for nitric oxide synthase in the guinea-pig small intestine. Neurosci Lett 1992;148(1–2): 121–125.
202. Murthy KS, Grider JR, Jin JG, et al. Interplay of VIP and nitric oxide in the regulation of neuromuscular activity in the gut. Arch Int Pharmacodyn Ther 1995;329(1):27–38.
203. Bishop-Bailey D, Larkin SW, Warner TD, et al. Characterization of the induction of nitric oxide synthase and cyclo-oxygenase in rat aorta in organ culture. Br J Pharmacol 1997;121(1):125–133.
204. Chartrain NA, Geller DA, Koty PP, et al. Molecular cloning, structure, and chromosomal localization of the human inducible nitric oxide synthase gene. J Biol Chem 1994;269(9):6765–6772.
205. Vazquez-Torres A, Jones-Carson J, Mastroeni P, et al. Antimicrobial actions of the NADPH phagocyte oxidase and inducible nitric oxide synthase in experimental salmonellosis. I. Effects on microbial killing by activated peritoneal macrophages in vitro. J Exp Med 2000;192(2):227–236.
206. Furchgott RF, Carvalho MH, Khan MT, et al. Evidence for endothelium-dependent vasodilation of resistance vessels by acetylcholine. Blood Vessels 1987;24(3):145–149.
207. Duncan C, Dougall H, Johnston P, et al. Chemical generation of nitric oxide in the mouth from the enterosalivary circulation of dietary nitrate. Nat Med 1995;1(6):546–551.
208. Benjamin N, O'Driscoll F, Dougall H, et al. Stomach NO synthesis. Nature 1994;368(6471):502.
209. Weitzberg E, Lundberg JO. Nonenzymatic nitric oxide production in humans. Nitric Oxide 1998;2(1):1–7.
210. Zweier JL, Samouilov A, Kuppusamy P. Non-enzymatic nitric oxide synthesis in biological systems. Biochim Biophys Acta 1999;1411(2–3):250–262.
211. Zhang Z, Naughton D, Winyard PG, et al. Generation of nitric oxide by a nitrite reductase activity of xanthine oxidase: a potential pathway for nitric oxide formation in the absence of nitric oxide synthase activity. Biochem Biophys Res Commun 1998;249(3):767–772.

212. Lovren F, Triggle CR. Involvement of nitrosothiols, nitric oxide and voltage-gated K^+ channels in photorelaxation of vascular smooth muscle. Eur J Pharmacol 1998;347(2–3):215–221.
213. Megson IL, Holmes SA, Magid KS, et al. Selective modifiers of glutathione biosynthesis and "repriming" of vascular smooth muscle photorelaxation. Br J Pharmacol 2000;130(7):1575–1580.
214. Matsunaga K, Furchgott RF. Interactions of light and sodium nitrite in producing relaxation of rabbit aorta. J Pharmacol Exp Ther 1989;248(2):687–695.
215. Andrews KL, McGuire JJ, Triggle CR. A photosensitive vascular smooth muscle store of nitric oxide in mouse aorta: no dependence on expression of endothelial nitric oxide synthase. Br J Pharmacol 2003;138(5):932–940.
216. Denninger JW, Marletta MA. Guanylate cyclase and the NO/cGMP signaling pathway. Biochim Biophys Acta 1999;1411(2–3):334–350.
217. Wedel B, Harteneck C, Foerster J, et al. Functional domains of soluble guanylyl cyclase. J Biol Chem 1995;270(42):24,871–24,875.
218. Robertson BE, Schubert R, Hescheler J, et al. cGMP-dependent protein kinase activates Ca-activated K channels in cerebral artery smooth muscle cells. Am J Physiol 1993;265(1 pt 1):C299–C303.
219. Archer SL, Huang JM, Hampl V, et al. Nitric oxide and cGMP cause vasorelaxation by activation of a charybdotoxin-sensitive K channel by cGMP-dependent protein kinase. Proc Natl Acad Sci USA 1994;91(16):7583–7587.
220. Bolotina VM, Najibi S, Palacino JJ, et al. Nitric oxide directly activates calcium-dependent potassium channels in vascular smooth muscle. Nature 1994;368:850–853.
221. Mistry DK, Garland CJ. Nitric oxide (NO)-induced activation of large conductance Ca^{2+}-dependent K^+ channels (BK(Ca)) in smooth muscle cells isolated from the rat mesenteric artery. Br J Pharmacol 1998;124(6):1131–1140.
222. Lang RJ, Harvey JR, McPhee GJ, et al. Nitric oxide and thiol reagent modulation of Ca^{2+}-activated K^+ (BK_{Ca}) channels in myocytes of the guinea-pig taenia caeci. J Physiol 2000;525(pt 2):363–376.
223. Haburcak M, Wei L, Viana F, et al. Calcium-activated potassium channels in cultured human endothelial cells are not directly modulated by nitric oxide. Cell Calcium 1997;21(4):291–300.
224. Jeong SY, Ha TS, Park CS, et al. Nitric oxide directly activates large conductance Ca^{2+}-activated K^+ channels (rSlo). Mol Cells 2001;12(1):97–102.
224a. Verma S, Lovren F, Dumont AS, et al. Tetrahydrobiopterin improves endothelial function in human saphenous veins. J Thorac Cardiovasc surg 2000;120:668–671.
224b. Pannirselvam M, Verma S, Anderson JJ, and Triggle CR. Cellular basis of endothelial dysfunction in small mesenteric arteries from spontaneously diabetic (db/db–/–) mice: role of decreased tetrahydrobiopterin bioavailaility. Br J Pharmacol 2002;136(2):255–263.
225. Feletou M, Vanhoutte PM. Activation of vascular smooth muscle K^+ channels by endothelium-derived factors. In: Archer SL, Rusch NJ, eds. Potassium Channels in Cardiovascular Biology. Kluwer Academic/Plenum: New York, 2001, pp. 691–723.
226. McGuire JJ, Ding H, Triggle CR. Endothelium-derived relaxing factors: a focus on endothelium-derived hyperpolarizing factor(s). Can J Physiol Pharmacol 2001;79 443–470.
227. Busse R, Edwards G, Feletou M, et al. EDHF: bringing the concepts together. Trends Pharmacol Sci 2002;23(8):374–380.
228. Triggle CR, Ding H. Endothelium-derived hyperpolarizing factor: is there a novel chemical mediator? Clin Exp Pharmacol Physiol 2002;29(3):153–160.
229. Ding H, Triggle CR. Relaxing blood vessels: are there novel endothelium-derived mediators to be found and can their discovery lead to the development of new therapeutic agents? Pharm News 2001;8(42):42–49.
230. Waldron GJ, Garland CJ. Contribution of both nitric oxide and a change in membrane potential to acetylcholine-induced relaxation in the rat small mesenteric artery. Br J Pharmacol 1994;112(3): 831–836.
231. Edwards G, Dora KA, Gardener MJ, et al. K^+ is an endothelium-derived hyperpolarizing factor in rat arteries. Nature 1998;396(6708):269–272.
232. Ding H, Triggle CR. The contribution of EDHF and the role of potassium channels in the regulation of vascular tone. Drug Discovery Res 2003;58:81–89.
233. Hinton JM, Langton PD. Inhibition of EDHF by two new combinations of K(+)-channel inhibitors in rat isolated mesenteric arteries. Br J Pharmacol 2003;138(6):1031–1035.
234. Dong H, Jiang Y, Cole WC, et al. Comparison of the pharmacological properties of EDHF-mediated vasorelaxation in guinea-pig cerebral and mesenteric resistance vessels. Br J Pharmacol 2000;130(8): 1983–1991.

235. Fisslthaler B, Popp R, Kiss L, et al. Cytochrome P450 2C is an EDHF synthase in coronary arteries. Nature 1999;401(6752):493–497.
236. Sandow SL, Hill CE. Incidence of myoendothelial gap junctions in the proximal and distal mesenteric arteries of the rat is suggestive of a role in endothelium-derived hyperpolarizing factor-mediated responses. Circ Res 2000;86(3):341–346.
237. Coleman HA, Tare M, Parkington HC. K^+ currents underlying the action of endothelium-derived hyperpolarizing factor in guinea-pig, rat and human blood vessels. J Physiol 2001;531(pt 2):359–373.
238. Chaytor AT, Evans WH, Griffith TM. Central role of heterocellular gap junctional communication in endothelium-dependent relaxations of rabbit arteries. J Physiol 1998;508 (pt 2):561–573.
239. Dora KA, Martin PE, Chaytor AT, et al. Role of heterocellular Gap junctional communication in endothelium-dependent smooth muscle hyperpolarization: inhibition by a connexin-mimetic peptide. Biochem Biophys Res Commun 1999;254(1):27–31.
240. Edwards G, Weston AH. EDHF—are there gaps in the pathway? J Physiol 2001;531(pt 2):299.
241. Ujiie H, Chaytor AT, Bakker LM, et al. Essential role of Gap junctions in NO- and prostanoid-independent relaxations evoked by acetylcholine in rabbit intracerebral arteries. Stroke 2003;34(2):544–550.
242. Emerson GG, Segal SS. Electrical coupling between endothelial cells and smooth muscle cells in hamster feed arteries: role in vasomotor control. Circ Res 2000;87(6):474–479.
243. Emerson GG, Segal SS. Endothelial cell pathway for conduction of hyperpolarization and vasodilation along hamster feed artery. Circ Res 2000;86(1):94–100.
244. Welsh DG, Segal SS. Role of EDHF in conduction of vasodilation along hamster cheek pouch arterioles in vivo. Am J Physiol Heart Circ Physiol 2000;278(6):H1832–H1839.
245. Matoba T, Shimokawa H, Kubota H, et al. Hydrogen peroxide is an endothelium-derived hyperpolarizing factor in human mesenteric arteries. Biochem Biophys Res Commun 2002;290(3):909–913.
246. Matoba T, Shimokawa H, Morikawa K, et al. Electron spin resonance detection of hydrogen peroxide as an endothelium-derived hyperpolarizing factor in porcine coronary microvessels. Arterioscl Thromb Vasc Biol 2003;23(7):1224–1230.
247. Matoba T, Shimokawa H. Hydrogen peroxide is an endothelium-derived hyperpolarizing factor in animals and humans. J Pharmacol Sci 2003;92(1):1–6.
248. Cohen RA, Plane F, Najibi S, et al. Nitric oxide is the mediator of both endothelium-dependent relaxation and hyperpolarization of the rabbit carotid artery. Proc Natl Acad Sci USA 1997;94(8):4193–4198.
249. Bauersachs J, Popp R, Hecker M, et al. Nitric oxide attenuates the release of endothelium-derived hyperpolarizing factor. Circulation 1996;94(12):3341–3347.
250. Deshpande NN, Sorescu D, Seshiah P, et al. Mechanism of hydrogen peroxide-induced cell cycle arrest in vascular smooth muscle. Antioxid Redox Signal 2002;4(5):845–854.
251. Rabelo LA, Cortez SF, Alvarez-Leite JI, et al. Endothelium dysfunction in LDL receptor knockout mice: a role for H_2O_2. Br J Pharmacol 2003;138(5):1215–1220.
252. Pomposiello S, Rhaleb NE, Alva M, et al. Reactive oxygen species: role in the relaxation induced by bradykinin or arachidonic acid via EDHF in isolated porcine coronary arteries. J Cardiovasc Pharmacol 1999;34(4):567–574.
253. Hamilton CA, McPhaden AR, Berg G, et al. Is hydrogen peroxide an EDHF in human radial arteries? Am J Physiol Heart Circ Physiol 2001;280(6):H2451–H2455.
254. Ellis A, Pannirselvam M, Anderson TJ, et al. Catalase has negligible inhibitory effects on endothelium-dependent relaxation in mouse isolated aorta and small mesenteric artery. Br J Pharmacol 2003;140(7): 1193–1200.
255 Ellis A, Triggle CR. Endothelium-derived reactive oxygen species: their relationship to endothelium-dependent hyperpolarization and the regulation of vascular tone. Can J. Physiol Pharmacol. 2003;81(11):1013–1028.
256. McCarron JG, Halpern W. Potassium dilates rat cerebral arteries by two independent mechanisms. Am J Physiol 1990;259(3 pt 2):H902–H908.
257. Knot HJ, Zimmermann PA, Nelson MT. Extracellular K(+)-induced hyperpolarizations and dilatations of rat coronary and cerebral arteries involve inward rectifier K(+) channels. J Physiol 1996; 492(pt 2):419–430.
258. Taddei S, Mattei P, Virdis A, et al. Effect of potassium on vasodilation to acetylcholine in essential hypertension. Hypertension 1994;23(4):485–490.
259. Pannirselvam M, Anderson TJ, Triggle CR. Endothelial cell dysfunction in type I and II diabetes—the cellular basis for dysfunction. Drug Discovery Res 2003;58:28–41.
260. Angulo J, Cuevas P, Fernandez A, et al. Calcium dobesilate potentiates endothelium-derived hyperpolarizing factor-mediated relaxation of human penile resistance arteries. Br J Pharmacol 2003;139(4):854–862.

4 Chemical Interaction of Nitric Oxide With Protein Thiols

S-Nitrosylation Signaling

Allan Doctor and Benjamin M. Gaston

CONTENTS

SUMMARY

Protein *S*-nitrosylation reactions signal physiological effects. These reactions are carefully regulated in cells under normal conditions to achieve specificity and to prevent nitrosative stress. This regulation is achieved through metabolic control of *S*-nitrosothiol (SNO) synthesis and catabolism, as well as through cellular localization. Conditions involving excessive SNO accumulation (i.e., nitrosative stress) and regional SNO depletion (i.e., asthma and cystic fibrosis) have been associated with the pathophysiology of specific diseases.

SNO signaling involves a covalent modification of protein thiolate groups and, as such, is distinctly different from NO signaling involving free-radical reactions with metal centers. The study of SNO signaling is a rapidly emerging discipline that is relevant to nearly every field of medicie. However, the field remains hampered by imprecise assays and biochemical controversies. This chapter reviews the bioactivities and metabolism of physiological SNO compounds as well as the interaction of these compounds with gasotransmitters.

Key Words: *S*-Nitrosothiol; *S*-nitrosoglutathione; *S*-nitrosohemoglobin; *S*-nitrosoglutathione lyase; signaling.

From: *Signal Transduction and the Gasotransmitters: NO CO, and H_2S in Biology and Medicine* Edited by: Rui Wang © Humana Press Inc., Totowa, NJ

1. INTRODUCTION

It is now appreciated that nitric oxide (NO) does not exclusively—or even usually—signal simply by gas-phase diffusion and interaction with the heme porphyrin center of guanylate cyclase *(1,2)*. Further, toxicity of NO is not always limited to its gas-phase diffusion and reactivity with either oxygen or superoxide. This chapter reviews a class of signaling reactions in which NO is transferred covalently from one peptide or protein to another. These are reactions of NO with cysteine thiol groups known as *S*-nitrosylation and transnitrosylation reactions *(1–4)*. The products of these *S*-nitrosylation reactions are a class of compounds known as SNOs, which are found endogenously in most cells and tissues.

SNO signaling contrasts substantially with NO (free radical) signaling in biology because of a small, but critically important, distinction in the redox state of nitrogen between NO and SNO. In biological systems, nitrogen exists in each of its possible redox states from –3 (ammonia) to +5 (nitrate) *(5)*. Classically, NO is viewed as a signaling molecule when nitrogen is in its +2 oxidation state (NO radical). However, in the presence of an electron acceptor, NO can be oxidized to an NO^+ (nitrosonium) equivalent; these NO^+ groups can be covalently added, removed, and transferred to signal changes in protein biology *(1–5)*. It is critical to understand, however, that NO^+ groups are not stable ions in solution but always exist in biological systems in complex with a relatively electronegative species such as thiolate anions (in SNOs) or transition metals.

This redox chemistry dramatically expands the signaling repertoire available to NO in physiology. This chapter reviews the bioactivities of these SNO compounds and their regulation in vivo. It also presents what we believe to be important interactions between these SNO compounds and other gasotransmitters, carbon monoxide (CO) and hydrogen sulfide (H_2S).

2. BIOACTIVITIES OF SNOS

2.1. Guanylate Cyclase-Independent Bioactivities

For more than a decade, it has been appreciated that many effects downstream of nitric oxide synthase (NOS) activation are independent of guanylate cyclase activation *(6,7)*. That is to say, they are not mimicked pharmacologically by cell-permeable cyclic guanosine 5'-monophosphate (cGMP) analogs, they are not enhanced by inhibitors of cGMP phosphodiesterases, they are not prevented by inhibitors of guanylate cyclase, and/or they are not associated with increased tissue cGMP levels.

Conventionally, cGMP-independent effects are often discussed as being toxicities mediated by reactions of NO with superoxide and carbon dioxide (CO_2) that result in posttranslational protein inactivation through tyrosine nitration. However, additional cGMP-independent mechanisms involved in physiological (as opposed to toxic) effects downstream of NOS activation have been well characterized. Examples of these biological processes include human airway smooth muscle relaxation *(6,7)* and normoxic upregulation of hypoxia inducible factor (HIF)-1-dependent gene transcription *(8)*. The most important of these mechanisms is posttranslational protein *S*-nitrosylation.

SNO bonds have been identified endogenously in proteins and/or peptides from virtually every organ system. Concentrations of *S*-nitrosoglutathione (GSNO) are approx 7 μ*M* in rat brain stem homogenates *(9)*. By contrast, GSNO levels are low, and *S*-nitroso-cysteinyl glycine (CGSNO) levels are higher, in the rat thalamus *(10)*. Total SNO concentration in human airway lining fluid is approx 500 n*M* under physiological conditions *(11)*. Cellular concentrations in lymphocytes, hepatocytes, erythrocytes, and other cells are on the order of 50–100 n*M*, depending on the protein type and the specific cellular localization *(12–15)*.

A broad array of different proteins have been shown to undergo posttranslation modification as a result of *S*-nitrosylation reactions. These proteins include circulating molecules such as albumin and hemoglobin (Hb) *(15,16)*; cell-surface molecules such as erythrocytic anion-exchange protein 1 (AE1) and neuronal *N*-methyl-D-aspartate receptor protein *(17,18)*; cytosolic/metabolic proteins such as glyceraldehyde phosphate dehydrogenate and methionine adenosyl transferase *(19,20)*; mitochondrial proteins such as caspases 3 and 9 *(13,14)*; and proteins involved in the regulation of gene transcription such as those involved in the activity of HIF-1, nuclear factor-κB (NF-κB), SP1, and SP3 *(8,21)*. Prokaryotic proteins are also known to be *S*-nitrosylated *(22)*. Of note, activity can be increased (such as that of thioredoxin *[23]*) or inhibited (such as that of caspases *[12,13]*) by *S*-nitrosylation. In certain proteins, such as the ryanodine-sensitive calcium channel, different degrees of *S*-nitrosylation under different redox conditions can have opposite effects *(24,25)*. More importantly, most of these posttranslational protein modifications appear to be carried out under physiological conditions by transnitrosation reactions in which an NO^+ group is exchanged for an H^+ between one thiolate and another according to **reaction A** *(26)*:

$$\text{RS-NO} + \text{R'SH} \leftrightarrow \text{RSH} + \text{R'S-NO} \tag{A}$$

2.2. Stereospecificity of SNO Bioactivities

Increasingly, it is appreciated that these cGMP-independent *S*-nitrosylation signaling processes are stereoselective. That is, they are replicated in experimental models by the L-isomer of *S*-nitrosocysteine (L-CSNO)—or peptides containing this moiety—but not by the D-isomer of CSNO. These effects include profound heart rate, blood pressure, and ventilatory effects caused by injection of L-CSNO, but not D-CSNO, into the brain stem nucleus tractus solitarius (**Fig. 1**; *27,28*). Additionally, the systemic peripheral vascular effects of CSNO are stereoselective *(29)*. The pharmacological actions of L-CSNO imply a specific protein receptor, particularly in neuronal and endothelial cells; however, these receptors have not yet been identified.

2.3. NO–Hemoglobin Interactions

Our understanding of the chemistry of NO and Hb interactions has reflected, and at times driven, more global models of NO biochemistry and physiology. The oxidative reaction of NO with oxyHb, yielding nitrate and methemoglobin, was long considered the major route of NO catabolism and constituted a method for assaying NO synthesis; additionally, the stereochemistry of NO and heme ligand interactions is important to our understanding of guanylate cyclase activation by NO. In examining *thiol*-based reactivity of Hb, specifically at a highly conserved, reactive cysteine (βcys^{93}), it was observed that NO bioactivity is preserved following *S*-nitrosylation at βcys^{93}, and that this reaction is under allosteric regulation *(15)*. Further, the conformation-dependent association of βcys^{93} with the amino-terminal cytoplasmic domain of AE1 in the red blood cell (RBC) membrane may provide a mechanism for nitrosative traffic between the erythrocyte and plasma thiols or cellular protein targets (**Fig. 2**; *17*). AE1 has not only been demonstrated to transport several NO_x species but also contains a classic *S*-nitrosylation motif at the site of Hb binding. Indeed, blockade of AE1 has been noted to abrogate (S)NOHb vasoactivity,

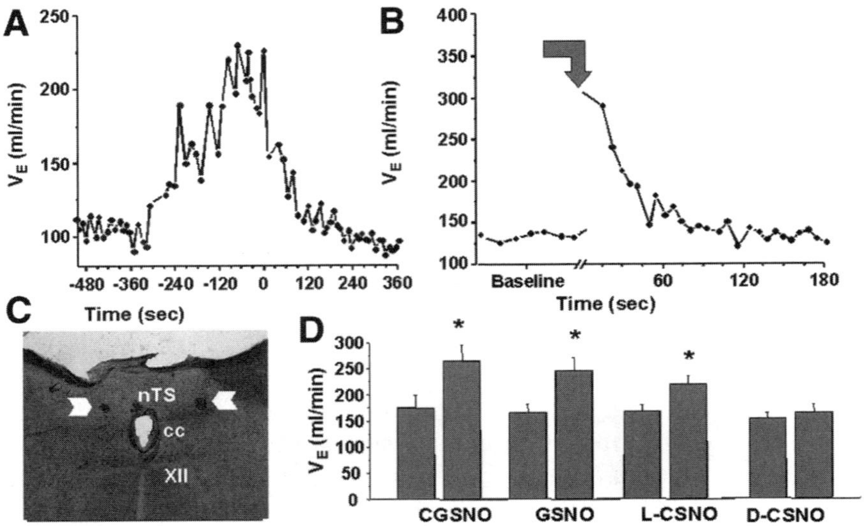

Fig. 1. Ventilatory effects of SNOs. **(A)** Minute ventilation (V_E) during (shaded) and following a short period of hypoxia. **(B)** Injection of 1 nmol of CGSNO into the nucleus tractus solitarius (nTS) resulted in a marked increase in V_E (injection indicated by arrow) with onset and decay characteristics identical to those observed during short exposure of the whole animal to hypoxia and return to normoxia. **(C)** Neuronal tissue section showing nTS. Arrows indicate the injection site. cc, central canal; XII, hypoglossal nucleus. Because conscious animals were used, baseline V_E varied considerably; however, a substantial increase was observed with each L-SNO isomer. **(D)** All L-SNO isomers caused increases in V_E (change from baseline for CGSNO: $*p < 0.001$, $n = 10$; GSNO: $*p < 0.0001$, $n = 14$; L-CSNO: $*p < 0.0001$, $n = 20$), whereas D-CSNO was without effect (p = NS; $n = 20$). (From ref. *27* with permission.)

and, notably, transnitrosation from *S*-nitrosylated Hb to AE1 is promoted on conversion from R- to T-state Hb, further relating RBC NO traffic to ambient pO_2.

Erythrocyte nitrosative flux is coupled to regional oxygen tensions by allosteric governance of (Heme) Fe-O_2 binding on (globin chain) thiol-NO binding. The stability of the βcys[93] *S*-nitrosothiol ([S]NO) bond varies with O_2-dependent Hb conformational change; it is stable in oxyHb and destabilized ondeoxygenation, allowing for NO egress from the red cell (through AE1 and other thiols) in response to falling oxygen tension. Thus, regional tissue Po2 and erythrocytic Hb-(S)NO affinity are allosterically linked ($\downarrow Po_2 \rightarrow \downarrow$[(S)NOHb]), working in concert with hypoxia-induced vasodilation and mobilizing vascular control by coupling NO release and capture with O_2 delivery within circulating erythrocytes and measures of regional perfusion sufficiency. Some researchers have argued that the arteriovenous gradient of SNO-Hb is inadequate to allow for SNO-Hb peripheral vasoactivity. However, these interpretations dismiss and/or do not account for the activity of AE1 *(17)*; the fact that allosteric transfer of low nanomolar quantities of NO from iron to AE1 (through SNO-Hb)—a minute fraction of the total NO-Hb pool—is sufficient to cause relaxation of resistance vessels *(30)*; the fact that conditions favoring NO transfer from SNO-Hb recapitulate the ventilatory and erythropoietic response to hypoxia in vivo (**Fig. 1**; *27,31*); and the fact several groups have now shown, using many methods—including protein crystallography—that NO-Hb interaction at the βcys[93] is like every other Hb chemical interaction in physiology: it is allosterically regulated *(27,32–34)*.

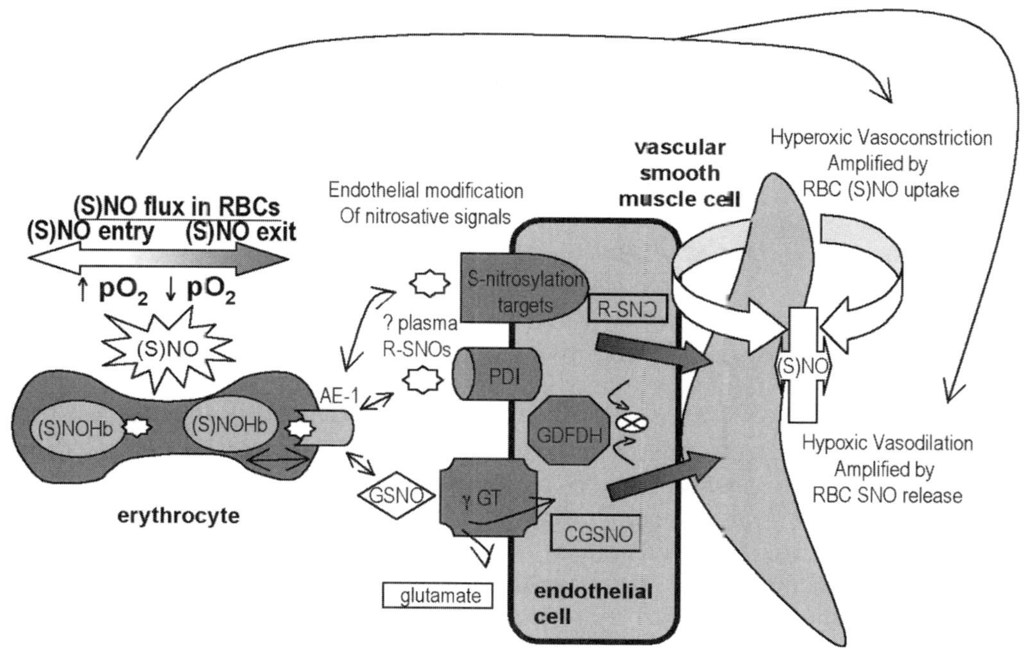

Fig. 2. Erythrocyte-endothelium communication via thiol-based nitrosative signaling in the microcirculation. Governed by regional Po_2 gradients, serial *S*-transnitrosation reactions from Hb may pass (S)NOs, through plasma thiols, to endothelial cell-surface targets, triggering signaling events or membrane traversal to effect change in underlying smooth muscle tension. Putative nodes in endothelial nitrosative signaling include protein disufide isomerase (PDI) (facilitating transmembrane transnitrosation reactions), γ glutamyl transpeptidase (γGT) (converting plasma *S*-nitrosoglutathione to CGSNO for intracellular transport and bioactivity), and glutathione-dependent formaldehyde dehydrogenase (GDFDH) (serves as an intracellular low-mass nitrosothiol lyase, quenching nitrosative signaling events).

2.4. SNO Signaling in Pathophysiology

Altered SNO metabolism has been implicated in conditions as diverse as diabetes, rheumatoid arthritis, malaria, atherosclerosis and thrombosis, multiple sclerosis, familial amyotrophic lateral sclerosis, and preeclampsia. Specifically in the lung, it appears to be relevant to primary pulmonary hypertension, pulmonary hypertension of the newborn, asthma, cystic fibrosis, and hypoxic signaling in pulmonary endothelium *(5,35–42)*. Vasoactivity of (S)NOHb; GSNO; GSNO-generating prodrug, *O*-nitrosoethanol; and *S*-nitrosocysteine has been demonstrated *(38,43,44)*.

In summary, *S*-nitrosylation signaling results from transnitrosation reactions that effect posttranslational modifications in a broad spectrum of proteins. These signaling reactions are increasingly appreciated to be stereoselective, cGMP-independent processes of biological importance.

3. REGULATION OF SNO BIOACTIVITIES

The potent signaling effects and other bioactivities of SNOs under physiological conditions are not simply the result of inorganic reactions associated with oxidative or nitrosative stress but, rather, are carefully regulated biochemical processes. In many ways, these reactions are analogous to phosphorylation. It is now appreciated that there are at least three levels of regulation for SNO bioactivity: synthesis, degradation, and cellular compartmentalization.

3.1. SNO Synthesis

NOS activation can produce NO. Additionally, however, it can produce hydroxylamine, peroxynitrite ($ONOO^-$), nitrate, and SNOs *(45)*. Indeed, it has been argued that these more versatile nitrogen oxides, in many circumstances, represent the principal products of NOS activation—products that can be enzymatically interconverted or converted to NO to achieve specific bioactivities.

When it is NO radical that is formed from NOS, it will only *S*-nitrosylate cysteine thiols— as a general rule—in the presence of an electron acceptor. In the past, it had been widely assumed that this electron acceptor must be oxygen in biological systems. Indeed, the third-order rate constant for the rate-limiting reaction of NO with oxygen to form an NO_2 intermediate—which is then followed by a reaction with another NO to form N_2O_3, a potent *S*-nitrosylating agent (**reactions B** and **C**)—is 200-fold higher in lipid membranes than in aqueous phase, suggesting that SNO synthesis can, and probably does, occur in membranes with oxygen as an electron acceptor *(4,46)*.

$$2NO + O_2 \rightarrow 2NO_2 \quad \text{(B)}$$

$$NO_2 + NO \rightarrow = N_2O_3 \leftrightarrow {}^+ON \ldots\ldots NO_2^- \quad \text{(C)}$$

$${}^+ON \ldots\ldots NO_2^- + RSH \rightarrow RS\text{-}NO + HNO_2^- \quad \text{(D)}$$

Similar N_2O_3-mediated *S*-nitrosylation has been proposed to occur in the hydrophobic pocket of albumin *(16)*, where a conserved cysteine residue is endogenously *S*-nitrosylated. Indeed, NAD^+ can serve as an inorganic electron acceptor and result in cysteine *S*-nitrosylation in biological and physiological conditions *(47)*. Finally, SNOs can be and probably are formed through intermediate inorganic iron-nitrosyl species *(48)*.

However, there is accumulating evidence that SNO formation is not exclusively—or even primarily—the result of inorganic reactions. Human ceruloplasmin recently has been shown to catalyze formation of GSNO *(49)*. Here, Cu^{2+} serves as the electron acceptor, forming a Cu^+-NO^+ complex. This complex reacts with glutathione (GSH) to form GSNO, and the electron is shuttled through the ceruloplasmin copper complexes, ultimately to form water from oxygen. Similarly, erythrocytic Hb/AE1 complex can serve, as already described, as an oxygen-sensitive GSNO synthase *(17)*.

Analogous to phosphorylation, there are consensus motifs that predict which specific cysteines will be *S*-nitrosylated in a particular protein *(2,50)*. These motifs are relevant to both the primary sequence and the tertiary structure of the protein *(20)*, such that redox-active amino acids will favor *S*-nitrosylation of a nearby cysteine. Remarkably, studies using site-directed mutagenesis have shown that it is generally only these highly specific, motif-predicted cysteines that are endogenously *S*-nitrosylated or experimentally *S*-nitrosylated under physiological conditions, and it is only *S*-nitrosylation of these specific cysteines that results in functional protein modification. This degree of specificity implies a more extensive level of enzymatic regulation than is currently understood.

3.2. SNO Catabolism

Substantially more is known about the enzymatic regulation of SNO catabolism than about SNO synthesis. At least six enzymes/enzyme systems have been shown to have activity in breaking down one or more SNOs. These include glutathione-dependent formaldehyde dehydrogenase (recently proposed to be renamed GSNO lyase *[14]*), Cu/Zn superoxide dismutase (SOD), γ-glutamyl transpeptidase (GGT), thioredoxin/thioredoxin reductase, xanthine/xanthine oxidase, and glutathione peroxidase. Additionally, membrane-based SNO signal-transducing peptides, including AE1 and protein disulfide isomerase, can, depending on conditions, serve as SNO lyases. Intriguingly, the products of these catabolic enzymes are extraordinarily diverse—ranging from ammonia to $ONOO^-$—suggesting that the activation and cellular localization of these enzymes may distribute or transduce nitrogen oxide bioactivity from an SNO pool.

GSNO lyase is relatively ubiquitous expressed as a "housekeeping" enzyme. Abundant expression in tissues not exposed to alcohol or formaldehyde had been difficult to rationalize in the context of its previous characterization as an alcohol dehydrogenase. However, as a GSNO lyase it appears to be critically important in preventing nonspecific cytosolic protein *S*-nitrosylation that can be associated with nitrosative stress in all tissues *(14,51)*. Indeed, levels of *S*-nitrosylated proteins more than double—and low molecular weight SNOs become detectable—in the cytosol of *Escherichia coli*, yeast ,and mouse hepatocytes from which the GSNO lyase gene has been deleted *(14)*. It also appears that GSNO lyase serves an important signaling function. For example, there is evidence that it is upregulated in allergic disease, augmenting bronchoconstriction *(14,39,52)*.

GGT, known for its role in GSH metabolism and transport, also appears to have a central role in SNO signaling. Like GSH, GSNO is not generally cell membrane permeable. Cleavage of the γ-glutamyl bond of GSNO – like that of GSH—to form CGSNO permits the dipeptide to enter the cell *(53)*. GGT is thus an important gatekeeper for transcellular GSNO signaling. The GGT inhibitor, acivicin, prevents the hypoxia-mimetic effect of GSNO at the level of the nucleus tractus solitarius *(27)*, the GSNO-mediated stabilization and activation of HIF-1 in normoxia *(8)*, and the GSNO-mediated expression of cystic fibrosis transmembrane regulatory protein (CFTR) *(40)*. Interestingly, each of these effects is overcome when the GGT product CGSNO, as opposed to GSNO, is used as the pharmacological agent in the presence of acivicin. Moreover, mice deficient in GGT lack the normal ventilatory and erythropoietic response to hypoxia, suggesting the possibility that erythrocyte deoxygenation-mediated GSNO synthesis is an important signaling pathway from deoxyhemoglobin that is regulated, in part, by GGT *(27)*. Of note, humans given exogenous *N*-acetylcysteine—which can also, through transnitrosation, deliver SNO signal intracellularly in a GGT-independent fashion—have dramatically augmented ventilatory and erythropoietin responses to hypoxia *(31)*.

Copper/zinc SOD catalyzes the breakdown of several different low-mass SNOs to NO in the presence of physiological concentrations of reducing agents such as glutathione and ascorbate *(41)*. Note that many of these low-mass SNOs are neuroprotective, whereas NO and $ONOO^-$ are neurotoxic *(54)*. Therefore, it is of interest that Cu/Zn SOD mutants associated with familial amyotrophic lateral sclerosis represent gain-of-function mutations that substantially increase the efficiency of low-mass SNO catabolism to NO, thus potentially converting neuroprotective to neurotoxic nitrogen oxides *(41)*.

It appears that thioredoxin is a critical intracellular reservoir of SNO bioactivity. Indeed, *S*-nitrosylation of thioredoxin cysteine 69 is required to permit the antioxidant function of separate cysteines (32 and 35) that are not *S*-nitrosylated in the protein *(23)*. As predicted by the surrounding consensus motif and confirmed by site-directed mutagenesis, it is only cysteine 69 that is *S*-nitrosylated. Thioredoxin reductase, in turn, appears to catalyze both denitrosylation of cysteine 69 and reduction of the redox-active thiols *(23,55)*. It is apparent from studies done on GSNO lyase that a transnitrosation equilibrium exists between GSNO and *S*-nitrosylated cytosolic proteins such as thioredoxin *(14)*. Signaling interactions among GSNO, SNO-thioredoxin, other regulated SNO proteins (such as caspases), thioredoxin reductase, and GSNO lyase are only beginning to be investigated.

There is evidence in vitro that the xanthine/xanthine oxidase system will enzymatically convert SNOs to $ONOO^-$ *(56)*. The relevance of this observation to intact cell and organ systems remains to be clarified. Similarly, the physiological relevance of in vitro evidence that glutathione peroxidase and other selenium-containing species catalyze GSNO catabolism is not clear *(57)*. In particular, unlike the other enzyme systems that we have described, the glutathione peroxidase K_M for GSNO is substantially higher than are levels of GSNO found in normal tissues.

Inorganic reactions also catalyze SNO decomposition but are likely to be of little relevance in physiology *(58)*. Indeed, Cu^+ is the most facile inorganic species in biology capable of carrying out SNO reduction, yet levels of free Cu^+ have been estimated to be as low as 1 atom per cell, largely because of stringent regulation of copper trafficking *(59)*. However, inorganic SNO breakdown may have important therapeutic implications. For example, SNOs are broken down to NO by light (photolyzed). The mechanism by which photophoresis attenuates adverse host/graft immune interactions in transplantation patients is not known *(60)* but may involve photolysis of SNO bonds. Specifically, Fas-Fas ligand binding triggers cleavage of SNO bonds in caspases 3 and 9, activating the caspases and leading to lymphocyte apoptosis. Treatment with Fas ligand in many ways mimics (but is more toxic than) photophoresis *(12,13)*. It could be speculated that photophoresis works—at least in part—by photolysis of SNO caspase bonds.

3.3. Compartmentalization of SNO Signaling

The bioactivities of SNO are also regulated by compartmentalization. As already noted, SNOs are relatively unstable in cytosol, though specific bonds—such as those buried in the protein structure and/or in hydrophilic pockets—are protected from transnitrosation and catabolism. However, SNO proteins are remarkably stable in membranes and at relatively low pH. Thus, the majority of caspase 3 and caspase 9 in the mitochondrial intermembrane space is *S*-nitrosylated in vivo *(13)*. There, specific active-site cysteine *S*-nitrosylation prevents autocatalytic protein degradation. With Fas-Fas ligand binding, these caspases are released into the cytosol, where they undergo rapid denitrosylation/activation *(13)*.

A great deal of SNO bioactivity appears to occur at or in the cell membrane. Stereoselective signaling by LCSNO-containing peptides and proteins seems to be cell membrane associated, likely involving various different proteins, including forms of protein disulfide isomerase *(61)*. Moreover, recent evidence suggests that *S*-nitrosylation signaling at the cell membrane can initiate downstream phosphorylation cascades, particularly involving Akt and phosphoinositol-3 (PI3) kinase, not to mention G-proteins such as $p21^{ras}$ *(62,63)*. Much work remains to be done on the interaction between cell membrane-based nitrosylation signaling and phosphorylation signaling cascades.

Additionally, the nucleus appears to be an important site for endogenous SNO bioactivity. Recent fluorescent studies suggest preferential nuclear localization of SNOs *(64)*. Indeed, SNO signaling has been shown to have dramatic effects on signaling of gene expression through HIF-1, specificity protein (SP)1, SP3, activation protein (AP)1, and NF-κB *(8,21,65–67)*. Interestingly, physiological levels of nitrosothiols (approx 500 n*M*) augment SP1 and SP3 binding (by electromobility shift assay), whereas nitrosative stress levels inhibit SP3 binding and augment SP1 binding (unpublished observation). Taken together, these data suggest that SNOs have important regulatory effects on gene expression, both at baseline and in the context of stress response.

S-Nitrosothiols also appear to be localized in vesicles. Immunofluorescent images by Gow and coworkers *(68)* showing apical localization of SNOs in what appear to be vesicles and epithelial cells are consistent with the work of Lewis and coworkers *(68a)*, who have measured millimolar concentrations of SNOs in vesicles isolated from endothelial cells. It is tempting to speculate that these epithelial vesicles are caveolae, whose membranes are rich in NOS3, but direct evidence is lacking. It is also tempting to speculate that lamellar bodies in type 2 airway epithelial cells, which have a pH of approx 3.0, because of the activity of vacuolar adenosine triphosphatases, may contain *S*-nitrosylated surfactant proteins. These SNO proteins could be secreted into the alveolus to augment host defense. Vesicular/vacuolar localization of SNOs for storage, trafficking, signaling, and host defense is an additional area of SNO research and is only just beginning to be explored.

In summary, SNO signaling and other bioactivities appear to be regulated at the level of synthesis, degradation, and cellular compartmentalization Each of these areas represents a novel and exciting field of research in cell biology.

4. INTERACTION OF SNOS WITH OTHER GASOTRANSMITTERS

4.1. Carbon Monoxide

Interactions between NO and CO signaling pathways are established *(69)*. Additionally, there appear to be interactions between SNO and CO signaling pathways. SNOs increase the stability and activity of HIF-1 in normoxia *(8)*. In turn, HIF-1 signals an increase in the transcriptional regulation of heme oxygenase-1 (HO-1), which produces CO. Thus, SNOs can signal upregulation of CO production in normoxia. Paradoxically, the NO radical actually inhibits HIF-1 binding in hypoxia *(8)*, underscoring the importance of the distinction between SNO- and NO-mediated bioactivities.

In asthma, exhaled levels of NO and CO are both high *(70,71)*. The conventional wisdom is that these levels represent increased expression of NOS2 and the CO-producing enzyme HO-1. However, there is also an interaction between NO and CO metabolic pathways that has not been considered previously. By way of background, levels of NO in expired air are 3 log orders too low to have any physiological effect *(71)*. In this sense, NO appears simply to be a reporter of nitrogen oxide bioactivities in the airways, rather than the effector molecule. Although NOS activation is ultimately important for NO generation, there are many capacitors in the circuit of NO production in the airway. Indeed, alkalinization of the ordinarily acidic asthmatic airway lining fluid decreases expired NO, suggesting that protonation of nitrite contributes to the net expired signal *(72)*. Similarly, inhalation of GSNO transiently increases expired NO; the NO signal decays depending on the rate of GSNO catabolism in the airways *(42)*. Of note, there is improved oxygenation of GSNO inhalation that is unrelated to the expired NO concentration *(42)*.

We have already seen that GSNO lyase (GSH-dependent formaldehyde dehydrogenase) is an important GSNO catabolic enzyme in the airway, and that airway GSNO catabolism is accelerated in asthma *(14,39,52,73)*. This enzyme catalyzes sequential reduction of GSNO through hydroxylamine to ammonia *(14,51)*. However, some of this product can be shunted away from complete reduction to ammonia by catalase, which oxidizes hydroxylamine to NO *(74)*. Thus, in the presence of catalase, increased expired NO can represent a reporter/biomarker for GSNO lyase activity. More classically, GSNO lyase catalyzes the oxidation of $GSCH_2OH$ to *S*-formyl glutathione, GSCHO *(51)*. In the presence of esterase D, GSCHO is further oxidized to formic acid, which in the presence of catalase, will form CO_2 *(75)*. However, GSCHO may also spontaneously dissociate to GSH and CO. Thus, increased expression of GSNO lyase in the asthmatic airway could lead both to increased NO production and increased CO production. Note that this process can represent a cycle involving sequential production and consumption of NADPH, all through one enzyme. This may be a model for integral processes occurring in other organ systems.

4.2. Hydrogen Sulfide

Recently we have observed another interaction between SNOs and the gasotransmitter H_2S. GSNO and H_2S are both present in the brain stem. The two compounds react inorganically (unpublished observation) to form $HSNO/SNO^-$ according to **reaction E**:

$$H_2S + GSNO \rightarrow HSNO + GSH \qquad \text{(E)}$$

Whether this reaction is regulated, and/or is physiologically relevant, remains to be established.

5. CONCLUSION

NO does not always signal as NO radical. Indeed, many significant effects of NOS activation may be signaled through posttranslational modifications of proteins at cysteine thiols. These bioactivities are cGMP independent and may be stereoselective. They occur under physiological conditions and are relevant to cell biology under physiological conditions, i.e., in the absence of any exogenous NO donor or nitrosative stress.

SNO signaling pathways appear to be responsible for a broad range of effects, ranging from inflammatory cell apoptosis to endothelial cell gene regulation. These processes are regulated at the level of SNO synthesis, which occurs at specific consensus motifs in the primary and tertiary structure of proteins; SNO breakdown, for which many selective enzymes have been described; and cellular compartmentalization. Each of these levels of regulation represents an exciting and novel field of research. Finally, NO may have important interaction with other gasotransmitters, CO and H_2S, through SNOs; these are exemplified by the implications of GSNO upregulation in the asthmatic airway.

REFERENCES

1. Arnelle D, Stamler J. NO+, NO, and NO– donation by S-nitrosothiols: implications for regulation of physiological functions by S-nitrosylation and acceleration of disulfide formation. Arch Biochem Biophys 1995;318:279–285.
2. Stamler JS, Lamas S, Fang FC. Nitrosylation: the prototypic redox-based signaling mechanism. Cell 2001;106:675–683.

3. Stamler JS, Singel DJ, Loscalzo J. Biochemistry of nitric oxide and its redox-activated forms. Science 1992;258:1898–1902.
4. Gaston B. Nitric oxide and thiol groups. Biochim Biophys Acta 1999;1411:323–333.
5. Gaston B, Stamler JS. Nitrogen oxides and lung function. In: Crystal R, West J, Weibel E, et al., eds. The Lung: Scientific Foundations. 2nd ed, Lippincott Raven: Philadelphia, 1997, pp. 239–253.
6. Gaston B, Drazen J, Jansen A, et al. Relaxation of human bronchial smooth muscle by S-nitrosothiols in vitro. J Pharmacol Exp Ther 1994;268:978–984.
7. Perkins WJ, Pabelick C, Warner DO, et al. cGMP-independent mechanism of airway smooth muscle relaxation induced by S-nitrosoglutathione. Am J Physiol Cell Physiol 1998;275:C468–C474.
8. Palmer L, Gaston B, Johns R. Normoxic stabilization of hypoxia-inducible factor 1 expression and activity: redox-dependent effect of nitrogen oxides. Mol Pharmacol 2000;58:1197–1203.
9. Kluge I, Gutteck-Amsler U, Zollinger M, et al. S-Nitrosoglutathione in rat cerebellum: identification and quantification by liquid chromatography-mass spectrometry. J Neurochem 1997;69:2599–2607.
10. Salt E, Zhang H, Mayer B, et al. Novel mode of nitric oxide neurotransmission mediated via S-nitroso-cysteinyl-glycine. Eur J Neurosci 2000;12:3919–3925.
11. Gaston B, Reilly J, Drazen J, et al. Endogenous nitrogen oxides and bronchodilator S-nitrosothiols in human airways. Proc Natl Acad Sci USA 1993;90:10,957–10,961.
12. Mannick JB, Hausladen A, Liu L, et al. Fas-induced caspase denitrosylation. Science 1999;284: 651–654.
13. Mannick J, Schonhoff C, Papeta N, et al. S-Nitrosylation of mitochondrial caspases. J Cell Biol 2001;154: 1111–1116.
14. Liu L, Hausladen A, Zeng M, et al. A metabolic enzyme for S-nitrosothiol conserved from bacteria to humans. Nature 2001;410:490–494.
15. Stamler JS, Jia L, Eu JP, et al. Blood flow regulation by S-nitrosohemoglobin in the physiological oxygen gradient. Science 1997;276:2034–2037.
16. Rafikova O, Rafikov R, Nudler E. Catalysis of S-nitrosothiols formation by serum albumin: the mechanism and implication in vascular control. Proc Natl Acad Sci USA 2002;99:5913–5918.
17. Pawloski J, Hess D, Stamler JS. Export by red blood cells of nitric oxide bioactivity. Nature 2001;409: 622–626.
18. Choi YB, Tenneti L, Le DA, et al. Molecular basis of NMDA receptor-coupled ion channel modulation by S-nitrosylation. Nat Neurosci 2000;3:15–21.
19. Mohr S, Stamler J, Brune B. Postranslational modification of glyceraldehyde-3-phosphate dehydrogenase by S-nitrosylation and subsequent NADH attachment. J Biol Chem 1996;271:4209–4214.
20. Perez-Mato I, Castro C, Ruiz FA, et al. Methionine adenosyltransferase S-nitrosylation is regulated by the basic and acidic amino acids surrounding the target thiol. J Biol Chem 1999;274:17,075–17,079.
21. Zaman K, Palmer LA, Doctor A, et al. Concentration-dependent effects of endogenous *S*-nitrosoglutathione on gene regulation by specificity proteins Sp3 and Sp1. Biochem J 2004;379:1–8.
22. Kim SO, Merchant K, Nudelman R, et al. OxyR:nA molecular code for redox-related signaling. Cell 2002;109:383–396.
23. Haendeler J, Hoffmann J, Tischler V, et al. Redox regulatory and anti-apoptotic functions of thioredoxin depend on S-nitrosylation at cysteine 69. Nat Cell Biol 2002;4:743–749.
24. Xu L, Eu JP, Meissner G, et al. Activation of the cardiac calcium release channel (ryanodine receptor) by poly-S-nitrosylation. Science 1998;279:234–237.
25. Sun J, Xin C, Eu J, et al. Cysteine-3635 is responsible for skeletal muscle ryanodine receptor modulation by NO. Proc Natl Acad Sci USA 2001;98:11,158–11,162.
26. Meyer D, Kramer H, Özer N, et al. Kinetics and equilibria of S-nitrosothiol-thiol exchange between glutathione, cysteine, penicillamines and serum albumin. FEBS Lett 1994;345:177–180.
27. Lipton A, Johnson M, Macdonald T, et al. S-Nitrosothiols signal the ventilatory response to hypoxia. Nature 2001;413:171–174.
28. Ohta H, Bates J, Lewis S, et al. Actions of S-nitrosocysteine in the nucleus tractus solitarii are unrelated to release of nitric oxide. Brain Res 1997;746:98–104.
29. Davisson R, Travis M, Bates J, et al. Hemodynamic effects of L- and D-S-nitrosocysteine in the rat. Circ Res 1996;79:256–262.
30. Jia L, Bonaventura C, Bonaventura J, et al. S-Nitrosohaemoglobin: a dynamic activity of blood involved in vascular control. Nature 1996;380:221–226.
31. Hildebrandt W, Alexander S, Bartsch P, et al. Effect of N-acetyl-cysteine on the hypoxic ventilatory response and erythropoietin production: linkage between plasma thiol redox state and O_2 chemosensitivity. Blood 2002;99:1552–1555.

32. Gow AJ, Stamler JS. Reactions between nitric oxide and haemoglobin under physiological conditions. Nature 1998;391:169–173.
33. Luchsinger BP, Rich EN, Gow AJ, et al. Routes to S-nitroso-hemoglobin formation with heme redox and preferential reactivity in the subunits. Proc Natl Acad Sci USA 2003;100:461–466.
34. Chan N, Rogers P, Arnone A. Crystal structure of the S-nitroso form of liganded human hemoglobin. Biochemistry 1998;37:16,459–16,464.
35. Milsom AB, Jones CJ, Goodfellow J, et al. Abnormal metabolic fate of nitric oxide in type I diabetes. Diabetologia 2002;45:1515–1522.
36. Hilliquin P, Borderie D, Hernvann A, et al. Nitric oxide as S-nitrosoproteins in rheumatoid arthritis. Arthritis Rheumatol 1997;40:1512–1517.
37. Lorch SA, Foust R 3rd, Gow A, et al. Immunohistochemical localization of protein 3-nitrotyrosine and S-nitrosocysteine in a murine model of inhaled nitric oxide therapy. Pediatr Res 2000;47:798–805.
38. Moya MP, Gow AJ, Califf RM, et al. Inhaled ethyl nitrite gas for persistent pulmonary hypertension of the newborn. Lancet 2002;360:141–143.
39. Gaston B, Sears S, Woods J, et al. Bronchodilator S-nitrosothiol deficiency in asthmatic respiratory failure. Lancet 1998;351:1317–1319.
40. Zaman K, McPherson M, Vaughan J, et al. S-Nitrosoglutathione increases cystic fibrosis transmembrane regulator maturation. Biochem Biophys Res Commun 2001;284:65–70.
41. Johnson MA, Macdonald TL, Mannick JB, et al. Accelerated S-nitrosothiol breakdown by amyotrophic lateral sclerosis mutant copper, zinc-superoxide dismutase. J Biol Chem 2001;276:39,872–39,878.
42. Snyder A, McPherson ME, Hunt JF, et al. Acute effects of aerosolized S-nitrosoglutathione in cystic fibrosis. Am J Respir Crit Care Med 2002;165:922–926.
43. Moya MP, Gow AJ, et al. S-Nitrosothiol repletion by an inhaled gas regulates pulmonary function. Proc Natl Acad Sci USA 2001;98:5792–5797.
44. McMahan T, Moon R, Luschinger B, et al. Nitric oxide in the human respiratory cycle. Nat Med 2002;8:711–717.
45. Schmidt H, Hofmann H, Schindler U, et al. No ·NO from NO synthase. Proc Natl Acad Sci USA 1996;93:14,492–14,497.
46. Liu X, Miller M, Joshi M, et al. Accelerated reaction of nitric oxide with O_2 within the hydrophobic interior of biological membranes. Proc Natl Acad Sci USA 1998;95:2175–2179.
47. Gow AJ, Buerk DG, Ischiropoulos H. A novel reaction mechanism for the formation of S-nitrosothiol in vivo. J Biol Chem 1997;272:2841–2845.
48. Vanin AF, Malenkova IV, Serezhenkov VA. Iron catalyzes both decomposition and synthesis of S-nitrosothiols: optical and electron paramagnetic resonance studies. Nitric Oxide 1997;1:191–203.
49. Inoue K, Akaike T, Miyamoto Y, et al. Nitrosothiol formation catalyzed by ceruloplasmin: implication for cytoprotective mechanism in vivo. J Biol Chem 1999;274:27,069–27,075.
50. Stamler JS, Toone E, Lipton S, et al. (S)NO signals: translocation, regulation, and a consensus motif. Neuron 1997;18:691–696.
51. Jensen D, Belka G, DuBois G. S-Nitrosoglutathione is a substrate for rat alcohol dehydrogenase class III isozyme. Biochem J 1998;331:659–668.
52. Fang K, Johns R, Macdonald T, et al. S-Nitrosoglutathione breakdown prevents airway smooth muscle relaxation in the guinea-pig. Am J Physiol Lung Cell Mol Biol 2000;279:L716–L721.
53. Hogg N, Singh RJ, Konorev E, et al. S-Nitrosoglutathione as a substrate for gamma-glutamyl transpeptidase. Biochem J 1997;323:477–481.
54. Lipton SA, Choi YB, Pan ZH, et al. A redox-based mechanism for the neuroprotective and neurodestructive effects of nitric oxide and related nitroso-compounds. Nature 1993;364:626–632.
55. Nikitovic D, Holmgren A. S-Nitrosoglutathione is cleaved by the thioredoxin system with liberation of glutathione and redox regulating nitric oxide. J Biol Chem 1996;271:19,180–19,185.
56. Trujillo M, Alvarez MN, Peluffo G, et al. Xanthine oxidase–mediated decomposition of S-nitrosothiols. J Biol Chem 1998;273:7828–7834.
57. Hou Y, Guo Z, Li J, et al. Seleno compounds and glutathione peroxidase catalyzed decomposition of S-nitrosothiols. Biochem Biophys Res Commun 1996;228:88–93.
58. McAninly J, Williams LH, Askew SC, et al. Metal ion catalysis of nitrosothiol (RSNO) decomposition. J Chem Soc Chem Commun 1993;1758, 1759.
59. Rae TD, Schmidt PJ, Pufahl RA, et al. Undetectable intracellular free copper: the requirement of a copper chaperone for superoxide dismutase. Science 1999;284:805–808.
60. Khuu HM, Desmond R, Huang ST, et al. Characteristics of photopheresis treatments for the management of rejection in heart and lung transplant recipients. J Clin Apheresis 2002;17:27–32.

61. Ramachandran N, Root P, Jiang XM, et al. Mechanism of transfer of NO from extracellular S-nitrosothiols into the cytosol by cell-surface protein disulfide isomerase. Proc Natl Acad Sci USA 2001;98:9539–9544.
62. Carver D, Gaston B, Palmer L. S-Nitrosyl glutathione increases Akt kinase activity in normoxia through γ-glutamyl transpeptidase. Am J Respir Crit Care Med 2003;167:A125.
63. Lander HM, Hajjar DP, Hempstead BL, et al. A molecular redox switch on p21ras: structural basis for the nitric oxide-p21ras interactions. J Biol Chem 1997;272:4323–4326.
64. Ckless K, Reynaert NL, Taatjes D, et al. S-Nitrosoproteins can be detected via chemical derivitization and immunocytochemistry in intact lun epithelial cells. Am J Respir Crit Care Med 2003;167:A52.
65. Peng HB, Libby P, Liao JK. Induction and stabilization of I-Kappa B alpha by nitric oxide mediates inhibition of NF-kappa B. J Biol Chem 1995;270:14,214–14,219.
66. Marshall HE, Stamler JS. Inhibition of NF-kappa B by S-nitrosylation. Biochemistry 2001;40:1688–1693.
67. Nikitovic D, Holmgren A, Spyrou G. Inhibition of AP-1 DNA binding by nitric oxide involving conserved cysteine residues in Jun and Fos. Biochem Biophys Res Commun 1998;242:109–112.
68. Gow AJ, Chen Q, Hess DT, et al. Basal and stimulated protein S-nitrosylation in multiple cell types and tissues. J Biol Chem 2002;277:9637–9640.
68a. Hoque A, Bates JN, Lewis SJ. Redox regulation of *S*-nitrosocysteine-mediated vasodilation in vivo. Eur J Pharmacol 2000;408:195–198.
69. Foresti R, Green CJ, Motterlini R. Nitric oxide and the heme oxygenase/carbon monoxide system: cooperation in the control of vascular function. In: Wang R, ed. Carbon Monoxide and Cardiovascular Functions, CRC Press: Boca Raton, FL, 2002, pp. 111–124.
70. Gaston B. Emissions testing II: carbon monoxide. J Pediatr 1999;135:537, 538.
71. Hunt J, Gaston B. Clinical implications of nitrogen oxide measurements in asthma and other pediatric respiratory diseases. J Pediatr 2000;137:14–21.
72. Hunt JF, Fang K, Malik R, et al. Endogenous airway acidification: implications for asthma pathophysiology. Am J Resp Crit Care Med 2000;161:694–699.
73. Dweik R, Comhair S, Gaston B, et al. NO chemical events in the human airway during the immediate and late antigen-induced asthmatic response. Proc Natl Acad Sci USA 2001;98:2622–2627.
74. Katsuki S, Arnold W, Mittal C, et al. Stimulation of guanylate cyclase by sodium nitroprusside, nitroglycerin and nitric oxide in various tissue preparations and comparison to the effects of sodium azide and hydroxylamine. J Cyclic Nucleotide Res 1977;3:23–35.
75. Koivusalo M, Lapatto R, Uotila L. Purification and characterization of S-formylglutathione hydrolase from human, rat and fish tissues. In: Weiner H, et al., eds. Enzymology and Molecular Biology of Carbonyl Metabolism 5. Plenum Press: New York, 1995, pp. 4427–4433.
??. Gaston B, Fry E, Sears S, et al. Umbilical arterial S-nitrosothiols in stressed newborns: Role in perinatal circulatory transition. Biochem Biophys Res Commun 1998;253:899–901.

5

Nitric Oxide and Adenosine Triphosphate-Sensitive Potassium Channels

Their Different Properties But Analogous Effects on Cellular Protection

Shoji Sanada, Jiyoong Kim, and Masafumi Kitakaze

CONTENTS

SUMMARY

Myocardial protection by ischemic preconditioning is effective in experimental studies, and ischemic preconditioning can also prevent cellular damage in many tissues and organs. This has encouraged investigators in various fields to study ischemic preconditioning intensively. In search of the essential cardioprotective factors, they have begun to clarify the major events during brief periods of ischemia. Ca^{2+} overload, free radicals, catecholamines, cytokines, and hormones have been proposed as candidate causes of ischemic damage but have also been identified as triggers for ischemic preconditioning-derived cardioprotection. Ischemic preconditioning leads to the activation of intracellular messengers, including nitric oxide (NO) and K_{ATP} channels, and other enzymes to produce a cardioprotective effect. These two agents have essentially different properties. However, they appear to use the analogous pathways to reduce the severity of both

From: *Signal Transduction and the Gasotransmitters: NO, CO, and H_2S in Biology and Medicine* Edited by: Rui Wang © Humana Press Inc., Totowa, NJ

myocardial infarction and myocardial dysfunction such as stunning, hibernating myocardium, and remodeling. Because preconditioning ischemia must precede lethal ischemia for these effects to occur, the underlying mechanisms such as NO and the opening of K_{ATP} channel should be effectively applied to strategies for protection after ischemic insults. Here, we summarize previous and current investigations related to the interaction of NO and K_{ATP} channels, especially in cardioprotection including controversial issues, and discuss the future directions of investigation, including some successfully proceeding clinical trials.

Key Words: Preconditioning; protein kinase C; K_{ATP} channel; adenosine; nitric oxide; infarction, cardioprotection; clinical study.

1. INTRODUCTION

It is critically important to consider how cardioprotection is achieved in diseased heart, because both mortality and morbidity resulting from heart diseases have increased worldwide, and there has been an increasing need for safe, effective, and efficient strategies or therapies to prevent cardiac diseases. In the clinical setting, the methods to treat patients with acute coronary syndrome (angina pectoris or acute myocardial infarction [MI]) dramatically progressed by the innovation and application of either percutaneous transluminal coronary angioplasty (PTCA) *(1)* or percutaneous transluminal coronary recanalization (PTCR) *(2)*. As a result of these developments, the mortality of patients with acute MI has decreased; however, the functional recovery of the reperfused heart is unfortunately less than expected, resulting in an increased number of patients with ischemic heart failure *(3)*.

2. IDENTIFICATION OF ISCHEMIC PRECONDITIONING: A POTENT, ENDOGENOUS PROTECTION AGAINST ISCHEMIC STRESSES

Some clinical cardiologists who have encountered patients with acute cardiac syndrome, including either severe unstable angina or acute MI, have occasionally observed the "cardiac warm-up phenomenon" *(4)*—the phenomenon that patients who have experienced at least one episode of prodromal angina paradoxically experience less ischemic damage, despite the increase in total ischemic duration. In 1986, Murry et al. *(5)* first documented this phenomenon experimentally and termed it *ischemic preconditioning*. Thereafter, numerous studies were conducted using various kinds of tissues and species, such as liver *(6)*, kidney *(7)*, brain *(8)*, and endothelial cells *(9)*, and all of the studies achieved the same result without any exception; that is, short period(s) of ischemic or anoxic insult(s) rescued many kinds of tissues from subsequent lethal damage. This strong, ubiquitous, and promising protection of ischemic preconditioning, which is rarely discovered in any experimental or clinical field, has been one of the major topics of study for the prevention of ischemic damage, not only acute reversible and irreversible injuries but also chronic cardiac disorders (hibernation, remodeling) *(10)*.

At present, these findings expand the definition of ischemic preconditioning to include any kind of protection afforded by brief periods of ischemia against ischemic damage caused by subsequent sustained ischemic insult. Finally, ischemic preconditioning has also been evidenced in the clinical setting *(11)*.

3. NITRIC OXIDE AND K_{ATP} CHANNEL AS PUTATIVE COMPONENTS OF ISCHEMIC PRECONDITIONING

3.1. Initial Approaches to Cellular Mechanisms of Ischemic Preconditioning

In 1991, the Downey laboratory first opened the door to pharmacological protections. In an in vivo experimental study, Liu et al. *(12)* found that pharmacological inhibition of adenosine A_1 receptor by 8-SPT prior to sustained ischemia abolished infarct limitation of ischemic preconditioning. They primarily proposed adenosine as a potent candidate of pharmacological preconditioning and the triggering factor of ischemic preconditioning. Their finding, along with the facts that (a) an adenosine A_1 receptor is one of the typical G-protein-coupled receptors that have seven transmembranal domains and that (b) the intracellular Ca^{2+} concentration can be elevated early in ischemia, led them to hypothesize that a potent subsequent cardioprotective mechanism following adenosine A_1 receptor-derived stimulus is related to both G-protein signals and Ca^{2+}. Ytrehus et al. *(13)* later found that the inhibition of protein kinase C (PKC) abolishes infarct size limitation by both pretreatment with adenosine and ischemic preconditioning, suggesting that PKC plays a crucial role in the infarct limitation of preconditioning. Because PKC could also be activated by either ischemia or some extracellular stimulators such as catecholamines, lipopolysaccharide (LPS), or phorbol 12-myristate 13-acetate (PMA) *(14,15)* and confer cardioprotection *(16)*, PKC has been recognized as a major mediator of ischemic preconditioning.

On the other hand, we have reported that transient Ca^{2+} overload prior to sustained ischemia also mimics ischemic preconditioning *(15)*. Furthermore, these cardioprotective effects are cancelled by GF109203X, a selective inhibitor of Ca^{2+}-dependent PKC (classic PKC) *(17)*. Because PKC-α is Ca^{2+} dependent, these data taken together led us to consider that the trigger of early phase preconditioning is also highly associated with transient changes in intracellular Ca^{2+} level in a short time period of ischemia. In search of responsible subtype(s) of PKC, we also found in a dog model *(18)* and a rat model *(19)* that PKC mediates cardioprotection of preconditioning, and some other candidates such as PKC-δ *(19–21)* or PKC-ε *(19,21–23)* have been raised in other models of smaller animals.

We have studied intensively the local metabolism of adenosine in normal and ischemic conditions *(24)*. As shown in **Fig. 1**, two enzymes (*S*-adenosylhomocysteine [SAH] hydrolase and 5'-nucleotidase) produce adenosine from SAH and adenosine monophosphate (AMP), respectively, whereas two other enzymes (adenosine deaminase and adenosine kinase) rapidly change adenosine into inosine and AMP, respectively. We reported for the first time *(17)* that PKC directly activates ecto-5'-nucleotidase, which is located on the surface membrane and maintains a low level of adenosine production under normal conditions but robustly increases adenosine production specifically under either ischemia or other extracellular stresses, in support of intracellular AMP increase, by phosphorylating the serine/threonine residue of ecto-5'-nucleotidase. Because adenosine *per se* is an endogenous bioactive factor *(25)*, which can have various effects on the cardiovascular system, such as negative inotropic effect, negative contractile effect, increasing effect of coronary blood flow, and platelet-deactivating effect, it could work as both a trigger and a mediator of ischemic preconditioning.

When all these data are considered, it is recognized that a circular pathway of adenosine-PKC- ecto-5'-nucleotidase-adenosine works as a booster that can enhance the activation

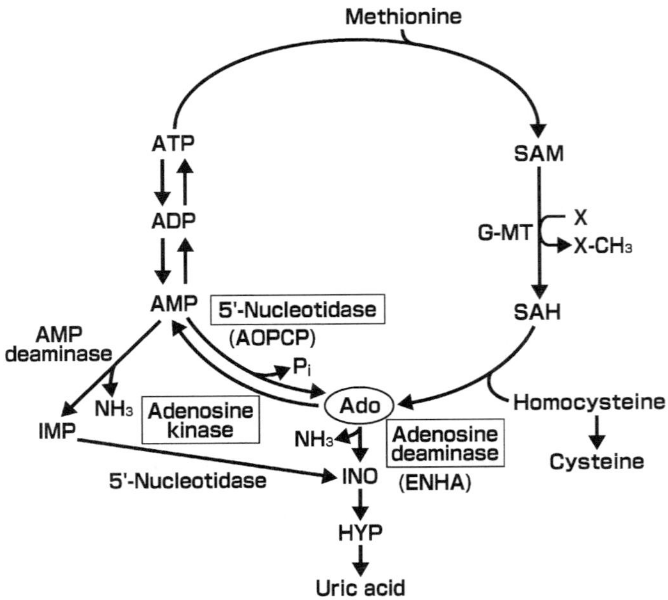

Fig. 1. Endogenous metabolic pathways of adenosine (ADO) and associated substances. Two enzymes (SAH hydrolase and 5'-nucleotidase) produce adenosine from SAH and AMP, respectively, while two other enzymes (adenosine deaminase and adenosine kinase) rapidly change adenosine into inosine and AMP, respectively. The specific pharmacological inhibitors of the respective enzymes are indicated in parentheses.

of all other components to build strong cardioprotection of ischemic preconditioning. Using the dog model in vivo, we found that treatment with AOP-CP, a specific ecto-5'-nucleotidase inhibitor, during either (a) the preconditioning period or (b) early reperfusion can equally and partly cancel the cardioprotection of ischemic preconditioning (**Fig. 2**) *(26)*, which supports the finding that adenosine triggers and mediates cardioprotection.

3.2. Adenosine Triphosphate-Sensitive Potassium Channels (K_{ATP} Channels)

Investigations of adenosine triphosphate (ATP-sensitive potassium channels (K_{ATP} channels) have been ongoing since Noma *(27)* reported its existence in cardiac muscle for the first time in 1983. At first, K_{ATP} channels were identified in cardiovascular physiology as releasers of vascular smooth muscles for either major or small arteries or a negative inotropic agent on the surface of cardiomyocytes *(28)*. These channels consist of the tetramer of identical units containing a Kir family subunit (located inward and forming a channel pore) and an SUR family subunit (located outward contacting with Kir subunits) on the membrane and the activity of K_{ATP} channels are modulated by the presence of Mg and ATP *(29)*. Among some known subtypes of either Kir or SUR subunits, there are two types of Kir subunit and three types of SUR subunit that form at least three important subtypes of K_{ATP} channel: pancreatic type (SUR1/Kir6.2, which enhances insulin secretion from pancreatic β-cells by its closure), cardiac type (SUR2A/Kir6.2, which hyperpolarizes cardiomyocytes by its opening), and vascular type (SUR2B/Kir6.1 or 6.2, which releases vascular smooth muscle cells and dilates vasculature by its opening). SUR subunits have 17 transmembrane domains (TMD1–17) and 2 nucleotide-binding domains (NBD1–2) in cytosolic domains. NBD1 is located between TMD11 and

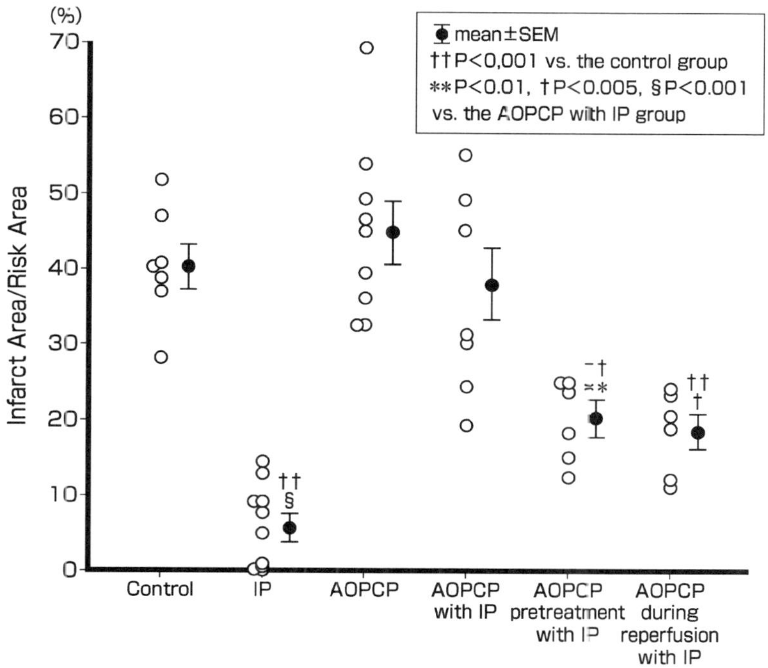

Fig. 2. Infarct size afforded by 90 min of ischemia followed by 6 h of reperfusion in anesthetized dog model. Ischemic preconditioning markedly reduces infarct size, and this protection is blunted by AOPCP (specific blocker of ecto-5'-nucleotidase) during either the preischemic or postischemic period. Open circles represent the infarct sizes in each individual.

TMD12 (cytosolic linker 11–12) and NBD2 is located between TMD17 and the C-terminus. Under the existence of Mg, ATP or adenosine 5'-diphosphate binds to NBDs, which decrease or increase the channel activity, respectively, indicating that a loss of ATP can enhance the probability of the opening of K_{ATP} channels. Furthermore, the C-terminus of either SURs *(30)* or Kirs *(31)* modulates the binding of nucleotides to NBDs. Because K_{ATP} channels work as an inward K^+ rectifier *(29)* when activated, the effects of K_{ATP} channels could be caused by the increase in the threshold of electronic depolarization, which downregulates the excitation of either vascular smooth muscles or cardiomyocytes, followed by vasorelaxation and decreased action potential duration, and finally results in intracellular Ca^{2+} unloading *(28,29)*. This phenomenon resembles the cardioprotective action afforded by ischemic preconditioning, which led investigators to hypothesize that K_{ATP} channels are one of the critical effectors of ischemic or pharmacological preconditioning *(32)*.

In 1992, pharmacological preconditioning afforded by K_{ATP} channel openers was evidenced for the first time *(33)*, and it was hypothesized to be downstream of A_1 adenosine receptor stimulation *(34)*. On the other hand, in 1991, Inoue et al. *(35)* found that not only the surface of the cell but also mitochondrial inner membrane have ATP-sensitive inward K^+ rectifier activity (but its structure was not identified), named "mitochondrial K_{ATP} channels," in contrast to the three types of "sarcolemmal K_{ATP} channels" just described. Thereafter, the contribution of mitochondrial and sarcolemmal K_{ATP} channels to cardioprotection of ischemic preconditioning was extensively studied. It was demonstrated mainly from the results of pharmacological studies that Ca^{2+} unloading and ATP preservation, which had been considered to be the effect of sarcolemmal K_{ATP} channels, were actually afforded by mitochondrial K_{ATP} channels *(36)*. Recent studies have primarily focused on the prevention of either mitochondrial swelling or leakage of apoptotic triggers such as caspases in the protective effects by mitochondrial K_{ATP} channels. Sato et al. *(37)* reported that these channels are directly activated by PKC, but it is also reported that they have a pathway further downstream that confers cardioprotection through the generation of oxygen radicals *(38)*. Because oxygen radicals can activate PKC, K_{ATP} channels also seem to form the "booster circuit" with PKC and oxygen radicals like adenosine, which is a rapid and strong protective mechanism.

However, there are some critical problems regarding mitochondrial K_{ATP} channels as the potent final effector of ischemic preconditioning. First, in more than 10 yr, mitochondrial K_{ATP} channels have not yet cloned out in spite of the dramatically innovated technology in cloning out the mitochondrial proteins. Now it is agreed that mitochondrial K_{ATP} channels might not include Kir 6.1 or Kir 6.2 subunits, which are common with other K_{ATP} channels *(39)*, but the whole structure of K_{ATP} channels on mitochondria remains largely unknown. Second, although some reports in vitro have shown the potency of the putative modulators of mitochondrial K_{ATP} channels *(37)* (diazoxide as an opener of K_{ATP} channels and 5-hydroxydecanoate [5-HD] as an inhibitor, which can be said to be basically the almost only tools to test the function of mitochondrial K_{ATP} channels), it turns out that these modulators have other important roles in modulating cellular metabolism directly; diazoxide is a direct and potent modulator of succinylate dehydrogenase in the tricarboxylic acid (TCA) cycle *(40)*, and 5-HD can be a competitive inhibitor of acyl-CoA, a major substrate of the TCA cycle *(41)*. Third, most recent reports using other models fail to modulate complete cardioprotection of ischemic preconditioning by these two drugs *(42,43)*.

In addition, a study with Kir 6.1 and Kir 6.2 knockout mice indicated in vivo that mitochondrial K_{ATP} channel activity does not contribute to the cardioprotection of ischemic preconditioning *(39)*. The reason for the discrepancy is not clear, but some of the investigators in this field focusing on mitochondrial K_{ATP} channels believe that the contribution of sarcolemmal K_{ATP} channel to cardioprotection is more critical when heartbeat (rapid pulsate contraction of myocardium) is present, whereas it is less when there is no heartbeat (i.e., in vitro). However, the cardioprotective mechanisms afforded by K_{ATP} channels remain to be further elucidated.

3.3. Nitric Oxide

Nitric Oxide (NO) is an endogenous cardiovascular relaxant that was first identified as endothelium-derived relaxing factor. The physiological roles of NO in the cardiovascular system are quite similar to those of adenosine, especially to those of adenosine A_1 receptor activation *(44)*, in terms of negative inotropic effect, vasodilating effect, and inhibitory effect on platelet aggregation or immunoreactive cytokine production, all through cyclic guanosine 5'-monophosphate (cGMP) as a messenger *(45)*. Furthermore, some reports documented that the presence of NO prior to sustained ischemia plays a part in either ischemic or pharmacological preconditioning *(46)*, indicating that NO could work as either a trigger or a final effector of preconditioning, which also resembles the role of adenosine. However, these two components are reported to be independently and differently involved in the triggering mechanisms of preconditioning *(47,48)*. Furthermore, NO-triggered early phase cardioprotection remains controversial, because a negative role of endogenous NO in exerting early phase preconditioning has also been reported *(49)*, which further implies the different involvement in the mechanism of preconditioning compared with adenosine. NO is reported to share the downstream pathways with other stimulants such as acetylcholine (ACh), bradykinin, opioids, and phenylephrine, which involves PKC, K_{ATP} channels, and the generation of reactive oxygen species (ROS) in the triggering phase *(48)*, but activities further downstream are unknown.

Although NO affords direct cardioprotective effects, it also coincidentally forms peroxynitrite ($ONOO^-$) and promptly loses its bioactivity as a stimulator of cGMP when NO is colocalized with a great deal of free radicals. It remains controversial whether $ONOO^-$ is beneficial or harmful to myocardium, because it is reported to violate vascular endothelial cells *(50)*, whereas other reports show that the formation of $ONOO^-$ is required to make NO function as a trigger of preconditioning *(51)*. In addition, some studies using gene-targeting mice show that endothelial nitrix oxide synthase (NOS) is responsible for cardioprotection *(52)*. It could be agreed that overall modulation of NO on the cardiovascular system is beneficial, although results might vary among species and experimental models, and the circumstances, such as the extent of colocalizing free radicals.

3.4. Interaction of NO and K_{ATP} Channels

We discussed earlier that PKC directly activates ecto-5'-nucleotidase *(17)* and increase adenosine production in response to either ischemia or other extracellular stresses by the phosphorylating serine/threonine residue of ecto-5'-nucleotidase. Furthermore, we documented that acute ischemia instantly increases regional production of NO in the heart induced by intracellular acidosis *(53)*. On the other hand, K_{ATP} channels also increase the probability of the opening in response to both ADP *(30)* and PI-(4,5)P2 *(54)*, a signal transmitter in the G-protein-coupled pathway that directly leads to PKC activation. Taken together, these data show that production or activity of all these transmitters (adenosine, K_{ATP} channels, and NO) rapidly and concurrently increases in response to acute metabolic stresses such as ischemia. In fact, the effects of these cardioprotectants on the cardiovascular system are quite analogous—negative inotropic effect, negative contractile effect, increasing effect of coronary blood flow, cytokine inhibitory effect, and platelet deactivating effect—in affording protection against cytotoxic stresses. However, the redundancy of these factors in practical cardioprotection is controversial; Mercus et al. *(54)* reports that the blockade of each element results in impairment of metabolic vasodilation in swine and human *(55)*, whereas we have documented that pre- and postischemic pharmacological inhibition of each element does not exacerbate infarct size afforded by ischemia-reperfusion injuries *(17,42,56)*. Therefore, we can say that these three components (adenosine, K_{ATP} channels, and NO) act at once in parallel to prevent a further increase in tissue damages by external stresses.

We mentioned earlier that either adenosine or K_{ATP} channels are linked with PKC and are separately involved in the "booster circuit" to respond promptly to stresses; the former is the PKC–5' nucleotidase–adenosine–PKC pathway, and the latter is the PKC–K_{ATP} channels–oxygen radical–PKC pathway. In the case of cardioprotection of ischemic preconditioning, because various stresses can activate PKC at once and either 5'-nucleotidase or K_{ATP} channel is directly activated by acute stresses, the existing booster circuit systems enable ubiquitous types of tissues to respond promptly to extracellular stresses and confer protective signals. On the other hand, although the contribution of endogenous

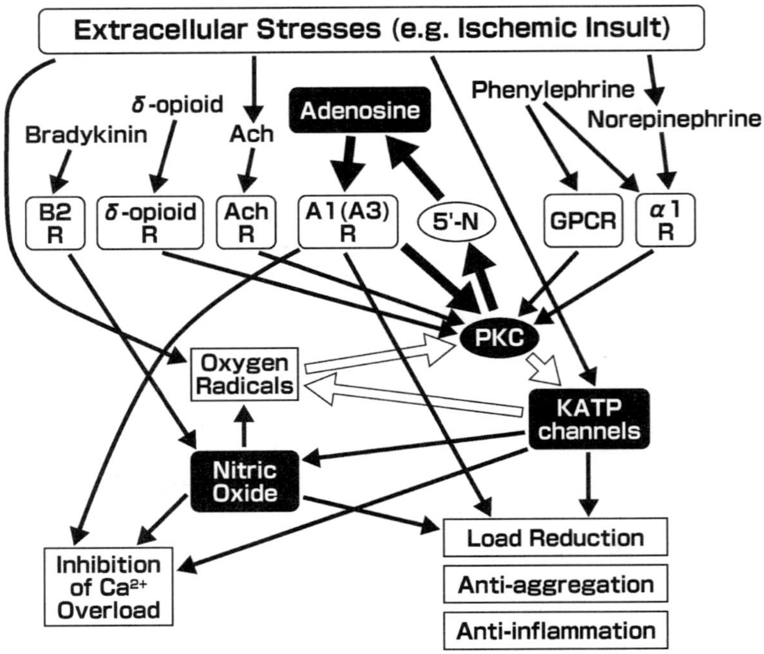

Fig. 3. Schematic illustration of mechanism of preconditioning and relative contribution of NO, ATP-sensitive potassium channel (K_{ATP} channel), and adenosine. Each trigger causing ischemic and pharmacological preconditioning is indicated at the top, and the major effects afforded by the early phase of preconditioning are indicated at the bottom. Bold arrows (solid or striped) indicate the evidenced or putative "booster circuits," respectively, which enhance themselves and each other. Ach, acetylcholine; R, receptors; GPCR, G-protein-coupled receptors; 5'-N, ecto-5'-nucleotidase. Very bold arrows indicate the existing booster circuit systems detailed in the text.

NO in this internal cardioprotective system is controversial *(47–49)*, NO can be induced by many extracellular factors (e.g., ACh, bradykinin, opioids, and phenylephrine) *(48)* and is reported to stimulate PKC *(49)* in harmony with them. Therefore, we can say that NO is also closely related to the triggering effect of cardioprotective pathway formed by either adenosine or K_{ATP} channels (**Fig. 3**).

After a brief triggering stimulus is applied, the "early window" of protection begins within minutes and lasts for several hours. However, the signal transduction sequence activated by these triggers is not fully understood. ROS have important roles in the signaling pathway involved in preconditioning triggering *(57–61)*. Some reports *(62,63)* indicate that exogenous H_2O_2 induces preconditioning in cardiomyocytes and that mitochondrial ROS are involved in the triggering by hypoxia or by administration of ACh. However, an interesting controversy has developed regarding the relationship between the mitochondrial K_{ATP} channel activation and the ROS signal during triggering. Pain et al. *(38)* and Forbes et al. *(64)* showed that activation of the mitochondrial K_{ATP} channel elicits an increase in ROS generation that is required for preconditioning protection. Furthermore, the K_{ATP} channel inhibitors abolish ROS generation during ACh triggering, indicating that ROS signaling occurs as a consequence of channel activation *(63)*. In contrast, inhibition of K_{ATP} channel abolished protection without effects for the ROS signaling *(65)*. These findings indicate that oxidant signals may participate at multiple steps during triggering.

NO has also been suggested to participate in the intracellular signaling during preconditioning triggering *(66–69)*. However, the requirement for NO in the triggering of preconditioning is controversial—some groups find evidence for the involvement of NO during the early window of protection *(70,71)* but others do not *(72)*. In cardiomyocytes, inhibition of NOS did not abolish the ROS signal during hypoxic triggering *(62)*, which suggests that NO is not required for ROS generation during hypoxic triggering. On the other hand, inhibition of NOS abolished the ROS signal during ACh triggering and they abrogated its protective effects *(73)*. Therefore, the role of NO in early preconditioning is not fully understood, and the relationships among NO, ROS, and the K_{ATP} channel are not fully clear. Lebuffe et al. *(74)* have developed a model, which is concomitant with ours (*see* **Fig. 3**), to elucidate these relationships during the triggering phase of early preconditioning *(36,75–79)*. The model proposes that ROS are required at two separate steps in the triggering process and that NO is required for events occurring downstream of the K_{ATP} channel. According to this model, preconditioning triggering requires both ROS and NO. ROS are required both upstream and downstream of mitochondrial K_{ATP} channel activation, whereas NO generation occurs downstream of the K_{ATP} channel.

As reported by Nakano et al. *(72)*, PKC activates nuclear factor-κB after several hours and finally leads to the increased expression of inducible NOS after 4–8 h, which plays a critical role in cardioprotection in distant phase, called "late phase preconditioning." Therefore, NO could act as the analogous cardioprotective factor in a different time window in response to acute stresses, compared with adenosine and K_{ATP} channels.

4. CARDIOPROTECTION: CURRENT CLINICAL TRIALS

Accordingly, recent clinical investigations also confirm that the presence of prodromal angina (transient anginal attack[s] prior to MI) limits infarct size, attenuates lethal side effects (i.e., ventricular fibrillation and acute heart failure), and improves the prognosis of patients with acute MI *(11,80)*. Although we cannot apply "preconditioning ischemia" directly to patients with heart disease because it should be done prior to acute cardiac damage, many trials have been conducted with the aim of translating the strong cardioprotection of ischemic preconditioning into effective therapeutic strategies against various kinds of heart diseases, such as by using some potent final effectors to treat acute diseases or by inducing preconditioning-like effects to prevent (or attenuate) future disease and chronic damage. Here, we discuss some of these trials and provide some encouraging results derived from them.

Currently, at least two strategies have been attempted. One is therapy to eliminate injuries, and the other is primary prevention, to build up the cardioprotective factors continuously. Adjunctive therapy following PTCA or PTCR is one of the typical examples of the former. Among some candidates raised by experimental studies, such as adenosine, the agents NO and the K_{ATP} channel opener nicorandil were also evidenced to attenuate reperfusion injuries in the clinical setting.

4.1. Adenosine and Its Mother Compounds

In the AMISTAD trial *(81)*, patients with acute MI in the anterior wall who underwent continuous intravenous infusion of adenosine together with PTCR showed smaller infarct size and better functional recovery than those not receiving adenosine infusion. Our group is now operating a clinical trial (Cooperative Osaka Adenosine Trial [COAT] study) to demonstrate that patients undergoing an intracoronary administration of ATP for a few minutes just after successful recanalization by PTCA experience smaller myocardial infarct size and better functional recovery.

4.2. NO and K_{ATP} Channel Openers

A marked reduction of regional myocardial perfusion with severe reduced contraction (hypokinesis) of the heart despite successful interventional recanalization of a major coronary artery, named the "no-reflow" phenomenon, occurs frequently after reperfusion that causes prolonged cardiac dysfunction. Some experimental *(82)* and clinical trials, by either intravenous *(83)* or intracoronary *(84)* administration, show that the severity of this phenomenon after successful PTCA is eliminated and that there is both limited infarct size and better recovery from severe hypokinesis, by continuous infusion of a K_{ATP} channel opener, nicorandil, just after successful PTCA. The pathophysiological mechanisms of the no-reflow phenomenon and the detailed mechanism of how nicorandil attenuates this phenomenon remain controversial, although some reports propose that the no-reflow phenomenon occurs because of vasospasm *(83)* or microembolization *(85)*. Furthermore, the effect of K_{ATP} channel (mitochondrial or sarcolemmal) on this phenomenon by nicorandil also remains controversial, because many reports confirm that nicorandil has a higher potency (10–100 times) than mitochondrial K_{ATP} channel than the sarcolemmal K_{ATP} channel in vitro *(86)*, whereas we have observed in the in vivo dog model that nicorandil has much less selectivity to the mitochondrial K_{ATP} channel *(42)*, and that a specific sarcolemmal K_{ATP} channel blocker, HMR1098, completely blunted the effect of nicorandil against the no-reflow phenomenon (unpublished data). In the case of severe ischemia, the IONA study *(87)* revealed that the long-term oral administration of nicorandil in patients with moderate to severe stable angina reduced the risk of cardiac incident and improved prognosis.

Several clinical trials targeting the prevention or reduction of reperfusion injury are now in progress. Nicorandil, a hybrid of K_{ATP} channel opener and nitrates, and atrial natriuretic peptide (ANP) are the promising candidates for an adjunctive therapy for acute MI. In animal models, several studies—including ours—demonstrated that nicorandil and ANP reduce myocardial infarct size and improve postischemic left ventricular function. In the clinical setting, however, the beneficial effects of nicorandil were only tested in single-center studies and the number of patients was relatively small. Thus, larger multicenter studies are needed to assess whether these effects can translate into clinical benefits. Japan-working groups of acute MI for the reduction of necrotic damage (J-WIND) is a prospective, randomized, multicenter study to evaluate the beneficial effects of nicorandil and calperitide (recombinant human ANP) as adjunctive therapy for acute MI. In the J-WIND trial (headquarters are at the National Cardiovascular Center; 54 institutes in Japan are participating), in addition to examining the effects of these treatments on clinical outcomes including infarct size and left ventricular regional function, the association between single nucleotide polymorphisms of genes that may influence either the function or metabolism of the drugs and the responsiveness of nicorandil or calperitide therapy is being analyzed. Further, by comparing the prevalence of single nucleotide polymorphisms of genes that may influence the occurrence of acute MI between healthy subjects and patients with acute MI enrolled in J-WIND, one can genetically predict patient populations who have a high risk of acute MI. Because we have documented that calperitide confers cardioprotection through the increase in NO production in experimental conditions in canine heart, we can expect, for the first time, very important clinical evidence of whether K_{ATP} channel opening or NO production is beneficial as an adjunctive therapy in acute MI.

Although some successful interventions derived from the cardioprotection by preconditioning have been proposed, they are still less effective than ischemic preconditioning itself. To establish safe and strong strategies, alone or in combination, that can be applicable in the clinical setting, further investigations are necessary.

REFERENCES

1. O'Neill WW. Primary percutaneous coronary angioplasty: a protagonist's view. Am J Cardiol 1988;62:15K–20K.
2. Ganz W, Geft I, Maddahi J, et al. Nonsurgical reperfusion in evolving myocardial infarction. J Am Coll Cardiol 1983;1:1247–1253.
3. Kloner RA, Jennings RB. Consequences of brief ischemia: stunning, preconditioning, and their clinical implications: part 1. Circulation 2001;104:2981–2989.
4. Tomai F. Warm up phenomenon and preconditioning in clinical practice. Heart 2002;87:99, 100.
5. Murry CE, Jennings RB, Reimer KA. Preconditioning with ischemia a delay of lethal cell injury in ischemic myocardium. Circulation 1986;74:1124–1136.
6. Jaeschke H. Molecular mechanisms of hepatic ischemia-reperfusion injury and preconditioning. Am J Physiol Gastrointest Liver Physiol 2003;284:G15–G26.
7. Bonventre JV. Kidney ischemic preconditioning. Curr Opin Nephrol Hypertens 2002;11:43–48.
8. Kirino T. Ischemic tolerance. J Cereb Blood Flow Metab 2002;22:1283–1296.
9. Pohlman TH, Harlan JM. Adaptive responses of the endothelium to stress. J Surg Res 2000;89:85–119.
10. Opie LH. The multifarious spectrum of ischemic left ventricular dysfunction: relevance of new ischemic syndromes. J Mol Cell Cardiol 1996;28:2403–2414.
11. Ishihara M, Sato H, Tateishi H, et al. Implications of prodromal angina pectoris in anterior wall acute myocardial infarction: acute angiographic findings and long-term prognosis. J Am Coll Cardiol 1997;30:970–975.
12. Liu GS, Thornton J, Van Winkle DM, et al. Protection against infarction afforded by preconditioning is mediated by A1 adenosine receptors in rabbit heart. Circulation 1991;84:350–356.
13. Ytrehus K, Liu Y, Downey JM. Preconditioning protects ischemic rabbit heart by protein kinase C activation. Am J Physiol 1994;266:H1145–H1152.
14. Kitakaze M, Hori M, Morioka T, et al. Alpha 1–adrenoceptor activation mediates the infarct size-limiting effect of ischemic preconditioning through augmentation of 5'-nucleotidase activity. J Clin Invest 1994;93:2197–2205.
15. Node K, Kitakaze M, Sato H, et al. Role of intracellular Ca^{2+} in activation of protein kinase C during ischemic preconditioning. Circulation 1997;96:1257–1265.
16. Kitakaze M, Hori M, Morioka T, et al. Alpha 1–adrenoceptor activation increases ecto-5'-nucleotidase activity and adenosine release in rat cardiomyocytes by activating protein kinase C. Circulation 1995;91:2226–2234.
17. Kitakaze M, Node K, Minamino T, et al. Role of activation of protein kinase C in the infarct size-limiting effect of ischemic preconditioning through activation of ecto-5'-nucleotidase. Circulation 1996;93:781–791.
18. Kitakaze M, Funaya H, Minamino T, et al. Role of protein kinase C-alpha in activation of ecto-5'-nucleotidase in the preconditioned canine myocardium. Biochem Biophys Res Commun 1997;239:171–175.
19. Yoshida K, Kawamura S, Mizukami Y, et al. Implication of protein kinase C-alpha, delta, and epsilon isoforms in ischemic preconditioning in perfused rat hearts. J Biochem (Tokyo) 1997;122:506–511.
20. Inagaki K, Kihara Y, Hayashida W, et al. Anti-ischemic effect of a novel cardioprotective agent, JTV519:is mediated through specific activation of delta-isoform of protein kinase C in rat ventricular myocardium. Circulation 2000;101:797–804.
21. Liu H, McPherson BC, Yao Z. Preconditioning attenuates apoptosis and necrosis: role of protein kinase C epsilon and -delta isoforms. Am J Physiol Heart Circ Physiol 2001;281:H404–H410.
22. Ping P, Song C, Zhang J, et al. Formation of protein kinase C(epsilon)-Lck signaling modules confers cardioprotection. J Clin Invest 2002;109:499–507.
23. Gray MO, Karliner JS, Mochly-Rosen D. A selective epsilon-protein kinase C antagonist inhibits protection of cardiac myocytes from hypoxia-induced cell death. J Biol Chem 1997;272:30,945–30,951.
24. Hori M, Kitakaze M. Adenosine, the heart, and coronary circulation. Hypertension 1991;18:565–574.

25. Kitakaze M, Hori M. It is time to ask what adenosine can do for cardioprotection. Heart Vessels 1998;13:211–228.
26. Kitakaze M, Hori M, Morioka T, et al. Infarct size-limiting effect of ischemic preconditioning is blunted by inhibition of 5'-nucleotidase activity and attenuation of adenosine release. Circulation 1994;89: 1237–1246.
27. Noma A. ATP-regulated K+ channels in cardiac muscle. Nature 1983;305:147–148.
28. Yokoshiki H, Sunagawa M, Seki T, et al. ATP-sensitive K+ channels in pancreatic, cardiac, and vascular smooth muscle cells. Am J Physiol 1998;274:C25–C37.
29. Snyders DJ. Structure and function of cardiac potassium channels. Cardiovasc Res 1999;42:377–390.
30. Matsuoka T, Matsushita K, Katayama Y, et al. C-terminal tails of sulfonylurea receptors control ADP-induced activation and diazoxide modulation of ATP-sensitive K(+) channels. Circ Res 2000;87:873–880.
31. Drain P, Li L, Wang J. KATP channel inhibition by ATP requires distinct functional domains of the cytoplasmic C terminus of the pore-forming subunit. Proc Natl Acad Sci USA 1998;95:13,953–13,958.
32. Gross GJ, Fryer RM. Sarcolemmal versus mitochondrial ATP-sensitive K+ channels and myocardial preconditioning. Circ Res 1999;84:973–979.
33. Gross GJ, Auchampach JA. Blockade of ATP-sensitive potassium channels prevents myocardial preconditioning in dogs. Circ Res 1992;70:223–233.
34. Grover GJ, Sleph PG, Dzwonczyk S. Role of myocardial ATP-sensitive potassium channels in mediating preconditioning in the dog heart and their possible interaction with adenosine A1–receptors. Circulation 1992;86:1310–1316.
35. Inoue I, Nagase H, Kishi K, et al. ATP-sensitive K+ channel in the mitochondrial inner membrane. Nature 1991;352:244–247.
36. Liu Y, Sato T, Seharaseyon J, et al. Mitochondrial ATP-dependent potassium channels: viable candidate effectors of ischemic preconditioning. Ann NY Acad Sci 1999;874:27–37.
37. Sato T, O'Rourke B, Marban E. Modulation of mitochondrial ATP-dependent K+ channels by protein kinase C. Circ Res 1998;83:110–114.
38. Pain T, Yang XM, Critz SD, et al. Opening of mitochondrial K(ATP) channels triggers the preconditioned state by generating free radicals. Circ Res 2000;87:460–466.
39. Suzuki M, Sasaki N, Miki T, et al. Role of sarcolemmal K(ATP) channels in cardioprotection against ischemia/reperfusion injury in mice. J Clin Invest 2002;109:509–516.
40. Dzeja PP, Bast P, Ozcan C, et al. Targeting nucleotide-requiring enzymes: implications for diazoxide-induced cardioprotection. Am J Physiol Heart Circ Physiol 2003;284:H1048–H1056.
41. Hanley PJ, Gopalan KV, Lareau RA, et al. Beta-oxidation of 5-hydroxydecanoate, a putative blocker of mitochondrial ATP-sensitive potassium channels. J Physiol 2003;547:387–393.
42. Sanada S, Kitakaze M, Asanuma H, et al. Role of mitochondrial and sarcolemmal K(ATP) channels in ischemic preconditioning of the canine heart. Am J Physiol Heart Circ Physiol 2001;280:H256–H263.
43. Schwartz LM, Welch TS, Crago MS. Cardioprotection by multiple preconditioning cycles does not require mitochondrial K(ATP) channels in pigs. Am J Physiol Heart Circ Physiol 2002;283: H1538–H1544.
44. Minamino T, Kitakaze M, Matsumura Y, et al. Impact of coronary risk factors on contribution of nitric oxide and adenosine to metabolic coronary vasodilation in humans. J Am Coll Cardiol 1998;31: 1274–1279.
45. Dawn B, Bolli R. Role of nitric oxide in myocardial preconditioning. Ann NY Acad Sci 2002;962: 18–41.
46. Kojda G, Kottenberg K. Regulation of basal myocardial function by NO. Cardiovasc Res 1999;41: 514–523.
47. Yoshida K, Mizukami Y, Kitakaze M. Nitric oxide mediates protein kinase C isoform translocation in rat heart during postischemic reperfusion. Biochim Biophys Acta 1999;1453:230–238.
48. Cohen MV, Yang XM, Liu GS, et al. Acetylcholine, bradykinin, opioids, and phenylephrine, but not adenosine, trigger preconditioning by generating free radicals and opening mitochondrial K(ATP) channels. Circ Res 2001;89:273–278.
49. Csonka C, Csont T, Onody A, et al. Preconditioning decreases ischemia/reperfusion-induced peroxynitrite formation. Biochem Biophys Res Commun 2001;285:1217–1219.
50. Altug S, Demiryurek AT, Kane KA, et al. Evidence for the involvement of peroxynitrite in ischaemic preconditioning in rat isolated hearts. Br J Pharmacol 2000;130:125–131.
51. Bell RM, Yellon DM. The contribution of endothelial nitric oxide synthase to early ischaemic preconditioning: the lowering of the preconditioning threshold. An investigation in eNOS knockout mice. Cardiovasc Res 2001;52:274–280.

52. Kitakaze M, Node K, Takashima S, Asanuma H, et al. Role of cellular acidosis in production of nitric oxide in canine ischemic myocardium. J Mol Cell Cardiol 2001;33:1727–1737.
53. Xie LH, Horie M, Takano M. Phospholipase C–linked receptors regulate the ATP-sensitive potassium channel by means of phosphatidylinositol 4,5-bisphosphate metabolism. Proc Natl Acad Sci USA 1999;96:15,292–15,297.
54. Merkus D, Haitsma DB, Fung TY, et al. Coronary blood flow regulation in exercising swine involves parallel rather than redundant vasodilator pathways. Am J Physiol Heart Circ Physiol 2003;285:H424–H433.
55. Asanuma H, Kitakaze M, Funaya H, et al. Nifedipine limits infarct size via NO-dependent mechanisms in dogs. Basic Res Cardiol 2001;96:497–505.
56. Baxter GF. Role of adenosine in delayed preconditioning of myocardium. Cardiovasc Res 2002;55: 483–494.
57. Ambrosio G, Tritto I, Chiariello M. The role of oxygen free radicals in preconditioning. J Mol Cell Cardiol 1995;27:1035–1039.
58. Baines CP, Goto M, Downey JM. Oxygen radicals released during ischemic preconditioning contribute to cardioprotection in the rabbit myocardium. J Mol Cell Cardiol 197;29:207–216.
59. Chen W, Gabel S, Steenbergen C, et al. A redox-based mechanism for cardioprotection induced by ischemic preconditioning in perfused rat heart. Circ Res 1995;77:424–429.
60. McPherson BC, Yao ZH. Morphine mimics preconditioning via free radical signals and mitochondrial KATP channels in myocytes. Circulation 2001;103:290–295.
61. Tritto I, D'Andrea D, Eramo N, et al. Oxygen radicals can induce preconditioning in rabbit hearts. Circ Res 1997;80:743–748.
62. Vanden Hoek TL, Becker LB, Shao Z, et al. Reactive oxygen species released from mitochondria during brief hypoxia induce preconditioning in cardiomyocytes. J Biol Chem 1998;273:18,092–18,098.
63. Yao Z, Tong J, Tan X, et al. Role of reactive oxygen species in acetylcholine-induced preconditioning in cardiomyocytes. Am J Physiol Heart Circ Physiol 1999;277:H2504–H2509.
64. Forbes RA, Steenbergen C, Murphy E. Diazoxide-induced cardioprotection requires signaling through a redox-sensitive mechanism. Circ Res 2001;88:802–809.
65. Vanden Hoek TL, Becker LB, Shao ZH, et al. Preconditioning in cardiomyocytes protects by attenuating oxidant stress at reperfusion. Circ Res 2000;86:541–548.
66. Lochner A, Marais E, Genade S, et al. Nitric oxide: a trigger for classic preconditioning? Am J Physiol Heart Circ Physiol 2000;279:H2752–H2765.
67. Ockaili R, Emani VR, Okubo S, et al. Opening of mitochondrial KATP channel induces early and delayed cardioprotective effect: role of nitric oxide. Am J Physiol Heart Circ Physiol 1999;277: H2425–H2434.
68. Rakhit RD, Edwards RJ, Mockridge JW, et al. Nitric oxide-induced cardioprotection in cultured rat ventricular myocytes. Am J Physiol Heart Circ Physiol 2000;278:H1211–H1217.
69. Yao Z, Gross GJ. Role of nitric oxide, muscarinic receptors, and the ATP-sensitive K channel in mediating the effects of acetylcholine to mimic preconditioning in dogs. Circ Res 1993;73:1193–1201.
70. Lu HR, Remeysen P, De Clerck F. Does the antiarrhythmic effect of ischemic preconditioning in rats involve the L-arginine nitric oxide pathway? J Cardiovasc Pharmacol 1995;25:524–530.
71. Ping PP, Takano H, Zhang J, et al. Isoform selective activation of protein kinase C by nitric oxide in the heart of conscious rabbits: a signaling mechanism for both nitric oxide-induced and ischemia-induced preconditioning. Circ Res, 1999;84:587–604.
72. Nakano A, Liu GS, Heusch G, et al. Exogenous nitric oxide can trigger a preconditioned state through a free radical mechanism, but endogenous nitric oxide is not a trigger of classical ischemic preconditioning. J Mol Cell Cardiol 2000;32:1159–1167.
73. Liu H, McPherson BC, Zhu X, et al. Role of nitric oxide and protein kinase C in ACh-induced cardioprotection. Am J Physiol Heart Circ Physiol 2001;281:H191–H197.
74. Lebuffe G, Schumacker PT, Shao ZH, et al. ROS and NO trigger early preconditioning: relationship to mitochondrial K_{ATP} channel. Am J Physiol Heart Circ Physiol 2003;284:H299–H308.
75. Das DK, Engelman RM, Maulik N. Oxygen free radical signaling in ischemic preconditioning. Ann NY Acad Sci 1999;874:49–65.
76. Gross GJ, Auchampach JA. Role of ATP dependent potassium channels in myocardial ischaemia. Cardiovasc Res 1992;26:1011–1016.
77. Marber MS. Ischemic preconditioning in isolated cells. Circ Res 2000;86:926–931.
78. O'Rourke B. Myocardial KATP channels in preconditioning. Circ Res 2000;87:845–855.
79. Okubo S, Xi L, Bernardo NL, et al. Myocardial preconditioning: basic concepts and potential mechanisms. Mol Cell Biochem 1999;196:3–12.

80. Noda T, Minatoguchi S, Fujii K, et al. Evidence for the delayed effect in human ischemic preconditioning: prospective multicenter study for preconditioning in acute myocardial infarction. J Am Coll Cardiol 1999;34:1966–1974.
81. Mahaffey KW, Puma JA, Barbagelata NA, et al. Adenosine as an adjunct to thrombolytic therapy for acute myocardial infarction: results of a multicenter, randomized, placebo-controlled trial: the Acute Myocardial Infarction STudy of ADenosine (AMISTAD) trial. J Am Coll Cardiol 1999;34:1711–1720.
82. Genda S, Miura T, Miki T, et al. K(ATP) channel opening is an endogenous mechanism of protection against the no-reflow phenomenon but its function is compromised by hypercholesterolemia. J Am Coll Cardiol 2002;40:1339–1346.
83. Ito H, Taniyama Y, Iwakura K, et al. Intravenous nicorandil can preserve microvascular integrity and myocardial viability in patients with reperfused anterior wall myocardial infarction. J Am Coll Cardiol 1999;33:654–660.
84. Sakata Y, Kodama K, Komamura K, et al. Salutary effect of adjunctive intracoronary nicorandil administration on restoration of myocardial blood flow and functional improvement in patients with acute myocardial infarction. Am Heart J 1997;133:616–621.
85. Kotani J, Nanto S, Mintz GS, et al. Plaque gruel of atheromatous coronary lesion may contribute to the no-reflow phenomenon in patients with acute coronary syndrome. Circulation 2002;106:1672–1677.
86. Sato T, Sasaki N, O'Rourke B, et al. Nicorandil, a potent cardioprotective agent, acts by opening mitochondrial ATP-dependent potassium channels. J Am Coll Cardiol 2000;35:514–518.
87. IONA study group. Effect of nicorandil on coronary events in patients with stable angina: the Impact of Nicorandil in Angina (IONA) randomized trial. Lancet 2002;359:1269–1275.

6 Interaction of Nitric Oxide and Related Radical Species With K_{Ca} Channels

Yanping Liu and David D. Gutterman

CONTENTS

SUMMARY

Nitric oxide (NO$^•$) is a biological-signaling molecule, which plays a fundamental role as a regulator of many physiological events. Substantial evidence points to the involvement of K_{Ca} channels in NO$^•$-induced responses. K_{Ca} channels are distributed in different types of smooth muscle. Opening of the channels causes cell membrane hyperpolarization and smooth muscle relaxation. In physiological conditions, the activation of K_{Ca} channels by NO$^•$ involves two major pathways: the cyclic guanosine 5'-monophosphate–protein kinase G pathway and *S*-nitrosation of thiols in the channel or in a closely associated regulatory protein. In some pathological conditions, where the production of superoxide ($O_2^{•-}$) is enhanced and NO$^•$ bioavailability is reduced, activation of K_{Ca} channels is mediated by endothelium-derived hyperpolarizing factor to compensate for loss of NO$^•$. On the other hand, NO$^•$ reacts with $O_2^{•-}$ to form peroxynitrite (ONOO$^-$). ONOO$^-$ inhibits large-conductance K_{Ca} channels and impairs smooth muscle relaxation. Therefore, in disease states associated with an increase in $O_2^{•-}$ generation, the effects of $O_2^{•-}$, loss of NO$^•$, and actions of ONOO$^-$ should all be considered when examining smooth muscle functional responses.

Key Words: Nitric oxide; K_{Ca} channels; smooth muscle; cyclic guanosine 5'-monophosphate; oxidant; oxidative stress.

From: *Signal Transduction and the Gasotransmitters: NO, CO, and H_2S in Biology and Medicine* Edited by: Rui Wang © Humana Press Inc , Totowa, NJ

1. INTRODUCTION

Nitric oxide ($NO^•$) and related radical species have emerged as ubiquitous cellular messengers. $NO^•$ is derived enzymatically from L-arginine by nitric oxide synthase (NOS) *(1–3)*, which exists in three isoforms: endothelial NOS (eNOS), neuronal NOS, and inducible NOS *(4–6)*. The formation of $NO^•$ by these enzymes requires five cofactors, including flavin adenine dinucleotide, flavin adenine mononucleotide, heme, calmodulin, and tetrahydrobiopterin, and three cosubstrates including L-arginine, NADPH, and O_2 *(7)*.

$NO^•$ synthesized by different isoforms of NOS regulates many physiological events including vasodilation *(8–12)*, inflammation *(13–15)*, vascular proliferation *(8,16)*, and synaptic plasticity *(4,17,18)*. The role of $NO^•$ in relaxation of smooth muscle cells (SMCs) involves several mechanisms, including opening Ca^{2+}-activated potassium channels (K_{Ca}), which have been identified in all SMCs *(19–23)*. K_{Ca} channels are highly expressed in vascular smooth muscle cells (VSMCs). Activation of vascular K_{Ca} channels has been identified as an important buffering mechanism to counteract vessel depolarization and constriction in response to physiological and pathophysiological stimuli *(24–27)* and to maintain sufficient blood flow to the tissue. A mechanistic understanding of how $NO^•$ interacts with K_{Ca} channels will have important implications for potential therapies involving restoration of NO in those disease states.

The effects of $NO^•$ on K_{Ca} channel function in physiological conditions have been studied extensively *(10,28,29)*. However, in pathological conditions, especially in situations when oxidative stress is enhanced and $NO^•$ bioavailability is reduced, $NO^•$-mediated smooth muscle relaxation is altered because of the interaction of $NO^•$ with other reactive oxygen species (ROS). These ROS may not only quench $NO^•$ but also interfere with $NO^•$-mediated signaling through effects on K_{Ca} channels. This chapter provides an overview of the effect of $NO^•$ on K_{Ca} channel function in physiological and pathological conditions.

2. EFFECTS OF NO ON K_{CA} CHANNELS IN DIFFERENT TYPES OF TISSUES

2.1. Vascular Smooth Muscle

In the vasculature, eNOS is localized to caveolae *(6,30,31)*, small invaginations in the plasma membrane. In the resting state, the interaction of eNOS with caveolin greatly attenuates eNOS activity. When endothelial cells are exposed to shear stress or receptor-dependent agonists, such as acetylcholine (ACh), adenosine 5'-diphosphate, bradykinin thrombin, and serotonin, intracellular calcium increases. Ca^{2+}-bound calmodulin competitively displaces caveolin from eNOS, thereby activating eNOS and leading to synthesis of $NO^•$.

The role of K_{Ca} channels in $NO^•$-induced vasodilation has been extensively studied. It was shown that in physiological conditions, specific inhibitors of K_{Ca} channels significantly reduce dilation to $NO^•$ and $NO^•$ donors, such as sodium nitroprusside, spermine NONOate, 3-morpholinosydnonimine (SIN-1), or ACh in various vascular beds *(32–35)* as well as in nonvascular smooth muscle *(35–38)*.

K_{Ca} channels also participate in humoral responses to $NO^•$. For example, in human coronary myocytes *(39)*, 17β-estradiol increases $NO^•$ production, which in turn enhances large-conductance K_{Ca} (BK_{Ca}) channel activity. The enhancement is reduced by iberiotoxin (IbTx), a specific BK_{Ca} channel blocker, or N^G-monomethyl-L-arginine, an NOS inhibitor *(39,40)*. In porcine coronary arteries, increased cyclic guanosine 5'-monophosphate (cGMP) production was observed in response to testosterone *(41–43)*. Testosterone induced both vasorelaxation and an increase in the open-state probability of BK_{Ca} channels. These responses can be eliminated with tetraethylammonium (TEA) *(41)*. Similarly, dilation to adrenomedullin in rat renal arteries *(44)* can be inhibited by TEA or N^{ω}-nitro-L-arginine-methyl ester (L-NAME).

Direct evidence that $NO^{\bullet}$ interacts with K_{Ca} channels comes from studies in *Xenopus* oocytes, in which human BK_{Ca} channel α- and studie-subunits genes are expressed. Spermine NONOate increased K_{Ca} currents in this model, in an IbTx-sensitive fashion *(45)*.

2.2. GastroIntestinal Smooth Muscle

It is well established that stimulation of nonadrenergic, noncholinergic (NANC) inhibitory neurons of gastrointestinal (GI) tract releases $NO^{\bullet}$, which plays an essential role in regulation of contractility of visceral smooth muscle. Similar to vascular smooth muscle, relaxation of visceral smooth muscle by $NO^{\bullet}$ involves K_{Ca} channels via cGMP-dependent and -independent pathways. There are two components to the $NO^{\bullet}$-induced relaxation and hyperpolarization of visceral smooth muscle. One is sensitive to apamin, the blocker of small-conductance K_{Ca} (SK_{Ca}), and the other is sensitive to NOS inhibitors. Apamin reduces the portion sensitive to inhibition of NOS in the human colon and dog pylorus *(46,47)*. Similarly, NOS inhibitors reduce the apamin-sensitive component in the dog ileocolonic sphincter *(48)*. In single myocytes of the opossum esophagus *(49)* and guinea pig proximal colon *(50)*, $NO^{\bullet}$ donors have been shown to increase whole-cell K^+ currents. This increased outward current is blocked by charybdotoxin (ChTx) or IbTx. At the single-channel level, both $NO^{\bullet}$ donors and membrane-permeable analogs of cGMP increase the open probability of BK_{Ca} channels recorded in cell-attached and inside-out patches of colonic myocytes in both guinea pig and dog models *(50,51)*, suggesting direct activation of BK_{Ca} by $NO^{\bullet}$. Therefore, K_{Ca} channels are critically involved in the response to $NO^{\bullet}$ released by NANC nerve fibers in the GI tract and, thus, may play a role in modulating the motility of visceral smooth muscle.

2.3. Myometrial Smooth Muscle

Myometrial smooth muscle cells are richly endowed with BK_{Ca} channels, which play a major role in regulating uterine contractility *(52,53)*. It has been reported that during pregnancy, the activity of cGMP is increased because of elevated levels of $NO^{\bullet}$, which stimulates activity of BK_{Ca} channels *(54,55)*. The enhanced BK_{Ca} channel activity is especially important in maintaining uterine quiescence in pregnancy. Patch-clamp studies using myometrial SMCs from pregnant women show an increased open-state probability of IbTx-inhibitable BK_{Ca} channels in excised patches, in response to cGMP analog or protein kinase G (PKG) stimulation *(39)*. More recently, studies using human nonpregnant myometrium demonstrated that diethylamine-NO induced dose-dependent relaxation of myometrium and that this relaxation was abolished by apamin or scyllatoxin, small-conductance K_{Ca} channel blockers, or by ChTx *(56)*. Therefore, the NO-K_{Ca} channel signal transduction pathway may also be important in myometrial conditioning during pregnancy.

3. MECHANISMS OF cGMP-DEPENDENT AND cGMP-INDEPENDENT EFFECTS OF NO ON K_{Ca} CHANNELS

The mechanism by which NO activates K_{Ca} channels is not clear. At least two distinct mechanisms have been proposed, one involving activation of cGMP, and the other a cGMP-independent pathway.

3.1. cGMP-Dependent Activation of K_{Ca} Channels

There is substantial evidence demonstrating cGMP-dependent activation of K_{Ca} channels in smooth cells *(57–63)*. Soluble guanylyl cyclase is a heme-containing enzyme existing in all cells. The enzyme has a high affinity for NO•, which serves to activate it in a dose-dependent fashion. When NO• reacts with the iron-containing heme moiety of the enzyme, enzymatic conversion of guanosine-triphosphate (GTP) to cGMP increases *(64–67)*.

In cultured bovine aortic cells, Williams et al. *(68)* first showed that sodium nitroprusside increased intracellular cGMP levels and activated BK_{Ca} channels in cell-attached membrane patches. Membrane-permeable cGMP analogs produced similar effects on BK_{Ca} channels. These findings have been confirmed by other investigators in various SMC types *(51,63,69–72)*.

The important target protein of cGMP is PKG *(61,62,71)*. Robertson et al. *(61)* first reported the role of PKG in cGMP-mediated dilation. They showed that in cerebral arterial smooth muscle cells, SIN-1, an NO• donor, and 8-pCPT-cGMP, a membrane-permeable analog of cGMP, both activate BK_{Ca} channels in cell-attached patch-clamp configuration. The enhanced BK_{Ca} channel activity was also seen when PKG was added to the bath solution of inside-out patches in the presence of cGMP and adenosine triphosphate. Similarly, in canine coronary arterial SMC *(62)*, application of G kinase to the cytoplasmic face of excised patches increased voltage sensitivity and open-state probability of BK_{Ca} channels. Similar findings have also been observed in multiple vascular beds *(57–60,62,63)*. The signaling cascade involving cGMP activation of PKG activity and subsequent phosphorylation of K_{Ca} channels provides an important amplifying system especially critical during situations in which levels of NO• are low.

3.2. cGMP-Independent Activation of K_{Ca} Channels

It was once thought that all of the actions of NO• were mediated by a rise in cGMP. However, this is clearly not the case, and at least some properties of NO• in smooth muscle are cGMP independent. This conclusion is derived from several studies in which the hyperpolarization of vascular smooth muscle was only partly blocked by specific inhibitors of cGMP, such as methylene blue or 1H-[1,2,4]oxadiazolo-[4,3-a]quinoxalin-1-one (ODQ) *(57,73,74)*. Interestingly, studies by Najibi and colleagues *(75,76)* demonstrated that L-NAME partially blocked the dilation to ACh in hypercholesterolemic rabbit carotid arteries without a measurable rise in cGMP levels. The mechanism of cGMP-independent activation of K_{Ca} has been studied. It involves *S*-nitrosation of low molecular weight thiols or cysteinyl side chains of proteins *(74)*. Cysteine residues are important for maintaining the native conformation of many proteins, and when located at active sites of an enzyme, nitrosation by NO• can alter enzymatic activity. Cysteine residues are also sites for covalent attachment of other regulatory molecules; therefore, any modification at this site may have implications for other signaling pathways. NO• reacts with thiols to form various oxidized thiol species, including sulfenic acids *(77)*, disulfides *(78)*, mixed disulfides *(79)*, and *S*-nitrosothiols *(80)*. Studies by Gow et al. *(81)* demonstrated an increased immunoreactivity of nitrosothiol protein in aortic rings treated with ACh. The enhanced immunostaining of nitrosothiol protein was completely eliminated in aortic rings exposed to both acetylcholine and L-NAME *(81)*. Similar findings were also observed in cultured endothelial cells treated with eNOS *(81,82)*. It remains to be confirmed whether nitosylation of cysteine residues on K_{Ca} channels may modulate the function of these channels in vascular tissue.

Thiol modification by NO• species may yield vasoactive intermediates. *S*-Nitrosoglutathione dilates rat aorta, mesenteric arteries, and porcine coronary arteries *(83)*. *S*-Nitrosoglutathione-induced dilation can be inhibited by tetrabutylammonium, a K_{Ca} channel blocker *(84)*. The *S*-nitrosation of tissue thiols has been postulated as a mechanism of formation of local NO• stores from which NO• can subsequently be released. This mechanism may be especially important as a buffering system when levels of NO• in the tissue are high.

4. EFFECT OF NO•-DERIVED FREE RADICALS ON K_{Ca} CHANNEL ACTIVITY

NO• has an unpaired electron, making it highly reactive. Chemically, NO• can exist in several redox forms, thus providing this species with a multitude of potential regulatory and chemical functions. One electron reduction produces nitroxyl anion (NO^-). NO^- is a reactive short-lived species that can react with another molecule of NO• to generate nitrous oxide, and possibly hydroxyl radicals. In aqueous aerobic solutions, NO• predominantly forms nitrite (NO_2^-). In the presence of oxyhemoglobin and oxymyoglobin, NO• is readily oxidized to nitrate (NO_3^-); whereas NO• reacts with $O_2^{\bullet-}$ to form peroxynitrite ($ONOO^-$), in situations in which $O_2^{\bullet-}$ production is increased, $ONOO^-$ has been proposed as an important intermediate radical species, since Beckman et al. *(85)* originally described its formation in 1990. $ONOO^-$ is a highly reactive and short-lived species that promotes oxidative and nitrosative molecular and tissue damage *(86)*. Generation of $ONOO^-$ is proposed as a contributing factor to the pathogenesis of several diseases, including acute and chronic inflammatory processes *(87,88)*, ischemia-reperfusion injury *(89–92)*, and diabetes *(93–95)*, among others *(96–98)*.

To understand the biology of NO• and other ROS such as ONOO, it is important to recognize the differential effect of NO• and $ONOO^-$ on cells and tissues. In particular, NO• is neither a strong oxidant nor a potent nitrating agent; it mostly participates in reversible interactions with iron containing moieties, radical-radical combination reactions, and nitrosylation reactions via intermediate formation of dinitrogen trioxide. In contrast to NO•, $ONOO^-$ is a strong oxidant and nitrating agent and a poor nitrosylating agent, which promotes nitration (incorporation of a nitro $-NO_2$ group) of aromatic and aliphatic residues. Most notably, protein tyrosine residues constitute key targets for $ONOO^-$- mediated nitration, and the presence of 3-nitrotyrosine in proteins has been used as pathopneumonic for the presence of $ONOO^-$ *(93,99,100)*. However, other ROS can also stimulate production of nitrotyrosine residues in cellular proteins *(101)*.

Growing evidence suggests that $ONOO^-$ has an inhibitory effect on BK_{Ca} channels *(102,103)*. In human coronary arterioles, incubation with 5 μ*M* authentic $ONOO^-$ significantly reduced dilation to bradykinin, which was previously shown to be dependent on ChTx-sensitive K_{Ca} channels. Inhibition of dilation to NS 1619, an opener of BK_{Ca} channels, was also reduced by $ONOO^-$. Application of $ONOO^-$ to the cytosolic face of excised patches of human coronary arteriolar SMCs reduced the open-state probability of BK_{Ca} channels, suggesting a direct inhibitory effect of $ONOO^-$ on BK_{Ca} channels *(102)*. Similar findings were also reported in rat cerebral arteries *(103)*. In the renal cortical collecting duct, Lu and Wang *(104)* demonstrated that $ONOO^-$ also decreased small-conductance K_{Ca} channel activities.

The mechanism by which $ONOO^-$ inhibits K_{Ca} channel activity is not fully understood. Future mechanistic studies will be required to explore how $ONOO^-$ impairs K_{Ca} channel function.

5. INTERACTIONS BETWEEN NO AND K_{Ca}-ACTIVATING ENDOTHELIAL FACTORS

In addition to NO•, the other endothelial-derived dilator substance that dilates the vessel by opening of K_{Ca} channels is endothelium-derived hyperpolarizing factor (EDHF). For more than a decade, the chemical nature of EDHF has been intensively investigated. The classic method used to identify a role for EDHF was the demonstration that the agonist produced an endothelium-dependent dilation in the presence of inhibitors of both NOS and cyclooxygenase. More specific proof of an EDHF mechanism requires demonstration of frank hyperpolarization, either indirectly by inhibiting the response with specific K^+ channel blockers, or directly by demonstrating vascular smooth muscle hyperpolarization using patch clamping or membrane potential recording techniques. Most investigations show that EDHF participates more in regulating resistance vessel function than conduit arterial tone *(105,106)*; thus, EDHF may play an important role in regulating tissue perfusion.

A prominent interaction between endothelial-derived NO• and EDHF has been described. NO• inhibits cytochrome P450, thereby limiting epoxyeicosatrienoic acid (EET) production *(107)*, a major candidate for EDHF. Even subthreshold concentrations of NO• inhibit EDHF-mediated dilation *(107)*. This interaction may be very important in atherosclerotic disease states and in the presence of its risk factors, where reduced NO• bioavailability is accompanied by increased contributions of EDHF to vasodilation *(75,108)*. Consistent with this notion, when endothelial NO• production is prevented, pathways generating EDHF can fully compensate for loss of NO• in the microcirculation *(75,109)*. Our laboratory has reported that in human coronary arterioles, flow-induced dilation was partially blocked by L-NAME in patients without coronary disease, but not in patients with coronary artery disease, in which NO• played no role in dilation *(110)*. However, in patients with coronary disease, flow-induced dilation was inhibited by ChTx and by inhibitors of cytochrome P450, but not L-NAME, indicating that EDHF is the key component in flow-mediated dilation in coronary disease *(110)*.

EETs, derived from AA through CYP450 are considered EDHFs in various circulations *(29,111–115)*. Evidence of support includes the following: First, cytochrome P450 enzymes are abundant in the endothelium *(116,117)* and mediate the formation of EETs *(118,119)*. Second, inhibitors of the cytochrome P450 monoxygenase and antisense oligonucleotides against specific isoforms suppress EDHF-mediated responses in coronary arteries from a variety of species *(111,120–122)*. Finally, EETs open large-conductance K_{Ca} channels and hyperpolarize SMCs in the cell-attached patch-clamp configuration *(123,124)*. However, evidence that does not support EETs as a candidate of EDHF includes the facts that EETs do not relax all blood vessels, such as hepatic arteries *(125)*, and that some cytochrome P450 blockers do not inhibit EDHF-induced dilation *(126)*. Furthermore, EDHF-mediated responses are observed in blood vessels, such as aorta and carotid arteries of the rabbit, which do not produce metabolites of arachidonic acid by the cytochrome P450 pathway *(127)*, thus suggesting that other factors may also be involved in endothelium-dependent hyperpolarization and relaxation.

More recently, H_2O_2 has been proposed as a candidate EDHF. Several studies have documented that exogenously applied H_2O_2 can produce vasodilation of isolated arteries from pulmonary *(128)*, coronary *(129)*, and cerebral beds *(130,131)*. In many cases, this dilation is associated with opening of K_{Ca} channels and hyperpolarization of cell membranes *(129,130,132)*. The second approach tests the effect of removing endogenously produced H_2O_2. Bradykinin, which generates ROS, dilates cerebral arterioles in rats *(132)*. This dilation is blocked by catalase, consistent with a critical role for H_2O_2 in the response *(132)*. Similarly, in mice lacking the gene for NOS, dilation to ACh is because of EDHF *(133)*. This dilation to ACh and the attendant hyperpolarization are inhibited by catalase *(134)*. Thus, under certain conditions, particularly those associated with enhanced oxidative stress in which $NO^{\bullet}$ production is reduced, H_2O_2 can act as an EDHF, which opens K_{Ca} channel and exerts effects similar to those of $NO^{\bullet}$.

The situation is more complex than simple compensation by EDHF for loss of NO. H_2O_2 can stimulate GTP cyclohydrolase-I and increase NO production through NOS *(135)*. Thus, in some cases, EDHF may act to stimulate the formation of $NO^{\bullet}$.

6. ROLE OF K_{Ca} CHANNELS IN PATHOLOGICAL CONDITIONS

Several studies suggest that a decrease in $NO^{\bullet}$ production and/or bioactivity contributes to the pathophysiology of many disease states, including hypercholesterolemia, hypertension, and diabetes. The role of K_{Ca} channels may also be altered under these conditions.

Although the magnitude of endothelium-dependent relaxation is preserved in hypercholesterolemia, compared with control rabbit carotid arteries, a significant ChTx-sensitive component is observed only in the carotid arteries of hypercholesterolemic animals *(76)*. Similarly, ChTx and TEA abolished $NO^{\bullet}$-dependent phenylephrine-induced rhythmic activity in aortic rings of atherosclerotic but not of healthy mice *(136)*. Subsequent studies revealed that the mechanism of the vasodilator response to exogenous $NO^{\bullet}$ becomes ChTx-sensitive, despite an impaired production of cGMP after cholesterol feeding, suggesting that the functional role of K_{Ca} channels in vascular muscle is markedly increased during hypercholesterolemia.

In hypertension, there is strong evidence that $NO^{\bullet}$ bioavailability is reduced, but the functional role of the K_{Ca} channel is enhanced in VSMCs. Pharmacological inhibitors of K_{Ca} channels cause enhanced depolarization and constriction of arteries from hypertensive animals. This phenomenon occurs similarly in several vascular beds, including aorta *(26,137,138)*; carotid artery *(139–141)*; and the mesenteric *(140)*, femoral *(140,142)*, and cerebral arteries *(25,143)*. Electrophysiological measurements obtained from patch-clamp studies in arterial myocytes isolated from hypertensive animals have confirmed that the whole-cell K^+ current through K_{Ca} channels is enhanced, in comparison with currents recorded from normotensive myocytes *(137,144)*. Based on a comprehensive molecular, electrophysiological, and functional study of the cerebral circulation of spontaneously hypertensive rats, the enhanced K_{Ca} currents can be explained by greater expression of K_{Ca} channels rather than a change in open-state probability or Ca^{2+} sensitivity of K_{Ca} channels *(25)*. The enhanced K_{Ca} channel function may serve as a protective mechanism against progressive increases in blood pressure and decreases in vasodilator responsiveness.

Diabetes, like other vascular disease, is recognized as a condition in which there is increased $O_2^{\bullet-}$ generation and reduced $NO^{\bullet}$ bioavailability. Although there is no direct evidence of alteration of K_{Ca} channel in diabetes, K_{Ca} channel opening appears to be refractory to $O_{22}^{\bullet-}$ *(102)*. Given et al. *(145)* compared dilator response to bradykinin in vivo in coronary circulation between euglycemic and diabetic rats. They showed that dilation to bradykinin was inhibited by either IbTx or clotrimazole, suggesting a role for EDHF. A similar dilation to bradykinin was observed in euglycemic and diabetic rats.

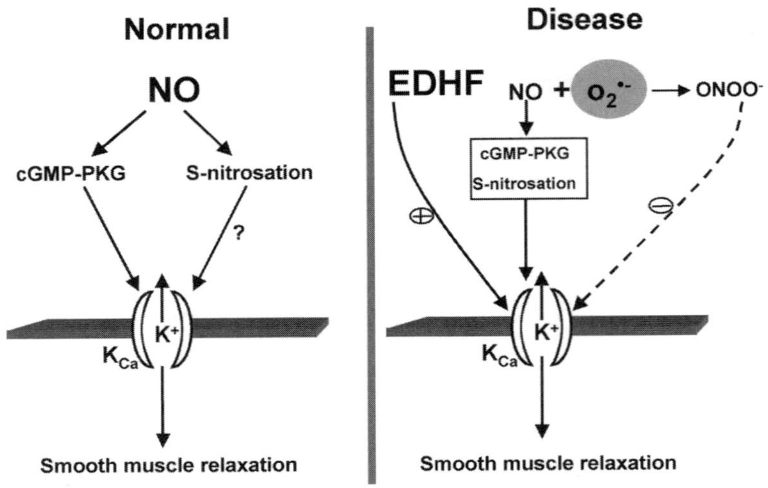

Fig. 1. The interaction of nitric oxide and KCa channels in smooth muscle membranes in normal and disease states.

These results are consistent with findings from our laboratory indicating that $O_2^{\bullet -}$ generated by the reaction of xanthine and xanthine oxidase, at a dose sufficient for inhibiting K_{ATP} and Kv channel function, had no effect on bradykinin-induced dilation in human coronary arterioles *(102)*. However, the redox situation in diabetes, as well as many other pathological conditions, is more complicated. In addition to $O_2^{\bullet -}$, other free-radical species, such as $ONOO^-$, can be generated *(93,94)*, depending on the state and severity of the disease. $ONOO^-$ has inhibitory effects on K_{Ca} channels, as discussed earlier *(102)*. Therefore K_{Ca} channel function may also be expected to be impaired in some chronic diabetic states. The effect of diabetes on K_{Ca} channel activity is a subject for future investigation.

7. CONCLUSION

Figure 1 summarizes the interaction between $NO^\bullet$ and K_{Ca} channels in regulating smooth muscle function under physiological and pathological conditions. In the normal situation, $NO^\bullet$ is synthesized from L-arginine by a different isoform of NOS. Once it reaches SMCs, $NO^\bullet$ stimulates hyperpolarization and relaxation by opening of K_{Ca} channels via activation of the cGMP-PKG pathway or via *S*-nitrosation of channel proteins.

$ONOO^-$, an important derivative radical formed by the interaction between superoxide and $NO^\bullet$, is generated in many pathological states. Unlike $O_2^{\bullet -}$, which has little effect on K_{Ca} channels, $ONOO^-$ impairs K_{Ca} channel function. Thus, on one hand, loss of $NO^\bullet$ may be compensated by EDHF or by an increase in protein expression of K_{Ca} channels. On the other hand, $ONOO^-$ often formed in this situation reduces K_{Ca} channel function and impairs smooth muscle relaxation. Thus, in disease states associated with enhanced $O_2^{\bullet -}$ production, the effects of $O_2^{\bullet -}$, loss of $NO^\bullet$, and actions of ONOO must all be considered when examining vasomotor and other functional responses.

REFERENCES

1. Ignarro LJ, Buga GM, Wood KS, et al. Endothelium-derived relaxing factor produced and released from artery and vein is nitric oxide. Proc Natl Acad Sci USA 1987;84:9265–9269.
2. Ignarro LJ, Byrns RE, Buga GM, et al. Endothelium-derived relaxing factor from pulmonary artery and vein possesses pharmacologic and chemical properties identical to those of nitric ocide radical. Circ Res 1987;61:866–879.
3. Ignarro LJ, Harbison RG, Wood KS, et al. Activation of purified soluble guanylate cyclase by endothelium-derived relaxing factor from intrapulmonary artrery and vein: stimulation by acetylcholine, bradykinin and arachidonic acid. J Pharmacol Exp Ther 1986;237:893–900.
4. Bredt DS, Hwang PM, Snyder SH. Localization of nitric oxide synthase indicating a neural role for nitric oxide. Nature 1990;347:768–770.
5. Lamas S, Marsden PA, Li GK, et al. Endothelial nitric oxide synthase: molecular cloning and characterization of a distinct constitutive enzyme isoform. Proc Natl Acad Sci USA 1992;89:6348–6352.
6. Michel T, Feron O. Nitric oxide synthases: which, where, how, and why? J Clin Invest 1997;100: 2146–2152.
7. Nathan C, Xie QW. Nitric oxide synthases: roles, tolls, and controls. Cell 1994;78:915–918.
8. Arnal JF, Dinh-Xuan AT, Pueyo M, et al. Endothelium-derived nitric oxide and vascular physiology and pathology. Cell Mol Life Sci 1999;55:1078–1087.
9. Chester AH, O'Neil GS, Tadjkarimi S, et al. The role of nitric oxide in mediating endothelium dependent relaxations in the human epicardial coronary artery. Int J Cardiol 1990;29:305–309.
10. Cooke JP, Rossitch E Jr., Andon NA, et al. Flow activates an endothelial potassium channel to release an endogenous nitrovasodilator. J Clin Invest 1991;88:1663–1671.
11. Costa F, Biaggioni I. Role of nitric oxide in adenosine-induced vasodilation in humans. Hypertension 1998;31:1061–1064.
12. Duffy SJ, Castle SF, Harper RW, et al. Contribution of vasodilator prostanoids and nitric oxide to resting flow, metabolic vasodilation, and flow-mediated dilation in human coronary circulation. Circulation 1999;100:1951–1957.
13. Fukumoto Y, Shimokawa H, Kozai T, , et al. Vasculoprotective role of inducible nitric oxide synthase at inflammatory coronary lesions induced by chronic treatment with interleukin-1 beta in pigs in vivo. Circulation 1997;96:3104–3111.
14. Ikeda U, Shimada K. Nitric oxide and cardiac failure. Clin Cardiol 1997;20:837–841.
15. Kelly RA, Balligand JL, Smith TW. Nitric oxide and cardiac function. Circ Res 1996;79:3:363–380.
16. Dusting GJ. Nitric oxide in coronary artery disease: roles in atherosclerosis, myocardial reperfusion and heart failure. EXS 1996;76:33–55.
17. Kano T, Shimizu-Sasamata M, Huang PL, et al. Effects of nitric oxide synthase gene knockout on neurotransmitter release in vivo. Neuroscience 1998;86:695–699.
18. Kurz T, Tolg R, Richardt G. Bradykinin B2-receptor-mediated stimulation of exocytotic noradrenaline release from cardiac sympathetic neurons. J Mol Cell Cardiol 1997;29:2561–2569.
19. Nelson MT, Quayle JM. Physiological roles and properties of potassium channels in arterial smooth muscle. Am J Physiol 1995;268:C799–C822.
20. Faraci FM, Sobey CG. Role of potassium channels in regulation of cerebral vascular tone. J Cereb Blood Flow Metab 1998;18:1047–1063.
21. Brayden JE. Potassium channels in vascular smooth muscle. Clin Exp Pharmacol Physiol 1996;23: 751–768.
22. Clapp LH, Tinker A. Potassium channels in the vasculature. Curr Opin Nephrol Hypertension 1998;7:91–98.
23. Nelson MT, Patlak JB, Worley JF III, et al. Calcium channels, potassium channels, and voltage dependence of arterial smooth muscle tone. Am J Physiol 1990;259:C3–C18.
24. Rusch NJ, Liu Y. Potassium channels in hypertension: homeostatic pathways to buffer arterial contraction. J Lab Clin Med 1997;130:245–251.
25. Liu Y, Hudetz AG, Knaus HG, et al. Increased expression of Ca^{2+}-sensitive K^+ channels in the cerebral microcirculation of genetically hypertensive rats: evidence for their protection against cerebral vasospasm. Circ Res 1998;82:729–737.
26. Liu Y, Pleyte K, Knaus HG, et al. Increased expression of Ca^{2+}-sensitive K^+ channels in aorta of hypertensive rats. Hypertension 1997;30:1403–1409.
27. Young DB, Lin H, McCabe RD. Potassium's cardiovascular protective mechanisms. Am J Physiol 1995;268:R825–R837.

28. Bialecki RA, Stinson-Fisher C. K_{Ca} channel antagonists reduce NO donor-mediated relaxation of vascular and tracheal smooth muscle. Am J Physiol 1995;268:L152–L159.
29. Li PL, Zou AP, Campbell WB. Regulation of potassium channels in coronary arterial smooth muscle by endothelium-derived vasodilators. Hypertension 1997;29:262–267.
30. Feron O, Belhassen L, Kobzik L, et al. Endothelial nitric oxide synthase targeting to caveolae: specific interactions with caveolin isoforms in cardiac myocytes and endothelial cells. J Biol Chem 1996;271:22,810–22,814.
31. Shaul PW, Smart EJ, Robinson LJ, et al. Acylation targets emdothelial nitric-oxide synthase to plasmalemmal caveolae. J Biol Chem 1996;271:6518–6522.
32. Fujimoto S, Ikegami Y, Isaka M, et al. K^+ channel blockers and cytochrome P450 inhibitors on acetylcholine-induced, endothelium-dependent relaxation in rabbit mesenteric artery. Eur J Pharmacol 1999;384:7–15.
33. Cohen RA, Plane F, Najibi S, et al. Nitric oxide is the mediator of both endothelium-dependent relaxation and hyperpolarization of the rabbit carotid artery. Proc Natl Acad Sci USA 1997;94:4193–4198.
34. Cowan CL, Palacino JJ, Najibi S, et al. Potassium channel-mediated relaxation to acetylcholine in rabbit arteries. J Pharmacol Exp Ther 1993;266:1482–1489.
35. Shikano K, Long CJ, Ohlstein EH, et al. Comparative pharmacology of endothelium-derived relaxing factor and nitric oxide. J Pharmacol Exp Ther 1988;247:873–881.
36. Watson N, Barnes PJ, Maclagan J. Actions of methoctramine, a muscarinic M2 receptor antagonist, on muscarinic and nicotonic cholinoceptors in guinea-pig airways in vivo and in vitro. Br J Pharmacol 1992;105:107–112.
37. Sun D, Huang A, Recchia FA, et al. Nitric oxide-mediated arteriolar dilation after endothelial deformation. Am J Physiol 2001;280:H714–H721.
38. Wigg SJ, Tare M, Tonta MA, et al. Comparison of effects of diabetes mellitus on an EDHF-dependent and an EDHF-independent artery. Am J Physiol 2001;281:H232–H240.
39. White RE, Han G, Maunz M, et al. Endothelium-independent effect of estrogen on Ca^{2+}-activated K^+ channels in human coronary artery smooth muscle cells. Cardiovasc Res 2002;53:650–661.
40. White RE, Darkow DJ, Lang JLF. Estrogen relaxes coronary arteries by opening bkca channels through a cgmp-dependent mechanism. Circ Res 1995;77:936–942.
41. Deenadayalu VP, White RE, Stallone JN, et al. Testosterone relaxes coronary arteries by opening the large-conductance, calcium-activated potassium channel. Am J Physiol 2001;281:H1720–H1727.
42. Chou TM, Sudhir K, Hutchison SJ, et al. Testosterone induces dilation of canine coronary conductance and resistance arteries in vivo. Circulation 1996;94:2614–2619.
43. Chute CG, Baron JA, Plymate SR, et al. Sex hormones and coronary artery disease. Am J Med 1987;83:853–859.
44. Wangensteen R, Quesada A, Sainz J, et al. Role of endothelium-derived relaxing factors in adrenomedullin-induced vasodilation in the rat kidney. Eur J Pharmacol 2002;444:97–102.
45. Nara M, Dhulipala PD, Ji GJ, et al. Guanylyl cyclase stimulatory coupling to K(Ca) channels. Am J Physiol Cell Physiol 2000;279:C1938–C1945.
46. Keef KD, Du C, Ward SM, et al. Enteric inhibitory neural regulation of human colonic circular muscle: role of nitric oxide. Gastroenterology 1993;105:1009–1016.
47. Bayguinov O, Sanders KM. Role of nitric oxide as an inhibitory neurotransmitter in the canine pyloric sphincter. Am J Physiol 1993;264:G975–G983.
48. Ward SM, McKeen ES, Sanders KM. Role of nitric oxide in non-adrenergic, non-cholinergic inhibitory junction potentials in canine ileocolonic sphincter. Br J Pharmacol 1992;105:776–782.
49. Murray JA, Shibata EF, Buresh TL, et al. Nitric oxide modulates a calcium-activated potassium current in muscle cells from opossum esophagus. Am J Physiol 1995;269:G606–G612.
50. Lang RJ, Watson MJ. Effects of nitric oxide donors, S-nitroso-L-cysteine and sodium nitroprusside, on the whole-cell and single channel currents in single myocytes of the guinea-pig proximal colon. Br J Pharmacol 1998;123:505–517.
51. Koh SD, Campbell JD, Carl A, et al. Nitric oxide activates multiple potassium channels in canine colonic smooth muscle. J Physiol 1995;489(pt 3):735–743.
52. Perez GJ, Toro L, Erulkar SD, et al. Characterization of large-conductance, calcium-activated potassium channels from human myometrium. Am J Obstet Gynecol 1993;168:652–660.
53. Anwer K, Oberti C, Perez GJ, et al. Calcium-activated K^+ channels as modulators of human myometrial contractile activity. Am J Physiol 1993;265:C976–C985.
54. Benkusky NA, Korovkina VP, Brainard AM, et al. Myometrial maxi-K channel beta1 subunit modulation during pregnancy and after 17 beta-estradiol stimulation. FEBS Lett 2002;524:97–102.

55. Zhou XB, Wang GX, Ruth P, et al. BKCa channel activation by membrane-associated cGMP kinase may contribute to uterine quiescence in pregnancy. Am J Physiol Cell Physiol 2000;279: C1751–C1759.
56. Modzelewska B, Kostrzewska A, Sipowicz M, et al. Apamin inhibits NO-induced relaxation of the spontaneous contractile activity of the myometrium from non-pregnant women. Reprod Biol Endocrinol 2003;1:8–15.
57. Plane F, Wiley KE, Jeremy JY, Cohen RA, et al. Evidence that different mechanisms underlie smooth muscle relaxation to nitric oxide and nitric oxide donors in the rabbit isolated carotid artery. Br J Pharmacol 1998;123:1351–1358.
58. Plane F, Hurrell A, Jeremy JY, et al. Evidence that potassium channels make a major contribution to SIN-1-evoked relaxation of rat isolated mesenteric artery. Br J Pharmacol 1996;119:1557–1562.
59. Archer SL, Huang JMC, Reeve HL, et al. Differential distribution of electrophysiologically distinct myocytes in conduit and resistance arteries determines their response to nitric oxide and hypoxia. Circ Res 1996;78:431–442.
60. Fujino K, Nakaya S, Wakatsuki T, Miyoshi Y, et al. Effects of nitroglycerin on ATP-induced Ca^{++}-mobilization, Ca^{++}-activated K channels and contraction of cultured smooth muscle cells of porcine coronary artery. J Pharmacol Exp Ther 1991;256:371–377.
61. Robertson BE, Schubert R, Hescheler J, et al. cGMP-dependent protein kinase activates Ca-activated K channels in cerebral artery smooth muscle cells. Am J Physiol 1993;265:C299–C303.
62. Taniguchi J, Furukawa KI, Shigekawa M. Maxi K^+ channels are stimulated by cyclic guanosine monophosphate-dependent protein kinase in canine coronary artery smooth muscle cells. Pflügers Arch 1993;423:167–172.
63. Carrier GO, Fuchs LC, Winecoff AP, et al. Nitrovasodilators relax mesenteric microvessels by cGMP-induced stimulation of Ca-activated K channels. Am J Physiol 1997;273:H76–H84.
64. Schini V, Grant NJ, Miller RC, et al. Morphological characterization of cultured bovine aortic endothelial cells and the effects of atriopeptin II and sodium nitroprusside on cellular and extracellular accumulation of cyclic GMP. Eur J Cell Biol 1988;47:53–61.
65. Brandes RP, Kim D, Schmitz-Winnenthal FH, et al. Increased nitrovasodilator sensitivity in endothelial nitric oxide synthase knockout mice: role of soluble guanylyl cyclase. Hypertension 2000;35: 231–236.
66. Cohen RA, Vanhoutte PM. Endothelium-dependent hyperpolarization. Beyond nitric oxide and cyclic GMP. Circulation 1995;92:3337–3349.
67. Furchgott RF, Vanhoutte PM. Endothelium-derived relaxing and contracting factors. FASBE J 1989;3:2007–2018.
68. Williams DL Jr., Katz GM, Roy-Contancin L, et al. Guanosine 5'-monophosphate modulates gating of high-conductance Ca^{2+}-activated K^+ channels in vascular smooth muscle cells. Proc Natl Acad Sci USA 1988;85:9360–9364.
69. Lander HM, Sehajpal PK, Novogrodsky A. Nitric oxide signaling: a possible role for G proteins. J Immunol 1993;151:7182–7187.
70. Li PL, Jin MW, Campbell WB. Effect of selective inhibition of soluble guanylyl cyclase on the K_{Ca} channel activity in coronary artery smooth muscle. Hypertens 1998;31:303–308.
71. White RE, Lee AB, Shcherbatko AD, et al. Potassium channel stimulation by natriuretic peptides through cGMP-dependent dephosphorylation. Nature 1993;361:263–266.
72. Sansom SC, Stockand JD, Hall D, et al. Regulation of large calcium-activated potassium channels by protein phosphatase 2A. J Biol Chem 1997;272:9902–9906.
73. Weisbrod RM, Griswold MC, Yaghoubi M, et al. Evidence that additional mechanisms to cyclic GMP mediate the decrease in intracellular calcium and relaxation of rabbit aortic smooth muscle to nitric oxide. Br J Pharmacol 1998;125:1695–1707.
74. Bolotina VM, Najibi S, Palacino JJ, et al. Nitric oxide directly activates calcium-dependent potassium channels in vascular smooth muscle. Nature 1994;368:850–853.
75. Najibi S, Cowan CL, Palacino JJ, et al. Enhanced role of potassium channels in relaxations to acetylcholine in hypercholesterolemic rabbit carotid artery. Am J Physiol 1994;266:H2061–H2067.
76. Najibi S, Cohen RA. Enhanced role of potassium channels in relaxation of hypercholesterolemic rabbit carotid artery to nitric oxide. Am J Physiol 1995;38:H805–H811.
77. Becker K, Savvides SN, Keese M, et al. Enzyme inactivation through sulfhydryl oxidation by physiologic NO-carriers. Nat Struct Biol 1998;5:267–271.
78. Lei SZ, Pan ZH, Aggarwal SK, et al. Effect of nitric oxide production on the redox modulatory site of the NMDA receptor-channel complex. Neuron 1992;8:1087–1099.

79. Luperchio S, Tamir S, Tannenbaum SR. NO-induced oxidative stress and glutathione metabolism in rodent and human cells. Free Radic Biol Med 1996;21:513–519.
80. Williams DLH. S-Nitrosation and the reaction of S-nitrosocompounds. Chem Soc Rev 1985;14:171–196.
81. Gow AJ, Chen Q, Hess DT, et al. Basal and stimulated protein S-nitrosylation in multiple cell types and tissues. J Biol Chem 2002;277:9637–9640.
82. Lang RJ, Harvey JR, et al. Sodium (2-sulfonatoethyl) methanethiosulfonate prevents S-nitroso-L-cysteine activation of Ca^{2+}-activated K^+ (BK_{Ca}) channels in myocytes of the guinea-pig taenia caeca. Br J Pharmacol 2003;139:1153–1163.
83. Alencar JL, Lobysheva I, Geffard M, et al. Role of S-nitrosation of cysteine residues in long-lasting inhibitory effect of nitric oxide on arterial tone. Mol Pharmacol 2003;63:1148–1158.
84. Muller B, Kleschyov AL, Malblanc S, et al. Nitric oxide–related cyclic GMP-independent relaxing effect of N-acetylcysteine in lipopolysaccharide-treated rat aorta. Br J Pharmacol 1998;123: 1221–1229.
85. Beckman JS, Beckman TW, Chen J, et al. Apparent hydroxyl radical production by peroxynitrite: implications for endothelial injury from nitric oxide and superoxide. Proc Natl Acad Sci USA 1990;87:1620–1624.
86. Beckman JS, Chen J, Ischiropoulos H, et al. Oxidative chemistry of peroxynitrite. Methods Enzymol 1994;233:229–240.
87. Beckmann JS, Ye YZ, Anderson PG, et al. Extensive nitration of protein tyrosines in human atherosclerosis detected by immunohistochemistry. Biol Chem Hoppe Seyler 1994;375:81–88.
88. Lefer DJ, Scalia R, Campbell B, et al. Peroxynitrite inhibits leukocyte-endothelial cell interactions and protects against ischemia-reperfusion injury in rats. J Clin Invest 1997;99:684–691.
89. Ma XL, Lopez BL, Liu GL, et al. Peroxynitrite aggravates myocardial reperfusion injury in the isolated perfused rat heart. Cardiovasc Res 1997;36:195–204.
90. Nossuli TO, Hayward R, Jensen D, Scalia R, et al. Mechanisms of cardioprotection by peroxynitrite in myocardial ischemia and reperfusion injury. Am J Physiol 1998;275:H509–H519.
91. Nossuli TO, Hayward R, Scalia R, et al. Peroxynitrite reduces myocardial infarct size and preserves coronary endothelium after ischemia and reperfusion in cats. Circulation 1997;96:2317–2324.
92. Yasmin W, Strynadka KD, Schulz R. Generation of peroxynitrite contributes to ischemia-reperfusion injury in isolated rat hearts. Cardiovasc Res 1997;33:422–432.
93. Zou MH, Shi C, Cohen RA. High glucose via peroxynitrite causes tyrosine nitration and inactivation of prostacyclin synthase that is associated with thromboxane/prostaglandin H(2) receptor-mediated apoptosis and adhesion molecule expression in cultured human aortic endothelial cells. Diabetes 2002;51:198–203.
94. Lyall F, Gibson JL, Greer IA, et al. Increased nitrotyrosine in the diabetic placenta: evidence for oxidative stress. Diabetes Care 1998;21:1753–1758.
95. Munzel T, Heitzer T, Harrison DG. The physiology and pathophysiology of the nitric oxide/superoxide system. Herz 1997;22:158–172.
96. Ma XL, Lopez BL, Liu GL, et al. Hypercholesterolemia impairs a detoxification mechanism against peroxynitrite and renders the vascular tissue more susceptible to oxidative injury. Circ Res 1999;80:894–901.
97. Noiri E, Nakao A, Uchida K, et al. Oxidative and nitrosative stress in acute renal ischemia. Am J Physiol Renal Physiol 2001;281:F948–F957.
98. Radi R, Peluffo G, Alvarez MN, et al. Unraveling peroxynitrite formation in biological systems. Free Radic Biol Med 2001;30:463–488.
99. Kooy NW, Lewis SJ, Royall JA, et al. Extensive tyrosine nitration in human myocardial inflammation: evidence for the presence of peroxynitrite. Crit Care Med 1997;25:812–819.
100. Ducrocq C, Blanchard B, Pignatelli B, et al. Peroxynitrite: an endogenous oxidizing and nitrating agent. Cell Mol Life Sci 1999;55:1068–1077.
101. Ogino K, Kodama N, Nakajima M, et al. Catalase catalyzes nitrotyrosine formation from sodium azide and hydrogen peroxide. Free Radic Res 2001;35:735–747.
102. Liu Y, Terata K, Chai Q, et al. Peroxynitrite inhibits Ca^{2+}-activated K^+ channel activity in smooth muscle of human coronary arterioles. Circ Res 2002;91:1070–1076.
103. Brzezinska AK, Gebremedhin D, Chilian WM, et al. Peroxynitrite reversibly inhibits Ca^{2+}-activated K^+ channels in rat cerebral artery smooth muscle cells. Am J Physiol Heart Circ Physiol 2000;278: H1883–H1890.
104. Lu M, Wang W. Reaction of nitric oxide with superoxide inhibits basolateral K^+ channels in the rat CCD. Am J Physiol 1998;275:C309–C316.

105. Urakami-Harasawa L, Shimokawa H, Nakashima M, et al. Importance of endothelium-derived hyperpolarizing factor in human arteries. J Clin Invest 1997;100:2793–2799
106. Shimokawa H, Yasutake H, Fujii K, et al. The importance of the hyperpolarizing mechanism increases as the vessel size decreases in endothelium-dependent relaxations in rat mesenteric circulation. J Cardiovasc Pharmacol 1996;28:703–711.
107. Bauersachs J, Popp RL, Hecker M, et al. Nitric oxide attenuates the release of endothelium-derived hyperpolarizing factor. Circ 1996;94:3341–3347.
108. Miura H, Liu Y, Gutterman DD. Human coronary arteriolar dilation to bradykinin depends on membrane hyperpolarization. Circulation 1999;99:3132–3138.
109. Thollon C, Fournet-Bourguignon MP, Saboureau D, et al. Consequences of reduced production of NO on vascular reactivity of porcine coronary arteries after angioplasty: importance of EDHF. Br J Pharmacol 2002;136:1153–1161.
110. Miura H, Wachtel RE, Liu Y, et al. Flow-induced dilation of human coronary arterioles: important role of Ca^{2+}-activated K^+ channels. Circulation 2001;103:1992–1998.
111. Campbell WB, Gebremedhin D, Pratt PF, et al. Identification of epoxyeicosatrienoic acids as endothelium-derived hyperpolarizing factors. Circ Res 1996;78:415–423.
112. Fukao M, Hattori Y, Kanno M, et al. Evidence against a role of cytochrome P450–derived arachidonic acid metabolites in endothelium-dependent hyperpolarization by acetylcholine in rat isolated mesenteric artery. Br J Pharmacol 1997;120:439–446.
113. Gebremedhin D, Ma Y-H, Falck JR, et al. Mechanism of action of cerebral epoxyeicosatrienoic acids on cerebral arterial smooth muscle. Am J Physiol 1992;263:H519–H525.
114. Hoebel BG, Kostner GM, Graier WF. Activation of microsomal cytochrome P450 mono-oxygenase by Ca^{2+} store depletion and its contribution to Ca^{2+} entry in porcine aortic endothelial cells. Br J Pharmacol 1997;121:1579–1588.
115. Oltman CL, Weintraub NL, VanRollins M, et al. Epoxyeicosatrienoic acids and dihydroxyeicosatrienoic acids are potent vasodilators in the canine coronary microcirculation. Circ Res 1998;83:932–939.
116. Singer HA, Peach MJ. Endothelium-dependent relaxation of rabbit aortas. I. Relaxation stimulated by arachidonic acid. J Pharmacol Exp Ther 1983;226:790–795.
117. Weintraub NL, Fang X, Kaduce TL, et al. Potentiation of endothelium-dependent relaxation by epoxyeicosatrienoic acids. Circ Res 1997;81:258–267.
118. HU S, Kim HS. Activation of K^+ channel in vascular smooth muscles by cytochrome P450 metabolites of arachidonic acid. Eur J Pharmacol 1993;230:215–221.
119. Oyekan AO, McGiff JC, Rosencrantz-Weiss P, et al. Relaxant responses of rabbit aorta: Influence of cytochrome P450 inhibitors. J Pharmacol Exp Ther 1994;268:262–269.
120. Fisslthaler B, Popp R, Kiss L, et al. Cytochrome P450 2C is an EDHF synthase in coronary arteries. Nature 1999;401:493–497.
121. Fisslthaler B, Popp R, Michaelis UR, et al. Cyclic stretch enhances the expression and activity of coronary endothelium-derived hyperpolarizing factor synthase. Hypertension 2001;38:1427–1432.
122. Dong H, Waldron GJ, Galipeau D, et al. NO/PGI_2-independent vasorelaxation and the cytochrome P450 pathway in rabbit carotid artery. Br J Pharmacol 1997;120:695–701.
123. Li PL, Campbell WB. Epoxyeicosatrienoic acids activate K^+ channels in coronary smooth muscle through a guanine nucleotide binding protein. Circ Res 1997;80:877–884.
124. Roman RJ, Maier KG, Sun CW, et al. Renal and cardiovascular actions of 20-hydroxyeicosatetraenoic acid and epoxyeicosatrienoic acids. Clin Exp Pharmacol Physiol 2000;27:855–865.
125. Zygmunt PM, Edwards G, Weston AH, et al. Effects of cytochrome P450 inhibitors on EDHF-mediated relaxation in the rat hepatic artery. Br J Pharmacol 1996;118:1147–1152.
126. Zygmunt PM, Högestätt ED. Role of potassium channels in endothelium-dependent relaxation resistant to nitroarginine in the rat hepatic artery. Br J Pharmacol 1996;117:1600–1606.
127. Pfister SL, Falck JR, Campbell WB. Enhanced synthesis of epoxyeicosatrienoic acids by cholesterol-fed rabbit aorta. Am J Physiol 1991;261:H843–H852.
128. Burke-Wolin T, Abate CJ, Wolin MS, et al. Hydrogen peroxide–induced pulmonary vasodilation: role of guanosine 3',5'-cyclic monophosphate. Am J Physiol 1991;261:L393–L398.
129. Beny JL, Von der Weid P-Y. Hydrogen peroxide: an endogenous smooth muscle cell hyperpolarizing factor. Biochem Biophys Res Commun 1991;176:378–384.
130. Iida Y, Katusic ZS. Mechanisms of cerebral arterial relaxations to hydrogen peroxide. Stroke 2000;31:2224–2230.
131. Wei EP, Kontos HA, Beckman JS. Mechanisms of cerebral vasodilation by superoxide, hydrogen peroxide, and peroxynitrite. Am J Physiol 1996;271:H1262–H1266.

132. Sobey CG, Heistad DD, Faraci FM. Mechanisms of bradykinin-induced cerebral vasodilatation in rats: evidence that reactive oxygen species activate K^+ channels. Stroke 1997;28:2290–2294.
133. Huang A, Sun D, Smith CJ, et al. In eNOS knockout mice skeletal muscle arteriolar dilation to acetylcholine is mediated by EDHF. Am J Physiol Heart Circ Physiol 2000;278:H762–H768.
134. Matoba T, Shimokawa H, Nakashima M, et al. Hydrogen peroxide is an endothelium-derived hyperpolarizing factor in mice. J Clin Invest 2000;106:1521–1530.
135. Shimizu S, Shiota K, Yamamoto S, et al. Hydrogen peroxide stimulates tetrahydrobiopterin synthesis through the induction of GTP-cyclohydrolase I and increases nitric oxide synthase activity in vascular endothelial cells. Free Radic Biol Med 2003;34:1343–1352.
136. Jiang J, Thoren P, Caligiuri G, et al. Enhanced phenylephrine-induced rhythmic activity in the atherosclerotic mouse aorta via an increase in opening of K_{Ca} channels: relation to Kv channels and nitric oxide. Br J Pharmacol 1999;128:637–646.
137. Rusch NJ, De Lucena RG, Wooldridge TA, et al. A Ca^{2+}-dependent K^+ current is enhanced in arterial membranes of hypertensive rats. Hypertension 1992;19:301–307.
138. Rusch NJ, Runnells AM. Remission of high blood pressure reverese arterial potassium channel alterations. Hypertension 1994;23:941–945.
139. Asano M, Masuzawa-Ito K, Matsuda T, et al. Functional role of Ca^{2+}-activated K^+ channels in resting state of carotid arteries from SHR. Am J Physiol 1993;265:H843–H851.
140. Asano M, Masuzawa-Ito K, Matsuda T. Charybdotoxin-sensitive K^+ channels regulate the myogenic tone in the resting state of arteries from spontaneously hypertensive rats. Br J Pharmacol 1993;108: 214–222.
141. Kolias TJ, Chai S, Webb RC. Potassium channel antagonists and vascular reactivity in stroke-prone spontaneously hypertensive rats. Am J Hypertension 1993;6:528–533.
142. Asano M, Nomura Y, Ito K, et al. Increased function of voltage-dependent Ca^{++} channels and Ca^{++}-activated K^+ channels in resting state of femoral arteries from spontaneously hypertensive rats at prehypertensive stage. J Pharmacol Exp Ther 1995;275:775–783.
143. Paterno R, Heistad DD, Faraci FM. Functional activity of Ca^{2+}-dependent K^+ channels is increased in basilar artery during chronic hypertension. Am J Physiol 1997;272:H1287–H1291.
144. Martens JR, Gelband CH. Alterations in rat interlobar artery membrane potential and K^+ channels in genetic and nongenetic hypertension. Circ Res 1996;79:295–301.
145. Given MB, Greenberg SS, Giles TD. Coronary vasodilator responses to bradykinin in euglycemic and diabetic rats. Am J Hypertension 2001;14:446–454.

7 Nitric Oxide and Voltage-Gated Ca^{2+} Channels

Claudio Grassi, Marcello D'Ascenzo, and Gian Battista Azzena

CONTENTS

SUMMARY

Nitric oxide (NO) markedly influences intracellular calcium homeostasis by affecting the influx of Ca^{2+} through the plasma membrane and its release from intracellular stores. There is a large body of experimental evidence indicating that all mechanisms controlling the intracellular Ca^{2+} concentrations are regulated by NO. In excitable cells, activation of the voltage-gated Ca^{2+} channels is certainly the most effective means of generating Ca^{2+} influx from the extracellular space in response to membrane depolarization, and Ca^{2+} passing through these channels is known to regulate fundamental cellular functions, including neurotransmitter release, heart and smooth muscle contraction, synthesis and modulation of intracellular enzymes, regulation of gene expression, cell proliferation, and apoptosis. This chapter reviews numerous studies highlighting direct and indirect modulatory effects of NO on various types of voltage-gated Ca^{2+} channels and discusses the functional implications of the interaction of NO with voltage-gated Ca^{2+} channels.

Key Words: Voltage-gated Ca^{2+} channels; Ca_v1; $Ca_v2.1$; $Ca_v2.2$; Ca_v3; nitric oxide; cyclic guanasine 5'-monophosphate; protein kinase A; protein kinase G; phosphodiesterase.

From: *Signal Transduction and the Gasotransmitters: NO, CO, and H_2S in Biology and Medicine* Edited by: Rui Wang © Humana Press Inc., Totowa, NJ

1. INTRODUCTION

The physiological stimuli leading to nitric oxide (NO) synthesis and the functional roles played by this gasotransmitter in different systems have been explored in previous chapters. As we have seen, NO is produced by the conversion of L-arginine to L-citrulline following the activation of nitric oxide synthase (NOS). The most effective stimulus for NO synthesis is the increase in intracellular Ca^{2+} levels *(1,2)* that occurs with the activation of *N*-methyl-D-aspartate (NMDA) and other ligand-operated receptors and that of voltage-gated calcium channels (VGCCs). This chapter reviews the literature data showing that NO produced in response to Ca^{2+} signals can, in turn, modulate Ca^{2+} influx through VGCCs: a feedback control loop that significantly affects numerous calcium-mediated cell functions.

1.1. Nitric Oxide and Ion Channels

Over the last decade, interest in the functional roles played by NO in the central and peripheral nervous systems, where this gasotransmitter acts as an unconventional intercellular messenger, has grown *(3,4)*. Numerous studies have documented the importance of NO in synaptic plasticity phenomena such as long-term potentiation and depression *(4,5)*. NO is also involved in the repair of nerve injury *(6)* as well as the transmission of sensory information *(7–10)*. The physiological effects produced by NO in the nervous system are mediated primarily by its modulatory effects on ion channel function. Because of its gaseous nature, NO can spread from its site of production and act on various proteins in the surrounding cells, including the ion channels. As described in detail in Chapters 3 and 4, NO can modify channel function either directly or through the effects of intracellular second messengers.

The direct effects depend on *S*-nitrosylation of the channel proteins *(11)*. In baroreceptors and hippocampal neurons, NO-induced *S*-nitrosylation activates Na^+ channels *(12,13)*, but in C-type dorsal root ganglion neurons it inhibits fast, slow, and persistent Na^+ channels (*[14]*; *see* also Chapter 8). Ca^{2+}-activated K^+-channels and cyclic nucleotide–gated channels are also activated by NO-induced *S*-nitrosylation (*[15–19]*; see also Chapters 6 and 9), and the same mechanism is responsible for modulation of cardiac and skeletal muscle calcium release channels *(20,21)* and NMDA receptors *(22–25)*.

NO's indirect influence on ion channels is mediated by activation of soluble guanylate cyclase (sGC), which leads to increases in intracellular levels of cyclic guanosine 5'-monophosphate (cGMP) *(26,27)*. cGMP, in turn, can produce various effects. Apart from its direct action on cGMP-gated ion channels, cGMP can also modify the activities of voltage-gated and ligand-operated channels by activating protein kinase G (PKG), which then phosphorylates the target channel or proteins closely associated with it. cGMP can also stimulate or inhibit the activity of various cAMP-specific phosphodiesterases (PDEs). In this case, channel modulation is the result of changes in the intracellular levels of cyclic adenosine monophosphate (cAMP) and protein kinase A (PKA) activity.

These indirect effects of NO have been demonstrated in various ion channels. Ca^{2+}-activated K^+ channels in vascular smooth muscle fibers and adenosine triphosphate (ATP)-sensitive K^+ channels in ventricular myocytes are known to be activated by NO via cGMP-dependent protein kinase (*[28–30]*; see also Chapters 5 and 6). cGMP-mediated inhibition of Ca^{2+}-dependent Cl^- currents has been observed in cat tracheal smooth muscle following NO donor application *(31)*, which activates cGMP-gated non-selective cation channels in retinal ganglion cells *(32)*. Store-operated Ca^{2+} entry is also regulated by NO via a cGMP/PKG-dependent mechanism *(33)*.

We next focus our attention on the effects induced by this gasotransmitter on the different classes of VGCCs.

1.2. Voltage-Gated Ca^{2+} Channels (VGCCs)

Calcium channels are a large family of transmembrane proteins that mediate Ca^{2+} influx (I_{Ca}) in response to membrane depolarization. They are composed of four to five subunits, including the pore-forming α_1-subunit; the β-subunit; the $\alpha_2\delta$ complex; and, in some cases, a γ-subunit. The α_1-subunit is the largest Ca^{2+} channel component and also the most important because it includes the ion-permeant pore; the voltage sensor; and the sites involved in gating modulation by second messengers, drugs, and toxins (*see* references in ref. *34*). Different α_1-subunits have been identified, each characterizing a different type of VGCC with distinct biophysical and pharmacological properties.

The VGCCs basically are classified as low-voltage-activated (LVA) or high-voltage-activated (HVA) channels, and the structural and biophysical differences between the two classes also correspond to different functional roles.

The LVA channels, also known as T-type or, more recently, Ca_v3 channels, are characterized by an activation threshold ranging from about –50 to –55 mV and small, single-channel conductance. They are blocked by Ni^{2+} and unaffected by the organic Ca^{2+} antagonists and polypeptide toxins from snail and spider venoms that block the HVA channels. The term *T-type* reflects the "transient" kinetics of their currents, which are characterized by a pronounced and rapid time-dependent inactivation. T-type channels are mainly involved in those oscillations in the resting membrane potential that are responsible for rhythmic spontaneous firing of neurons and other excitable cells. For example, these channels are well known for the roles they play in thalamic relay neuron firing during sleep and in the prepotential of pacemaker cells that generate the rhythmic activity of the heart.

The HVA channels include different types: L-type, N-type, P/Q-type, and R-type. L-type channels (Ca_v1) are expressed in different variants in skeletal muscle, cardiac, smooth muscle, brain, and endocrine cells. They are characterized by a high threshold of activation; large single-channel conductance; and slow voltage-dependent inactivation. The latter property gave rise to the denomination L channels, which refers to their "long-lasting" activity. They are blocked by different classes of organic Ca^{2+} channel antagonists, including the dihydropyridines (DHPs), phenylalkylamines, and benzothiazepines. Evaluation of Ca^{2+} channel sensitivity to the DHPs is a very useful pharmacological method for distinguishing L-type channels from the other HVA channels, which often exhibit overlapping biophysical properties.

The N-type channel ($Ca_v2.2$) was the first DHP-resistant HVA channel identified in neurons, and the letter *N* was chosen to indicate that this channel is "neither T nor L." It is activated by strong depolarization (threshold of about –20 mV). Unaffected by DHPs, the N-type channels are selectively blocked by ω-conotoxin-GVIA (CTx-GVIA), a toxin obtained from the venom of the cone snail, *Conus geographus*. Their activity is also markedly modulated by numerous neurotransmitters.

The P-type channel ($Ca_v2.1$) was first identified in Purkinje cells (hence its name). It activates at relatively high thresholds and exhibits slow inactivation that is not easily distinguishable from that of L- and N-type channels. P channels are insensitive to either DHPs or CTx-GVIA. They can be blocked by low concentrations (i.e., in the nanomolar range) of ω-agatoxin-IVA, a toxin present in the venom of the funnel web spider.

R-type channels ($Ca_v2.3$) were identified as rapidly inactivating HVA channels that were "resistant" to the blocking effects of all of the agents discussed, i.e., the organic Ca^{2+} antagonists and the venom-based polypeptide toxins. They differ from the previously identified HVA channels in their rapid inactivation and more negative half-inactivation voltage. R currents are blocked by the spider toxin SNX-482.

Collectively, the HVA Ca^{2+} channels contribute to the regulation of numerous physiological functions, including neurotransmitter release, neurite outgrowth during development, heart and smooth muscle contraction, synthesis and modulation of intracellular enzymes, regulation of gene expression, cell proliferation, and apoptosis (*see* references in ref. *34*).

Neurotransmitter release is usually mediated by cooperation among multiple HVA channel types. The relative contribution of each varies depending on specific locations in the nervous system and the neurotransmitter being released *(34–36)*. In most mammalian nerve terminals, the major role is usually played by P/Q-type channels although N-type channels contribute between 20 and 30%, as also shown by our studies on noradrenaline release from cortical synaptosomes *(37)*. N-type channel activity is fundamental in the release of catecholamines from postganglionic sympathetic fibers in mammals and in neurotransmitter release from amphibian nerve terminals *(38)*. R-type channels have been suggested to provide transient surges of Ca^{2+} influx in response to brief depolarizations, and, at certain presynaptic terminals, they may contribute significantly to the Ca^{2+} increase that triggers neurotransmitter release. The L-type channels, whose involvement in neurotransmitter release is quite limited, play a very important role in the release of hormones from endocrine cells *(39–43)*, and their contribution to the regulation of gene expression is also well recognized *(44)*.

2. EFFECTS OF NO ON HVA CHANNELS

Although the action of NO on L- and N-type VGCCs has been widely investigated and clearly defined, its role in P/Q-type channel modulation has received less attention, and the picture that has emerged is still unclear. No information is available on the possible influence of NO on R-type VGCC.

2.1. L-Type Channels (Ca_v1)

The effects of NO on L-type Ca^{2+} channels have been explored in numerous experimental models of cardiac, smooth muscle, and neuroendocrine cells. NO appears to modulate the function of these channels by *direct* and *indirect* actions, and the second-messenger pathways involved in the latter mechanisms seem to vary with the cell types being studied.

A cGMP-independent, i.e., direct NO action similar to that exerted on NMDA receptors and Na^+ and K^+ channels *(12,16,25,45)*, has been described for L-type channels *(46–49)*. In glomus cells of the rabbit carotid body, application of the NO donors sodium nitroprusside (SNP) and spermine NO was found to provoke a short-latency, voltage-independent reduction in macroscopic Ba^{2+} currents *(48)*. This action could be prevented by blocking L-type channels with nisoldipine, whereas CTx-GVIA, which blocks the N-type channels, was ineffective. The SNP-induced current reduction could not be reproduced by exposing the cells to the cGMP analog 8-Br-cGMP, and it was prevented by application of *N*-ethylmaleimide, which produces covalent modifications of sulfhydryl groups that render them resistant to nitrosylation. These data were interpreted to suggest that glomus cell L-type channels are inhibited by NO via a cGMP-independent mechanism involving the modification of channel-protein sulfhydryl groups. Cardiovascular and smooth muscle L-type channels expressed in human embryonic kidney cells are similarly inhibited by NO donors *(47,49)*. At the single-channel level, the inhibition of L-type channels induced by *S*-nitrosothiol nitrosoglutathione in smooth muscle cells (SMCs) is characterized primarily by a reduction in the open probability that has been attributed to the nitrosation of critical thiol groups in the α_1-subunit *(49)*.

In other preparations, NO regulates Ca^{2+} channel activity through the activation of sGC and the subsequent elevation of cGMP levels. The latter compound can affect the function of L-type channels through various second-messenger pathways. In frog cardiac myocytes, NO/cGMP can either inhibit or activate L-type channels by modulating the activities of distinct cGMP-dependent PDEs *(50,51)*. Consequent changes in cAMP levels affect the cAMP-PKA pathway, which is a well-known upregulator of L-type channel activity. At low concentrations, cGMP may target the cGMP-inhibited PDE, PDE3, provoking an increase in cAMP levels that activates L-type channels. More frequently, however, cGMP-stimulated PDE2 is activated, and the consequent reduction in the cAMP concentration inhibits L-type channels *(50,52,53)*. The latter action has little or no effect on basal I_{Ca}: it seems to occur almost exclusively under conditions in which the current is activated by physiological stimuli that increase cAMP levels *(50,51)*. Recently, it has been suggested that the functional interaction between PDE2 and L-type channels occurs within a tight intracellular microdomain, and this finding might partially explain some of the contradictory data concerning regulation of L-type channel by NO and cGMP in cardiac cells *(54)*. Recent studies indicate that L-type channels in guinea pig cardiomyocytes are also subjected to tonic inhibition by NO via its effects on cGMP and PDE2 *(55,56)*.

NO-induced increases in cGMP levels have also been shown to modulate L-type channel function by activating PKG. PKG-mediated inhibition of L-type channel activity has been reported in chick and rabbit cardiac cells *(57,58)*, guinea pig SMCs *(59)*, and rat pinealocytes *(60)*. It might also be implicated in the cGMP-mediated reduction of high-threshold Ca^{2+} currents observed in rat dorsal root ganglion cells *(61)*.

We have investigated the NO effects on L-type channels of neuroendocrine cells, by studying the changes of both macroscopic currents and single-channel properties *(62–64)*. The experimental models used in these studies, rat insulinoma RINm5F cells and bovine chromaffin cells, are characterized by an abundant expression of L-type channels, whose activation is essential for the release of insulin and catecholamines, respectively *(39,40,65)*.

In RINm5F cells, extracellular application of the NO donor SNP produced dose-dependent reductions in macroscopic Ba^{2+} currents with an IC_{50} at 45 μM. At a concentration of 200 μM, SNP reduced these currents by approx 40% (39 ± 4%) with respect to control values, and similar results were observed following application of a different NO donor, (±)*S*-nitroso-*N*-acetylpenicillamine (SNAP). The NO-induced inhibition was voltage independent, with similar decreases observed at voltages ranging from –20 to +40 mV.

SNP was then tested on L-type currents alone (**Fig.** 1), which had been isolated from the DHP-insensitive HVA currents by treatment of cells with 2.5 μM ω-conotoxin-MVIIC *(65)*, and a similar reduction (30 ± 3%) was seen. The effects of SNP were prevented by the selective inhibitor of guanylate cyclase activity, 1H-[1,2,4]oxadiazolo[4,3-a]quinoxalin-1-one ([ODQ]; 10 μM), and they were mimicked by the membrane-permeant analog of cGMP, 8-Br-cGMP. These findings demonstrate that, in our experimental model, the action of NO on L-type channels is mediated by activation of guanylate cyclase and, consequently, by increased cGMP levels *(62)*.

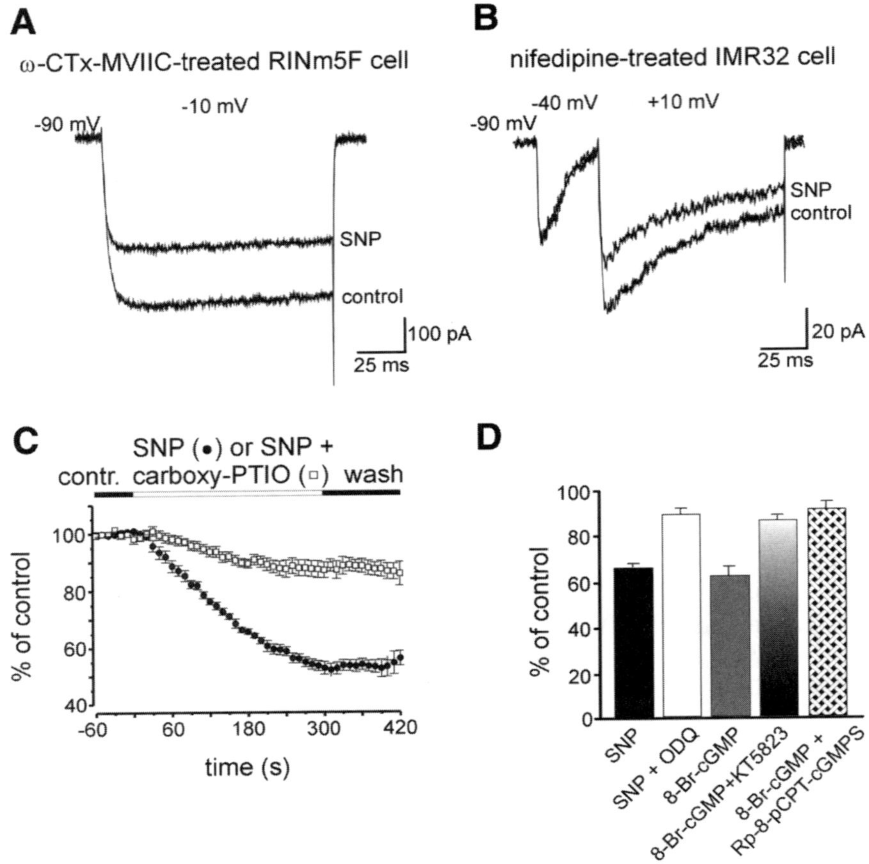

Fig. 1. The NO-donor SNP markedly reduces macroscopic L- and N-type currents via cGMP and PKG. **(A)** Inhibitory effect of 200 μ*M* SNP on L-type currents recorded during membrane depolarization at –10 mV from hoding potential (V_h) = –90 mV in rat insulinoma RINm5F cells. Current recordings were performed in the presence of 2.5 μ*M* ω-conotoxin-MVIIC blocking the non-L-type HVA currents. **(B)** Representative traces showing SNP effects on LVA and HVA currents elicited by step depolarization at +10 mV preceded by a 30-ms pre-pulse at –40 mV from V_h = –90 mV in IMR32 cells. Recordings were performed in the presence of 5 μ*M* nifedipine blocking L-type currents. **(C)** Time course of SNP effect on N-type current. The percentage changes in current amplitude are plotted against time during application of either 200 μ*M* SNP (●) or 200 μ*M* SNP together with the NO scavenger carboxy-PTIO (300 μ*M*) (□). (Data shown are averages (±SEM) of currents normalized with respect to the control amplitude (*n* = 9 in each group). **(D)** Percentage decrease in peak-current amplitude measured at third minute of cell exposure to 200 μ*M* SNP alone (*n* = 21), 200 μ*M* SNP in presence of 10 μ*M* ODQ (*n* = 5), 400 μ*M* 8-Br-cGMP (*n* = 7), and 400 μ*M* 8-Br-cGMP after cell treatment with either 1 μ*M* KT5823 (*n* = 9) or 20 μ*M* Rp-8-pCPT-cGMPS (*n* = 5).

In a more recent study conducted on bovine chromaffin cells, we used cell-attached recordings to identify the L-type channel gating properties that are influenced by NO *(64)*. In the presence of SNP, the channel open probability (P_o) was markedly reduced (–60%), and there was a considerable increase in the number of null sweeps **(Fig. 2)**. Again, similar effects were seen at all voltages tested (from 0 to +20 mV). The mean

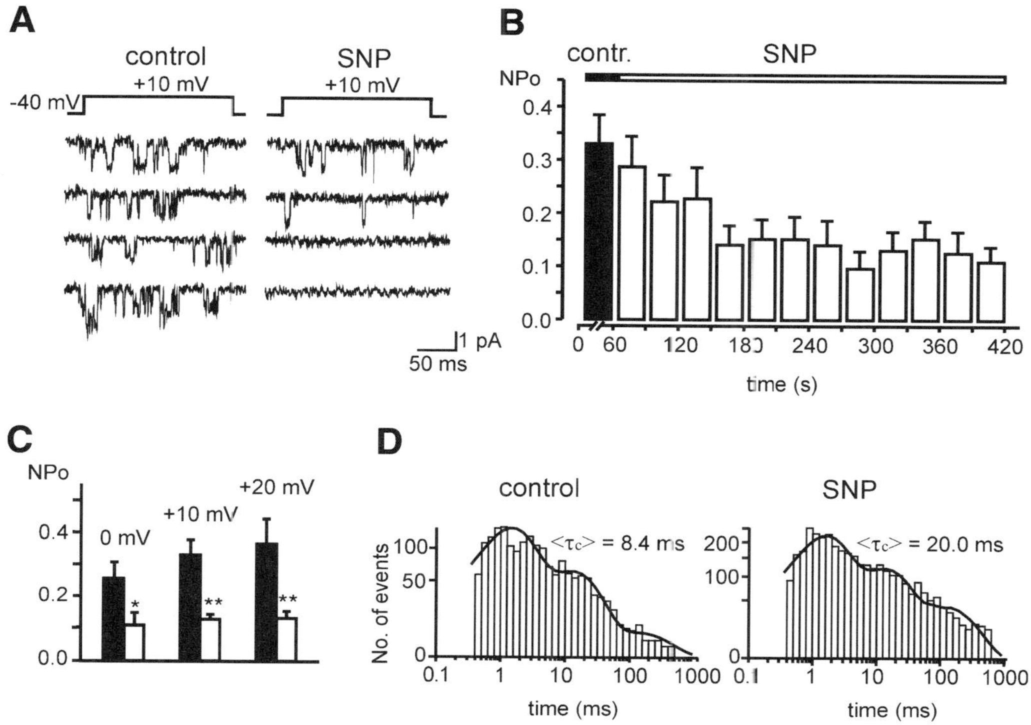

Fig. 2. SNP-induced inhibition of single L-type channel activity in bovine chromaffin cells. **(A)** Representative traces of L-type channel activity recorded in a cell-attached patch containing more than one channel under control conditions and during exposure to 200 μ*M* SNP. Ten micromolar ω-conotoxin-MVIIC, 5 μ*M* Bay K 8644, and a mixture of purinergic/opioidergic receptor antagonists (100 μ*M* suramin, 10 μ*M* naloxone) were present in the pipet solution. **(B)** Mean NP_o vs time before and during SNP application. In patches containing more than one channel, NP_o was calculated by adding the time duration of single, double, and even triple openings and dividing the sum by the duration of the analyzed time interval (for more details, *see* ref. *64*). The solid bar indicates the mean value of data collected during 1 min of recording from 13 patches under control conditions. Open bars are values during SNP application obtained at 30-s intervals. **(C)** Mean values of NP_o at 0, +10, and +20 mV in controls and during SNP application, with the reductions induced by the NO donor being 57.7, 60.6, and 64.9%, respectively. **(D)** Closed time distribution at +10 mV in control and during SNP application. The data were collected from five patches displaying single channel openings and depolarized with pulses of 600 ms to +10 mV. The distributions were fitted with a three exponential function with the following time constants: τ_{C1} = 1.3 ms (65%), τ_{C2} = 12.7 ms (32%), and τ_{C3} = 127 ms (3%) in controls and τ_{C1} = 1.4 ms (56%), and τ_{C2} = 12.5 ms (32%) and τ_{C3} = 127 ms (12%) with SNP. Average mean closed times ($<\tau_C>$) derived from the fit are given at the top right of each distribution and are similar to values obtained from the arithmetic mean of all data (21.9 ms vs 8.42 ms in controls). * $p < 0.05$; ** $p < 0.01$.

closed time was also appreciably prolonged by cell exposure to the NO donor (21.9 vs 8.42 ms in controls), whereas channel conductance and mean open time were unaffected.

As in the previously cited study, the effects of SNP were mimicked by 8-Br-cGMP, which reduced the L-type channel open probability by 56%. However, in cells pretreated with the PKG inhibitor KT5823, the P_o was not significantly modified by 8-Br-cGMP, which demonstrates that the observed changes in channel gating are indeed mediated by a cGMP/PKG pathway.

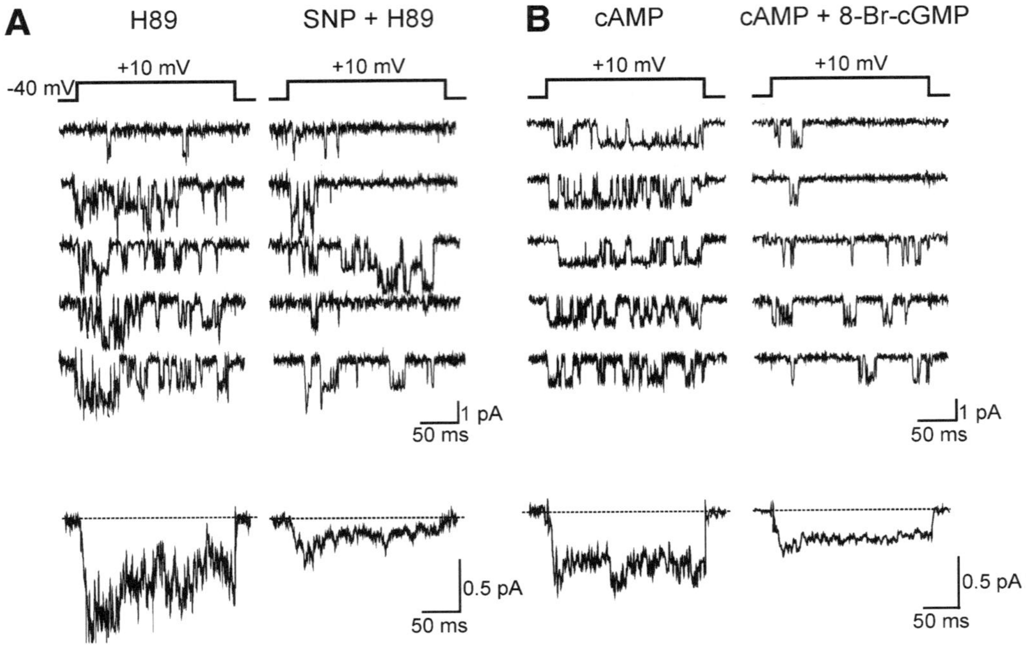

Fig. 3. The effects of SNP and 8-Br-cGMP on NP_0 of L-type channels are not influenced by blockade or activation of PKA. **(A)** Control multichannel activity in a bovine chromaffin cell incubated for 20 min with 1 μ*M* H89 (left) is markedly inhibited by exposure to 200 μ*M* SNP (right). The strong depression induced by SNP is comparable with that shown in **Fig. 2A**. **(B)** Single L-type channel activity recorded from a chromaffin cell incubated for 30 min with 8-CPT-cAMP (1 m*M*) (left) exhibiting high P_o activity. The addition of 400 μ*M* 8-Br-cGMP (right) produced a marked inhibition, which is comparable with that observed in the absence of cAMP application. Lower traces in **(A)** and **(B)** are averaged currents obtained from 10 and 40 traces respectively recorded in control and during application of the test agents from the same patches shown in the upper panels.

As noted, there is a large body of experimental evidence indicating that NO can also influence L-type channel function by modulating the channel-activating cAMP/PKA pathway. To determine whether the observed inhibition of L-type channel was related to any modulation of the PKA-mediated regulatory pathway, we repeated our single-channel recordings in cells treated with the PKA inhibitor H89 (1 μ*M*). The SNP-induced decrease in P_o in these cells was almost identical (–59%) to that observed in the absence of PKA blockade **(Fig. 3A)**. Moreover, although the membrane-permeant cAMP analog 8-CPT-cAMP (1 m*M*) markedly upregulated L-type channel activity, it did not prevent the inhibitory effects of 8-Br-cGMP **(Fig. 3B)**. The cAMP/PKA and cGMP/PKG pathways thus appear to represent two distinct modalities for the regulation of L-type channel gating. However, both can be influenced by NO synthesis, and it is possible that in some cell systems they also function in a cooperative manner to modulate the activity of L-type channels.

The PKG-mediated downregulation of neuroendocrine L-type channel gating is very similar to the voltage-independent inhibition produced on the same channels by opioids and ATP *(66)*. This effect, which is mediated by pertussis toxin (PTx)-sensitive G-proteins, consists, in fact, of a marked decrease in P_o that is primarily because of increases in closed times and the number of null sweeps. This similarity prompted us to investigate the possibility that the two signaling pathways might be operating at closely related binding sites. The two effects proved to be additive, however, and 8-Br-cGMP reduced the *Po* by approx 50% in patches in which the G-proteins had been activated by purinergic and opioidergic receptor agonists *(64)*. These data seem to suggest that $G_{i/o}$ protein subunit binding occurs in a channel region that is distinct from the PKG phosphorylation site, which, in the cardiac α_{1C}-subunit, is located at position Ser^{533} in the I-II cytoplasmic linker *(67)*.

2.2. N-Type Channels ($Ca_v2.2$)

More recently, we demonstrated that the inhibitory pathway underlying NO's modulation of neuroendocrine L-type channels is also involved in the regulation of N-type channels in human neuronal cells *(68)*. We investigated the functional roles of NO, cGMP, and PKG in nifedipine-treated human neuroblastoma IMR32 cells. After the blockade of L-type channels, the HVA currents recorded in these cells are, in fact, almost completely (>90%) attributable to N-type channel activity *(68–70)*. Cell exposure to the NO donor SNP consistently reduced the amplitude of N-type currents without affecting their activation and inactivation kinetics. The effect produced by NO was observed after a latency of 20–40 s and reached maximal intensity between min 4 and 5 of drug application. As shown in **Fig. 1C**, the current was reduced by 34.1 ± 1.5% (with respect to control values) 3 min after the addition of SNP and by 46.9 ± 1.6% 2 min later. When SNP was applied together with the NO scavenger carboxy-PTIO no significant decrease in N-type current amplitude was observed.

The N-type channel inhibition produced by SNP resembled that exerted on L-type channel activity in that it was prevented by blockade of guanylate cyclase (10 μ*M* ODQ) and reproduced by application of 400 μ*M* 8-Br-cGMP (**Fig. 1D**). Moreover, the inhibitory effects of 8-Br-cGMP were almost completely abolished by cell treatment with the specific PKG inhibitors KT5823 (1 μ*M*) or Rp-8-pCPT-cGMPS (20 μ*M*).

The inhibition of macroscopic N-type current produced by NO via cGMP/PKG activation was paralleled at the single-channel level by a marked voltage-independent reduction (–59%) in channel open probability (**Fig. 4**). The mean closed time was also significantly increased (16.08 ± 0.94 vs 9.44 ± 0.67 ms), as was the null sweep probability, but no significant changes were observed in channel conductance, mean open time, or latency of the first opening.

Yoshimura et al. *(71)* have also described NO's inhibition of N-type channels via the cGMP pathway in dorsal root ganglion neurons from the rat urinary bladder. They found that macroscopic HVA currents were reduced by 23–27% in neurons exposed to 500 μ*M* SNAP. The experimental model they used expressed both nimodipine-sensitive (L-type) and CTx-GVIA-sensitive (N-type) channels, but the effect of NO seemed to be mediated almost exclusively by the latter channels. In fact, the SNAP-induced inhibition of current was almost completely abolished by CTx-GVIA and unaffected by nimodipine.

By contrast, in rat sympathetic neurons, NO and cGMP reportedly produce a mild increase in Ca^{2+} influx. When Chen and Schofield *(72)* exposed these cells to high concentrations of NO donors (500 μ*M* SNP or SNAP), macroscopic HVA currents increased slightly (approx +10 and +17%, respectively, compared with controls). The channel-activating effects of SNP were significantly diminished by application of the guanylate cyclase inhibitor methylene blue, suggesting that the NO-induced Ca^{2+} channel activation was mediated by cGMP.

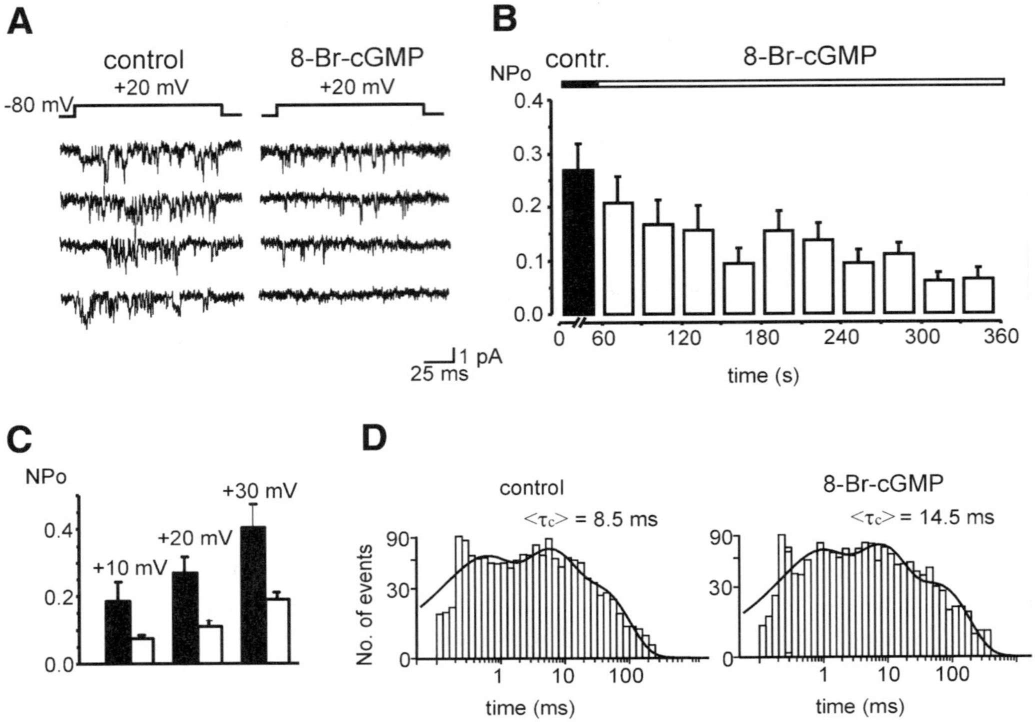

Fig. 4. 8-Br-cGMP markedly reduces the open probability of N-type channels in human neuroblastoma cells. **(A)** Representative traces of N-type channel activity recorded in a cell-attached patch containing more than one channel under control conditions and during exposure to 400 μ*M* 8-Br-cGMP. Nifedipine (5 μ*M*) was present in the pipette solution to block L-type channels, and depolarization at +20 mV was delivered from $V_h = -80$ mV. **(B)** Mean changes in NP_o induced by 8-Br-cGMP in 7 patches containing two or three N-type channels. The solid bar shows the NP_o value obtained by averaging data collected during 1 min of recording under control conditions before application of the test agent. The open bars indicate mean NP_o obtained by averaging the data collected in the seven studied patches over 30-s periods. **(C)** The effect of 8-Br-cGMP on NP_o is voltage independent, with the percentage decrease found at +10, +20, and +30 mV being 60.3, 59.3, and 52.9%, respectively. **(D)** Closed time distribution at +20 mV is fitted with a three-exponential function with the following time constants: τ_{C1} = 0.45 ms (34.1%), τ_{C2} = 4.86 ms (43.1%), and τ_{C3} = 27.51 ms (22.8%) in controls and τ_{C1} = 0.66 ms (33.1%), τ_{C2} = 6.55 ms (45.1%), and τ_{C3} = 51.83 ms (21.8%) in the presence of 8-Br-cGMP. The mean ($\langle\tau_C\rangle$) values derived from the fit are given at the top of each distribution and compare well with those derived by the arithmetic mean of the collected data (τ_C = 9.44 ± 0.67 ms in controls and 16.08 ± 0.94 ms with 8-Br-cGMP).

Studies performed on primary cell cultures from lower vertebrates have also revealed modest increases in macroscopic Ba^{2+} currents in response to NO donors *(73,74)*. In salamander retinal ganglion cells, the enhancement in current induced by high concentrations of SNAP (1 m*M*) is abolished by CTx-GVIA, suggesting that N-type channels are the targets of this activation. This effect was mimicked by the cGMP analog CPT-cGMP, and it was blocked by PKG inhibitors. However, in a previous study one of the same research groups found that Ca^{2+} influx in salamander rod photoreceptors is enhanced by NO donors through a cGMP-*independent* mechanism *(73)*.

The data just reviewed indicate that NO-induced modulation of N-type channels varies considerably depending on the experimental model used, with effects ranging from marked depression to moderate activation. The heterogeneity of these findings can be attributed to numerous factors. First, the function of N-type channels in mammalian cells and the mechanisms underlying their modulation are probably different from those of lower vertebrate cells *(74)*. Second, it is also important to recall that variants of the N-type channel α_{1B}-subunit have been identified, each exhibiting different functional properties *(75)*. Finally, there are also several functionally distinct cGMP-dependent protein kinases, which reportedly mediate different physiological effects *(76,77)*. Therefore, the type of modulation exerted on these channels by NO in a given experimental model probably depends on the specific N-type channel variant and/or G-kinase isoforms expressed in that preparation. Similar heterogeneity has been observed for L-type channels, which, as we have seen, are subject to different mechanisms of modulation in the various cell types. However, as far as mammalian neurons are concerned, our data and those of Yoshimura et al. *(71)* suggest that the predominant effect of NO on the N-type channels is inhibitory, and this action would be consistent with the principal effects of this nonconventional transmitter on other HVA Ca^{2+}-channel types.

2.3. P/Q-Type Channels ($Ca_v2.1$)

The action exerted by NO on P/Q-type channels has been investigated less extensively, and the picture that has emerged is by no means complete. Chen et al. *(78)* studied the effects of H_2O_2 in baby hamster kidney (BHK) cells transfected with $Ca_v2.1/\beta_{1a}/\alpha_2\delta$ cDNAs to produce stable expression of a VGCC identified as a P/Q-type *(79)*. The oxidant agent produced an irreversible increase in Ca^{2+} influx. Similar results were observed in transfected *Xenopus* oocytes, and the kinetics of the H_2O_2-potentiated currents were different when the β_3- rather than β_{2a}-subunit was present in the channel complex *(80)*. The current-enhancing effect of H_2O_2 in BHK cells was mimicked by the NO donors SNAP and diethylamine NONOate. However, it did not appear to be cGMP dependent, because it was not antagonized by the guanylate cyclase inhibitor ODQ, but was abolished by application of the reducing agent dithiothreitol. These observations led the investigators to suggest that the increased Ca^{2+} influx might be the result of cysteine-residue oxidation. In an earlier study, however, cysteine modification by DTBNP had failed to produce any appreciable increase in P/Q-type currents *(80)*. The effects of the oxidant agents were voltage dependent and, at voltages higher than +25 mV, the increase in current amplitude was often replaced by a slight decrease. Enhancement of current similar to that caused by H_2O_2 and the NO donors was also achieved through overexpression of NOS (obtained by cell incubation with endothelial NOS–adenovirus particles) and its subsequent activation with the Ca^{2+} ionophore A23187.

Collectively, these findings have been interpreted to suggest that P/Q-type channel oxidation by different agents, including NO, is capable of upregulating the activity of these channels although, as the investigators themselves noted, other explanations cannot be excluded (e.g., H_2O_2-induced oxidation of intracellular enzymes or membrane lipids).

Our own data regarding the effects of NO on P/Q-type channels are limited, but they pave the way for an alternative to the aforementioned hypothesized scenario. In **Subheading 2.1.1.**, we described the inhibitory action exerted by NO and cGMP on Ca^{2+} influx in rat insulinoma RINm5F cells *(62)*. In this experimental model, HVA currents were generated mainly by L-type channels and by a nifedipine-insensitive channel that was partially and reversibly inhibited by ω-agatoxin-IVA and ω-conotoxin-MVIIC. The latter channel has been considered to be an anomalous P/Q-type channel and defined as "Q-like" *(65,81)*. Although the main focus of our study was the effect of NO on the total HVA current, we also conducted some experiments in the presence of L-type channel blocker (5 μ*M* nifedipine). The residual HVA current that could be recorded under these experimental conditions—i.e., the nifedipine-resistant HVA current generated mainly by the so-called Q-like channels—was also significantly diminished (–36%) by NO. This is a preliminary observation, but it suggests that the P/Q-type channel of rat insulinoma cells might be downregulated by NO via a cGMP-dependent mechanism similar to that demonstrated for L- and N-type channels.

More detailed studies are required to fully elucidate the functional roles of NO and other oxidizing agents, as well as those of the NO-activated second-messenger cascade, in regulating the activity of P/Q-type channels. If our knowledge of L- and N-type channels is any indication, multiple mechanisms (direct and indirect) could be used by NO to modulate the activity of P/Q-type channels. Cell types may differ in terms of the predominant mechanism used to achieve this regulation, or, in certain cells and/or under certain physiological conditions, channel modulation might be achieved through a combination of two or more mechanisms. These channels play a fundamental role in the control of neurotransmitter release in the mammalian nervous system *(36,82,83)*, and for this reason their modulation by NO merits much more active investigation.

3. NO AND LVA, T-TYPE VGCCs (Ca_v3)

To our knowledge there is no clear evidence in the literature that T-type channels (Ca_v3) are influenced by NO. As shown in **Fig. 1B**, the stimulation protocol we used to activate VGCCs of IMR32 cells allowed us to segregate LVA from HVA currents by delivery of a pre-pulse depolarizing stimulus at –40 mV prior to the test depolarization. In these human neuroblastoma cells, NO donor application that significantly reduced HVA currents flowing through N-type channels had no significant effects on the LVA currents *(68)*. Similar results have been described in human coronary myocytes *(84)*, in which the NO donor SNAP and cGMP dose-dependently inhibited L-type channels without significantly altering T-type currents. However, in newt olfactory receptor cells, application of cGMP or blockade of cGMP PDE increased a transient inward ion current flowing through Na^+ channels and T-type and L-type Ca^{2+} channels *(85)*. The effect of cGMP, which was mediated by PKG activation, was confined for the most part to the Na^+ and T-type Ca^{2+} components of this current. The enhancement of T-type current was characterized primarily by a negative shift of the activation curve, whereas no marked changes were observed in the peak amplitude of the intensity-to-voltage curve.

It is clear that the action of NO on T-type channels needs to be explored in greater detail. It is important to consider that multiple T-type channel isoforms have been identified (i.e., $Ca_v3.1$, $Ca_v3.2$, and $Ca_v3.3$), and they are encoded by three different genes. Future studies will probably clarify whether NO exerts any physiological action on the different Ca_v3 variants.

4. FUNCTIONAL IMPLICATIONS OF NO-INDUCED MODULATION OF VGCCs

In light of the data reviewed thus far (**Fig. 5**), it seems quite likely that NO is capable of influencing numerous cell functions regulated by Ca^{2+} signals in excitable cells. The

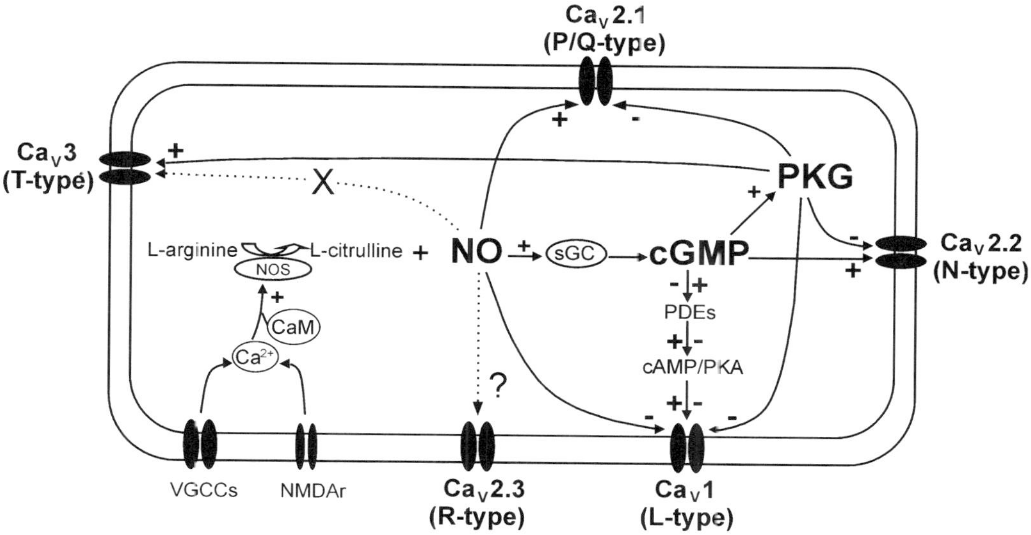

Fig. 5. Different actions of NO and the related second messengers on the different VGCC types.

spectrum of functions potentially influenced by NO-induced VGCC modulation is immense, and they cannot be explored exhaustively in this setting. Therefore, we will limit our discussion to a few of the more relevant examples.

In neural and endocrine cells, the most widely studied function regulated by HVA Ca^{2+} channel activation is exocytosis of neurotransmitters and hormones, respectively. With a few exceptions *(72,74,78)*, the vast majority of the studies conducted thus far indicate that NO produces direct or cGMP-mediated *inhibition* of neuronal HVA channel activity. These findings are suggestive of a downregulation of Ca^{2+} entry in response to physiological stimuli that trigger NO synthesis, and they are compatible with diminished neurotransmitter release and depression of synaptic transmission. Indeed, several studies have shown that NO and cGMP inhibit synaptic transmission through actions exerted at the presynaptic level. At glutamatergic synapses in rat visual cortex slices, the cGMP analog 8-Br-cGMP and the specific activator of PKG, Sp-8-Br-PET-cGMPS, have been shown to reduce the stimulus-evoked EPSPs *(86)*. In cultured visual cortex neurons, the same agents induce a reduction in spontaneous EPSC frequency that is associated with a decrease in VGCC currents. NO has been reported to inhibit both the release of noradrenaline from sympathetic nerve terminals and the vasoconstrictor response to adrenergic nerve stimulation *(38,87)*. Moreover, the cGMP/PKG pathway has been considered responsible for the SNP-induced inhibition of the release of glutamate from rat hippocampal nerve terminals *(88)*.

On the other hand, numerous studies have revealed NO-induced enhancement of neurotransmitter release and synaptic transmission *(89–92)*. One must recall, however, that different steps in the neurotransmitter release process are potentially affected by the NO. In addition to its effects on Ca^{2+} entry into nerve terminals, NO has been found to promote neurotransmitter vesicle docking/fusion by stimulating the formation of the VAMP/SNAP-25/syntaxin 1a core complex *(93)*. This effect involves *S*-nitrosylation of synaptic proteins, and it is not dependent on Ca^{2+} influx. Furthermore, inhibition of Ca^{2+} influx through VGCCs is only one of the mechanisms used by NO to control Ca^{2+} signals in neurons. Increases in intracellular Ca^{2+} levels can be the result of other NO-mediated effects, including the activation of NMDA receptors, store- and second-messenger-operated Ca^{2+} channels, and release from ryanodine-sensitive and inositol 1,4,5-triphosphate–sensitive Ca^{2+} stores *(94–96)*. NO and cGMP also activate a voltage-independent Ca^{2+} current that probably mediates the effects of muscarinic receptor agonists *(97)*. The actions exerted by NO on Na^+ and K^+ channels (see Chapters 5, 6, and 8) also have obvious repercussions on membrane potential and neuronal excitability and, therefore, on VGCC activation. The net result of these multiple, and in some cases contrasting, effects can be expected to vary with the preparation being examined, and it is also likely to be influenced by different physiological conditions. However, the NO-induced modulation of VGCCs described in this chapter must be considered a very important component of the complex mechanism controlling neurotransmitter release and synaptic plasticity.

As for hormone release by neuroendocrine cells, it should be significantly influenced by the inhibitory cGMP/PKG-mediated effect of NO that we demonstrated in L-type channels of rat insulinoma and bovine chromaffin cells. In the latter cells, marked inhibition of acetylcholine- and KCl-stimulated catecholamine secretion can be induced with NO, SNP, or 8-Br-cGMP *(98)*. NO has also been reported to inhibit neuropeptide secretion from posterior pituitary nerve terminals *(99,100)* and glucose-induced insulin secretion *(101)*. However, other studies have shown that NO might actually facilitate insulin secretion from pancreatic β-cells by enhancing the release of Ca^{2+} from mitochondria and endoplasmic reticulum stores *(102–103)*. The same mechanisms may be involved in NO's reported enhancement of the release of catecholamine from chromaffin cells under basal conditions *(98–104)*.

The NO-induced inhibition of cAMP-stimulated L-type channels observed in cardiomyocytes has important functional implications for heart function and its regulation by the autonomic nervous system and hormones *(105,106)*. It is well documented that the positive inotropic effect induced by sympathetic command is mediated by β-adrenergic receptor activation, which results in increased intracellular levels of cAMP and PKA activity and, consequently, the phosphorylation of numerous regulatory proteins, including the L-type Ca^{2+} channels *(105,107)*. Therefore, modulation of the channel-activating cAMP/PKA pathway produced by NO and cGMP via modifications in PDE activity is a fundamental element in the regulation of cardiac function.

The examples just provided give researchers some idea of the important physiological consequences of NO's modulatory effects on VGCCs, but the list is by no means complete. The actions of NO that we have discussed in this chapter can potentially affect all cell functions that are Ca^{2+} regulated.

5. CONCLUSION

NO, as being an unconventional gasotransmitter *(108)*, is capable of markedly influencing intracellular Ca^{2+} homeostasis in both excitable and nonexcitable cells by multiple mechanisms. Ca^{2+} influx through VGCCs and ligand-operated receptor channels, and capacitative Ca^{2+} entry and Ca^{2+} release from intracellular stores, are all influenced by NO and/or by the intracellular second messengers produced in response to NO synthesis. Different mechanisms involved in this complex control system can produce opposite effects on intracellular Ca^{2+} levels. However, not all of the regulatory pathways are operative in every cell type, and the net result of NO will thus depend on the mechanism that prevails in the given model. In excitable cells, modulation of VGCCs can be expected to be the predominant or at least one of the most important pathways of NO-mediated Ca^{2+} regulation. In this manner, NO can potentially modify all of the calcium-mediated functions in these cells. NO synthesized in neurons and glial cells can have potent effects on neurotransmitter release and, therefore, synaptic transmission. In all probability, it can also play a crucial role in the interplay between neurons and glial cells, which modulate synaptic strength by sequestering neurotransmitters from the synaptic cleft and releasing them through calcium-dependent and -independent mechanisms.

ACKNOWLEDGMENTS

This research was supported by grants from Ministero dell' Istruzione, dell' Università" e della Ricerca (MIUR) and local funds from Catholic University.

REFERENCES

1. Garthwaite J, Charles SL, Chess-Williams R. Endothelium-derived relaxing factor release on activation of NMDA receptors suggests role as intercellular messenger in the brain. Nature 1988;336: 385–388.
2. Bredt DS, Snyder SH. Isolation of nitric oxide synthetase, a calmodulin-requiring enzyme. Proc Natl Acad Sci USA 1990;87:682–685.
3. Rand MJ, Li CG. Nitric oxide as a neurotransmitter in peripheral nerves: nature and mechanism of transmission. Annu Rev Physiol 1995;57:659–682.
4. Garthwaite J, Boulton CL. Nitric oxide signaling in the central nervous system. Annu Rev Physiol 1995;57:683–706.
5. Shuman EM, Madison DV. Nitric oxide and synaptic function. Annu Rev Neurosci 1994;17:153–183.
6. Kumar SM, Portefield M, Muller KJ, et al. Nerve injury induces a rapid efflux of nitric oxide (NO) detected with a novel NO microsensor. J Neurosci 2001;21:215–220.
7. Haley JE, Dickenson AH, Schachter M. Electrophysiological evidence for a role of nitric oxide in prolonged chemical nociception in the rat. Neuroscience 1992;31:251–258.
8. Sousa AM, Prado WA. The dual effect of a nitric oxide donor in nociception. Brain Res 2001;897:9–19.
9. Grassi C, Santarelli R, Nisticò S, et al. Possible modulation of auditory middle latency responses by nitric oxide in the inferior colliculus of anaesthetized rats. Neurosci Lett 1995;196:213–217.
10. Azzena GB, Ferraresi A, Filippi GM, et al. Proprioceptive afferents from extraocular muscles and oculomotor control: functional role of nitric oxide. Pflügers Arch 2000;439:R263:38.
11. Jaffrey SR, Erdjument-Bromage H, Ferris CD, et al. Protein S-nitrosylation: a physiological signal for neuronal nitric oxide. Nat Cell Biol 2001;3:193–196.
12. Li Z, Chapleau MW, Bates JN, et al. Nitric oxide as an autocrine regulator of sodium currents in baroreceptor neurons. Neuron 1998;20:1039–1049.
13. Hammarström AKM, Gage, PW. Nitric oxide increases persistent sodium current in rat hippocampal neurons. J Physiol (Lond) 1999;520:451–461.
14. Renganathan M, Cummins TR, Waxman SG. Nitric oxide blocks fast, slow and persistent Na^+ channels in C-type DRG neurons by S-nitrosylation. J Neurophysiol 2002;87:761–775.
15. Kitamura K, Lian Q, Carl A, et al. S-Nitrosocysteine, but not sodium nitroprusside produces apamin-sensitive hyperpolarization in rat gastric fundus. Br J Pharmacol 1993;109:415–423.
16. Bolotina VM, Najibi S, Palacino JJ, et al. Nitric oxide directly activates calcium-dependent potassium channels in vascular smooth muscle. Nature 1994;368:850–853.
17. Ahern GP, Hsu S-F, Jackson MB. Direct actions of nitric oxide on rat neurohypophysial K^+ channels. J Physiol (Lond) 1999;520:165–176.
18. Lang RJ, Harvey JR, McPhee GJ, et al. Nitric oxide and thiol reagent modulation of Ca^{2+}-activated K^+ (BK_{Ca}) channels in myocytes of the guinea-pig taenia caeci. J Physiol (Lond) 2000;525:363–376.
19. Broillet MC, Firestein S. Direct activation of the olfactory cyclic nucleotide–gated channel through modification of sulfhydryl groups by NO compunds. Neuron 1996;16:377–385.
20. Xu L, Eu JP, Meissner G, et al. Activation of the cardiac calcium release channel (ryanodine receptor) by poly-S-nitrosylation. Science 1998;279:234–237.

21. Suko J, Drobny H, Hellmann G. Activation and inhibition of purified skeletal muscle calcium release channel by NO donors in single channel current recordings. Biochim Biophys Acta 1999;1451:271–287.
22. Hoyt Kr, Tang LH, Aizenman E, et al. Nitric oxide modulates NMDA-induced increases in intracellular Ca^{2+} in cultured rat forebrain neurons. Brain Res 1992;592:310–316.
23. Manzoni O, Prezeau L, Marin P, et al. Nitric oxide-induced blockade of NMDA receptor. Neuron 1992;8:653–662.
24. Lei SZ, Pan ZH, Aggarwal SK, et al. Effect of nitric oxide production on the redox modulatory site of the NMDA receptor-channel complex. Neuron 1992;8:1087–1099.
25. Choi YB, Tenneti L, Le DA, et al. Molecular basis of NMDA receptor–coupled ion channel modulation by S-nitrosylation. Nat Neurosci 2000;3:15–21.
26. Moncada S, Palmer RM, Higgs EA. Nitric oxide: physiology, pathophysiology, and pharmacology. Pharmacol Rev 1991;43:109–142.
27. Snyder SH. Nitric oxide: first in a new class of neurotransmitters. Science 1992;257:494–496.
28. Archer SL, Huang JMC, Hampl V, et al. Nitric oxide and cGMP cause vasorelaxation by activation of a charybdotoxin-sensitive K channel by cGMP-dependent protein kinase. Proc Natl Acad Sci USA 1994;91:7583–7587.
29. Swayze RD, Braun AP. A catalytically inactive mutant of type I cGMP-dependent protein kinase prevents enhancement of large conductance, calcium-sensitive K^+ channels by sodium nitroprusside and cGMP. J Biol Chem 2001;276:19729–19737.
30. Han J, Kim N, Kim E, et al. Modulation of ATP-sensitive potassium channels by cGMP-dependent protein kinase in rabbit ventricular myocytes. J Biol Chem 2001;276:22,140–22,147.
31. Waniishi Y, Inoue R, Morita H, et al. Cyclic GMP-dependent but G-kinase-independent inhibition of Ca^{2+}-dependent Cl^- currents by NO donors in cat tracheal smooth muscle. J Physiol (Lond) 1998;511:719–731.
32. Ahmad I, Leinders-Zufall T, Kocsis JD, et al. Retinal ganglion cells express a cGMP-gated cation conductance activatable by nitric oxide donors. Neuron 1994;12:155–165.
33. Kwan H-Y, Huang Y, Yao X. Store-operated calcium entry in vascular endothelial cells is inhibited by cGMP via protein kinase G–dependent mechanism. J Biol Chem 2000;275:6758–6763.
34. Hofmann F, Lacinova L, Klugbauer N. Voltage-dependent calcium channels: from structure to function. Rev Physiol Biochem Pharmacol 1999;139:33–87.
35. Dunlap K, Luebke JI, Turner TJ. Exocytotic Ca^{2+} channels in mammalian central neurons. Trends Neurosci 1995;18:89–98.
36. Catterall WA. Structure and function of neuronal Ca^{2+}-channels and their role in neurotransmitter release. Cell Calcium 1998;24:307–323.
37. Grassi C, Martire M, Altobelli D, et al. Characterization of Ca^{2+}-channels responsible for K^+-evoked [^{3}H]noradrenaline release from rat brain cortex synaptosomes and their response to amyotrophic lateral sclerosis IgGs. Exp Neurol 1999;159:520–527.
38. Tesfamariam B, Weisbrod RM, Cohen RA. Endothelium inhibits responses of rabbit carotid artery to adrenergic nerve stimulation. Am J Physiol 1987;253:H792–H798.
39. García AG, Sala F, Reig JA, et al. Dihydropyridine BAY-K-8644 activates chromaffin cell calcium channels. Nature 1984;309:69–71.
40. Prentki M, Matschinsky FM. Ca^{2+}, cAMP and phospholipid-derived messengers in coupling mechanisms of insulin secretion. Physiol Rev 1987;67:1185–1248.
41. Lopez MG, Villaroya M, Lara B, et al. Q- and L-type Ca^{2+} channels dominate the control of secretion in bovine chromaffin cells. FEBS Lett 1994;349:331–337.
42. Kim SJ, Lim W, Kim J. Contribution of L- and N-type calcium currents to exocytosis in rat adrenal medullary chromaffin cells. Brain Res 1995;675:289–296.
43. Lomax RB, Michelena P, Nunez L, García-Sancho J, et al. Different contributions of L- and Q-type Ca^{2+} channels to Ca^{2+} signals and secretion in chromaffin cell subtypes. Am J Physiol 1997;272:C476–C484.
44. Dolmetsch RE, Pajvani U, Fife K, et al. Signaling to the nucleus by an L-type calcium channel-calmodulin complex through the MAP kinase pathway. Science 2001;294:333–339.
45. Aizenman E, Brimecombe JC, Potthoff WK, et al. Why is the role of nitric oxide in NMDA receptor function and dysfunction so controversial? Prog Brain Res 1998;118:53–71.
46. Campbell DL, Stamler JS, Strauss HC. Redox modulation of L-type calcium channels in ferret ventricular myocytes: dual mechanism regulation by nitric oxide and *S*-nitrosothiols. J Gen Physiol 1996;108:277–293.
47. Hu H, Chiamvimonvat N, Yamagishi T, et al. Direct inhibition of expressed cardiac L-type Ca^{2+} channels by *S*-nitrosothiol nitric oxide donors. Circ Res 1997,81:742–752.

48. Summers BA, Overholt JL, Prabhakar NR. Nitric oxide inhibits L-type Ca^{2+} current in glomus cells of the rabbit carotid body via a cGMP-independent mechanism. J Neurophysiol 1999;81:1449–1457.
49. Poteser M, Romanin C, Schreibmayer W, et al. *S*-Nitrosation controls gating and conductance of the α_1 subunit of class C L-type Ca^{2+} channels. J Biol Chem 2001;276:14,797–14,803.
50. Méry P, Pavoine C, Belhassen L, et al. Nitric oxide regulates cardiac Ca^{2+} current: involvement of cGMP-inhibited and cGMP-stimulated phosphodiesterases through guanylyl cyclase activation. J Biol Chem 1993;268:26,286–26,295.
51. Levi RC, Alloatti G, Penna C, et al. Guanylate-cyclase-mediated inhibition of cardiac I_{Ca} by carbachol and sodium nitroprusside. Pflügers Arch 1994;426:419–426.
52. Fischmeister R, Hartzell HC. Cyclic guanosine 3':5'-monophosphate regulates the calcium current in single cell from frog ventricule. J Physiol (Lond) 1987;387:453–472.
53. Méry P, Pavoine C, Pecker F, et al. Erythro-9-(2-hydroxy-3-nonyl)adenine inhibits cyclic GMP-stimulated phosphodiesterases in isolated cardiac myocytes. Mol Pharmacol 1995;48:121–130.
54. Dittrich M, Jurevieius J, Georget M, et al. Local response of L-type Ca^{2+} current to nitric oxide in frog ventricular myocytes. J Physiol (Lond) 2001;534:109–121.
55. Gallo MP, Ghigo D, Bosia A, et al. Modulation of guinea-pig cardiac L-type calcium current by nitric oxide synthase inhibitors. J Physiol (Lond) 1998;506:639–654.
56. Gallo MP, Malan D, Bedendi I, et al. Regulation of cardiac calcium current by NO and cGMP-modulating agents. Pflügers Arch 2001;441:621–628.
57. Tohse N, Sperelakis N. cGMP inhibits the activity of single calcium channels in embryonic chick heart cells. Circ Res 1991;69:325–331.
58. Tohse N, Nakaya H, Takeda Y, et al. Cyclic GMP–mediated inhibition of L-type Ca^{2+} channel activity by human natriuretic peptide in rabbit heart cells. Br J Pharmacol 1995;114:1076–1082.
59. Tewari K, Simard JM. Sodium nitroprusside and cGMP decrease Ca^{2+} channel availability in basilar artery smooth muscle cells. Pflügers Arch 1997;433:304–311.
60. Chik CL, Kiu Q, Li B, et al. cGMP inhibits L-type channel current through protein phosphorylation in rat pinealocytes. J Neurosci 1995;15:3104–3109.
61. Kim SJ, Song SK, Kim J. Inhibitory effect of nitric oxide on voltage-dependent calcium currents in rat dorsal root ganglion cells. Biochem Biophys Res Commun 2000;271:509–514.
62. Grassi C, D'Ascenzo M, Valente A, et al. Ca^{2+} channel inhibition induced by nitric oxide in rat insulinoma RINm5F cells. Pflügers Arch 1999;437:241–247.
63. D'Ascenzo M, Azzena GB, Grassi C. Effect of nitric oxide on high-voltage activated Ca^{2+}-channels. In: Morad M, Kostyuk P, eds. Calcium Signaling. IOS Press: Amsterdam, 2001, pp. 117–123.
64. Carabelli V, D'Ascenzo M, Carbone E, et al. Nitric oxide inhibits neuroendocrine Ca_v1 L-channel gating via cGMP-dependent protein kinase in cell-attached patches of bovine chromaffin cells. J Physiol (Lond) 2002;541:351–366.
65. Magnelli V, Pollo A, Sher E, et al. Block on non-L-, non-N-type Ca^{2+} channels in rat insulinoma RINm5F cells by ω-agatoxin IVA and ω-conotoxin MVIIC. Pflügers Arch 1995;429:762–771.
66. Carabelli V, Hernández-Guijo JM. Baldelli P, et al. Direct autocrine inhibition and cAMP-dependent potentiation of single L-type Ca^{2+} channels in bovine chromaffin cells. J Physiol 2001;532:73–90.
67. Jiang LH, Gawler DJ, Hodson N, et al. Regulation of cloned cardiac L-type calcium channels by cGMP-dependent protein kinase. J Biol Chem 2000;275:6135–6143.
68. D'Ascenzo M, Martinotti G, Azzena GB, et al. c-GMP/PKG-dependent inhibition of N-type Ca^{2+}-channels induced by nitric oxide in human neuroblastoma IMR32 cells. J Neurosci 2002;22: 7485–7492.
69. Carbone E, Sher E, Clementi F. Ca currents in human neuroblastoma IMR32 cells: kinetics, permeability and pharmacology. Pflügers Arch 1990;416:170–179.
70. Grassi C, Magnelli V, Carabelli V, et al. Inhibition of low- and high-threshold Ca^{2+}-channels of human neuroblastoma IMR32 cells by Lamber-Eaton myasthenic syndrome (LEMS) IgGs. Neurosci Lett 1994;181:50–56.
71. Yoshimura N, Seki S, de Groat WC. Nitric oxide modulates Ca^{2+} channels in dorsal root ganglion neurons innervating rat urinary bladder. J Neurophysiol 2001;86:304–311.
72. Chen C, Schofield GG. Nitric oxide enhanced Ca^{2+} currents and blocked noradrenaline-induced Ca^{2+} current inhibition in rat sympathetic neurons. J Physiol (Lond) 1995;482:521–531.
73. Kurenny DE, Moroz LL, Turner RW, et al. Modulation of ion channels in rod photoreceptors by nitric oxide. Neuron 1994;13:315–324.
74. Hirooka K, Kourennyi DE, Barnes S. Calcium channel activation facilitated by nitric oxide in retinal ganglion cells. J Neurophysiol 2000;83:198–206.

75. Lin Z, Lin Y, Schorge S, et al. Alternative splicing of a short cassette exon in α_{1B} generates functionally distinct N-type calcium channels in central and peripheral neurons. J Neurosci 1999;19: 5322–5331.
76. Lohman SM, Vaandrager AB, Smolenski A, et al. Distinct and specific funtions of cGMP-dependent protein kinases. Trends Biochem Sci 1997;22:307–312.
77. Hofmann F, Ammendola A, Schlossmann J. Rising behind NO: cGMP-dependent protein kinases. J Cell Sci 2000;113:1671–1676.
78. Chen J, Daggett H, De Waard M, et al. Nitric oxide augments voltage-gated P/Q-type Ca^{2+} channels constituting a putative positive feedback loop. Free Radic Biol Med 2002;32:638–649.
79. Niidome T, Teramoto T, Murata Y, et al. Stable expression of the neuronal BI (class A) calcium channel in baby hamster kidney cells. Biochem Biophys Res Commun 1994;203:1821–1827.
80. Li A, Segui J, Heinemann SH, et al. Oxidation regulates cloned neuronal voltage-dependent Ca^{2+} channels expressed in *Xenopus* oocytes. J Neurosci 1998;18:6740–6747.
81. Magnelli V, Avaltroni A, Carbone E. A single non-L-, non-N-type Ca^{2+} channel in rat insulin secreting RINm5F cells. Pflügers Arch 1996;431:341–352.
82. Uchitel O, Protti DA, Sanchez V, et al. P-type voltage-dependent calcium channel mediates presynaptic calcium influx and transmitter release in mammalian synapses. Proc Natl Acad Sci USA 1992;89:3330–3333.
83. Catterall WA. Interaction of presynaptic Ca^{2+}-channels and snare proteins in neurotransmitter release. Ann NY Acad Sci 1999;868:144–159.
84. Quignard JF, Frapier JM, Harricane MC, et al. Voltage-gated calcium channel currents in human coronary myocytes: regulation by cyclic GMP and nitric oxide. J Clin Invest 1997;99:185–193.
85. Kawai F, Miyachi E-I. Modulation of the voltage-gated currents in newt olfactory receptor cells. Neurosci Res 2001;39:327–337.
86. Wei J-Y, Ethan J, Cohen D, et al. cGMP-induced presynaptic depression and postsynaptic facilitation at glutamatergic synapses in visual cortex. Brain Res 2002;927:42–54.
87. Greenberg S, Diecke FPJ, Peevy K, et al. The endothelium modulates adrenergic neurotransmission to canine pulmonary arteries and veins. Eur J Pharmacol 1989;162:57–80.
88. Sequeira SM, Carvalho AP, Carvalho CM. Both protein kinase G dependent and independent mechanisms are involved in the modulation of glutamate release by nitric oxide in rat hippocampal nerve terminals. Neurosci Lett 261;1999:29–32.
89. Prast H, Philippu A. Nitric oxide releases acetylcholine in the basal forebrain. Eur J Pharmacol 1992;216:139–140.
90. Hawkins RD, Son H, Arancio O. Nitric oxide as a retrograde messenger during long-term potentiation in hippocampus. Prog Brain Res 1998;118:155–172.
91. Herring N, Paterson JD. Nitric oxide–cGMP pathway facilitates acetylcholine release and bradycardia during vagal nerve stimulation in the guinea-pig in vitro. J Physiol (Lond) 2001;535:507–518.
92. Ahern GP, Klyachko VA, Jackson MB. cGMP and S-nitrosylation: two routes for modulation of neuronal excitability by NO. Trends Neurosci 2002;25:510–517.
93. Meffert MK, Calakos NC, Scheller RH, et al. Nitric oxide modulates synaptic vesicle docking/fusion reactions. Neuron 1996;16:1229–1236.
94. Clementi E. Role of nitric oxide and its intracellular signalling pathways in the control of Ca^{2+} homeostasis. Biochem Pharmacol 1998;55:713–718.
95. Hart JDE, Dulhunty AF. Nitric oxide activates or inhibits skeletal muscle ryanodine receptors depending on its concentration, membrane potential and ligand binding. J Membr Biol 2000;173:227–236.
96. Lu Y-F, Hawkins RD. Ryanodine receptors contribute to cGMP-induced late-phase LTP and CREB phosphorylation in the hippocampus. J Neurophysiol 2002;88:1270–1278.
97. Mathes C, Thompson SH. The nitric oxide/cGMP pathway couples muscarinic receptor to activation of Ca^{2+} influx. J Neurosci 1996;16:1702–1709.
98. Oset-Gasque MJ, Parramon M, Hortelano S, et al. Nitric oxide implication in the control of neurosecretion by chromaffin cells. J Neurochem 1994;63:1693–1700.
99. Chiodera P, Volpi R, Coiro V. Inhibitory control of nitric oxide on the arginine-vasopressin and oxytocin response to hypoglycaemia in normal men. NeuroReport 1994;5:1822–1824.
100. Lutz-Bucher B, Koch B. Evidence for an inhibitory effect of nitric oxide on neuropeptide secretion from isolated neural lobe of the rat pituitary gland. Neurosci Lett 1994;165:48–50.
101. Antoine MH, Ouedraogo R, Hermann M,et al. 3-Morpholinosydnonimine as instigator of a glibenclamide-sensitive reduction in the insulin secretory rate. Biochem Pharmacol 1997;53: 1211–1213.

102. Laffranchi R, Gogvadze V, Richter C, et al. Nitric oxide (nitrogen monoxide, NO) stimulates insulin secretion by inducing calcium release from mitochondria. Biochem Biophys Res Commun 1995;217:584–591.
103. Willmott NJ, Galione A, Smith PA. Nitric oxide induces Ca^{2+} mobilization and increases secretion of incorporated 5-hydroxytryptamine in rat pancreatic β-cells. FEBS Lett 1985;371:1981–1992.
104. O'Sullivan AJ, Burgoyne RD. Cyclic GMP regulates nicotine-induced secretion from cultured bovine adrenal chromaffin cells: effects of 8–bromo-cyclic GMP, atrial natriuretic peptide, and nitroprusside (nitric oxide). J Neurochem 1990;54:1805–1808.
105. McDonald TF, Pelzer S, Trautwein W, et al. Regulation and modulation of calcium channels in cardiac, skeletal and smooth muscle cells. Physiol Rev 1994;74:365–507.
106. Shimoni Y. Hormonal control of cardiac ion channels and transporters. Prog Biophys Mol Biol 1999;72:67–108.
107. Striessnig J. Pharmacology, structure and function of cardiac L-type Ca^{2+} channels. Cell Physiol Biochem 1999;9:242–269.
108. Wang R. Two's company, three's a crowd—can H_2S be the third endogenous gaseous transmitter? FASEB J 2002;16:1792–1798.

8 Interactions of Nitric Oxide and Cardiac Ion Channels

Zhao Zhang, Kathryn A. Glatter, and Nipavan Chiamvimonvat

CONTENTS

SUMMARY

Nitric oxide (NO) is a uniquely diffusible and reactive molecular messenger that is found in abundance and plays important regulatory roles in the cardiovascular system. NO modulates a wide variety of ion channels in different systems as diverse as neurons, vascular smooth muscles, carotid bodies, pancreatic cells, and hair cells in the inner ear. Indeed, the modulation of ion channels represents one of the important functional effects of NO. In the cardiovascular system, NO significantly modulates the cardiac ryanodine receptor channel, L-type Ca^{2+} channel, and Na^+ channel. The actions of NO are exceedingly multifaceted. There are at least two distinct downstream signaling actions for NO: an indirect pathway via cyclic guanosine 5'-monophosphate (cGMP) production and a direct pathway via protein thiol nitrosylation (*S*-nitrosylation). In addition, a low level of cGMP can mediate the inactivation of phosphodiesterase type 3, leading to an increased level of cyclic adenosine monophosphate. For example, Ca^{2+} channels can be stimulated or inhibited under different conditions by different concentrations of NO via indirect or direct pathways. Furthermore, NO modulations can be biphasic and highly sensitive to experimental conditions, such as redox state of the cells, concentrations of NO, temperature, and oxygen tension. More recently, it has been suggested that spatial confinement of different NO synthase (NOS) isoforms may allow NO signaling to have independent,

From: *Signal Transduction and the Gasotransmitters: NO, CO, and H_2S in Biology and Medicine* Edited by: Rui Wang © Humana Press Inc., Totowa, NJ

and even opposite, effects on cardiac function. Therefore, a precise knowledge of various pathways and multiple effectors of different NOS enzymes is critical to the development of diagnostic and therapeutic strategies for heart diseases.

Key Words: Nitric oxide; Ca^{2+} channel; Na^+ channel; ryanodine receptor; nitric oxide synthase.

1. INTRODUCTION

Nitric oxide (NO) is a uniquely diffusible and reactive molecular messenger that is found in abundance and plays important regulatory roles in different systems throughout the body *(1)*, including the nervous, immune, respiratory, gastrointestinal, and cardiovascular systems. In the cardiovascular system, NO is the major endothelium-derived relaxing factor, and it causes vasodilation and reduces blood pressure *(2)*. In addition, NO functions as an important endogenous inhibitor of vascular lesion formation *(3)*. NO significantly modulates the excitation-contraction (EC) coupling in the heart *(4–6)*. The coronary endothelium is responsible for the bulk of the endogenous, physiological production of NO *(7)*. However, NO can also be produced within the cardiac myocytes themselves by the constitutive NO synthase (NOS) *(8)*.

There is accumulating evidence that NO modulates cardiac contractility both in vitro and in vivo *(4–6)*, by participating in the regulation of many key ion channels involved in cardiac EC coupling *(4–6)*. NO influences over EC coupling are mediated by precise spatial localization of NOS isoforms with effector ion channels *(6,9)*. The two well-characterized ion channels modulated by NO are the L-type Ca^{2+} channel (modulated by NOS3) *(10–15)*, and the ryanodine receptor (RYR) (modulated by NOS1) *(9)*.

Intracellular Ca^{2+} homeostasis of cardiac myocytes is maintained by Ca^{2+} release and uptake by the sarcoplasmic reticulum (SR) and the Ca^{2+} flux across the sarcolemma *(16)*. Cardiac myocyte contraction is initiated by membrane depolarization, which leads to Ca^{2+} entry via voltage-gated L-type Ca^{2+} channels *(16)*. This Ca^{2+} entry results in a larger Ca^{2+} release from the SR through the RYR, which activates myofilament contraction, a process known as Ca^{2+}-induced Ca^{2+} release *(16)*. The Ca^{2+} release from the SR via RYR is largely responsible for tension development in the heart and accounts for approx 80% of the Ca^{2+} flux involved in EC coupling. Twenty percent of the Ca^{2+} is removed by two sarcolemmal transport systems: the Na^+-Ca^{2+} exchanger and the plasma membrane Ca^{2+} pump. The Na^+-Ca^{2+} exchanger is the principal mechanism for Ca^{2+} extrusion from myocytes. Myocyte relaxation requires Ca^{2+} removal from the cytoplasm, which is mediated by the SR Ca^{2+} adenosine triphosphatase (ATPase) (SERCA) and the sarcolemmal Na^+-Ca^{2+} exchanger *(16–18)*. The plasma membrane Ca^{2+} ATPase also plays a minor role in cytoplasmic Ca^{2+} removal *(19)*. NO interacts with SERCA, phospholamban, and the Na^+-Ca^{2+} exchanger; however, the precise mechanisms have not been fully identified.

1.1. Nitric Oxide Synthase

NO is produced by oxidation of the terminal guanidino nitrogen of L-arginine to form NO and the amino acid L-citrulline by NOS. Three isoforms of NOS enzymes have been described in mammalian systems *(20–23)*; neuronal NOS (or NOS1), inducible NOS (or NOS2), and endothelial NOS (or NOS3). NOS1 and NOS3 are activated by Ca^{2+} and calmodulin, whereas NOS2 is known to be Ca^{2+} independent because of its high basal Ca^{2+}/calmodulin affinity. NOS2 also has other distinct properties. Whereas NOS1 and 3 are constitutively present in cardiac myocytes *(24,25)*, NOS2 requires induction by cytokines *(26)*. Recently, it has been shown that spatial confinement of different NOS isoforms may allow NO signals to have independent, and even opposite, effects on cardiac functions *(6,9)*. In other words, precise local regulation of different effector molecules may represent the key mechanism by which NO exerts its biological activity *(6,27–29)*. Therefore, NO can provide exquisite fine-tuning of organ function by recruiting different downstream NO signaling pathways within distinct microdomains of the same cell.

1.2. Cardiac NOS3

The endothelial isoform of NOS (NOS3) is expressed both in the vascular endothelium and in cardiac myocytes, and the cellular regulation of NOS3 may represent an important determinant of cardiovascular homeostasis. Cardiac NOS3 is coupled to numerous receptors, including the muscarinic, β-adrenergic, and bradykinin receptors. NOS3 localizes to the caveolae of the sarcolemma and t-tubules and is inactivated by the scaffolding protein caveolin-3. The activity of NOS3, a Ca^{2+}/calmodulin-dependent enzyme, is markedly attenuated by its interaction with caveolin *(30,31)*. NOS3 is activated via an agonist-stimulated increase in intracellular Ca^{2+} leading to dissociation of the enzyme from caveolin-3. NOS3 can also be activated directly by Akt phosphorylation independent of intracellular increases in Ca^{2+} *(32,33)*.

1.3. Cardiac NOS1

NOS3 is known to be the predominant isoform that is constitutively present in cardiac endothelial cells and cardiomyocytes. However, NOS1 also contributes importantly to the regulation of myocardial function. In particular, it has recently been shown that NOS1 regulates cardiac function by its influences on myocyte Ca^{2+} handling that may be different or even opposite to NOS3 activities *(6,9)*.

NOS1 is expressed in cardiac myocytes *(9)*, skeletal muscle *(27)*, and neurons *(34)*. Like NOS3, NOS1 is a Ca^{2+}/calmodulin-activated enzyme. However, unlike NOS3, tissue-specific expression of NOS1 occurs via alternative splicing. NOS1 localizes to cardiac SR and influences SR Ca^{2+} cycling. In addition, it has recently been shown that NOS1 coimmunoprecipitates with the RYR *(9)* and likely stimulates the SR Ca^{2+} release.

2. CELLULAR ACTIONS OF NO

The actions of NO are exceedingly multifaceted. There are at least two distinct downstream signaling actions for NO: an indirect pathway via cyclic guanosine 5'-monophosphate (cGMP) production *(18)* and a direct pathway via protein thiol nitrosylation (*S*-nitrosylation) *(35)*. The free radical NO can exert many of its effects through an indirect pathway involving activation of guanylyl cyclase (GC) and increased levels of cGMP. NO can activate soluble GC by binding to its heme moiety, forming an Fe-nitrosyl complex *(36)*, leading to the production of cGMP, which in turn activates protein kinase G (PKG) and a cascade of biological signaling events *(37)*. Activation of soluble GC by NO requires low levels of NO concentrations (EC_{50} of 100 n*M*) *(36)*. NO that is produced by NOS3 can readily diffuse to adjacent cells and activate soluble GC. Thus, NO produced in myocytes as well as in endothelial cells has the potential to contribute to myocyte cGMP production.

A second biologically important signaling mechanism for NO is redox-regulated covalent modification of proteins *(35,38)*. NO reacts with nucleophilic centers in a nitrosylation reaction *(39)*, occurring with a broad array of low molecular weight compounds or proteins at thiol residues *(40–42)*. In addition, the free radical NO can result in a multitude of interrelated redox forms with distinct properties and reactivities. These molecules encompass the actions of several naturally occurring nitrogen (N)-oxides, which display reactivity profiles that are different from NO itself *(1)*.

Protein nitrosylation is involved in the modulatory actions of various proteins involved in cardiac EC coupling, including the L-type Ca^{2+} channel *(14,15)* and the RYR *(43,44)*. S-NO reactions are likely regulated in biological systems by the enzyme glutathione-dependent formaldehyde dehydrogenase, which breaks down the S-NO bond. There is ample evidence to suggest that both cyclic cGMP-dependent *(18,37)* and -independent *(45)* mechanisms contribute to NO influences of myocardial contractility.

Finally, it has recently been shown that NO can regulate both adenylyl cyclase (AC) and GC in cardiac myocytes. High levels of NO induce large increases in cGMP and a negative inotropic effect, while low levels of NO increase adenosine monophosphate (cAMP) and induce a positive contractile response *(46)*.

3. EFFECTS OF NO ON ION CHANNELS

NO modulates a wide variety of ion channels in different systems as diverse as neurons, vascular smooth muscles, carotid bodies, pancreatic cells, and hair cells in the inner ear *(47,48)*. Indeed, the modulation of ion channels represents one of the important functional effects of NO. In the cardiac systems, NO can inactivate the cardiac RYR channel *(49)* and modulate the cardiac Ca^{2+} and Na^{+} channels *(10–15,50)*. These actions can lead to significant effects on cardiac functions. In addition, NO plays an important role in cardiac pacemaking cells by mediating a muscarinic cholinergic attenuation of the L-type Ca^{2+} current in mammalian sinoatrial and atrioventricular nodes *(51)*.

3.1. Modulation of L-Type Ca^{2+} Channel By NO

The functions of NO in the cardiovascular system appear to be mediated, at least in part, by modulation of Ca^{2+} channels. In cardiomyocytes, NO has no effect on basal Ca^{2+} current *(10–13)* but exerts inhibitory *(11,13)* or biphasic effects on cAMP-stimulated Ca^{2+} current *(10,12)*. In frog ventricular myocytes, the NO donor 3-morpholinosydnonimine (SIN-1) induces a pronounced stimulation of Ca^{2+} current at low concentrations, whereas at higher concentrations, SIN-1 inhibits Ca^{2+} current *(10)*. The stimulatory effects are attributed to the activation of GC by NO, resulting in accumulation of intracellular cGMP *(10,12)*, which in turn suppresses cGMP-inhibited phosphodiesterases (PDEs) and thus elevates cAMP level and stimulates Ca^{2+} current *(10,12)*. On the other hand, inhibition of the current is secondary to activation of cGMP-stimulated PDEs *(10,52)* or cGMP-dependent protein kinases *(12,53,54)*. In addition, apart from the indirect mechanisms via cGMP stimulation, NO has been shown to modulate Ca^{2+} current via an additional direct modification of the Ca^{2+} channel by *S*-nitrosation *(14,15)* (*see also* **Fig. 1**).

3.1.1. Direct Modification of Ca^{2+} Channels

The effects of NO on defined molecular components of the Ca^{2+} channel itself have been directly examined using a heterologous expression system *(15,55)*. Such an approach enables the examination of each particular subunit of interest in isolation from other subunits. Ca^{2+} channels are complexes of a pore-forming, transmembrane-spanning α_1-subunit of about 190–250 kDa, a disulfide-linked complex of α_2- and δ-subunits; and an

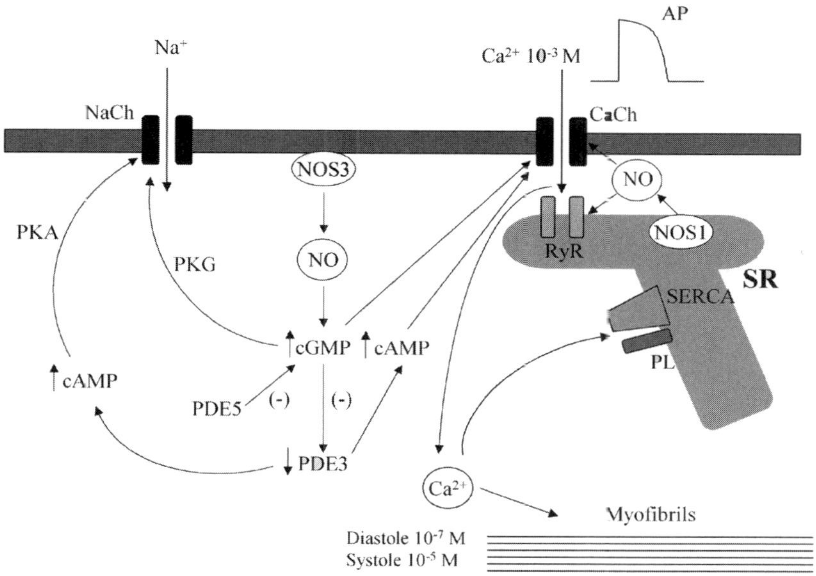

Fig. 1. NO modulations of different cardiac ion channels including Na^+ channel (NaCh), L-type Ca^{2+} channel (CaCh) on the sarcolemma, and RYR. PL, phospholamban; AP, action potential; (–) inhibitory effects.

intracellular β-subunit and a γ-subunit. The auxiliary subunits modulate the properties of the channel complex *(56–58)*. However, the pharmacological and permeation properties of Ca^{2+} channels arise primarily from the α_1-subunits.

NO donors from the nitrosothiol (RSNO) class have been shown to modulate expressed cardiac L-type Ca^{2+} channels (α_{C1}-subunit) coexpressed with auxiliary subunits (β- and δ-subunits) *(15)*. NO donors were found to inhibit the Ca^{2+} current in a dose-dependent manner, and the inhibitory effects were cGMP independent but occurred via a direct inhibition of the Ca^{2+} channel by redox chemical reactions. The mechanisms of the modulation of Ca^{2+} current by NO were found to be similar to the effects of various cysteine-oxidizing reagents *(15)*. In addition, redox modifications have previously been shown for rabbit smooth muscle Ca^{2+} channel ($\alpha_{1C\text{-}b}$ subunit alone without auxiliary subunits) in Chinese hamster ovary cells using thiol-specific modifying reagents *(55)*, suggesting that this redox property of the α_{1C}-subunit may be generalizable between splice variants and remains intact after coexpression of auxiliary subunits. Sulfhydryl modification of L-type Ca^{2+} channels resulted in a reduction in whole-cell Ca^{2+} current, which could be readily reversed by disulfide reduction *(55)*. At the single-channel level, this reduction in macroscopic current was mediated by a decrease in open probability and open time and an apparent decrease in the number of functional channels, with no change in single-channel conductance, consistent with changes in gating but not permeation of the channel. The effects of sulfhydryl modification were Ca^{2+}-channel specific, with no detectable changes in Na^+ current *(55)*.

3.1.2. Presence of Accessible Free Sulfhydryl Groups in Pore-Forming Subunit of Ca^{2+} Channel

Where within the Ca^{2+} channel molecule are the likely sites for redox modulation? Hu et al. *(15)* suggested that the more hydrophobic the modifying reagents, the more potent the inhibition. Thus, there seem to be one or more hydrophobically accessible redox sites in the Ca^{2+} channel. NO molecules would likely have access to the same site(s), given that they are small molecules and move freely through cellular membranes. Nevertheless, even if the targets for NO donors are the same as those for thiol-specific modifying reagents, the chemical reactions might not be identical. The RSNOs can modify thiol side chains by different mechanisms. They may undergo a transnitrosation reaction with thiols in the channel to form SNOs (R_x'SNOs, in which R_x' refers to cysteine residue in the channel protein) or form mixed disulfide bonds with cysteines (RSSR_x') *(38,59)*. In addition, if two cysteine residues are close to each other, RSNOs may facilitate the formation of a disulfide bond between the two residues *(38,59)* (R_x'SSR_y'). Furthermore, NO released by RSNOs may modify channel thiol side chains in similar ways, with the exception of forming mixed disulfide bonds. Finally, RSNOs release not only $NO^{\bullet}$ but also NO^+ and NO^- *(60)* and may contribute to further modification of the channel.

Voltage-activated Ca^{2+} channels can be viewed as having three modes of gating behavior: no openings (mode 0), brief repetitive openings (mode 1), and long-lasting openings with brief closures (mode 2) *(61,62)*. The dihydropyridine agonists (e.g., Bay K 8644) enhance Ca^{2+} current by promoting mode 2, whereas the antagonists favor mode 0. The mechanism of transitions between modes is not known. It has previously been shown that sulfhydryl modification of L-type Ca^{2+} channel mimics the effects of dihydropyridine antagonists, promoting transition of the channel to mode 0 and mode 1, with a resultant reduction in open time and open probability and an apparent decrease in functional channel number *(55)*. The results are consistent with the presence of free sulfhydryl groups on the Ca^{2+} channel, which are accessible from the extracellular side and are important in the gating of the channel. Previous biochemical studies have suggested the involvement of disulfide bonds and free sulfhydryl groups in the binding of dihydropyridine to the L-type Ca^{2+} channel in heart muscle *(63)*. The findings that oxidation of free sulfhydryl groups of the Ca^{2+} channel leads to a channel with characteristics similar to those after treatment with dihydropyridine antagonists are consistent with this interpretation.

3.1.3. Absence of Effects of Disulfide-Reducing Agent on Pore-Forming Subunit of Ca^{2+} Channel

In contrast to the marked effects of cysteine-specific oxidizing agent, it was shown that dithiothreitol (a disulfide-reducing agent) has no effect on Ca^{2+} current *(55)*. Although results are consistent with the idea that there are no accessible disulfide bonds in the native α_1-subunit of Ca^{2+} channels, reduction of a disulfide bond will not necessarily affect the whole-cell current. Nonetheless, this finding may not come as a surprise. Although the presence of disulfide bonds between subunits has been documented previously (namely, those between α_2- and-subunits), the presence of disulfide bonds within the α_1-subunit has not been proposed.

3.2. Modulation of Na^+ Channel by NO Via Indirect Pathway

It has been shown using heterologous expression systems that whereas the Ca^{2+} channel can be directly modulated by NO, Na^+ channels are unaffected by direct NO modulation *(15)*. By contrast, NO can modulate Na^+ channels in native cardiac myocytes *(50)*. There is further evidence to demonstrate that NO modulates Na^+ channels via second-messenger pathways through activation of protein kinase G (PKG) and PKA. Specifically, NO was shown to inhibit Na^+ current in isolated guinea pig and mouse ventricular myocytes *(50)*. NO modification of Na^+ channels resulted in a reduction in Na^+ current with no changes in the steady-state or time-dependent kinetics. At the single-channel level, the reduction in macroscopic current was mediated by a decrease in open probability, and/or a decrease in the number of functional channels with no change in single-channel conductance. Taken together with the macroscopic current findings, it is likely that the inhibitory effects of NO most likely result from a change in the channel number rather than changes in permeation or gating. The inhibitory effects of NO on Na^+ current involve the activation of both cGMP- and cAMP-dependent protein kinases and cannot be reversed by sulfhydryl reducing agents, as would be expected for a direct modulation. This is in contrast to the cardiac Ca^{2+} channels that can be modulated by both indirect (cGMP-dependent) and direct (*S*-nitrosylation/oxidation) pathways *(14,15)*.

3.2.1. Involvement of cGMP and cAMP

Previous studies have demonstrated that exogenously applied NO at high concentrations can produce a negative inotropic effect on cardiac contraction that is mediated by a cGMP-dependent PKG activation. By contrast, low concentrations of NO evoked a positive inotropic effect by a novel mechanism via a cGMP-independent activation of AC *(51)*. An elevation of the intracellular levels of cAMP (and PKA activation) could occur via a cGMP-dependent inhibition of the PDE3 (i.e., cGMP-inhibited PDE) *(64)*. Low concentrations of cGMP (0.1–10 μmol/L) were found to have a stimulatory effect on L-type Ca^{2+} current likely resulting from the inhibition of cAMP degradation, mediated by the inactivation of PDE3 by low levels of cGMP *(10)*. An alternative mechanism involves the activation of AC by NO either directly or via a G-protein. Previous studies have provided evidence suggesting that NO can directly or indirectly activate AC in a cGMP-independent manner *(65)*. The exact mechanism of NO activation of AC is uncertain. However, recent reports have demonstrated that NO can modulate G-protein function *(66)*. In addition, in endothelial cells, NO has been shown to selectively inhibit G-proteins of the G_i and G_q family but not those of the G_s family, and that this modulation of G-proteins could have a permissive action on the G_s-AC pathway *(67)*. Therefore, it is possible that NO can activate AC via the potential modulation of G-protein.

The aforementioned results from the study by Ahmmed et al. *(50)* support a direct involvement of AC. Even when a basal cGMP increase was completely abolished by the presence of the selective inhibitor of GC, NO was still able to induce an inhibitory effect on Na^+ current. In these settings, the additional block induced by NO must occur by a mechanism other than cGMP-mediated PDE inhibition, possibly by a cGMP-independent activation of AC. Indeed, these data suggest the direct involvement of both GC and AC as well as PKA and PKG on the NO modulation of the cardiac Na^+ channel (*see also* **Fig. 1**).

3.3. Direct Modulations of Other Ion Channels

Regulation by redox state and NO/RSNOs has been described for numerous intra- and extracellular proteins including ion channels *(68,69)*. Direct effects of N-oxides appear to derive from reactions of vicinal thiols that serve as allosteric regulators of channel function *(68)*. The oxidation state of sulfhydryl groups has been shown to be important in the function of the rat brain I_A K^+ channels (rapidly inactivating K^+ channels) expressed in oocytes. Sulfhydryl oxidation of the channel led to the abolition of fast inactivation of the channel *(68)*. This loss of inactivation was shown to result from the oxidation of a critical cysteine residue located near the inactivation domain of the channel *(68)*.

Sulfhydryl oxidation induced a rapid and reversible closure of the adenosine triphosphate-regulated K^+ channel in pancreatic β-cells *(70)*.

Direct modulation by NO has been described in the Ca^{2+}-activated K^+ channel of rabbit aortic smooth muscle *(71)*. As in the study by Hu et al. *(15)*, such modulation was only partially removed upon washout of NO donors, indicating possible covalent modification of the channels. In addition, redox reaction on thiol groups by NO donors has been proposed to be responsible for the downregulation of *N*-methyl-D-aspartate receptor activity *(59)* and for the stimulatory (but not the inhibitory) effect of NO donors on Ca^{2+} channels in ferret ventricular cells *(14)*. These studies, as well as ours, confirm the idea that the structure and function of many proteins, including ion channels, are critically dependent on the oxidative state of the sulfhydryl groups.

3.3.1. Ryanodine Receptor

NO has been shown to increase the open probability of the cardiac RYR *(43)*. Based on this finding and the fact that NOS1 resides in the SR, it was predicted that NOS1 would facilitate myocardial contractility by enhancing Ca^{2+} cycling *(9)* (*see* **Fig. 1**). Indeed, it has been shown that NOS1 exerts stimulatory influences on Ca^{2+} transients ($[Ca^{2+}]_i$) *(9)* and facilitates the force-frequency relationship, likely by increasing the SR Ca^{2+} release. Sulfhydryl oxidation of the Ca^{2+} release channel triggered Ca^{2+} release from the SR, whereas disulfide reduction led to a rapid reuptake of Ca^{2+} *(72)*. The effects were mediated by an increase in SR Ca^{2+} release channel open probability with no change in channel conductance *(73)*. It was hypothesized that a free sulfhydryl group on the Ca^{2+} release channel can be oxidized, leading to the opening of the channel, and that the coupling of this channel to the voltage-gated Ca^{2+} channel on the membrane is critically dependent on the oxidation of this sulfhydryl group *(74)*.

4. PHYSIOLOGICAL AND PATHOLOGICAL IMPLICATIONS

The physiological concentrations of NO are in the submicromolar range *(75,76)*. A high concentration of NO has been documented during the pathological state (e.g., sepsis) *(77)*. Excessive amounts of NO could be toxic to the heart. Submillimolar concentrations of NO donor have been reported to reduce the contractility of both cardiomyocytes and cardiac muscles *(64,78)*. NO has also been shown to have negative inotropic effects on isolated ventricular myocytes *(8)*. NO signals both by direct protein modification (S-nitrosylation; the likely mode of influence over the RYR and cardiac Ca^{2+} channel) and by cGMP production (the likely mode of influence over the cardiac Ca^{2+} channel and cardiac Na^+ channel). NO also participates in mitochondrial respiration. Also critical to NO signaling pathways are mechanisms that inactivate downstream messengers, such as the action of PDE5 on cGMP *(18)* and glutathione-dependent formaldehyde dehydrogenase in the case of *S*-nitrosylation *(79)*. In addition, superoxide has the potential to disrupt S-NO signaling *(80)*. It seems reasonable to speculate that the dependence of L-type Ca^{2+} channels on the redox state of the cellular environment may exert protective effects during cell injury in ischemia and reperfusion, because oxidative stress would favor a reduction in voltage-dependent Ca^{2+} entry.

5. CONCLUSION

There has been much controversy in the literature regarding the exact influences of NOS1 on cardiac excitation coupling as well as the direction and magnitude of NOS3-

related cardiac effects *(81)*. On the other hand, it is well accepted that NO plays important modulatory roles in several key proteins that are critical to cardiac function in both normal and disease states. NO signaling may lead to paradoxical effects on EC coupling, stimulating or inhibiting, depending on which isoform and the corresponding downstream signal are affected. Recent discoveries that distinct NOS enzymes are localized in the different compartments of the cardiac myocytes have helped to shed new light on our understanding of the roles of NO in the heart. The compartmentalization of the NOS enzymes may allow exquisite fine-tuning of the different NO signaling pathways. Furthermore, it is important to realize that NO modulations can be biphasic and highly sensitive to experimental conditions, such as redox state of the cells, concentrations of NO, temperature, and oxygen tension *(6,81)*. The same protein (e.g., Ca^{2+} channel) can be stimulated or inhibited under different conditions by different concentrations of NO via indirect or direct pathways. A precise and complete knowledge of the multitude of pathways and effectors of different NOS enzymes is critical to the development of diagnostic and therapeutic strategies for heart diseases.

ACKNOWLEDGMENTS

We are grateful to Dr. E. N. Yamoah (University of California, Davis) for helpful comments and suggestions. This work was supported in part by National Institutes of Health grants RO1 HL-68507 and HL-67737 and the Veteran Administration Merit Review Grant to N.Chiamvimonvat, and the Pfizer/SWHR Scholar Grant and American Heart Association Western Affiliate Beginning Grant-in-Aid to K. A. Glatter.

REFERENCES

1. Stamler JS, Singel DJ, Loscalzo J. Biochemistry of nitric oxide and its redox-activated forms. Science 1992;258:1898–1902.
2. Stamler JS, Loscalzo J. Endothelium-derived relaxing factor modulates the atherothrombogenic effects of homocysteine. J Cardiovasc Pharmacol 1992;20:S202–S204.
3. von der Leyen HE, Gibbons GH, Morishita R, et al. Gene therapy inhibiting neointimal vascular lesion: in vivo transfer of endothelial cell nitric oxide synthase gene. Proc Natl Acad Sci USA 1995;92: 1137–1141.
4. Brady AJ, Warren JB, Poole-Wilson PA, et al. Nitric oxide attenuates cardiac myocyte contraction. Am J Physiol 1993;265:H176–H182.
5. Hare JM, Colucci WS. Role of nitric oxide in the regulation of myocardial function. Prog Cardiovasc Dis 1995;38:155–166.
6. Hare JM. Nitric oxide and excitation-contraction coupling. J Mol Cell Cardiol 2003;35:719–729.
7. Shah AM. Paracrine modulation of heart cell function by endothelial cells. Cardiovasc Res 1996;31: 847–867.
8. Balligand JL, Kelly RA, Marsden PA, et al. Control of cardiac muscle cell function by an endogenous nitric oxide signaling system. Proc Natl Acad Sci USA 1993;90:347–351.
9. Barouch LA, Harrison RW, Skaf MW, et al. Nitric oxide regulates the heart by spatial confinement of nitric oxide synthase isoforms. Nature 2002;416:337–339.
10. Mery PF, Pavoine C, Belhassen L, et al. Nitric oxide regulates cardiac Ca^{2+} current: involvement of cGMP-inhibited and cGMP-stimulated phosphodiesterases through guanylyl cyclase activation. J Biol Chem 1993;268:26,286–26,295.
11. Han X, Kobzik L, Balligand JL, et al. Nitric oxide synthase (NOS3)-mediated cholinergic modulation of Ca^{2+} current in adult rabbit atrioventricular nodal cells. Circ Res 1996;78:998–1008.
12. Wahler GM; Dollinger SJ. Nitric oxide donor SIN-1 inhibits mammalian cardiac calcium current through cGMP-dependent protein kinase. Am J Physiol 1995;268:C45–C54.
13. Levi RC, Alloatti G, Penna C, et al. Guanylate-cyclase-mediated inhibition of cardiac ICa by carbachol and sodium nitroprusside. Pflugers Arch 1994;426:419–426.

14. Campbell DL, Stamler JS, Strauss HC. Redox modulation of L-type calcium channels in ferret ventricular myocytes: dual mechanism regulation by nitric oxide and S-nitrosothiols. J Gen Physiol 1996;108:277–293.
15. Hu H, Chiamvimonvat N, Yamagishi T, et al. Direct inhibition of expressed cardiac L-type Ca^{2+} channels by S-nitrosothiol nitric oxide donors. Circ Res 1997;81:742–752.
16. Bers DM. Cardiac excitation-contraction coupling. Nature 2002;415:198–205.
17. Bers DM. Calcium and cardiac rhythms: physiological and pathophysiological. Circ Res 2002;90:14–17.
18. Senzaki H, Smith CJ, Juang GJ, et al. Cardiac phosphodiesterase 5 (cGMP-specific) modulates beta-adrenergic signaling in vivo and is down-regulated in heart failure. FASEB J 2001;15:1718–1726.
19. Schuh K, Uldrijan S, Telkamp M, et al. The plasmamembrane calmodulin-dependent calcium pump: a major regulator of nitric oxide synthase I. J Cell Biol 2001;155:201–205.
20. Nathan C. Nitric oxide as a secretory product of mammalian cells. FASEB J 1992;6:3051–3064.
21. Michel T, Xie Q-W, Nathan C. Molecular biological analysis of nitric oxide synthase. John Wiley: Chichester, UK, 1996.
22. Michel T, Feron O. Nitric oxide synthases: which, where, how, and why? J Clin Invest 1997;100: 2146–2152.
23. Michel T, Smith TW. Nitric oxide synthases and cardiovascular signaling. Am J Cardiol 1993;72: 33C-38C.
24. Xu KY, Huso DL, Dawson TM, et al. Nitric oxide synthase in cardiac sarcoplasmic reticulum. Proc Natl Acad Sci USA 1999;96:657–662.
25. de Belder AJ, Radomski MW, Why HJ, et al. Nitric oxide synthase activities in human myocardium. Lancet 1993;341:84, 85.
26. Balligand JL, Ungureanu-Longrois D, Simmons WW, et al. Cytokine-inducible nitric oxide synthase (iNOS) expression in cardiac myocytes. Characterization and regulation of iNOS expression and detection of iNOS activity in single cardiac myocytes in vitro. J Biol Chem 1994;269:27,580–27,588.
27. Kobzik L, Reid MB, Bredt DS, et al. Nitric oxide in skeletal muscle. Nature 1994;372:546–548.
28. Brenman JE, Chao DS, Gee SH, et al. Interaction of nitric oxide synthase with the postsynaptic density protein PSD-95 and alpha1-syntrophin mediated by PDZ domains. Cell 1996;84:757–767.
29. Brenman JE, Chao DS, Xia H, et al. Nitric oxide synthase complexed with dystrophin and absent from skeletal muscle sarcolemma in Duchenne muscular dystrophy. Cell 1995;82:743–752.
30. Feron O, Dessy C, Opel DJ, et al. Modulation of the endothelial nitric-oxide synthase-caveolin interaction in cardiac myocytes: implications for the autonomic regulation of heart rate. J Biol Chem 1998;273:30,249–30,254.
31. Feron O, Smith TW, Michel T, et al. Dynamic targeting of the agonist-stimulated m2 muscarinic acetylcholine receptor to caveolae in cardiac myocytes. J Biol Chem 1997;272:17,744–17,748.
32. Fulton D, Gratton JP, McCabe TJ, et al. Regulation of endothelium-derived nitric oxide production by the protein kinase Akt. Nature 1999;399:597–601.
33. Dimmeler S, Fleming I, Fisslthaler B, et al. Activation of nitric oxide synthase in endothelial cells by Akt-dependent phosphorylation. Nature 1999;399:601–605.
34. Bredt DS, Snyder SH. Isolation of nitric oxide synthetase, a calmodulin-requiring enzyme. Proc Natl Acad Sci USA 1990;87:682–685.
35. Stamler JS, Lamas S, Fang FC. Nitrosylation: the prototypic redox-based signaling mechanism. Cell 2001;106:675–683.
36. Wink DA, Mitchell JB. Chemical biology of nitric oxide: insights into regulatory, cytotoxic, and cytoprotective mechanisms of nitric oxide. Free Radic Biol Med 1998;25:434–456.
37. Layland J, Li JM, Shah AM. Role of cyclic GMP-dependent protein kinase in the contractile response to exogenous nitric oxide in rat cardiac myocytes. J Physiol 2002;540:457–467.
38. Feelisch M, Stamler JS. Donors of nitrogen oxides. In: Feelisch M, Stamler JS, eds. Methods in Nitric Oxide Research. John Wiley: New York, 1996, pp. 69–114.
39. Wink DA, Cook JA, Kim SY, et al. Superoxide modulates the oxidation and nitrosation of thiols by nitric oxide-derived reactive intermediates: chemical aspects involved in the balance between oxidative and nitrosative stress. J Biol Chem 1997;272:11,147–11,151.
40. Scharfstein JS, Keaney JF Jr., Slivka A, et al. In vivo transfer of nitric oxide between a plasma protein-bound reservoir and low molecular weight thiols. J Clin Invest 1994;94:1432–1439.
41. Sun J, Xin C, Eu JP, et al. Cysteine-3635 is responsible for skeletal muscle ryanodine receptor modulation by NO. Proc Natl Acad Sci USA 2001;98:11158–11162.
42. Simon DI, Mullins ME, Jia L, et al. Polynitrosylated proteins: characterization, bioactivity, and functional consequences. Proc Natl Acad Sci USA 1996;93:4736–4741.

43. Xu L, Eu JP, Meissner G, et al. Activation of the cardiac calcium release channel (ryanodine receptor) by poly-S-nitrosylation. Science 1998;279:234–237.
44. Eu JP, Sun J, Xu L, et al. The skeletal muscle calcium release channel: coupled O2 sensor and NO signaling functions. Cell 2000;102:499–509.
45. Chesnais JM, Fischmeister R, Mery PF. Positive and negative inotropic effects of NO donors in atrial and ventricular fibres of the frog heart. J Physiol 1999;518(pt 2):449–461.
46. Vila-Petroff MG, Younes A, Egan J, et al. Activation of distinct cAMP-dependent and cGMP-dependent pathways by nitric oxide in cardiac myocytes. Circ Res 1999;84:1020–1031.
47. Bang L, Boesgaard S, Nielsen-Kudsk JE, et al. Nitroglycerin-mediated vasorelaxation is modulated by endothelial calcium-activated potassium channels. Cardiovasc Res 1999;43:772–778.
48. Li Z, Chapleau MW, Bates JN, et al. Nitric oxide as an autocrine regulator of sodium currents in baroreceptor neurons. Neuron 1998;20:1039–1049.
49. Zahradnikova A, Minarovic I, Venema RC, et al. Inactivation of the cardiac ryanodine receptor calcium release channel by nitric oxide. Cell Calcium 1997,22:447–454.
50. Ahmmed GU, Xu Y, Hong Dong P, et al. Nitric oxide modulates cardiac Na^+ channel *via* protein kinase A and protein kinase G. Circ Res 2001;89:1005–1013.
51. Han X, Shimoni Y, Giles WR. A cellular mechanism for nitric oxide–mediated cholineric control of mammalian heart rate. J Gen Physiol 1995;106:45–65.
52. Hartzell HC, Fischmeister R. Opposite effects of cyclic GMP and cyclic AMP on Ca^{2+} current in single heart cells. Nature 1986;323:273–275.
53. Levi RC, Alloatti G, Fischmeister R. Cyclic GMP regulates the Ca-channel current in guinea pig ventricular myocytes. Pflügers Arch 1989;413:685–687.
54. Mery PF, Lohmann SM, Walter U, et al. Ca^{2+} current is regulated by cyclic GMP-dependent protein kinase in mammalian cardiac myocytes. Proc Natl Acad Sci USA 1991;88:1197–1201.
55. Chiamvimonvat N, O'Rourke B, et al. Functional consequences of sulfhydryl modification in the pore-forming subunits of cardiovascular Ca^{2+} and Na^+ channels. Circ Res 1995;76:325–334.
56. Catterall WA. Structure and function of voltage-gated ion channels. Annu Rev Biochem 1995;64: 493–531.
57. Catterall WA. Structure and function of neuronal Ca^{2+} channels and their role in neurotransmitter release. Cell Calcium 1998;24:307–323.
58. Walker D, De Waard M. Subunit interaction sites in voltage-dependent Ca^{2+} channels: role in channel function. Trends Neurosci 1998;21:148–154.
59. Lipton SA, Choi YB, Pan ZH, et al. A redox-based mechanism for the neuroprotective and neurodestructive effects of nitric oxide and related nitroso-compounds. Nature 1993;364:626–632.
60. Arnelle DR, Stamler JS. NO+, NO, and NO^- donation by S-nitrosothiols: implications for regulation of physiological functions by S-nitrosylation and acceleration of disulfide formation. Arch Biochem Biophys 1995;318:279–285.
61. Hess P, Lansman JB, Tsien RW. Different modes of Ca channel gating behaviour favoured by dihydropyridine Ca agonists and antagonists. Nature 1984;311:538–544.
62. Pietrobon D, Hess P. Novel mechanism of voltage-dependent gating in L-type calcium channels. Nature 1990;346:651–655.
63. Murphy BJ, Washkurak AW, Tuana BS. Dihydropyridine binding to the L-type Ca^{2+} channel in rabbit heart sarcolemma and skeletal muscle transverse-tubules: role of disulfide, sulfhydryl and phosphate groups. Biochim Biophys Acta 1990;1052:333–339.
64. Mohan P, Brutsaert DL, Paulus WJ, et al. Myocardial contractile response to nitric oxide and cGMP. Circulation 1996;93:1223–1229.
65. Heuze-Joubert I, Mennecier P, Simonet S, et al. Effect of vasodilators, including nitric oxide, on the release of cGMP and cAMP in the isolated perfused rat kidney. Eur J Pharmacol 1992;220:161–171.
66. Lander HM, Sehajpal PK, Novogrodsky A. Nitric oxide signaling: a possible role for G proteins. J Immunol 1993;151:7182–7187.
67. Miyamoto A, Laufs U, Pardo C, et al. Modulation of bradykinin receptor ligand binding affinity and its coupled G-proteins by nitric oxide. J Biol Chem 1997;272:19,601–19,608.
68. Ruppersberg JP, Stocker M, Pongs O, et al. Regulation of fast inactivation of cloned mammalian IK(A) channels by cysteine oxidation. Nature 1991;352:711–714.
69. Stamler JS. Redox signaling: nitrosylation and related target interactions of nitric oxide. Cell 1994;78:931–936.
70. Islam MS, Berggren PO, Larsson O. Sulfhydryl oxidation induces rapid and reversible closure of the ATP-regulated K^+ channel in the pancreatic beta-cell. FEBS Lett 1993;319:128–132.

71. Bolotina VM, Najibi S, Palacino JJ, et al. Nitric oxide directly activates calcium-dependent potassium channels in vascular smooth muscle. Nature 1994;368:850–853.
72. Zaidi NF, Lagenaur CF, Abramson JJ, et al. Reactive disulfides trigger Ca^{2+} release from sarcoplasmic reticulum via an oxidation reaction. J Biol Chem 1989;264:21,725–21,736.
73. Stoyanovsky DA, Salama G, Kagan VE. Ascorbate/iron activates Ca^{2+}-release channels of skeletal sarcoplasmic reticulum vesicles reconstituted in lipid bilayers. Arch Biochem Biophys 1994;308: 214–221.
74. Salama G, Abramson JJ, Pike GK. Sulphydryl reagents trigger Ca^{2+} release from the sarcoplasmic reticulum of skinned rabbit psoas fibres. J Physiol 1992;454:389–420.
75. Malinski T, Taha Z, Grunfeld S, et al. Diffusion of nitric oxide in the aorta wall monitored in situ by porphyrinic microsensors. Biochem Biophys Res Commun 1993;193:1076–1082.
76. Malinski T, Radomski MW, Taha Z, et al. Direct electrochemical measurement of nitric oxide released from human platelets. Biochem Biophys Res Commun 1993;194:960–965.
77. MacKenzie IM, Garrard CS, Young JD. Indices of nitric oxide synthesis and outcome in critically ill patients. Anaesthesia 2001;56:326–330.
78. Kojda G, Kottenberg K, Nix P, et al. Low increase in cGMP induced by organic nitrates and nitrovasodilators improves contractile response of rat ventricular myocytes. Circ Res 1996;78:91–101.
79. Liu L, Hausladen A, Zeng M, et al. A metabolic enzyme for S-nitrosothiol conserved from bacteria to humans. Nature 2001;410:490–494.
80. Saavedra WF, Paolocci N, St John ME, et al. Imbalance between xanthine oxidase and nitric oxide synthase signaling pathways underlies mechanoenergetic uncoupling in the failing heart. Circ Res 2002;90:297–304.
81. Hare JM, Stamler JS. NOS: modulator, not mediator of cardiac performance. Nat Med 1999;5:273, 274.

9 *S*-Nitrosylation of Cyclic Nucleotide-Gated Channels

Marie-Christine Broillet

CONTENTS

SUMMARY

The activation of cyclic nucleotide-gated (CNG) channels is the final step in olfactory and visual transduction. Over the past several years, CNG channels have been found in various other cell types where they might fulfill various physiological functions. CNG channels rely on the binding of at least two molecules of cyclic adenosine monophosphate or cyclic guanosine 5'-monophosphate at intracellular sites on the channel protein to open a nonspecific cation conductance with a significant permeability to Ca ions. In addition to their activation by cyclic nucleotides, nitric oxide (NO)-generating compounds can directly open the olfactory CNG channels through a redox reaction that results in the *S*-nitrosylation of a free SH group on a cysteine residue. This cysteine is located in the C-linker region of the channel, which is known to be important in channel gating. Kinetic analyses suggest that at least two of these cysteine residues on different channel subunits are involved in the direct activation by NO.

Key Words: Cyclic nucleotide-gated channel; nitric oxide; *S*-nitrosylation; channel activation; olfaction; vision; ion channel.

1. INTRODUCTION

Cyclic nucleotide-gated (CNG) channels are a family of ligand-gated channels that are activated by the binding of at least two molecules of cyclic adenosine monophosphate (cAMP) or cyclic guanosine 5'-monophosphate (cGMP) at intracellular sites on the channel protein. They are nonselective cation channels conducting both mono- and divalent cations, and they belong to the superfamily of cation channels with six transmembrane segments (**Fig. 1**). This superfamily includes voltage-gated K^+, Na^+, and Ca^{2+} channels,

From: *Signal Transduction and the Gasotransmitters: NO, CO, and H_2S in Biology and Medicine* Edited by: Rui Wang © Humana Press Inc., Totowa, NJ

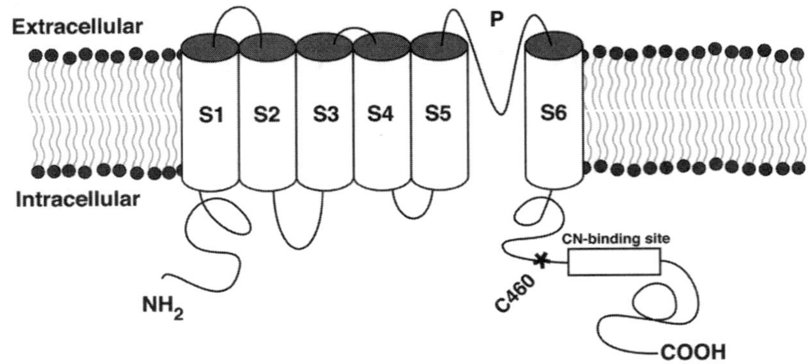

Fig. 1. Hypothetical model of two-dimensional architecture of a CNG channel subunit. S1–S6 are the putative transmembrane domains, and P is the putative pore region. The cyclic nucleotide (CN)–binding site is defined by homology to the sequences of cAMP- and cGMP-binding proteins. The cysteine C460 is the nitric oxide (NO) target site.

hyperpolarization-activated CNG channels, transient receptor potential channels, and the polycystins *(1)*.

CNG channels were originally identified in vertebrate photoreceptor cells *(2)* and olfactory receptor neurons (ORNs) *(3)*, where they mediate calcium entry, providing an intracellular calcium signal that is important for both excitation and adaptation *(4–7)* (**Fig. 2**). Because of the strong calcium permeability of CNG channels, their activation by the ubiquitous cyclic nucleotide second messengers leads not only to membrane depolarization but also to a significant calcium influx into the cells *(8)*.

2. CNG CHANNELS

2.1. Structure and Nomenclature

The genes encoding CNG channels have been cloned, and their transmembrane structures have been deduced from the primary amino acid sequences (for a review, *see* ref. *9*). CNG channels are constructed from different but highly homologous subunits (**Fig. 1**) that are similar in structure to voltage-gated K^+ channels except that they possess a cyclic nucleotide-binding site on the intracellular C-terminal tail and have no apparent voltage sensitivity *(6)*. Indeed, CNG channels possess a voltage-sensor motif in S4 with a reduced number of positively charged amino acids. This domain might be the ancestral S4 segment that has evolved into the voltage sensor of voltage-activated cation channels *(10)*.

A series of elegant experiments with cloned channels and chimeric constructs has revealed significant information regarding the binding and gating reactions that lead to CNG channel activation *(6,11–21)*. These studies have identified several regions as well as specific residues distributed throughout the approx 500 amino acids of the rod or the olfactory proteins that play a key role in channel regulation *(22)*.

The different CNG channel subunits can be grouped into two main types, called CNGA and CNGB. The CNGA subunits (or principal subunits) can form functional homomeric channels activatable by cyclic nucleotides when expressed in heterologous systems (such as HEK293 cells or *Xenopus laevis* oocytes). Three types of different CNGA subunits have been identified. The CNGA1 subunit was first identified in the rod

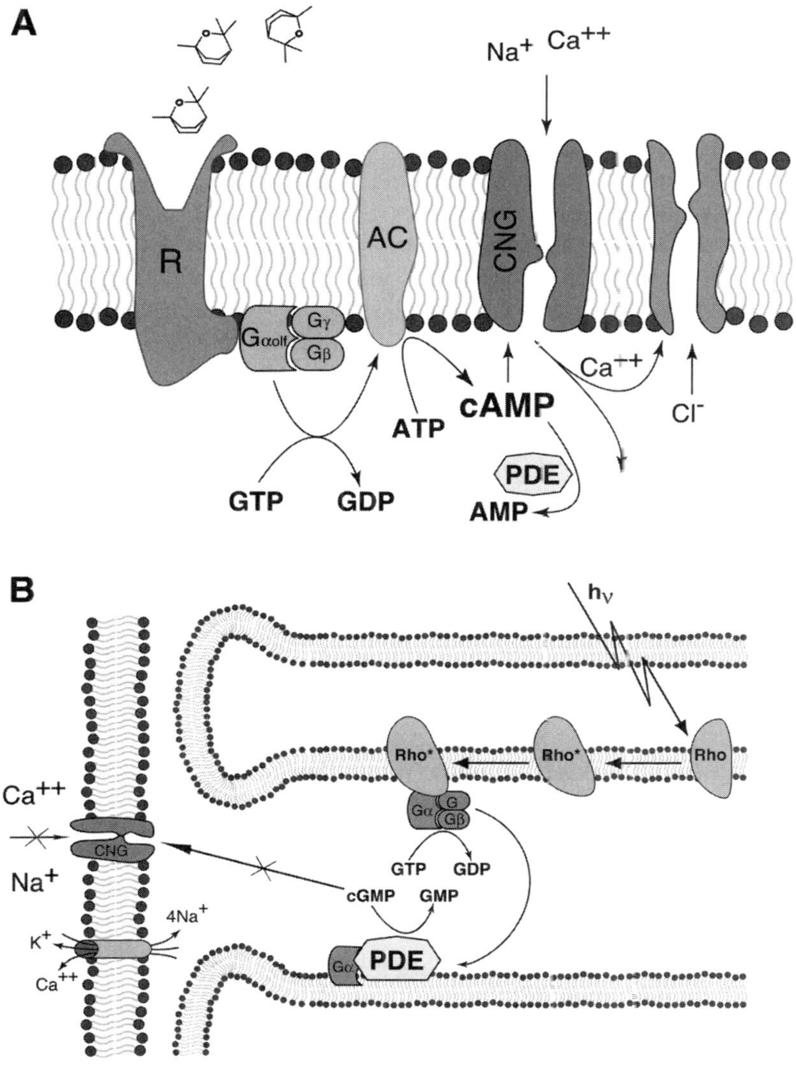

Fig. 2. **(A)** Olfactory signal transduction cascade. In this pathway, the binding of an odorant molecule (cineole is represented here) carried through the mucus layer via an odor-binding protein to the odorant receptor leads to the interaction of the receptor to a GTP-binding protein (G-protein). This interaction in turn leads to the release of the GTP-coupled α-subunit of the G-protein, which then stimulates the adenylyl cyclase (AC) to produce elevated levels of cAMP. The increase in cAMP opens CNG channels, causing an alteration in the membrane potential. **(B)** Visual signal transduction cascade. In the dark, the CNG channel is cooperatively kept open on binding of four cGMP molecules to its CNGB subunits. This causes an exchange of cations (Ca^{2+}, Na^+, K^+) between the cytoplasm and the surrounding interphotoreceptor space. To keep the ion gradients active, the cations are actively pumped across the plasma membrane by $Na^+/Ca^{2+}/K^+$ exchanger. In the light, cGMP is hydrolyzed to 5'GMP by the phosphodiesterases (PDEs). With decreased cGMP concentration, cGMP is removed from the CNG channel subunits and the channel is closed. This blocks the flow of Ca^{2+} and Na^+ inside the rod outer segment (ROS). In darkness, the inward flow of charges (Ca^{2+}, Na^+, K^+) is equal to the outward flow. This is obtained by an $Na^+/Ca^{2+}/K^+$ exchanger in the ROS membrane. At illumination, the $Na^+/Ca^{2+}/K^+$ exchanger is still active, but the inward Ca^{2+} and Na^+ flow through the CNG is blocked and the plasma membrane is hyperpolarized because the charge flow rates have become unequal.

photoreceptor cells *(23)*, the CNGA2 is the corresponding subunit in ORNs *(24)*, and the CNGA3 subunit has been cloned in the testis and cone photoreceptor cells *(25,26)*. These CNGA subunits share common structural features (60–70% homology) such as six transmembrane domains (S1–S6), an S4-like voltage sensor motif, a pore region, and a cyclic nucleotide-binding site in the intracellular N-terminal region (**Fig. 1**).

Additional subunits have been discovered, sharing the same structural properties as the CNGA subunits but with some added diversity; they are called CNGB subunits. They are considered to be modulatory subunits, because they cannot form functional CNG channels on their own in expression systems. The CNGB1a is the rod modulatory subunit *(27)*. The CNGA4 subunit is found in ORNs *(28,29)*; although it shares more sequence homology with the CNGA subunits, it cannot form homomeric cyclic nucleotide-activated ion channels and, thus, from a functional standpoint is classified with the modulatory CNGB subunits. The CNGB1b subunit, a splice variant of the rod modulatory subunit (CNGB1a), is also a component of the olfactory channel *(30)*. A modulatory subunit of the cone CNG channel (CNGB3) has also been cloned *(31)*.

The subunit composition of the native rod CNG channels recently has been determined in parallel by fluorescence resonance energy transfer *(32)*, by chemical crosslinking *(33)* and by analysis of intersubunit interactions *(34)*. The rod CNG channel is a tetramer composed of three CNGA1 and one CNGB1 subunits and represents an example of violation of symmetry in tetrameric channels.

The stoichiometry of the olfactory CNG channel is not known yet, but it is probably a tetramer composed of a mix of three different subunits: CNGA2 *(24)*, CNGA4 *(28,29)*, and CNGB1b *(30,35)*. Indeed, the coexpression of CNGA2/CNGA4/CNGB1b olfactory subunits in heterologous systems leads to the formation of a channel whose properties resemble the native olfactory channel and differ from those observed for homomeric CNGA2 channels. The native olfactory channel is sensitive to both cGMP and cAMP (with a higher affinity for cGMP; $K_{1/2}$ for cAMP = 4 μ*M*, $K_{1/2}$ for cGMP = 1.8 μ*M* *(30)*). The native channel subtypes also differ in their relative permeability to physiological concentrations of calcium such that the fractional current carried by calcium in the olfactory channel is greater than in the rod channel *(36)*. In heterologous expression systems, it has been shown that calcium permeation is determined by the subunit composition of the channel *(37)*.

In summary, CNG channels are heterotetrameric and their subunit composition clearly determines both their ligand sensitivity and electrophysiological properties, and therefore calcium entry in the cells.

2.2. Role in Olfactory Transduction Cascade

One important model system to study the different functions of CNG channels is the olfactory system *(38)*. The remarkable capacity to discriminate among a wide range of odor molecules begins at the level of the ORNs. These particular neurons perform the complex task of converting the chemical information contained in the odor molecules into changes in membrane potential *(39)*. In most vertebrates, the ORNs form a sensory epithelium within the nasal cavity. They are true neurons, sending an axon to the central nervous system (CNS). They have a bipolar morphology and a single dendrite extended to the epithelial surface bearing 10–12 cilia. The sensory transduction occurs at the level of the ciliary membrane. Odorant recognition involves membrane protein receptors and transduction components analogous to those that mediate the specific responses to hormones, growth factors, and neurotransmitters *(40)*. Every molecular element of the olfactory transduction cascade has been isolated, cloned, and expressed, allowing establishment of the scheme presented in **Fig. 2A**. The different steps of the transduction cascade can be summarized as follows: When a receptor molecule is occupied by an odorant, it activates a specific GTP-binding protein (G_{olf}), which modulates the activity of an adenylyl cyclase (AC type III), an enzyme producing the second-messenger cAMP. cAMP directly activates a CNG channel representing the final step in the biochemical cascade and the first step in the generation of the electrical response. An additional, unique membrane conductance, a Ca^{2+}-activated chloride current, is also involved in the electrical response to odors *(41)*.

This cascade of events results in the cell membrane shifting the resting potential from –65 to –45 mV. This depolarization spreads by passive current flow through the dendrite to the soma where it activates voltage-gated Na^+ channels, initiating impulse generation. The combination of Na^+ currents, voltage-dependent K^+ currents, and small Ca^{2+} currents acts to produce one or more action potentials that can propagate via the axon to the olfactory bulb of the brain.

In summary, a high density of CNG channels is present on the ciliary membrane of ORNs. These channels are selective for cations, and their activation plays a key role in cell membrane depolarization because the olfactory stimulus.

2.3. Role in Visual Transduction Cascade

Phototransduction is mediated by an enzymatic cascade that ultimately leads to the hydrolysis of cGMP (**Fig. 2B**). The photoreceptor cells, rods, and cones integrate and respond to cGMP hydrolysis via a CNG channel in the plasma membrane of the outer segment. The last step in phototransduction is the creation of a change in membrane potential that is mediated by CNG channels. In the dark, the CNG channel is cooperatively kept open by the binding of four cGMP molecules to its CNGB subunits *(42)*. This causes an exchange of cations (Ca^{2+}, Na^+, K^+) between the cytoplasm and the surrounding interphotoreceptor space. To keep the ion gradients active, the cations are actively pumped out across the plasma membrane by an $Na^+/Ca^{2+}/K^+$ exchanger *(43)*. In the light, cGMP is hydrolyzed to 5'GMP by the phosphodiesterases (PDEA-PDEB). With decreased cGMP concentration, cGMP is released from the CNG channel subunits and the channel is closed (**Fig. 2B**). This blocks the flow of Ca^{2+} and Na^+ into the outer segment of the rod (ROS).

In darkness, the inward flow of charges (Ca^{2+}, Na^+, K^+) is equal to the outward flow. This is obtained by a $Na^+/Ca^{2+}/K^+$ exchanger in the ROS membrane. At illumination the $Na^+/Ca^{2+}/K^+$-exchanger is still active but the inward Ca^{2+} and Na^+ flow through the CNG channel is blocked and the plasma membrane is hyperpolarized because the charge flow rates have become unequal. This hyperpolarization leads to a change in synaptic activity and ultimately alters the nerve impulse pattern that is sent to the brain *(44)*. The CNG channel activity is also susceptible to high Ca^{2+} calmodulin that leads to closure of the channel to reduce the Ca^{2+} influx *(27,45)*.

2.4. Roles in Other Systems

Over the past several years, CNG channels have been found in various other tissues, including kidney, testis, and heart (for a review, *see* ref. *9*), where they might fulfill various physiological functions. More recently, these channels have been found in the

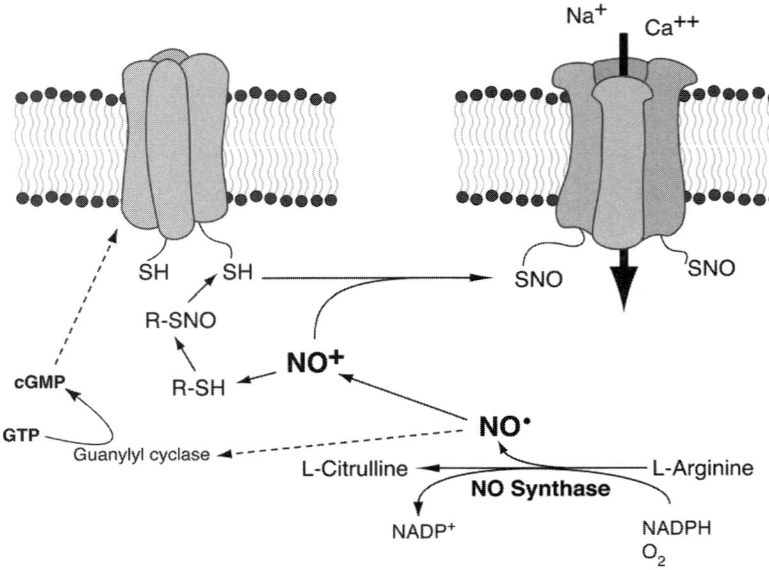

Fig. 3. Biosynthesis pathway for NO and nitrosothiols. Nitric oxide (NO·) is synthesized from L-arginine by NO synthases (NOSs). One NO· redox form, the nitrosonium ion (NO^+), can activate by *S*-nitrosylation an array of target proteins (such as the CNG channel represented here) either directly or via the formation of intermediate nitrosothiols (RSNOs). The CNG channel can also be activated by cGMP produced after stimulation of the guanylyl cyclase activity by NO·.

CNS *(7,46,47)* and have been implicated in processes as diverse as synaptic modulation, central communication, plasticity, and axon outgrowth in animals ranging from the nematode to mammals *(48,49)*. CNG channel subunits, in particular the olfactory CNG channel subunits, have been identified in the brain *(46,47,50)*. Specific subsets of neurons such as the CA1 and CA3 neurons of the hippocampus express CNG channel subunits, suggesting that these channels have a particular function in the CNS that is related specifically to certain cell types, rather than being of a general housekeeping nature *(49)*.

In the heart, the I_h channel in the sinoatrial node controls pacemaking activity and is regulated by the binding of cAMP *(51)*. Another CNG channel, similar to the olfactory CNGA2, is expressed throughout the heart of the mouse but its function remains unclear *(52)*. As sensors of cyclic nucleotide concentrations and conduits for Ca^{2+} entry, these CNG channels may play a role in regulating the heart rate and contraction.

2.5. Multiple Ligand Sensitivity

Olfactory CNG channels can be activated by either cAMP or cGMP, although it is generally believed that under normal physiological conditions it is a rise in intracellular cAMP that is responsible for channel activation *(39)*. However, because of the sensitivity of CNG channels to cGMP, NO has been proposed as a signaling molecule in the olfactory system *(53)* and in the visual system *(54)*. In this model, NO activates a soluble guanylyl cyclase, producing cGMP, which then activates the ion channel (**Fig. 3**). This interaction between NO and guanylyl cyclase represents a widespread signaling mechanism that links extracellular stimuli to the biosynthesis of cGMP in adjacent cells *(55–57)*.

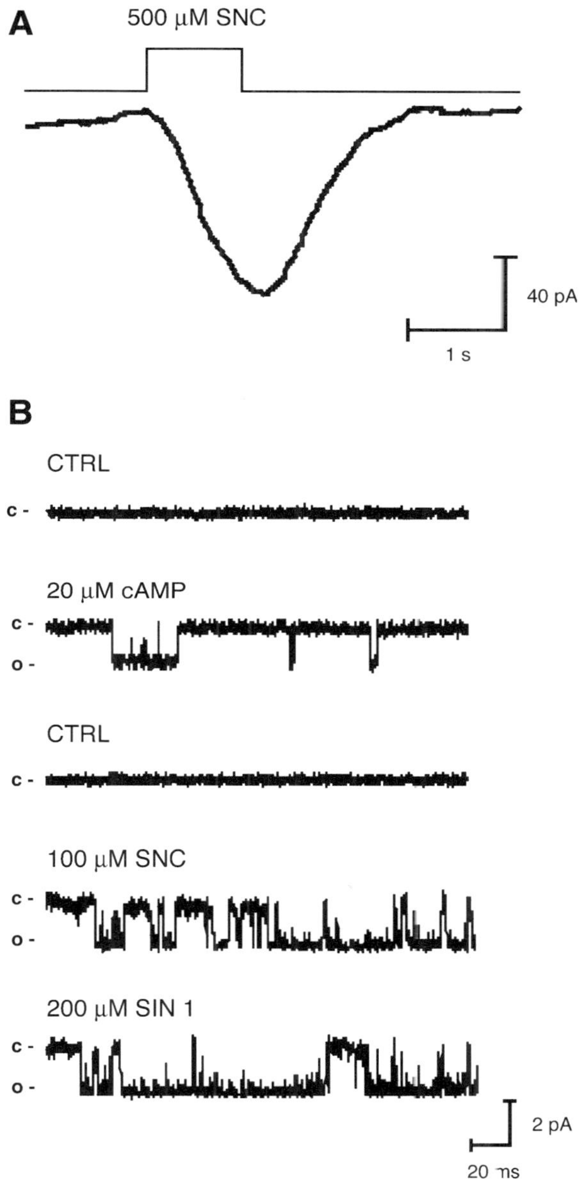

Fig. 4. NO activates olfactory CNG channels. **(A)** Response of voltage-clamped ORN to 1-s pulse of NO donor (SNC, 500 μ*M*). The intracellular pipet solution contained 110 m*M* CsCl, 1 m*M* $CaCl_2$, 2 m*M* $MgCl_2$, 10 m*M* EGTA, 4 m*M* HEPES, 1 m*M* Mg-ATP; pH 7.6; no GTP was added. The holding potential was –70 mV. **(B)** Direct effect of NO donors on CNG channels. An inside-out patch recording from the dendrite of an olfactory neuron contained a single channel activated either by cAMP (20 μ*M*), SNC (100 μ*M*) or SIN-1 (200 μ*M*). (Adapted from ref. *60*.)

In electrophysiological experiments with intact olfactory neurons from different animal species recorded under whole-cell voltage clamp, it was observed that application of NO donors such as sodium nitroprusside *(58,59)* or *S*-nitrosocysteine (SNC) **(Fig. 4A)** *(60)* induced the immediate appearance of a depolarizing current similar to the one observed when a cGMP membrane-permeant analog was perfused. This current was not dependent on the presence of GTP in the intracellular pipet solution and was therefore declared cGMP independent, providing a possible alternate pathway for CNG channel activation *(59,60)*.

2.6. Activation via S-Nitrosylation

Characteristic single CNG channel activity recorded from inside-out membrane patches from the soma and dendritic membrane of olfactory receptor neurons is shown in **Fig. 4B**. The control recording (top trace, CTRL), in the presence of symmetrical Ca^{2+}-free solutions, showed no channel openings. When cAMP (20 μ*M*) was applied, single channel openings and bursts of openings occurred. The effect of cAMP was fully and rapidly reversible after removal of agonist (CTRL). On addition of the NO donors, SNC at 100 μ*M*, or 3-morpholinosydnon-mine (SIN-1) at 200 μ*M*, long bursts of channel openings and clusters of channel openings appeared on the current trace. When no channels were activated by cAMP in a patch, SIN-1 or SNC also failed to activate channels, or to increase the patch permeability (data not shown), indicating that SIN-1 or SNC specifically activated a cAMP-gated conductance. Conversely, whenever a patch contained a channel activated by cAMP, SNC or SIN-1 was also able to activate a channel. Inactive SIN-1 or SNC (i.e., solutions more than 24 h old) had no effect on patches containing CNG channels, nor did the byproduct cystine (100 μ*M*) or 100 μ*M* cysteine alone, indicating that the effects were dependent on the production of NO groups. Channel activity induced by either SNC or SIN-1 was only slowly reversible, with occasional channel openings occurring for up to 30 min after removal of the drugs, suggesting that a persistent modification of the channel had been caused by NO, in one of its redox states *(60)*. Application of NO donors did not affect the single-channel conductance level or the mean amplitudes. Thus, it appears that SNC and cAMP are acting on the same population of channels.

One contradictory study using macropatches excised from the olfactory knob of rat olfactory neurons presented inhibition of the CNG conductance instead of activation through NO stimulation *(61)*.

NO action was independent of the presence of the normal ligand (cAMP or cGMP) and did not involve the cyclic nucleotide-binding site, suggesting an alternate site on the molecule that is critical in channel gating (**Fig. 5**) *(60)*.

In summary, in addition to their activation by cyclic nucleotides, NO-generating compounds can directly open the olfactory CNG channels through a redox reaction that results in the *S*-nitrosylation of a free SH group on a cysteine residue *(60)* (**Fig. 3**). This posttranslational modification, comparable to phosphorylation, has been shown to modulate the activity of other proteins, including caspases and *N*-methyl-D-aspartate receptor channels *(62)*.

It is not uncommon for several cysteine residues on a given protein to be candidates for nitrosylation. In the ryanodine receptor, of a total of 364 cysteines, 84 provide free SH groups, but only 12 are thought to undergo nitrosylation *(63)*. Although the precise parameters governing accessibility by NO are unknown, the existence of a consensus nitrosylation acid-base motif has been postulated based on large database screenings *(64)*. The proposed motif is XYCZ, in which X can be any of G, S, T, C, Y, N, or Q; Y can be K, R, H, D, or E; and Z can be D or E. The most important element of the sequence is believed to be the Asp/Glu residues following the cysteine. In spite of this rather

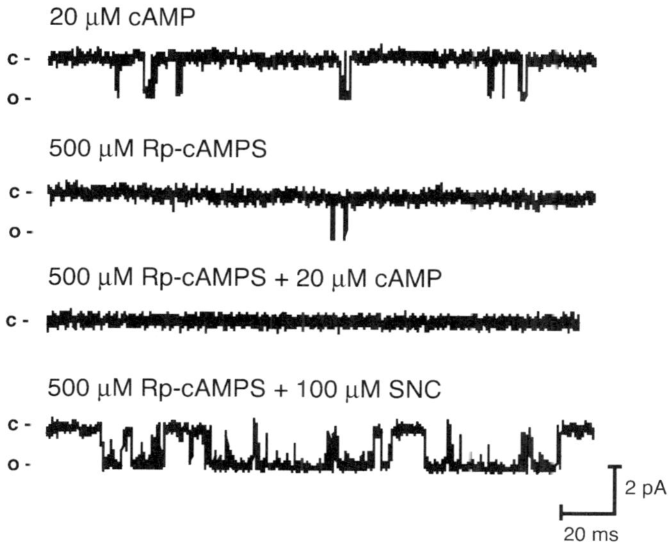

Fig. 5. NO activation is independent of cyclic nucleotides. Original recordings from inside-out patches of salamander olfactory neuron show the antagonistic effects of Rp-cAMPS (500 µ*M*) in the presence of cAMP (20 µ*M*) or SNC (100 µ*M*). The holding potential was –40 mV. (Adapted from ref. *60*.)

degenerate motif, in the olfactory CNG channel, only the cysteine C460 possesses the required motif (i.e., Q, D, C, E) (**Fig. 1**). Biochemical and mutation experiments (described hereafter) confirmed that C460 was indeed the NO target site *(60,65)*.

The NO target site on the CNG channel has been identified by mutating the four candidate intracellular cysteine residues Cys-460, Cys-484, Cys-520, and Cys-552 of the rat olfactory CNGA2 channel into serine residues. All mutant channels continue to be activated by cyclic nucleotides, but only one of them, the C460S mutant channel, exhibits a total loss of NO sensitivity *(65)* (**Fig. 6**). This result is consistent with the lack of NO sensitivity of the CNG channel expressed in *Drosophila melanogaster* (DmCNG), which does not have this specific cysteine residue *(65,66)* (**Fig. 6**). Cys-460 is located in the C-linker region of the channel known to be important in channel gating (**Fig. 1**). Kinetic analyses suggest that at least two of these Cys-460 residues on different channel subunits are involved in the activation by NO. These results show that one single cysteine residue is responsible for NO sensitivity but that several channel subunits need to be activated for channel opening by NO *(67)*.

Because a functional CNG channel is most probably made up of four subunits *(68)*, there are four potential nitrosylation sites per channel. However, factors other than those noted earlier may also determine the likelihood of NO activity at particular cysteines. Different degrees of accessibility to NO resulting from protein conformation, different reaction rates with NO at different cysteines because of redox status of the immediate environment, or cysteines in positions that may have no functional consequences on nitrosylation could also account for the observation that in most proteins a relatively few free thiols are in fact involved in nitrosylation-induced activity *(64)*. In the CNG channel, our concentration-response data indicate a Hill coefficient of <2, suggesting that as few as two of the four target cysteines may actually interact with NO. However, the Hill coefficient does not preclude activity at all four sites.

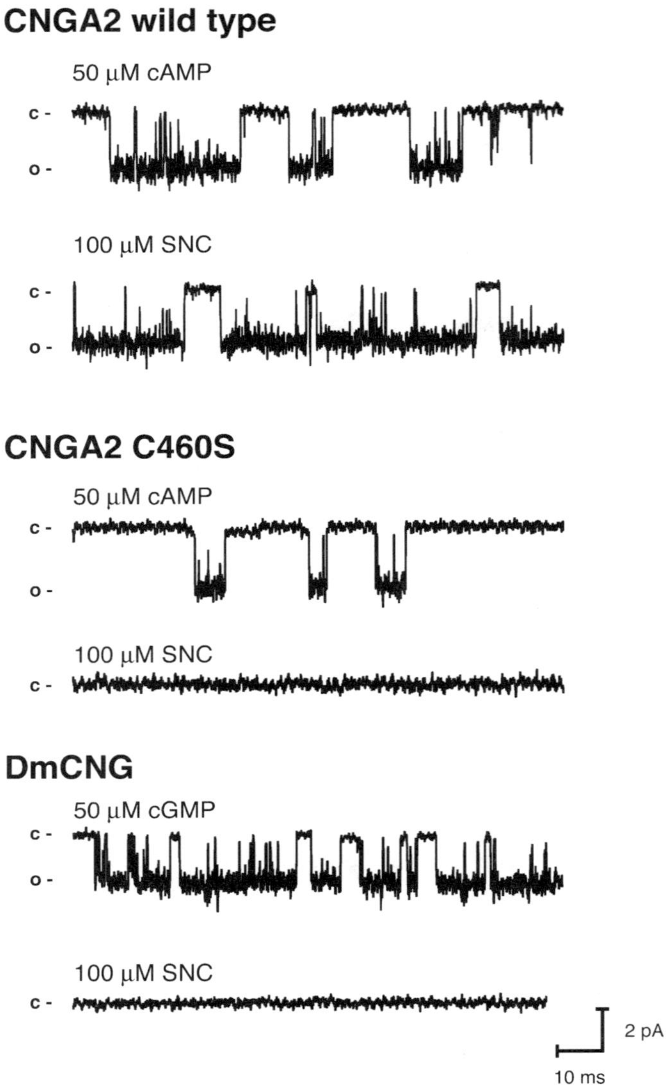

Fig. 6. C460 is the NO target site. A comparison of single-channel current recordings from channels composed of either wild-type CNGA2 subunits or CNGA2 C460S mutant subunits expressed in HEK 293 cells is presented. Channels were activated respectively by cAMP (50 μ*M*) and SNC (100 μ*M*), holding potential of –60 mV. In the case of the C460S mutant channel, no activation could be observed after SNC treatment and a decrease in open probability was observed after cAMP treatment compared with the wild-type channel. In inside-out patches from HEK 293 cells transfected with the cloned *Drosophila* CNG channel (DmCNG), treatment with cGMP (50 μ*M*) led to immediate channel activation. However, treatment with the NO donor SNC (1–1000 μ*M*) failed to induce channel activity. All recordings were performed at a holding potential of –60 mV. (Adapted from ref. *65*.)

Gordon et al. *(69)* have proposed an important role for cysteine residues in CNG channel activation. They have shown, working on the rod $CNGA_1$ subunit of the CNG channel, that the N- and C-terminal regions of each subunit interact in the tetrameric channel. This interaction involves particular cysteine residues with the formation of a disulfide bond between Cys 35 (N-terminal) and Cys 481 (C-terminal). Recently, Brown et al. *(20)* have also found that C481 is necessary for the potentiation of the rod CNGA1 CNG channel's response to cAMP and cGMP. This cysteine residue, located in the C-terminal region, corresponds exactly to the NO target site in the olfactory CNG channel (Cys 460), confirming that this amino acid—which is highly conserved among the different cloned CNG channels—likely plays a critical role in channel gating and in the potentiation of cyclic nucleotide action.

The fact that the Cys 460 is a potent activator of channel gating by NO and also affects channel gating by cAMP indicates the importance of this residue for channel function.

The corresponding cysteine on the olfactory CNGA4 subunit is the cysteine 352. This subunit cannot form homomeric channels that can be activated by cyclic nucleotides, but they can be activated by NO. There are striking differences between the properties of these NO-gated channels and either native CNG channels in rat olfactory neurons or homomeric CNGA2 channels expressed in heterologous systems, including the amount of open channel noise and the calcium permeability *(67)*. Thus, these homomeric CNGA4 channels might be expressed at different cellular locations and fulfill different physiological roles than the heteromeric CNG channels.

Although the significance of CNG channels being directly NO gated for olfactory transduction is not yet clear, the demonstration that NOS is expressed transiently by newly developing ORNs suggests a possible role for NO during neuronal regeneration *(70,71)*.

The NO-target site is also conserved in the CNGA1 and CNGB1 subunits of the rod CNG channels, but there is no experimental evidence yet whether or not NO has a direct effect on photoreceptor CNG channels *(72)*.

2.7. Activation Via NO/cGMP Pathway

CNG channels in outer segments of vertebrate photoreceptors generate electrical signals in response to changes in cGMP concentration during phototransduction. CNG channels also allow the influx of Ca^{2+}, which is essential for photoreceptor adaptation. In cone photoreceptors, cGMP triggers an increase in membrane capacitance indicative of exocytosis, suggesting that CNG channels are also involved in synaptic function. CNG channels occur in clusters and are indirectly activated by the NO donor SNC as the NO-induced transmitter release is suppressed by guanylyl cyclase inhibitors and prevented by direct activation of CNG channels, indicating that their activation is required for NO to elicit release *(73)*.

In the retina, in addition to their central role in phototransduction, CNG channels may be involved in NO signaling in bipolar neurons or in the hyperpolarizing synaptic response to glutamate in ON-type (depolarizing) bipolar cells. Several forms of particulate and soluble guanylyl cyclases are expressed in a variety of cells in the retina *(74,75)*. These cyclases can be activated by NO or by natriuretic peptides, resulting in elevated cGMP in selected cell types, including bipolar cells *(76)*. Therefore, the signaling pathways involving NO-sensitive and NO-insensitive guanylyl cyclases offer a possible link between physiological stimuli and CNG channels in bipolar cells. It is interesting that, in addition to bipolar cells, both amacrine cells and ganglion cells have been implicated in NO signaling *(77)*. This raises the possibility that cone CNG channels may mediate responses to signals such as NO that activate guanylyl cyclase in a variety of retinal neurons in addition to bipolar cells *(78)*.

3. CNG CHANNELS AND NO: THE FUTURE

CNG channels first identified in rods and olfactory receptors are now known to be distributed throughout many different cells of the body and may have important roles in processes such as cell motility, secretion, and development. Future work is likely to reveal new functions for CNG channels as cyclic nucleotide or NO-gated channels as well as more information about the functional significance of particular structural features and the modulation of these channels.

Important physiological roles for CNG channels might be found in different tissues such as in the heart, where they might be involved in regulating heart rate and contraction as sensors of cyclic nucleotide concentrations and conduits for Ca^{2+} entry. In the hippocampus or other parts of the CNS, they might play a role in neuronal pathfinding, synapse formation, and plasticity with cyclic nucleotides or NO as activators.

Significant advances in our understanding of the chemosensory systems have occurred at a rapid pace over the last several years. Although this knowledge has provided some very critical insights, many fundamental questions remain to be answered. For example, taste and pheromone transduction cascades have not been identified yet. Vertebrate taste transduction has recently received a lot of attention with the cloning of two new multigene families of G-protein-coupled receptors *(79–81)*. The screening of the newly published human genome library allowed the discovery of one gene sharing strong homology with the mouse and rat vomeronasal genes *(82)*, reopening the debate as to whether humans can detect pheromones. The presence of CNG channels in both systems as well as the presence of neuronal NOS may indicate a significant importance of these channels and NO for these two sensory systems.

Given the significant Ca permeability of the CNG channel, long-lasting activation of this channel by NO-produced nitrosothiols could be the initiating step in many cellular processes. Moreover, it has been recently shown that exposure to an NO donor also leads to increased protein and mRNA levels of the CNGA2 subunit contributing to enhanced channel activity and calcium elevation in endothelial cells *(83)*.

ACKNOWLEDGMENTS

I gratefully thank Olivier Randin for assistance in producing the figures. This work was supported by grants from the Novartis, Roche, and Leenaards Foundations and by the Fonds National Suisse de la Recherche.

REFERENCES

1. Chen XZ, Vassilev PM, Basora N, et al. Polycystin-L is a calcium-regulated cation channel permeable to calcium ions. Nature 1999;401:383–386.
2. Fesenko EE, Kolesnikov SS, Lyubarsky AL. Induction by cyclic GMP of cationic conductance in plasma membrane of retinal rod outer segment. Nature 1985;313:310–313.
3. Nakamura T, Gold GH. A cyclic-nucleotide gated conductance in olfactory receptor cilia. Nature 1987;325:442–444.
4. Kaupp UB. Family of cyclic nucleotide gated ion channels. Curr Opin Neurobiol 1995;5:434–442.
5. Finn JT, Yau KW. Cyclic nucleotide-gated channels: an extended family with diverse functions. Annu Rev Physiol 1996;58:395–426.
6. Zagotta WN, Siegelbaum SA. Structure and function of cyclic nucleotide-gated channels. Annu Rev Neurosci 1996;19:235–263.
7. Wei JY, Roy DS, Leconte L, et al. Molecular and pharmacological analysis of cyclic nucleotide-gated channel function in the central nervous system. Prog Neurobiol 1998;56:37–64.

8. Leinders-Zufall T, Rand MN, Shepherd GM, et al. Calcium entry through cyclic nucleotide-gated channels in individual cilia of olfactory receptor cells: spatiotemporal dynamics. J Neurosci 1997;17:4136–4148.
9. Kaupp UB, Seifert R. Cyclic nucleotide-gated ion channels. Physiol Rev 2002;82:769–824.
10. Jan LY, Jan YN. A superfamily of ion channels. Nature 1990;345:672.
11. Goulding EH, Tibbs GR, Liu D, et al. Role of H5 domain in determining pore diameter and ion permeation through cyclic nucleotide-gated channels. Nature 1993;364:61–64.
12. Goulding EH, Tibbs GR, Siegelbaum SA. Molecular mechanism of cyclic nucleotide-gated channel activation. Nature 1994;372:369–374.
13. Gordon SE, Zagotta WN. A histidine residue associated with the gate of the cyclic nucleotide–activated channels in rod photoreceptors. Neuron 1995;14:177–183.
14. Gordon SE, Zagotta WN. Localization of regions affecting an allosteric transition in cyclic nucleotide–activated channels. Neuron 1995;14:857–864.
15. Root MJ, MacKinnon R. Identification of an external divalent cation-binding site in the pore of a cGMP-activated chanel. Neuron 1993;11:459–466.
16. Root MJ, MacKinnon R. Two identical noninteracting sites on an ion channel revealed by proton transfer. Science 1994;265:1852–1856.
17. Park CS, MacKinnon R. Divalent cation selectivity in a cyclic nucleotide-gated ion channel. Biochemistry 1995;34:13,328–13,333.
18. Sun Z, Akabas MH, Goulding EH, et al. Exposure of residues in the cyclic nucleotide-gated channel pore: P-region structure and function in gating. Neuron 1996;16:141–149.
19. Liu DT, Tibbs GR, Paoletti P, et al. Constraining ligand-binding site stoichiometry suggests that a cyclic nucleotide-gated channel is composed of two functional dimers. Neuron 1998;21:235–248.
20. Brown LA, Snow SD, Haley TL. Movement of gating machinery during the activation of rod cyclic nucleotide-gated channels. Biophys J 1998;75:825–833.
21. Gavazzo P, Picco C, Eismann E, et al. A point mutation in the pore region alters gating, Ca^{2+} blockage, and permeation of olfactory cyclic nucleotide-gated channels. J Gen Physiol 2000;116:311–326.
22. Broillet M-C, Firestein S. Cyclic nucleotide-gated channel: molecular mechanisms of activation. In: Rudy B, Seeburg P, eds. Molecular and Functional Diversity of Ion Channels and Receptors. vol. 868. New York: The New York Academy of Sciences, 1999, pp. 730–740
23. Kaupp UB, Niidome T, Tanabe T, et al. Primary structure and functional expression from complementary DNA of the rod photoreceptor cyclic GMP–gated channel. Nature 1989;342:762–766.
24. Dhallan RS, Yau KW, Schrader KA, et al. Primary structure and functional expression of a cyclic nucleotide–activated channel from olfactory neurons. Nature 1990;347:184–187.
25. Bonigk W, Altenhofen W, Muller F, et al. Rod and cone photoreceptor cells express distinct genes for cGMP-gated channels. Neuron 1993;10:865–877.
26. Weyand I, Godde M, Frings S, et al. Cloning and functional expression of a cyclic-nucleotide-gated channel from mammalian sperm. Nature 1994;368:859–863.
27. Chen T-Y, Illing M, Molday LL, et al. Subunit 2 (or beta) of retinal rod cGMP-gated cation channel is a component of the 240–kDa channel-associated protein and mediates Ca^{2+}-calmodulin modulation. PNAS 1994;91:11,757–11,761.
28. Bradley J, Li J, Davidson N, et al. Heteromeric olfactory cyclic nucleotide-gated channels: a subunit that confers increased sensitivity to cAMP. PNAS 1994;91:8890–8894.
29. Liman ER, Buck LB. A second subunit of the olfactory cyclic nucleotide-gated channel confers high sensitivity to cAMP. Neuron 1994;13:611–621.
30. Bonigk W, Bradley J, Muller F, et al. The native rat olfactory cyclic nucleotide -gated channel is composed of three distinct subunits. J Neuroscie 1999;19:5332–5347.
31. Gerstner A, Zong X, Hofmann F, et al. Molecular cloning and functional characterization of a new modulatory cyclic nucleotide-gated subunit from mouse retina. J Neurosci 2000;20:1324–1332.
32. Zheng J, Trudeau MC, Zagotta WN. Rod cyclic nucleotide-gated channels have a stoichiometry of three CNGA1 subunits and one CNGB1 subunit. Neuron 2002;36:891–896.
33. Weitz D, Ficek N, Kremmer E, et al. Subunit stoichiometry of the CNG channel of rod photoreceptors. Neuron 2002;36:881–889.
34. Zhong H, Molday LL, Molday RS, et al. The heteromeric cyclic nucleotide-gated channel adopts a 3A:1B stoichiometry. Nature 2002;420:193–198.
35. Sautter A, Zong X, Hofmann F, et al. An isoform of the rod photoreceptor cyclic nucleotide-gated channel β subunit expressed in olfactory neurons. Proc Natl Acad Sci USA 1998;95:4696–4701.
36. Frings S, Seifert R, Godde M, et al. Profoundly different calcium permeation and blockage determine the specific function of distinct cyclic nucleotide-gated channels. Neuron 1995;15:169–179.

37. Dzeja C, Hagen V, Kaupp UB, et al. Ca^{2+} permeation in cyclic nucleotide-gated channels. EMBO J 1999;18:131–144.
38. Firestein S, Zufall F. The cyclic nucleotide gated channel of olfactory receptor neurons. Semin Cell Biol 1994;5:39–46.
39. Shepherd GM. Discrimination of molecular signals by the olfactory receptor neuron. Neuron 1994;13:771–790.
40. Hibert MF, Trump-Kallmeyer S, Bruinvels A, et al. Three-dimensional models of neurotransmitter G-binding protein coupled receptors. Mol Pharmacol 1991;40:8–15.
41. Lowe G, Gold GH. Nonlinear amplification by calcium-dependent chloride channels in olfactory receptor cells. Nature 1993;366:283–286.
42. Brown RL, Gramling R, Bert RJ, et al. Identification by photoaffinity labeling of peptide regions within retinal rod cGMP-activated channel subunits involved in cGMP binding. Invest Ophthalmol Vis Sci 1994;35:1473–1477.
43. Tucker JE, Winkfein RJ, Cooper CB, et al. cDNA cloning of the human retinal rod Na-Ca^{+} K exchanger: comparison with a revised bovine sequence. Invest Ophthalmol Vis Sci 1998;39:435–440.
44. Yau KW. Phototransduction mechanism in retinal rods and cones: The Friedenwald Lecture. Invest Ophthalmol Vis Sci 1994;35:9–32.
45. Hsu YT, Molday RS. Interaction of calmodulin with the cyclic GMP–gated channel of rod photoreceptor cells: modulation of activity, affinity purification, and localization. J Biol Chem 1994;269: 29,765–29,770.
46. Bradley J, Zhang Y, Bakin R, et al. Functional expression of the heteromeric "olfactory" cyclic nucleotide-gated channel in the hippocampus: A potential effector of synaptic plasticity in brain neurons. J Neurosci 1997;17:1993–2005.
47. Kingston PA, Zufall F, Barnstable CJ. Rat hippocampal neurons express genes for both rod retinal and olfactory cyclic nucleotide-gated channels: novel targets for cAMP/cGMP function. Proc Natl Acad Sci USA 1996;93:10,440–10,445.
48. Coburn CM, Bargmann CI. A putative cyclic nucleotide-gated channel is required for sensory development and function in C. elegans. Neuron 1996;17:695–706.
49. Zufall F, Shepherd GM, Barnstable CJ. Cyclic nucleotide gated channels as regulators of CNS development and plasticity. Curr Opin Neurobiol 1997;7:404–412.
50. Strijbos PJ, Pratt GD, Khan S, et al. Molecular characterization and in situ localization of a full-length cyclic nucleotide-gated channel in rat brain. Eur J Neurosci 1999;11:4463–4467.
51. DiFrancesco D, Tortora P. Direct activation of cardiac pacemaker channels by intracellular cyclic AMP. Nature 1991;351:145–147.
52. Ruiz ML, London B, Nadal-Ginard B. Cloning and characterization of an olfactory cyclic nucleotide-gated channel expressed in mouse heart. J Mol Cell Cardiol 1996;28:1453–1461.
53. Breer H, Shepherd GM. Implications of the NO/cGMP system for olfaction. TINS 1993;16:5–9.
54. Kurenny DE, Moroz LL, Turner RW, Set al. Modulation of ion channels in rod photoreceptors by nitric oxide. Neuron 1994;13:315–324.
55. Ignarro LJ. Nitric oxide. Hypertension 1990;16:477–483.
56. Moncada S, Palmer RMJ, Higgs EA. Nitric oxide: physiology, pathophysiology and pharmacology. Pharmacol. Rev. 1991;43:109–142.
57. Snyder SH. Nitric oxide: first in a new class of neurotransmitters? Science 1992;257:494–496.
58. Lischka FW, Schild D. Effects of nitric oxide upon olfactory receptor neurones in Xenopus laevis. NeuroReport 1993;4:582–584.
59. Schmachtenberg O, Diaz J, Bacigalupo J. NO activates the olfactory cyclic nucleotide-gated conductance independent from cGMP in isolated rat olfactory receptor neurons. Brain Res 2003;980:146–150.
60. Broillet M-C, Firestein S. Direct activation of the olfactory cyclic nucleotide-gated channel through modification of sulfhydryl groups by NO compounds. Neuron 1996;16:377–385.
61. Lynch JW. Nitric oxide inhibition of the rat olfactory cyclic nucleotide-gated cation channel. J Membr Biol 1998;65:227–234.
62. Broillet M-C. S-Nitrosylation of proteins. Cell Mol Life Sci 1999;55:1036–1042.
63. Xu L, Eu JP, Meissner G, et al. Activation of the cardiac calcium release channel (ryanodine receptor) by poly-S-nitrosylation. Science 1998;279:234–236.
64. Stamler JS, Toone EJ, Lipton SA, et al. (S)NO signals: translocation, regulation, and a consensus motif. Neuron 1997;18:691–696.
65. Broillet M-C. A single intracellular cysteine residue is responsible for the activation of the olfactory cyclic nucleotide-gated channel by NO. J Biol Chem 2000;275:15,135–15,141.

66. Baumann A, Frings S, Godde M, et al. Primary structure and functional expression of a drosophila cyclic nucleotide-gated channel present in eyes and antennae. EMBO J 1994;13:5040–5050.
67. Broillet M-C, Firestein S. β subunits of the olfactory cyclic nucleotide-gated channel form a nitric oxide activated Ca^{2+} channel. Neuron 1997;18:951–958.
68. Liu DT, Tibbs GR, Siegelbaum SA. Subunit stoichiometry of cyclic nucleotide-gated channels and effects of subunit order on channel function. Neuron 1996;16 983–990.
69. Gordon SE, Varnum MD, Zagotta WN. Direct interaction between amino- and carboxyl-terminal domains of cyclic nucleotide-gated channels. Neuron 1997;19:431–441.
70. Roskams AJ, Bredt DS, Dawson TM, et al. Nitric oxide mediates the formation of synaptic connections in developing and regenerating olfactory receptor neurons. Neuron 1994;13:289–299.
71. Zhao H, Firestein S, Greer CA. NADPH-diaphorase localization in the olfactory system. NeuroReport 1994;6:149–152.
72. Trivedi B, Kramer RH. Patch cramming reveals the mechanism of long-term suppression of cyclic nucleotides in intact neurons. J Neurosci 2002;22:8819–8826.
73. Savchenko A, Barnes S, Kramer RH. Cyclic-nucleotide-gated channels mediate synaptic feedback by nitric oxide. Nature 1997;390:694–698.
74. Ahmad I, Barnstable CJ. Differential laminar expression of particulate and soluble guanylate cyclase genes in rat retina. Exp Eye Res 1993;56:51–62.
75. Blute TA, Velasco P, Eldred WD. Functional localization of soluble guanylate cyclase in turtle retina: modulation of cGMP by nitric oxide donors. Vis Neurosci 1998;15:485–498.
76. Blute TA, Lee HK, Huffmaster T, et al. Localization of natriuretic peptides and their activation of particulate guanylate cyclase and nitric oxide synthase in the retina. J Comp Neurol 2000;424:689–700.
77. Ahmad I, Leinders-Zufall T, Kocsis JD, et al. Retinal ganglion cells express a cGMP-gated cation conductance activatable by nitric oxide donors. Neuron 1994;12:155–165.
78. Henry D, Burke S. Shishido E, et al. Retinal bipolar neurons express the cyclic nucleotide-gated channel of cone photoreceptors. J Neurophysiol 2003;89:754–761.
79. Hoon MA, Adler E, Lindemeier J, et al. Putative mammalian taste receptors: a class of taste-specific GPCRs with distinct topographic selectivity. Cell 1999;96:541–551.
80. Adler E, Hoon MA, Mueller KL, et al. A novel family of mammalian taste receptors. Cell 2000;100: 693–702.
81. Chandrashekar J, Mueller KL, Hoon MA, et al. T2Rs function as bitter taste receptor. Cell 2000;100: 703–711.
82. Rodriguez I, Greer CA, Mok MY, et al. A putative pheromone receptor gene expressed in human olfactory mucosa. Nat Genet 2000;26:18–19.
83. Zhang J, Xia SL, Block ER, et al. NO upregulation of a cyclic nucleotide-gated channel contributes to calcium elevation in endothelial cells. Am J Physiol Cell Physiol 2002;283:C1080–C1089.

III

Story of a Silent Killer: The Resurgence of Carbon Monoxide as the Second Gasotransmitter

10 Synthesis and Metabolism of Carbon Monoxide

Stefan W. Ryter and Augustine M. K. Choi

CONTENTS

SUMMARY

The biosynthesis, biodistribution, and physiological consequences of carbon monoxide (CO) in humans are intimately linked to the utilization of the heme molecule in vital cellular processes. CO, though chemically inert, acts as a heme iron ligand. Living organisms acquire CO by inhalation, by endogenous enzymatic heme degradation catalyzed by heme oxygenases (HOs), or as a byproduct of lipid and xenobiotic metabolism. The HO enzymes exist in both constitutive (HO-2, HO-3) and inducible (HO-1) isoforms, the latter classified as a mammalian stress protein involved in cellular defense. HO quantitatively converts heme to CO, biliverdin, and iron, and these metabolites may all participate in HO-mediated cytoprotection. Although extremely toxic at elevated concentrations, CO may exert anti-inflammatory and anti-apoptotic properties at low concentrations. Endogenously produced CO is redistributed throughout the body as carboxyhemoglobin and excreted through the lungs by diffusion. Possible medical applications for CO include its

From: *Signal Transduction and the Gasotransmitters: NO, CO, and H_2S in Biology and Medicine* Edited by: Rui Wang © Humana Press Inc., Totowa, NJ

therapeutic delivery as an anti-inflammatory agent, and its measurement in exhaled breath as a general biomarker of inflammatory states.

Key Words: Bilirubin; carbon monoxide; gasotransmitter; heme; heme oxygenase; iron.

1. INTRODUCTION

Carbon monoxide (CO), a simple gaseous molecule, arises from the oxidation of organic materials *(1)*. The observations that CO exists in human blood and originates in part from the endogenous degradation of hemoglobin (Hb), the oxygen carrier of the blood date back to the late 19th century, and mid-20th century, respectively *(2–4)*. The association between the degradation of erythrocyte Hb and heme with the formation of the principle colored pigment of the bile, bilirubin, has also been known since the mid-twentieth century *(5,6)*.

In 1968–1969, Tenhunen et al. *(7,8)* described a microsomal monooxygenase enzyme system, distinct from cytochrome P450, that catalyzes the rate-determining step in the degradation of heme. This enzyme, heme oxygenase (HO), releases stoichiometric amounts of CO during the cleavage of heme to the open-chain tetrapyrrole biliverdin-IXα(BV) while releasing the heme-iron *(8)*. A soluble enzyme NAD(P)H:biliverdin reductase (BVR) completes heme metabolism by converting biliverdin-IXα(BV) to bilirubin-IXα(BR) *(9)*. At the time of this discovery, the significance of HO was believed to lie primarily in its role in erythrocyte Hb turnover, in reticuloendothelial tissues such as the spleen, kidney, and liver *(8,10)*. CO was regarded as an inevitable, but toxic, waste product of heme utilization *(8,11)*.

The landmark discoveries of Maines and colleagues *(12–14)* (ca. 1986–1988) demonstrated the existence of HOs as both inducible (HO-1) and constitutive isozymes (HO-2), and their widespread distribution in extrahepatic tissues. In 1988–1989, HO attained a new significance when Keyse et al. *(15)* and others *(16)* established the identity of HO as the 32-kDa mammalian stress protein (p32), also known as the 32-kDa heat-shock protein HSP32, a species dramatically induced by multiple forms of cellular stress. These studies first suggested a role for heme metabolic activity in cellular adaptation to environmental stress and implied a cytoprotective function for HO-1. Since then, numerous studies have established that HO-1 serves as a general cytoprotectant against oxidative stress in cell culture *(17–23)*, and in animal models of inflammatory or oxidative tissue injury *(23–28)*.

The search for mechanisms of action of HO-mediated cytoprotection led to investigations of the biological activities of the heme metabolites BV, BR, CO, and iron *(29)*. Both BV and BR possess potent antioxidant properties in vitro *(30)*. Investigations into the fate of HO-derived heme-iron led to the suggestion that ferritin, an iron storage molecule, served as a co-cytoprotectant by sequestering and detoxifying iron released from HO activity *(31,32)*, and later led to the discovery of an intracellular iron pump associated with HO-1 *(33,34)*. The toxic waste paradigm surrounding CO has been challenged by recent discoveries that this byproduct of HO activity at low concentrations may exert considerable vasoregulatory, anti-inflammatory, and anti-apoptotic activity in cell culture and in vivo, by modulating intracellular signaling pathways *(22)* (*see* Chapter 14). The protective effects of exogenously applied CO have been demonstrated in organ transplant, inflammatory and oxidative lung pathologies; ischemia/reperfusion injury; and, most recently, vascular injury *(23,35–38)*. This chapter addresses the systems that govern the endogenous metabolic generation of CO, with an emphasis on the enzymology and regulation of the HO system, and other minor sources. The toxicology of environmental CO acquired by accidental inhalation exposure has been reviewed elsewhere *(1,39,40)*.

2. PHYSIOCHEMICAL CHARACTERISTICS OF CO

CO is a diatomic molecule with a mol wt of 28.01. CO has no odor, color, or taste, properties that increase the hazardous potential of this flammable and noxious substance *(1)*. CO exists in gaseous form at ambient temperature and pressure, melting at –205°C, and boiling at –191.5°C *(1,41)*. CO dissolves in water (2.3 mL/100 mL at 20°C) and in some organic solvents *(1)*. Other useful constants include a specific gravity in gaseous form of 1.250 g/L at 0°C. For the purpose of calculating molarity in the gas phase, 1 ppm = 1.25 mg/m^{-3} (44.6 n*M*) at 25°C *(1,41)*.

In comparison with its cognate gas nitric oxide (NO), CO is relatively inert. CO and NO share an important similarity in biochemical reactivity: they both act as heme ligands and form complexes with most known hemoproteins at the heme-iron center *(11)*. CO will bind only to reduced (ferrous) iron centers, whereas NO may bind to both ferrous and ferric hemes *(42)*. Unlike the free radical NO, CO is an inherently stable nonradical with little biochemical reactivity with respect to non-iron compounds. By contrast, the radical character of NO allows it to react with thiol groups to form nitrosothiols, or with superoxide anion radical (O_2^-) to form peroxynitrite ($ONOO^-$), and participate in other redox reactions as described elsewhere *(43)*.

2.1. Association of CO With Hemoproteins

The best known biochemical reaction of CO involves its association with the heme center of Hb, to form carboxyhemoglobin (COHb), with an affinity 245 times that of oxygen, thus competing for the four binding sites for oxygen *(44)*. Partial occupation of CO binding sites inhibits the release of O_2 from the remaining heme groups, resulting in the leftward shift of the dissociation curve. This property of CO reduces the O_2-carrying capacity of the blood to deliver O_2, leading to tissue hypoxia *(1,44)*. The lethality of CO usually is associated with hypoxia, though cellular targets cannot be discounted *(40)*. Spectral studies have revealed the formation of CO complexes with other hemoproteins such as myoglobin, soluble guanylate cyclase (sGC), inducible nitric oxide synthase, cytochrome P450, cytochrome-*c* oxidase, NADPH:oxidase, and the heme-HO complex *(45–52)*. The interaction of CO with enzymes such as cytochrome P450 or cytochrome-*c* oxidase results in inhibition of enzymatic activity *(51,53)*. CO binding to sGC results in a weak activation of enzyme activity relative to the classical agonist NO *(11,45,46)*. The interaction of CO with the heme of sGC differs from that of NO, in that a hexacoordinate, rather than pentacoordinate, complex is formed without axial ligand displacement *(45,46)*. The relevance of these interactions to intracellular signal transduction is discussed in Chapter 14.

3. ENDOGENOUS PRODUCTION OF CO BY METABOLIC PATHWAYS

CO arises in cells and tissues, in part, through endogenous metabolic processes *(2–4,8)*. The rate of endogenous CO production in humans has been estimated (0.42 mL/h) *(4)*. In the absence of significant environmental pollution, most of the COHb in the blood is accounted for by this endogenous production, corresponding to blood CO levels of 0.4–1% *(1)*. Blood COHb will increase above this endogenous background in response to environmental CO exposure and may reach levels of 10% or higher in smokers *(44)*. An estimated 86% of endogenous CO production originates in heme metabolism whereas the remaining fraction arises from less well-defined sources, such as cytochrome P450-dependent metabolism of xenobiotics or lipid oxidation processes *(54)*.

3.1. Heme Degradation

The origination of endogenous CO from oxidation of the α-methene bridge carbon of heme was first elucidated by Sjostrand *(2)*, and associated with HO activity by Tenhunen et al. *(8)*. The HO reaction proceeds through three oxidation cycles, each requiring molecular oxygen (O_2), whereby the heme molecule serves as both the substrate and the catalytic cofactor in its own destruction *(55,56)*. Although distinct from the cytochrome P450 monooxygenases involved in the hydroxylation of xenobiotics, HO requires the reductase component of cytochrome P450 (P450 reductase) *(57)*. In each cycle, P450 reductase reduces the iron center of the bound heme molecule at the expense of NADPH, allowing the binding of O_2 to the reduced iron *(57,58)*. The bound O_2 is activated by a second electron from NADPH to form a peroxo-intermediate that attacks the heme ring *(59)*. The first oxidation cycle forms α-*meso*-hydroxy-heme. The CO is released during the second oxidation cycle prior to the formation of a verdoheme intermediate *(60)*. The third cycle forms the (FeIII) biliverdin IXα, which dissociates to free BV and Fe(II), following an additional NADPH-dependent reduction step *(55,56)*.

3.2. Cytochrome P450

CO may form endogenously as a byproduct of the cytochrome P450–dependent metabolism of certain xenobiotics. The formation of CO and an increase in blood COHb levels have been observed in the metabolism of dihalomethanes such as dichloromethane (DCM), as well as of methylene chloride *(61,62)*. CO production occurs during the oxidative dechlorination of DCM by cytochrome P450 2E1 (CYT2E1), which competes with a second major metabolic pathway dependent on GSH *(63,64)*. The stimulation of CYT2E1 with aromatic hydrocarbons or ethanol increased the production of CO from DCM metabolism and the detectable amount of blood COHb in rats *(63,65,66)*. Hepatic glutathione (GSH) depletion also increased CO production from DCM metabolism *(64)*. The CO that evolves from cytochrome P450–dependent DCM metabolism will complex and inactivate cytochrome P450 *(67)*.

3.3. Lipid Peroxidation

Evidence from toxicological studies suggests that CO stimulates lipid peroxidation in the rat brain following inhalation exposure, possibly by neutrophil recruitment *(68,69)*. In vitro studies suggest that CO can be formed as a byproduct of lipid peroxidation processes *(70,71)*. Using the formation of carboxy–cytochrome P450 as an assay system, Arachov et al. demonstrated that CO originates during microsomal NADPH and Fe(II)-dependent lipid oxidation in vitro. This CO production depended on membrane fractions, was abolished by inhibitors of cytochrome P450 2B4, but occurred in the presence of HO inhibitors, distinguishing it from heme metabolism *(70)*. In similar experiments using chromatographic CO detection methods, CO was found to evolve in tissue homogenates incubated with Fe(II)/ascorbate and paralleled the formation of malondialdehyde, a lipid peroxidation end product *(71)*.

4. METABOLIC FATE OF THE HEME RING: BILIVERDIN/BILIRUBIN AND IRON/FERRITIN

4.1. Bile Pigments

In 1936, Lemberg and Wyndham *(72)* proposed that the water-soluble pigment BV was the biological precursor of BR. Using isotopically labeled BV, Goldstein and Lester *(73)* confirmed that injected BV appears as conjugated BR in the bile. Similar experiments using radiolabeled Hb confirmed that Hb is quantitatively converted to biliary BR *(74)*. The HO reaction, which generates BV, is regioselective in that only the α-isomer of BV is produced *(75)*. By contrast, P450 reductase alone degrades heme to propentdyopents *(58)*. In 1971, Tenhunen et al. *(9)* and later Maines and Trakshel *(76)* described the purification of BVR, the cytosolic NAD(P)H-dependent reductase responsible for BV to BR conversion in vivo.

The work of Stocker and colleagues *(30,77–82)* demonstrating that both BV and BR have potent antioxidant properties in vitro, has suggested a possible antioxidative role for HO activity *(83)*. BR may directly react with or quench reactive oxygen species (ROS) including O_2^-, HOCl, and $^1O_2$ *(77,84)*. At micromolar concentrations, BV and BR displayed antioxidant activity comparable to α-tocopherol in polyunsaturated fatty acid liposomes *(30)*. Stocker et al. *(30)* proposed that BR acts as a chain-breaking antioxidant, while BV directly traps peroxyl radicals. BR acted as a plasma antioxidant after the consumption of endogenous circulating antioxidants *(81)*. Albumin-bound BR prevented oxidative damage to both the albumin and associated fatty acids *(80,82)*. Conjugated BR inhibited lipid peroxidation in liposomes, possibly by recycling α-tocopherol, but also displayed prooxidant activity in its copper-bound form *(78,79)*. Free and albumin-bound BR also inhibited low-density lipoprotein oxidation by acting as coantioxidants with α-tocopherol *(81)*. In neural cell culture, nanomolar concentrations of BR conferred cytoprotection against oxidative damage *(85)*. The depletion of cellular BR by RNA interference against BVR markedly augmented tissue levels of ROS and caused apoptotic cell death *(86)*.

4.2. Heme Iron

The immediate metabolic fate of the heme iron is unknown but is thought to be complexed with low molecular weight chelators *(87)*. The iron released from HO activity may stimulate the biosynthesis of ferritin, by complexing the iron-binding protein and relieving translational repression of ferritin synthesis *(31,88,89)*. Ferritin may bind and sequester up to 4500 iron molecules as a major sink for the storage of reactive iron. An intrinsic H-chain ferroxidase activity maintains the iron in the oxidized Fe(III) form in a crystalline core. Thus, the sequestration of intracellular iron by ferritin is thought to limit its potential catalysis of deleterious reactions *(89)*. The possible role of ferritin in cytoprotection linked to HO-1 induction has been examined in cell culture models *(17,90–92)*.

In vivo studies describe abnormal accumulations of tissue iron and serum anemia in *ho-1*$^{-/-}$ and *ho-2*$^{-/-}$ transgenic mice *(93–95)*. The *ho-1*$^{-/-}$ mice accumulated non-heme iron in the kidney and liver but decreased total iron content in the lung, whereas *ho-2*$^{-/-}$ mice accumulated total lung iron without compensatory increase in ferritin levels *(93–95)*.

Recent studies have suggested that iron released from HO activity also stimulates increased iron efflux from cells, such that HO activity is inversely related to intracellular iron accumulation *(33)*. The cytoprotective effect of HO-1 was likewise found to inversely correlate to intracellular iron accumulation *(33)*. However, in some models of HO overexpression, excess HO activity has caused short-term hypersensitivity to oxidative stress, owing to transient intracellular iron elevation *(29)*. An adenosine triphosphate-dependent iron pump has been identified that colocalizes with HO-1 in microsomal membranes. This pump may facilitate the transfer of iron across the lumen of the endoplasmic reticulum (ER), following its generation by HO activity, leading to its exocytosis *(34)*.

5. THE HOS: THREE ISOZYMES

Three genetically distinct isozymes of HO have been characterized—an inducible form, HO-1, and two constitutively expressed forms, HO-2 and HO-3 *(12–14,96)*, representing the products of distinct genes *(96–98)*. HO-1 and HO-2 map to separate chromatin regions (HO-1: 22q12; HO-2: 16p13.3) *(99)*. The entire gene sequences of human rat, and murine HO-1, have been described *(100–102)*.

5.1. Heme Oxygenase-1

The inducible form of HO, HO-1, occurs at a high level of expression in tissues normally active in erythrocyte turnover, comprising the reticuloendothelial system of the liver, spleen, and kidney *(14)*. In tissue not specialized in erythrocyte turnover, HO-1 typically occurs at low to undetectable levels under basal conditions but responds to rapid transcriptional upregulation by diverse chemical and physical stimuli (*see* **Subheading 6.**). HO-1 (approx 32 kDa) has been purified to homogeneity from human liver microsomes *(103)*, as well as liver or spleen of various animals (32–34 kDa) *(104–107)*, and corresponding cDNA clones have been described *(16,108)*. HO-1 contains a conserved heme-binding catalytic domain, and a hydrophobic domain in the extreme carboxylterminus that facilitates its association with cellular membranes *(108,109)*. To date, the HO enzymes have been characterized as ER-associated proteins, but recently we have observed a functional association of HOs with plasma membrane caveolae *(110)*. Variants of HO-1 have been identified in the red algae *Cyanidium caldarium* and in the bacterial strains Hmu O, *Corynebacterium diphtheriae;* HemO, *Neisseria meningitides;* and PigA, *Pseudomonas aeruginosa*) *(111–114)*.

5.2. Heme Oxygenase-2

A distinct isozyme of HO, HO-2, has been isolated from rat liver, spleen, brain, and testes *(12,13,115,116)*, and its corresponding cDNA cloned *(117,118)*. The highest expression of HO-2 occurs in the testes, but the protein is also found abundantly in the brain and vasculature *(13,14,119)*. Rat HO-2 has an apparent mol wt of 36 kDa *(13)*. HO-1 and HO-2 proteins differ in physical properties including molecular weight, sequence, and K_m for heme *(13)*. Furthermore, the expression of HO-2 normally does not respond to induction by chemicals (i.e., $CdCl_2$) that induce HO-1 *(12)*. However, the expression of HO-2 may be modulated by glucocorticoid hormones *(120)*. HO-1 and HO-2 share less than 50% amino acid homology, yet they share a conserved domain of 24 amino acid residues, which may correspond to a common "distal helix" domain *(55,118)*. HO-2 contains two heme-binding motifs termed *heme regulatory domains* (HRD) *(121)*. The function of HRDs remains uncertain, but their presence suggests that HO-2 may have a regulatory role in heme-dependent cellular processes, or act as a cellular sink for small gas molecules *(121,122)*.

5.3. Heme Oxygenase-3

McCoubrey et al. *(96)* isolated a gene encoding a third HO isozyme (HO-3) with a high amino acid sequence homology to HO-2 (approx 90%). The HO-3 protein (approx 33 kDa) has little HO activity, and, therefore, the function of this protein remains uncertain. Like HO-2, HO-3 has two HRD motifs, suggesting that HO-2 and HO-3 share a heme-dependent function unrelated to heme degradation. HO-3 transcripts can be detected in the spleen, liver, thymus, prostate, heart, kidney, testis, and brain *(96)*, with a distribution in the latter organ mainly in the hippocampus, cerebellum, and cortex *(123)*.

6. INDUCTION OF HO: A GENERAL CELLULAR AND TISSUE RESPONSE TO SYSTEMIC OR ENVIRONMENTAL STRESS

HO-1 induction represents a general transcriptional response to oxidative cellular stress, imposed by stimulation from diverse chemical and physical agents. The oxidative stress response can be divided into four broad categories (a) agents that directly generate ROS, (b) cytokines and growth factors that modulate intracellular ROS production downstream of receptor activation, (c) agents that compromise intracellular antioxidant potential by reacting with or depleting intracellular reduced GSH, and (d) the modulation of oxygen (O_2) tension in excess (hyperoxia) or deficit (hypoxia) of normal physiological levels. The HO substrate heme, also a potent inducer of the gene, may promote prooxidant states if allowed to accumulate, by catalyzing iron-dependent reactions such as membrane lipid peroxidation *(29)*. Finally, HO also responds to other physical stress conditions such as vascular hemodynamic or shear stress, and heat stress. Analysis of the *ho-1* gene promoter region (mouse) has revealed two upstream enhancer sequences (E1, E2) that occur respectively at –4 and –10 kb of the transcriptional start site *(124–126)*. These enhancers mediate the transcriptional induction of the *ho-1* gene by multiple agents, including endotoxins, heavy metal salts, phorbol esters, oxidants, and heme. E1 and E2 contain repeated stress-responsive elements consisting of recognition sequences for several transcription factors, including Cap'n'collar/basic-leucine zipper proteins, v-maf oncoprotein, and activator protein-1 (AP-1) *(126)*.

6.1. Reactive Oxygen Species

The activation of HO-1 by ultraviolet A (UVA) radiation (320–380 nm) represents a well-studied example of the oxidative regulation of the HO-1 gene *(15)*. UVA imposes an oxidative cellular stress, mediated by the generation of singlet molecular oxygen ($^1O_2$) from the photoexcitation of intracellular chromophores *(127)*. Porphyrin-mediated photodynamic therapy, another potent inducer of HO-1, mimics the UVA effect by generating $^1O_2$ from the visible irradiation of synthetic chromophores *(128)*. The direct application of H_2O_2 and treatment with redox-cycling compounds that generate H_2O_2 and O_2^- moderately activate HO-1 expression *(15,129,130)*. The chemical depletion of intracellular GSH with buthionine-*S*-*R*-sulfoximine, an inhibitor of GSH synthesis, enhances the activation of HO-1 by oxidants, such as UVA radiation exposure *(131)*. Metal chelators such as desferral typically inhibit the oxidative induction of the gene *(130)*.

6.2. Cytokines and Growth Factors

HO-1 appears to be a general response to oxidative stress associated with inflammatory states. HO-1 responds to cell stimulation with proinflammatory mediators, such as lipopolysaccharide (LPS), cytokines (interleukin [IL]-1, IL-6, tumor necrosis factor [TNF]-α), and tumor promoters *(16,132–134)*. The transcriptional induction of HO-1 by LPS was enhanced by GSH depletion and inhibited by the antioxidant *N*-acetyl cysteine, suggesting a role for ROS in the induction mechanism *(133,134)*. Several growth factors also elicit an HO-1 response in cell-specific fashion, including TGF-β and platelet-derived growth factor *(135,136)*. Receptor-mediated induction of HO by such agents has also been associated with downstream intracellular generation of ROS.

6.3. Thiol-Reactive Substances

HO-1 activation responds to thiol (-SH)-reactive substances that complex intracellular reduced glutathione (GSH) including sodium *m*-arsenite, diethylmaleate, and heavy metal salts (i.e., $CdCl_2$) *(15,16,137,138)*. The activation of HO-1 by sodium arsenite appears to represent a general response in numerous animal cell types *(139)*. NO reacts with GSH to form *S*-nitrosothiol and may also generate oxidative stress by formation of $ONOO^-$ *(43)*. NO donors or NO gas activate HO-1 in cell culture models, involving both transcriptional and posttranscriptional regulation of the gene *(140–146)*.

6.4. Oxygen Tension

Increases or decreases in physiological oxygen tension cause metabolic stress that may induce HO-1. Hyperoxia (increased Po_2) causes an oxidative stress related to the increased mitochondrial generation of ROS as well as causes inflammatory lung damage that mimics the condition of adult respiratory distress syndrome *(147)*. Hypoxia (decreased Po_2) also causes metabolic stress related to mitochondrial dysfunction, but the reported increases in ROS production during hypoxia remain controversial. Several pathophysiological states are associated with hypoxia, including high-altitude sickness, cardiovascular diseases, and cancer, whereas hyperoxia arises only in critical care situations by forced delivery of O_2 *(148)*. Hypoxia triggers the synthesis of numerous stress response proteins in various animal cell types *(149–152)*, including Chinese hamster ovary cells, and pulmonary or systemic vascular endothelial cells or vascular smooth muscle cells *(153–156)*. Acute or chronic hypoxia triggers HO-1 transcription in vivo in various rat organs including the heart *(154,157)*. Nakayama et al. *(158)* report a lack of HO-1 response to hypoxia in several cell types of human origin. Hyperoxia (>95% O_2), the other extreme of O_2 tension, also activated *ho-1* transcription in cultures of lung origin (epithelial cells, fibroblasts, macrophages, and smooth muscle cells *(159)*, and increased HO enzymatic activity in the adult rat lung *(159,160)*. The induction of HO-1 by either hyperoxia or hypoxia has been associated with GSH depletion and may be inhibited by metal-chelating agents, indicating a role for redox processes and endogenous metal ions in either response *(156,161–163)*. The transcriptional induction of HO-1 by hypoxia may involve both hypoxia-inducible factor-1, and AP-1 transcription factors, with apparent tissue-specific variation between vascular cell types *(154,164)*. On the other hand, the hyperoxia-induced activation of HO-1 proceeds through a distinct mechanism involving the AP-1 transcription factor and STAT proteins *(165)*.

7. HO ACTIVITY AND ITS REGULATION BY PROTOPORPHYRINS/MESOPORPHYRINS

Fe-protoporphyrin-IX (heme-*b*) is the preferred substrate for HO-1 and HO-2 activity, but oxidation of heme-*c* has also been demonstrated *(166,167)*. Metalloporphyrin analogs of heme, with a central metal chelate other than iron, typically serve as competitive inhibitors of HO activity in vitro *(168)*. Examples of these compounds are tin-, cobalt-, and zinc-substituted protoporphyrin IX (Sn-PP, Co-PP, and Zn-PP, respectively) or tin-substituted mesoporphyrin (Sn-MP). Co-PP induces hepatic HO activity following injection in vivo, whereas Sn-PP and Zn-PP are inhibitors of in vivo enzyme activity *(168,169)*. In contrast to inhibition of activity, some metalloporphyrins (Co-PP and Zn-PP) induce HO-1 mRNA transcription *(170,173)*.

8. ELIMINATION OF CO AND POSSIBLE MEDICAL APPLICATIONS OF EXHALED GASES

The COHb complex is dissociable by O_2, and thus endogenous or inhaled CO is excreted through the lung by diffusion *(174)*. In humans, very little CO is converted to CO_2 by oxidation. In Gram-negative bacteria (i.e., *Carboxydothermus hydrogenoformans*), a CO-oxidizing enzyme system has been described *(175)*. Following prolonged CO inhalation, the half-life of COHb on return to room air is within the range of 30–180 min *(176)*. The amount of exhaled CO (E-CO) detectable on the breath of healthy subjects falls within the range of 1–5 ppm. E-CO likely includes the sum of inspired CO and endogenous CO from heme metabolism in various tissues including the lung and airways, as a result of inducible HO-1 and constitutive HO-2 activity. Elevation of E-CO may reflect an increase in exogenous background sources such as smoking or air pollution. In addition to changes in environmental factors, elevations of E-CO in lung diseases may parallel an increase in blood COHb levels in response to systemic inflammation or may involve an increase in pulmonary HO-1 expression in response to local inflammation *(177–180)*. Accumulating evidence suggests that elevated E-CO may occur as a general marker of inflammatory diseases such as asthma, cystic fibrosis, and chronic obstructive pulmonary disease, but conflicting reports indicate that a consensus has not yet been reached on the diagnostic value of this biomarker *(181,182)*.

9. FUTURE PERSPECTIVES

Centuries of observations have defined the toxicological consequences of CO exposure in humans, with the conclusion that environmental CO is a deadly asphyxiant *(40)*. The last century of research has also revealed that CO arises during the course of normal metabolism. The discovery of the HO enzymes identified the principal source of biological CO. The finding that HO-1, and later HO-1-derived CO, have cytoprotective properties points to the possible delivery of either HO-1 or CO for therapeutic gain in the treatment of inflammatory diseases. The administration of HO-1 would likely be achieved through gene therapy approaches using retroviral vectors, but pharmacological manipulation of HO-1 expression with natural nontoxic inducers also remains possible. The delivery of CO would likely be achieved by inhalation of mixed gas, or placement of organs in chambers for ex vivo applications. The synthetic CO-releasing compounds may serve as an alternate means for the pharmacological delivery of CO *(183)*. Progress in this area requires a thorough understanding of the toxicological sequelae, if any, of low-dose

CO exposure in humans. Furthermore, barriers to human experimentation with CO inhalation therapies remain. Thus, the future of this field lies in the possible therapeutic delivery of CO for anti-inflammatory potential in advanced lung disease, and organ transplantation.

ACKNOWLEDGMENTS

This work was supported by an award from the American Heart Association (AHA #0335035N) to S. W. Ryter, and National Institutes of Health grants R01-HL60234, R01-AI42365, and R01-HL55330 to A. M. K. Choi. Because of the vast nature of this field, we have cited selected representative or historical references, and therefore we regret the inevitable omission of important works.

REFERENCES

1. Von Berg R. Toxicology update: carbon monoxide. J Appl Toxicol 1999;19: 379–386.
2. Sjostrand T. The formation of carbon monoxide by the decomposition of hemoglobin in vivo. Acta Physiol Scand 1952;26:338–344.
3. Sjostrand T. Endogenous production of carbon monoxide in man under normal and pathophysiological conditions. Scand J Clin Lab Invest 1949;1:201–214.
4. Coburn RF, Blakemore WS, Forster RE. Endogenous carbon monoxide production in man. J Clin Invest 1963;42:1172–1178.
5. London IM, West R, Shemin D, et al. On the origin of bile pigment in normal man. J Biol Chem 1950;184:351–358.
6. Ostrow JD, Jandl JH, Schmid R. The formation of bilirubin from hemoglobin in vivo. J Clin Invest 1968;41:1628–1637.
7. Tenhunen R, Marver HS, Schmid R. The enzymatic conversion of heme to bilirubin by microsomal heme oxygenase. Proc Natl Acad Sci USA 1968;61:748–755.
8. Tenhunen R, Marver H, Schmid R. Microsomal heme oxygenase: characterization of the enzyme. J Biol Chem 1969;244:6388–6394.
9. Tenhunen R, Ross ME, Marver HS, et al. Reduced nicotinamide adenine dinucleotide phosphate dependent biliverdin reductase: partial purification and characterization. Biochemistry 1970;9:298–323.
10. Tenhunen R, Marver H, Schmid R. The enzymatic catabolism of hemoglobin: stimulation of microsomal heme oxygenase by hemin. J Lab Clin Med 1970;75:410–421.
11. Maines MD. The heme oxygenase system: a regulator of second messenger gases. Annu Rev Pharmacol Toxicol 1997;37:517–554.
12. Maines MD, Trakshel GM, Kutty RK. Characterization of two constitutive forms of rat liver microsomal heme oxygenase. J Biol Chem 1986;261:411–419.
13. Trakshel GM, Kutty RK, Maines MD. Purification and characterization of the major constitutive form of testicular heme oxygenase. J Biol Chem 1986;261:11,131–11,137.
14. Maines MD. Heme Oxygenase: Clinical Applications and Functions.CRC Press: Boca Raton, FL, 1992.
15. Keyse SM, Tyrrell RM. Heme oxygenase is the major 32-kDa stress protein induced in human skin fibroblasts by UVA radiation, hydrogen peroxide, and sodium arsenite. Proc Natl Acad Sci USA 1989;86:99–103.
16. Kageyama H, Hiwasa T, Tokunaga K, et al. Isolation and characterization of a complementary DNA clone for a Mr 32,000 protein which is induced with tumor promoters in BALB/c 3T3 cells. Cancer Res 1988;48: 4795–4798.
17. Vile GF, Basu-Modak S, Waltner C, et al. Heme oxygenase 1 mediates an adaptive response to oxidative stress in human skin fibroblasts. Proc Natl Acad Sci USA 1994;91:2607–2610.
18. Abraham NG, Lavrovsky Y, Schwartzman ML, et al. Transfection of the human heme oxygenase gene into rabbit coronary microvessel endothelial cells: protective effect against heme and hemoglobin toxicity. Proc Natl Acad Sci USA 1995;92:6798–6802.
19. Lee PJ, Alam J, Wiegand GW, et al. Overexpression of heme oxygenase-1 in human pulmonary epithelial cells results in cell growth arrest and increased resistance to hyperoxia. Proc Natl Acad Sci USA 1996;93:10,393–10,398.

20. Chen K, Gunter K, Maines MD. Neurons overexpressing heme oxygenase-1 resist oxidative stress-mediated cell death. J Neurochem 2000;75:304–313.
21. Petrache I, Otterbein LE, Alam J, et al. Heme oxygenase-1 inhibits TNF-alpha-induced apoptosis in cultured fibroblasts. Am J Physiol Lung Cell Mol Physiol 2000;278:L312–L319.
22. Ryter S, Otterbein LE, Morse D, et al. Heme oxygenase/carbon monoxide signaling pathways: regulation and functional significance. Mol Cell Biochem 2002;234/235:249–263.
23. Otterbein LE, Bach FH, Alam J, et al. Carbon monoxide has anti-inflammatory effects involving the mitogen-activated protein kinase pathway. Nat Med 2000;6:422–428.
24. Otterbein LE, Kolls JK, Mantell LL, et al. Exogenous administration of heme oxygenase-1 by gene transfer provides protection against hyperoxia-induced lung injury. J Clin Invest 1999;103: 1047–1054.
25. Minamino T, Christou H, Hsieh CM, et al. Targeted expression of heme oxygenase-1 prevents the pulmonary inflammatory and vascular responses to hypoxia. Proc Natl Acad Sci USA 2001;98:8798–8803.
26. Juan SH, Lee TS, Tseng KW, et al. Adenovirus-mediated heme oxygenase-1 gene transfer inhibits the development of atherosclerosis in apolipoprotein E-deficient mice. Circulation 2001;104:1519–1525.
27. Panahian N, Yoshiura M, Maines MD. Overexpression of heme oxygenase-1 is neuroprotective in a model of permanent middle cerebral artery occlusion in transgenic mice. J Neurochem 1999;72: 1187–1203.
28. Wagner M, Cadetg P, Ruf R, et al. Heme oxygenase-1 attenuates ischemia/reperfusion-induced apoptosis and improves survival in rat renal allografts. Kidney Int 2003;63:1564–1573.
29. Ryter S, Tyrrell RM. The heme synthesis and degradation pathways: role in oxidant sensitivity: heme oxygenase has both pro- and anti-oxidant properties. Free Radic Biol Med 2000;28:289–309.
30. Stocker R, Yamamoto Y, McDonagh A, et al. Bilirubin is an antioxidant of possible physiological importance. Science 1987;235:1043–1045.
31. Vile GF, Tyrrell RM. Oxidative stress resulting from ultraviolet A irradiation of human skin fibroblasts leads to a heme oxygenase–dependent increase in ferritin. J Biol Chem 1993;268:14,678–14,681.
32. Vile GF, Basu-Modak S, Waltner C, et al. Heme oxygenase 1 mediates an adaptive response to oxidative stress in human skin fibroblasts. Proc Natl Acad Sci USA 1994;91:2607–2610.
33. Ferris CD, Jaffrey SR, Sawa A, et al. Haem oxygenase-1 prevents cell death by regulating cellular iron. Nat Cell Biol 1999;1:152–157.
34. Baranano DE, Wolosker H, Bae BI, et al. A mammalian iron ATPase induced by iron. J Biol Chem 2000;275:15,166–15,173.
35. Chapman JT, Otterbein LE, Elias JA, et al. Carbon monoxide attenuates aeroallergen-induced inflammation in mice. Am J Physiol Lung Cell Mol Physiol 2001;281:L209–L216.
36. Otterbein LE, Zuckerbraun BS, Haga M, et al. Carbon monoxide suppresses arteriosclerotic lesions associated with chronic graft rejection and with balloon injury. Nat Med 2003;9:183–190.
37. Otterbein LE, Mantell LL, Choi AM. Carbon monoxide provides protection against hyperoxic lung injury. Am J Physiol 1999;276:L688–L694.
38. Sato K, Balla J, Otterbein L, et al. Carbon monoxide generated by heme oxygenase-1 suppresses the rejection of mouse-to-rat cardiac transplants. J Immunol 2001;166 4185–4194.
39. Weaver LK. Carbon monoxide poisoning. Crit Care Clin 1999;15:297–317.
40. Gorman D, Drewry A, Huang YL, et al. The clinical toxicology of carbon monoxide. Toxicology 2003;187:25–38.
41. Budavari S, O'Neill M, Smith A, et al., eds. The Merck Index. An Encyclopedia of Chemicals, Drugs, and Biologicals. 11th ed. Merck: Rahway, NJ, 1989.
42. Hartsfield CL. Cross talk between carbon monoxide and nitric oxide. Antioxid Redox Signal 2002;4:301–307.
43. Wink DA, Mitchell JB. Chemical biology of nitric oxide: insights into regulatory, cytotoxic, and cytoprotective mechanisms of nitric oxide. Free Radic Biol Med 1998;25:434–456.
44. Smith RP. Toxic responses of the blood. In: Klaassen CD, Amdur MO, Doull J., eds. Casarett and Doull's Toxicology, the Basic Science of Poisons. 3rd ed. MacMillan: New York, 1986, pp. 223–244.
45. Furchgott RF, Jothianandan D. Endothelium-dependent and -independent vasodilation involving cyclic GMP: relaxation induced by nitric oxide, carbon monoxide and light. Blood Vessels 1991;28:52–61.
46. Stone JR, Marletta MA. Soluble guanylate cyclase from bovine lung: activation with nitric oxide and carbon monoxide and spectral characterization of the ferrous and ferric states. Biochemistry 1994;33:5636–5640.
47. Stevenson TH, Gutierrez AF, Alderton WK, et al. Kinetics of CO binding to the haem domain of murine inducible nitric oxide synthase: differential effects of haem domain ligands. Biochem J 2001;358:201–208.

48. Volpe JA, O'Toole MC, Caughey WS. Quantitative infrared spectroscopy of CO complexes of cytochrome c oxidase, hemoglobin and myoglobin: evidence for one CO per heme. Biochem Biophys Res Commun 1975;62:48–53.
49. Guengerich FP, Ballou DP, Coon MJ. Purified liver microsomal cytochrome P-450: electron-accepting properties and oxidation-reduction potential. J Biol Chem 1975;250:7405–7414.
50. Cross AR, Higson FK, Jones OTG, et al. The enzymic reduction and kinetics of oxidation of cytochrome b_{-245} of neutrophils. Biochem J 1982;204:479–485.
51. Estabrook RW, Franklin MR, Hildebrandt AG. Factors influencing the inhibitory effect of carbon monoxide on cytochrome P450–catalyzed mixed function oxidation reactions. Ann NY Acad Sci 1970;174:218–232.
52. Migita CT, Matera KM, Ikeda-Saito M, et al. The oxygen and carbon monoxide reactions of heme oxygenase. J Biol Chem 1998;273:945–949.
53. Brown SD, Piantadosi CA. Reversal of carbon monoxide–cytochrome c oxidase binding by hyperbaric oxygen in vivo. Adv Exp Med Biol 1989;248:747–754.
54. Vremen HJ, Wong RJ, Stevenson DK. Carbon monoxide in breath, blood, and other tissues. In Penney DG, ed., Carbon Monoxide Toxicity. CRC Press: Boca Raton, FL, 2000, pp. 19–60.
55. Wilks A. Heme oxygenase: evolution, structure, mechanism. Antiox Redox Signal 2002;4:603–614.
56. Yoshida T, Kikuchi G. Features of the reaction of heme degradation catalyzed by the reconstituted microsomal heme oxygenase system. J Biol Chem 1978;253:4230–4236.
57. Yoshida T, Noguchi M, Kikuchi G. Oxygenated form of heme–heme oxygenase complex and the requirement for a second electron to initiate heme degradation from the oxygenated complex. J Biol Chem 1980;255:4418–4420.
58. Yoshinaga T, Sassa S, Kappas A. A comparative study of heme degradation by NADPH–cytochrome c reductase alone and by the complete heme oxygenase system: distinctive aspects of heme degradation by NADPH–cytochrome c reductase. J Biol Chem 1982;257:7794–7802.
59. Wilks A, Ortiz de Montellano, PR. Rat liver heme oxygenase: high level expression of a truncated soluble form and nature of the *meso*-hydroxylating species. J Biol Chem 1993;268:22,357–22,362.
60. Yoshida T, Noguchi M, Kikuchi G. The step of carbon monoxide liberation in the sequence of heme degradation catalyzed by the reconstituted microsomal heme oxygenase system. J Biol Chem 1982;257:9345–9348.
61. Stewart RD, Fisher TN, Hosko MJ, et al. Carboxyhemoglobin elevation after exposure to dichloromethane. Science 1972;176:295, 296.
62. Amsel J, Soden KJ, Sielken RL, et al. Observed versus predicted carboxyhemoglobin levels in cellulose triacetate workers exposed to methylene chloride. Am J Ind Med 2001;40:180–191.
63. Guengerich FP, Shimada T. Oxidation of toxic and carcinogenic chemicals by human cytochrome P-450 enzymes. Chem Res Toxicol 1991;4:391–407.
64. Oh SJ, Kim SK, Kim YC. Role of glutathione in metabolic degradation of dichloro-methane in rats. Toxicol Lett 2002;129:107–114.
65. Kim SK, Kim YC. Effect of a single administration of benzene, toluene or m-xylene on carboxyhaemoglobin elevation and metabolism of dichloromethane in rats. J Appl Toxicol 1996;16:437–444.
66. Wirkner K, Damme B, Poelchen W, et al. Effect of long-term ethanol pretreatment on the metabolism of dichloromethane to carbon monoxide in rats. Toxicol Appl Pharmacol 1997;143:83–88.
67. Takano T, Miyazaki Y. Metabolism of dichloromethane and the subsequent binding of its product, carbon monoxide, to cytochrome P-450 in perfused rat liver. Toxicol Lett 1988;40:93–96.
68. Thom SR. Carbon monoxide mediated brain lipid peroxidation in the rat. J Appl Physiol 1990;68:997–1003.
69. Thom, SR. Leucocytes in carbon monoxide mediated brain oxidative injury. Toxicol Appl Pharmacol 1993;123:234–247.
70. Archakov AI, Karuzina II, Petushkova NA, et al. Production of carbon monoxide by cytochrome P450 during iron-dependent lipid peroxidation. Toxicol In Vitro 2002;16:1–10.
71. Vreman HJ, Wong RJ, Sanesi CA, et al. Simultaneous production of carbon monoxide and thiobarbituric acid reactive substances in rat tissue preparations by an iron-ascorbate system. Can J Physiol Pharmacol 1998;76:1057–1065.
72. Lemberg R, Wyndham RA. Reduction of biliverdin to bilirubin in tissues. Biochem J 1936;30: 1147–1170.
73. Goldstein GW, Lester R. Reduction of biliverdin-C^{14} to bilirubin C^{14} in vivo. Proc Soc Exp Biol Med 1964;117:681–683.
74. Ostrow JD, Jandl JH, Schmid R. The formation of bilirubin from hemoglobin in vivo. J Clin Invest 1962;41:1628–1637.

75. Noguchi M, Yoshida T, Kikuchi G. Identification of the product of heme degradation catalyzed by the heme oxygenase system as biliverdin-IXα by reverse phase high performance liquid chromatography. J Biochem (Tokyo) 1882;91:1479–1483.
76. Maines MD, Trakshel GM. Purification and characterization of human biliverdin reductase. Arch Biochem Biophys 1993;300:320–326.
77. Stocker R, Peterhans E. Antioxidant properties of conjugated bilirubin and biliverdin: biologically relevant scavenging of hypochlorous acid. Free Radic Res Commun 1989;6:57–66.
78. Stocker R, Peterhans E. Synergistic interaction between vitamin E and the bile pigments bilirubin and biliverdin. Biochim Biophys Acta 1989;1002:238–244.
79. Stocker R, Ames B. Potential role of conjugated bilirubin and copper in the metabolism of lipid peroxides in the bile. Proc Natl Acad Sci USA 1987;84: 8130–8134.
80. Stocker R, Glazer AN, Ames BN. Antioxidant activity of albumin bound bilirubin. Proc Natl Acad Sci USA 1987;84:5918–5922.
81. Neuzil J, Stocker, R. Free and albumin-bound bilirubin are efficient co-antioxidants for alpha-tocopherol, inhibiting plasma and low density lipoprotein lipid peroxidation. J Biol Chem 1994;269:16,712–16,719.
82. Neuzil J, Stocker R. Bilirubin attenuates radical-mediated damage to serum albumin. FEBS Lett 1993;331:281–284.
83. Stocker R. Induction of haem oxygenase as a defense against oxidative stress. Free Radic Res Commun 1990;9:101–112.
84. Stevens B, Small RD. The photoperoxidation of unsaturated organic molecules-XV: $O_2\ ^1\Delta_g$ quenching by bilirubin and biliverdin. Photochem Photobiol 1976;23:33–36.
85. Doré S, Takahashi M, Ferris CD, et al. Bilirubin, formed by activation of heme oxygenase-2, protects neurons against oxidative stress injury. Proc Natl Acad Sci USA 1999;96:2445–2450.
86. Baranano DE, Rao M, Ferris CD, et al. Biliverdin reductase: a major physiologic cytoprotectant. Proc Natl Acad Sci USA 2002;99;16,093–16,098.
87. Richardson DR, Ponka P. The molecular mechanisms of the metabolism and transport of iron in normal and neoplastic cells. Biochim Biophys Acta 1997;1331:1–40.
88. Eisenstein RS, Garcia-Mayol D, Pettingel W, et al. Regulation of ferritin and heme oxygenase synthesis in rat fibroblasts by different forms of iron. Proc Natl Acad Sci USA 1991;88:688–692.
89. Theil E. Ferritin: structure, gene regulation, and cellular function in animals, plants, and microorganisms. Annu Rev Biochem 1987;56:289–315.
90. Lin F, Girotti AW. Hemin-enhanced resistance of human leukemia cells to oxidative killing: antisense determination of ferritin involvement. Arch Biochem Biophys 1998;352:51–58.
91. Balla GH, Jacob J, Balla J, et al. Ferritin: a cytoprotective stratagem of endothelium. J Biol Chem 1992;267:18,148–18,153.
92. Balla J, Jacob HS, Balla G, et al. 1993. Endothelial-cell heme uptake from heme proteins: induction of sensitization and desensitization to oxidant damage. Proc Natl Acad Sci USA 1993;90:9285–9289.
93. Dennery PA, Spitz DR, Yang G, et al. Oxygen toxicity and iron accumulation in the lungs of mice lacking heme oxygenase-2. J Clin Invest 1998;101:1001–1011.
94. Dennery PA, Visner G, Weng Y-H, et al. Resistance to hyperoxia with heme oxygenase-1 disruption: role of iron. Free Radic Biol Med 2003;34:124–133.
95. Poss KD, Tonegawa S. Heme oxygenase-1 is required for mammalian iron reutilization. Proc Natl Acad Sci USA 1997;94:10,919–10,924.
96. McCoubrey WK, Huang TJ, Maines MD. Isolation and characterization of a cDNA from the rat brain that encodes hemoprotein heme oxygenase-3. Eur J Biochem 1997;247:725–732.
97. Cruse I, Maines MD. Evidence suggesting that the two forms of heme oxygenase are products of different genes. J Biol Chem 1988;263:3348–3353.
98. Trakshel GM, Ewing JF, Maines MD. Heterogeneity of heme oxygenase 1 and 2 isoenzymes. Biochem J 1991;275:159–164.
99. Kutty RK, Kutty G, Rodriguez IR, et al. Chromosomal localization of the human heme oxygenase genes: heme oxygenase-1 (HMOX1) maps to chromosome 22q12 and heme oxygenase-2 (HMOX2) maps to chromosome 16p13.3. Genomics 1994;20:513–516.
100. Müller RM, Taguchi H, Shibahara S. Nucleotide sequence and organization of the rat heme oxygenase gene. J Biol Chem 1987;262:6795–6802.
101. Alam J, Cai J, Smith A. Isolation and characterization of the mouse heme oxygenase-1 gene. J Biol Chem 1994;269:1001–1009.
102. Shibahara S, Sato M, Müller RM, et al. Structural organization of the human heme oxygenase gene and the function of its promoter. Eur J Biochem 1989;179:557–563.

103. Abraham N, Mitrione S, Levere R. Kinetics and properties of human fetal and adult liver heme oxygenase. Biochem Arch 1986;2:253–259.
104. Yoshida T, Kikuchi G. Purification and properties of heme oxygenase from rat liver microsomes. J Biol Chem 1979;254:4487–4491.
105. Yoshida T, Kikuchi G. Purification and properties of heme oxygenase from pig spleen microsomes. J Biol Chem 1978;253:4224–4229.
106. Yoshinaga T, Sassa S, Kappas A. Purification and properties of bovine spleen heme oxygenase: amino acid composition and sites of action of inhibitors of heme oxidation. J Biol Chem 1982;257: 7778–7785.
107. Bonkovsky HL, Healey JF, Pohl J. Purification and characterization of heme oxygenase from chick liver: comparison of the avian and mammalian enzymes. Eur J Biochem 1990;189:155–166.
108. Shibahara S, Müller RM, Taguchi H, et al. Cloning and expression of cDNA for rat heme oxygenase. Proc Natl Acad Sci USA 1985;82:7865–7869.
109. Ishikawa K, Sato M, Yoshida T. Expression of rat heme oxygenase in *Esherichia coli* as a catalytically active, full length form that binds to membranes. Eur J Biochem 1991;202:161–165.
110. Kim HP, Wang X, Galbiati F, et al. Caveolae compartmentalization of oxygenase-1 in pulmonary artery endothelial cells. Faseb J 2004, in press.
111. Beale SI, Cornejo J. Enymatic heme oxygenase activity in soluble extracts of the unicellular red alga, *Cyanidium caldarium.* Arch Biochem Biophys 1984;235:371–384.
112. Wilks A, Schmitt MP. Expression and characterization of a heme oxygenase (Hmu O) from Corynebacterium diphtheriae: iron acquisition requires oxidative cleavage of the heme macrocycle. J Biol Chem 1998;273:837–841.
113. Zhu W, Wilks A, Stojiljkovic I. Degradation of heme gram-negative bacteria: the product of the *hemO* gene of neisseriae is a heme oxygenase. J Bacteriol 2000;182:6783–6790.
114. Ratliff M, Zhu W, Deshmukh R, et al. Homologues of neisserial heme oxygenase in Gram-negative bacteria: degradation of heme by the product of the *pigA* gene of *Pseudomonas aeruginosa.* J Bacteriol 2001;183:6394–6403.
115. Braggins PE, Trakshel GM, Kutty RK, et al. Characterization of two heme oxygenase isoforms in rat spleen: comparison with the hematin-induced and constitutive isoforms of the liver. Biochem Biophys Res Commun 1986;141:528–533.
116. Trakshel GM, Kutty RK, Maines MD. Resolution of rat brain heme oxygenase activity: absence of a detectable amount of the inducible form (HO-1). Arch Biochem Biophys 1988;260:732–739.
117. Rotenberg MO, Maines MD. Isolation, characterization, and expression in Escherichia coli of a cDNA encoding rat heme oxygenase-2. J Biol Chem 1990;265:7501–7506.
118. Rotenberg MO, Maines MD. Characterization of a cDNA-encoding rabbit brain heme oxygenase-2 and identification of a conserved domain among mammalian heme oxygenase isozymes: possible heme-binding site? Arch Biochem Biophys 1991;290:336–344.
119. Zakhary R, Gaine SP, Dinerman JL, et al. Heme oxygenase 2: endothelial and neuronal localization and role in endothelium-dependent relaxation. Proc Natl Acad Sci USA 1996;93:795–798.
120. Raju VS, McCoubrey WK Jr., Maines MD. Regulation of heme oxygenase-2 by glucocorticoids in neonatal rat brain: characterization of a functional glucocorticoid response element. Biochim Biophys Acta 1997;1351:89–104.
121. McCoubrey WK, Huang TJ, Maines MD. Heme oxygenase-2 is a hemoprotein and binds heme through heme regulatory motifs that are not involved in heme catalysis. J Biol Chem 1997;272: 12,568–12,575.
122. Maines MD, Panahian N. The heme oxygenase system and cellular defense mechanisms: do HO-1 and HO-2 have different functions? Adv Exp Med Biol 2001;502:249–72
123. Scapagnini G, D'Agata V, Calabrese V, et al. Gene expression profiles of heme oxygenase isoforms in the rat brain. Brain Res 2002;954:51–59.
124. Alam J. Multiple elements within the 5' distal enhancer of the mouse heme oxygenase-1 gene mediate induction by heavy metals. J Biol Chem 1994; 269:25,049–25,056.
125. Alam J, Camhi S, Choi AM. Identification of a second region upstream of the mouse heme oxygenase-1 gene that functions as a basal level and inducer-dependent transcription enhancer. J Biol Chem 1995;270:11,977–11,984.
126. Choi AM, Alam J. Heme oxygenase-1: function, regulation, and implication of a novel stress-inducible protein in oxidant-induced lung injury. Am J Respir Cell Mol Biol 1996;15:9–19.
127. Basu-Modak S, Tyrrell RM. Singlet oxygen: a primary effector in the ultraviolet A/near-visible light induction of the human heme oxygenase gene. Cancer Res 1993;53:4505–4510.

128. Gomer CJ, Luna M, Ferrario A, et al. Increased transcription and translation of heme oxygenase in chinese hamster fibroblasts following photodynamic stress or photofrin II incubation. Photochem Photobiol 1991;53:275–279.
129. Keyse SM, Tyrrell RM. Both near ultraviolet radiation and the oxidizing agent hydrogen peroxide induce a 32-kDa stress protein in normal human skin fibroblasts. J Biol Chem 1987;262:14,821–14,825.
130. Keyse SM, Tyrrell RM. Induction of the heme oxygenase gene in human skin fibroblasts by hydrogen peroxide and UVA (365nm) radiation: evidence for the involvement of hydroxyl radical. Carcinogenesis 1990;11:787–791.
131. Lautier D, Luscher P, Tyrrell RM. Endogenous glutathione levels moculate both constitutive and UVA radiation/hydrogen peroxide inducible expression of the human heme oxygenase gene. Carcinogenesis 1992;13:227–232.
132. Camhi SL, Alam J, Otterbein L, et al. Induction of heme oxygenase-1 gene expression by lipopolysaccharide is mediated by AP-1 activation. Am J Respir Cell Mol Biol 1995;13:387–398.
133. Rizzardini M, Carelli M, Cabello Porras MR, et al. Mechanisms of endotoxin-induced haem oxygenase mRNA accumulation in mouse liver: synergism by glutathione depletion and protection by *N*-acetylcysteine. Biochem J 1994;304:477–483.
134. Camhi SL, Alam J, Wiegand GW, et al. Transcriptional activation of the HO-1 gene by lipopolysaccharide is mediated by 5' distal enhancers: role of reactive oxygen intermediates and AP-1. Am J Respir Cell Mol Biol 1998;18:226–234.
135. Durante W, Peyton KJ, Schafer AI. Platelet-derived growth factor stimulates heme oxygenase-1 gene expression and carbon monoxide production in vascular smooth muscle cells. Arterioscl Thromb Vasc Biol 1999;19:2666–2672.
136. Kutty RK, Nagineni CN, Kutty G, et al. Increased expression of heme oxygenase-1 in human retinal pigment epithelial cells by transforming growth factor-beta. J Cell Physiol 1994;159:371–378.
137. Saunders EL, Maines MD, Meredith MJ, et al. Enhancement of heme oxygenase-1 synthesis by glutathione depletion in chinese hamster ovary cells. Arch Biochem Biophys 1991;288:368–373.
138. Maines MD, Kappas A. Metals as regulators of heme metabolism. Science 1977;198:1215–1221.
139. Applegate LA, Luescher P, Tyrrell RM. Induction of heme oxygenase: a general response to oxidant stress in cultured mammalian cells. Cancer Res 1991;51:974–978.
140. Foresti R, Clark JE, Green CJ, et al. Thiol compounds interact with nitric oxide in regulating heme oxygenase-1 induction in endothelial cells: involvement of superoxide and peroxynitrite anions. J Biol Chem 1997;272:18,411–18,417.
141. Hartsfield CL, Alam J, Cook JL, et al. Regulation of heme oxygenase-1 gene expression in vascular smooth muscle cells by nitric oxide. Am J Physiol 1997;273:L980–L988.
142. Motterlini R, Foresti R, Intaglietta M, et al. NO-mediated activation of heme oxygenase: endogenous cytoprotection against oxidative stress to endothelium. Am J Physiol 1996;270:H107–H114.
143. Yee EL, Pitt BR, Billiar TR, et al. Effect of nitric oxide on heme metabolism in pulmonary artery endothelial cells. Am J Physiol 1996;271:L512–L518.
144. Foresti R, Sarathchandra P, Clark JE, et al. Peroxynitrite induces haem oxygenase-1 in vascular endothelial cells: a link to apoptosis. Biochem J 1999; 339:729–736.
145. Marquis JC, Demple B. Complex genetic response of human cells to sublethal levels of pure nitric oxide. Cancer Res 1988;58:3435–3440.
146. Bouton C, Demple B. Nitric oxide–inducible expression of heme oxygenase-1 in human cells: translation-independent stabilization of the mRNA and evidence for direct action of nitric oxide. J Biol Chem 2000;275:32,688–32,693.
147. Freeman BA, Crapo JD. Hyperoxia increases oxygen radical production in rat lungs and lung mitochondria. J Biol Chem 1991;256:10,986–10,992.
148. Haddad JJ. Oxygen-sensing mechanisms and the regulation of redox-responsive transcription factors in the development and pathophysiology. Respir Res 2002;3:26.
149. Heacock CS, Sutherland RM. Induction characteristics of oxygen regulated proteins. Int J Radiat Oncol Biol Phys 1986;12:1287–1290.
150. Graven KK, Farber HW. Endothelial cell hypoxic stress proteins. J Lab Clin Med 1998;132: 456–463.
151. Helfman T, Falanga V. Gene expression in low oxygen tension. Am J Med Sci 1993;306:37–41.
152. Zimmerman LH, Levine RA, Farber HW. Hypoxia induces a specific set of stress proteins in cultured endothelial cells. J Clin Invest 1991;87:908–914.
153. Morita T, Perrella MA, Lee ME, et al. Smooth muscle cell–derived carbon monoxide is a regulator of vascular cGMP. Proc Natl Acad Sci USA 1995;92:1475–1479.

154. Lee PJ, Jiang BH, Chin BY, et al. Hypoxia-inducible factor-1 mediates transcriptional activation of the heme oxygenase-1 gene in response to hypoxia. J Biol Chem 1997;272:5375–5381.
155. Murphy BJ, Laderoute KR, Short SM, et al. The identification of heme oxygenase as a major hypoxic stress protein in Chinese hamster ovary cells. Br J Cancer 1991;64:69–73.
156. Ryter S, Si ML, Lai C-C, et al. Regulation of endothelial heme oxygenase activity during hypoxia is dependent on intracellular chelatable iron. Am J Physiol Heart Circ Physiol 2000;279:H2889–H2897.
157. Katayose D, Isoyama S, Fujita H, et al. Separate regulation of heme oxygenase and heat shock protein 70 mRNA expression in the rat heart by hemodynamic stress. Biochem Biophys Res Commun 1993;191:587–594.
158. Nakayama M, Takahashi K, Kitamuro T, et al. Repression of heme oxygenase-1 by hypoxia in vascular endothelial cells. Biochem Biophys Res Commun 2000;271:665–671.
159. Lee PJ, Alam J, Sylvester SL, et al. Regulation of heme oxygenase-1 expression in vivo and in vitro in hyperoxic lung injury. Am J Respir Cell Mol Biol 1996;14:556–568.
160. Dennery PA, Rodgers PA, Lum MA, et al. Hyperoxic regulation of lung heme oxygenase in neonatal rats. Pediatr Res 1996;40:815–821.
161. Motterlini R, Foresti R, Bassi R, et al. Endothelial heme oxygenase-1 induction by hypoxia: modulation by inducible nitric-oxide synthase and S-nitrosothiols. J Biol Chem 2000;275:13,613–13,620.
162. Fogg S, Agarwal A, Nick HS, et al. Iron regulates hyperoxia-dependent human heme oxygenase 1 gene expression in pulmonary endothelial cells. Am J Respir Cell Mol Biol 1999;20:797–804.
163. Takahashi S, Takahashi Y, Yoshimi T, et al. Oxygen tension regulates heme oxygenase-1 gene expression in mammalian cell lines. Cell Biochem Funct 1998;16:183–193.
164. Hartsfield CL, Alam J, Choi AM. Differential signaling pathways of HO-1 gene expression in pulmonary and systemic vascular cells. Am J Physiol 1999;277:L1133–L1141.
165. Lee PJ, Camhi SL, Chin BY, et al. AP-1 and STAT mediate hyperoxia-induced gene transcription of heme oxygenase-1. Am J Physiol Lung Cell Mol Physiol 2000;279:L175–L182.
166. Kutty RK, Maines MD. Oxidation of heme *c* derivatives by purified heme oxygenase: evidence for the presence of one molecular species of heme oxygenase in the rat liver. J Biol Chem 1982;257: 9944–9952.
167. Yoshinaga T, Sassa S, Kappas A. The oxidative degradation of heme *c* by the microsomal heme oxygenase system. J Biol Chem 1982;257:7803–7807.
168. Kappas A, Drummond GS. Control of heme and cytochrome P-450 metabolism by inorganic metals, organometals, and synthetic metalloporphyrins. Environ Health Perspect 1984;57:301–306.
169. Sardana MK, Kappas A. Dual control mechanism for heme oxygenase: tin(IV)-protoporphyrin potently inhibits enzyme activity while markedly increasing content of enzyme protein in liver. Proc Natl Acad Sci USA 1987;84:2464–2468.
170. Smith A, Alam J, Escriba PV, et al. Regulation of heme oxygenase and metallothionein gene expression by the heme analogs, cobalt, and tin-protoporphyrin. J Biol Chem 1993;268:7365–7371.
171. Lin JH, Villalon CP, Martasek P, et al. Regulation of heme oxygenase gene expression by cobalt in rat liver and kidney. Eur J Biochem 1990;192:577–582.
172. Mitani K, Fujita H, Fukuda Y, et al. The role of inorganic metals and metalloporphyrins in the induction of haem oxygenase and heat-shock protein 70 in human hepatoma cells. Biochem J 1993;290:819–825.
173. Shan Y, Pepe J, Lu TH, et al. Induction of the heme oxygenase-1 gene by metalloporphyrins. Arch Biochem Biophys 2000;380:219–227.
174. Wagner JA, Horvath SM, Dahms TE. Carbon monoxide elimination. Respir Physiol 1975;23:41–47.
175. Soboh B, Linder D, Hedderich R. Purification and catalytic properties of a CO-oxidizing: H2-evolving enzyme complex from Carboxydothermus hydrogenoformans. Eur J Biochem 2002;269:5712–5721.
176. Haddad LM. Clinical Management of Poisoning and Drug Overdose. 2nd ed. WB Saunders: Philadelphia, 1990.
177. Donnelly LE, Barnes PJ. Expression of heme oxygenase in human airway epithelial cells. Am J Respir Cell Mol Biol 2001;24:295–303.
178. Maestrelli P, El Messlemani AH, De Fina O, et al. Increased expression of heme oxygenase (HO)-1 in alveolar spaces and HO-2 in alveolar walls of smokers. Am J Respir Crit Care Med 2001;164: 1508–1513.
179. Yasuda H, Yamaya M, Yanai M, et al. Increased blood carboxyhaemoglobin concentrations in inflammatory pulmonary diseases. Thorax 2002;57:779–783.
180. Horvath I, Donnelly LE, Kiss A, et al. Raised levels of exhaled carbon monoxide are associated with an increased expression of heme oxygenase-1 in airway macrophages in asthma: a new marker of oxidative stress. Thorax 1998;53:668–672.

181. Slebos DJ, Ryter SW, Choi AMK. Heme oxygenase-1 and carbon monoxide in pulmonary medicine. Respir Res 2003;4:7.
182. Horvath I, MacNee W, Kelly FJ, et al. "Haemoxygenase-1 induction and exhaled markers of oxidative stress in lung diseases," summary of the ERS Research Seminar in Budapest, Hungary, September, 1999. Eur Respir J 2001;18:420–430.
183. Motterlini R, Clark JE, Foresti R, et al. Carbon monoxide-releasing molecules: characterization of biochemical and vascular activities. Circ Res 2002;90:E17–E24.

11 Interaction of Carbon Monoxide With K^+ Channels in Vascular Smooth Muscle Cells

Rui Wang

CONTENTS

SUMMARY

Being one of the gasotransmitters, carbon monoxide (CO) fulfills an important modulatory role in the cardiovascular system. It relaxes blood vessels and lowers peripheral resistance, thus influencing the homeostatic control of blood pressure. Stimulation of various types of K^+ channels in vascular smooth muscle cells (VSMCs) is one of the mechanisms for the CO-induced vasorelaxation. These K^+ channels include voltage-dependent Kv, adenosine triphosphate-sensitive K_{ATP}, and calcium-activated K_{Ca} channels. The stimulation of big-conductance K_{Ca} channels by CO from both exogenous and endogenous sources has been mostly documented in VSMCs. Calcium-spark-activated transient K_{Ca} channels are also activated by CO. Increased calcium sensitivity and/or improved coupling efficiency between calcium spark and K_{Ca} channel activities potentially underline the stimulatory effect of CO on K_{Ca} channels. The interaction of CO and K^+ channels may be the dominant force in driving the CO-induced vasorelaxation in specific types of blood vessels, especially peripheral resistant and cerebral arterioles. In other types of blood vessels, the effect of CO on K^+ channels may become less important in comparison to the activation of the cyclic guanosine 5'-monophosphate pathway by

From: *Signal Transduction and the Gasotransmitters: NO, CO, and H_2S in Biology and Medicine* Edited by: Rui Wang © Humana Press Inc., Totowa, NJ

CO. Altered cardiovascular functions under pathophysiological conditions may also be related to the abnormal interaction of CO and K^+ channels in VSMCs. Endogenous CO production may be lower such as in hypertension. K^+ channel response to CO may be reduced such as in diabetes. For the latter case, severe glycation of K_{Ca} channel proteins renders their insensitivity to CO. Elucidation of the molecular mechanisms for the interaction of CO and K^+ channels, tissue type-dependent selective effect of CO on K^+ channels, and variation in the production and function of CO under pathological conditions will undoubtedly improve the understanding of the pathogenesis and maintenance of many CO-related cardiovascular disorders. Subsequently, novel strategies targeting to the interaction of CO and K^+ channels for these diseases may be devised.

Key Words: Kv channel; K_{ATP} channel; K_{Ca} channel; vascular smooth muscle cells; carbon monoxide.

1. INTRODUCTION

Among many ion channels expressed in excitable cells, K^+ channels comprise the most versatile and complex superfamily. At least four types of K^+ channels have been identified in vascular smooth muscle cells (VSMCs): Kv, voltage-dependent; K_{ATP}, adenosine triphosphate (ATP)-sensitive; K_{Ca}, Ca^{2+}-activated; and Kir, inward rectifier. The resting membrane potential of VSMCs is controlled by these K^+ channels. Opening of K^+ channels increases K^+ efflux, resulting in membrane hyperpolarization. Decreased Ca^{2+} influx resulting from the closure of voltage-dependent Ca^{2+} channels, and vasodilation ensue. Beyond their physiological importance in modulating excitability of VSMCs under resting conditions, K^+ channels mediate physiological functions of many endogenous substances including gasotransmitters. Under pathophysiological conditions, K^+ channels serve as therapeutic targets for pharmaceutical interventions in dealing with various types of cardiovascular diseases.

Different types of K^+ channels have different responses to different gasotransmitters. The effects of nitric oxide (NO) (Chapters 5 and 6) and hydrogen sulfide (H_2S) (Chapter 21) on K^+ channels are discussed in detail in other chapters of this book. The interaction of carbon monoxide (CO) and K^+ channels, the focus of this chapter, leads to changes in cellular functions. This interaction can be mediated by known second messengers, such as cyclic guanosine 5'-monophosphate (cGMP), or result from structural changes in ion channel proteins induced by CO. The effect of CO on ion channels is also subject to the health status of the cells. In diabetes, severe glycation of cellular proteins may alter the interaction of CO with K^+ channels. Moreover, chronic exposure to elevated CO levels may affect the expression of K^+ channels. Readers are referred to Chapter 15 for a detailed discussion of the chronic effect of CO on K^+ channel expression.

2. PHYSIOLOGICAL IMPORTANCE OF K^+ CHANNELS IN THE REGULATION OF VSMC FUNCTION

2.1. K_{ATP} Channels in VSMCs and Their Endogenous Modulators

ATP-sensitive K^+ (K_{ATP}) channels are inhibited by intracellular ATP and extracellular sulfonylureas *(1)* but stimulated by K_{ATP} channel openers (KCOs) *(2)*. K_{ATP} channels were originally discovered in cardiac muscle *(3)* and later identified in many other tissues, including pancreatic β-cells, skeletal muscle cells, and many types of VSMCs *(4–8)*. Activation of K_{ATP} channels leads to membrane hyperpolarization and relaxation of

VSMCs. Glibenclamide and tolbutamine are representative sulfonylureas that block K_{ATP} channels. Pinacidil, cromakalim, nicorandil, and diazoxide act as KCOs to stimulate K_{ATP} channels. The pharmacological sensitivities of K_{ATP} channels to different agents are largely determined by the molecular composition of the K_{ATP} channel complex. The target site of KCOs in the K_{ATP} channel complex is also assumed to be the sulfonylurea receptor (SUR) subunit, considering that the response of reconstituted K_{ATP} channels to either diazoxide or pinacidil is correlated with the presence of SUR subtypes. The C-terminal end of the SUR appears to be a critical determinant for KCO pharmacology.

In response to a decrease in intracellular ATP level, plasma membrane K_{ATP} channels open to allow transmembrane movement of K^+ ions. Electrophysiology and pharmacology studies in the last 10 yr have revealed an isoform of K_{ATP} channels in the mitochondrial inner membrane, termed *mitoK$_{ATP}$ channels (9–11)*. Compared with plasma membrane K_{ATP} channels, mitoK_{ATP} channels are specifically sensitive to blockade by 5-hydroxydecanoate, and to opening by diazoxide. Glibenclamide (blocker) and pinacidil (opener) have similar efficacies on surfaceK_{ATP} and mitoK_{ATP} channels *(12)*.

K_{ATP} channels are a heterooctamer assembly of four pore-forming subunits (Kir6.x) and four regulatory SUR subunits $(Kir6.x/SUR)_4$ *(13)*. Kir6.1 and Kir6.2 belong to a class of inwardly rectifying K^+ channels with two membrane-spanning regions. SURs belong to the ATP-binding cassette superfamily. Both the C- and N-termini of Kir6.1 and Kir6.2 are located inside the cell and are important for intracellular ATP binding and interactions with SUR subunits *(14,15)*. SURs are large proteins with 17 putative transmembrane domains, having an extracellular N-terminus and an intracellular C-terminus. To date, five SUR subunits have been identified in various mammalian tissues: SUR1, SUR1B, SUR2A, SUR2B, and SUR2C. Binding sites for sulfonylureas and KCOs are on SURs *(16)*. Different combinations of Kir6x and SURs yield tissue-specific K_{ATP} channels with different electrophysiological and pharmacological features. Thus, Kir6.2/SUR1 constitutes K_{ATP} channels in pancreatic β-cells and some neurons, as does Kir6.2/SUR2A in cardiac and skeletal muscles. Kir6.2/SUR2B is the K_{ATP} isoform in non-VSMCs and some neurons. It is generally accepted that the Kir6.1/SUR2B channel may be specific for VSMCs *(17,18)*. My colleagues and I detected the transcripts of Kir6.1, Kir6.2, SUR2B, and SUR1 in rat mesenteric artery smooth muscle cells (SMCs) *(19)*. Furthermore, we cloned four K_{ATP} subunit genes from mesenteric artery SMCs and accordingly referred to them as rvKir6.1, rvKir6.2, rvSUR1, and rvSUR2B. Their GenBank access numbers are AB043636, AB043638, AB045281, and AB052294, respectively. It is possible that VSMCs possess multiple types of K_{ATP} channels constructed by Kir6.1 with either SUR1 or SUR2B being the regulatory subunit, because Kir6.1 confers relative ATP insensitivity (not inhibited by ATP at concentrations lower than 1 m*M*), which is one of the fingerprints of K_{ATP} channels in VSMCs.

K_{ATP} channels are regulated by intracellular ATP, ADP, or the ATP/ADP ratio. When binding to the pore-forming Kir6.x subunit, intracellular ATP inhibits channel opening (ligand action). By contrast, when associated with SUR subunits, ATP stimulates the channel. ADP and many other nucleoside diphosphates in the absence of Mg^{2+} also inhibit the activity of K_{ATP} channels. In the presence of Mg^{2+}, the inhibitory effect of ADP reversed to a stimulatory effect on K_{ATP} channels. Among known endogenous K_{ATP} channel modulators is endothelin, which inhibits K_{ATP} channels of VSMCs *(20)*. NO hyperpolarizes SMCs from rabbit mesenteric arteries by indirectly activating K_{ATP} channels with cGMP as the intermediate factor *(21)*. However, another study found that sodium nitroprusside, an NO donor, had no effect on K_{ATP} channel currents in porcine coronary artery SMCs *(22)*. Atrial natriuretic factor also activated K_{ATP} channels in rat aortic SMCs via the stimulation of particulate guanylate cyclase *(23)*. Calcitonin gene-related peptide activated K_{ATP} channels in VSMCs mediated by a cyclic adenoside monophosphate pathway *(22)*. Adenosine is also an endogenous modulator of K_{ATP} channels in VSMCs *(24)*. Recent studies have demonstrated that H_2S is an endogenous opener of K_{ATP} channels in VSMCs *(25)*. The interaction of H_2S and K_{ATP} channels is discussed specifically in Chapter 21.

2.2. Kv Channels in VSMCs

Voltage-gated K^+, Kv, channels in VSMCs are pivotal in transforming the membrane excitability into cellular contractility. When membrane potential is depolarized to more positive than –40 mV, a delayed outward rectifier Kv channel opens and the yielded current, I_K, inactivates slowly. I_K has been identified in almost all types of VSMCs. Within the physiological range of membrane potential and normal intracellular Ca^{2+} level, I_K represents the dominant repolarizing conductance *(26,27)*. For example, previous studies by my colleagues and I have shown that the predominant Kv current in rat mesenteric artery SMCs or tail artery SMCs is I_K *(28,29)*. Another type of Kv current, I_A, in VSMCs activates fast but quickly inactivated. I_A coexists with I_K in several types of VSMCs, but its actual function is still not clear. This ambiguity stems from both the voltage dependence and time dependency of I_A. Only residual I_A channels are available for opening at resting membrane potential level in VSMCs (around –50 mV). The fast inactivation kinetics also nullify a physiologically meaningful contribution of I_A to the regulation of membrane potentials of VSMCs, which do not usually undergo transient pulse changes as occurred in action potential.

There are at least nine subfamilies of Kv channels cloned in mammals, including Kv1–Kv4 subfamilies (corresponding to Shaker, Shab, Shaw, and Shal) and Kv5–Kv9 *(30)*. At the molecular level, the functional Kv channel is composed of four α-subunits and four smaller cytoplasmic β-subunits that act mainly as a regulatory moiety ($\alpha_4\beta_4$) *(30,31)*. The core region of each α-subunit is formed by six putative transmembrane domains (S1–S6) and an H5 segment between S5 and S6 that constitutes the channel pore. The voltage sensor of Kv channels is related to the charged S4 region. In line with the biophysical properties of Kv channels, Kv channel genes can be further divided into two groups: Kv1.1, 1.2, 1.3, 1.5, 1.6, 2.1, 3x, and 4.1 may underlie the functional I_K; Kv1.4, 1.7 *(32)*, 4.2, and 4.3 *(33)* may encode I_A. Kv1.1, 1.2, 1.4, 1.5, 4.2, and 2.1 have been identified in rat aorta at the mRNA level *(34)*. Transcripts of Kv1.5 *(35)* and another unidentified Kv channel clone *(36)* have also been discovered in various vascular tissues. Kv2.1 and 1.5 proteins were immunolocalized in rat cerebral, coronary, and renal arteries *(37)*, and human aorta (Kv1.5) *(38)*. In addition, three subfamilies of Kvβ-subunits (Kvβ1, Kvβ2, and Kvβ3) have also been identified *(39)*. In previous studies by my colleagues and I *(40,41)*, 11 Kv transcripts were detected in endothelium-free rat tail artery and mesenteric artery tissues. Among them are five I_K-encoding genes (Kv1.2, 1.3, 1.5, 2.1, and 3.2) and six I_A-encoding genes (Kv1.4, 3.3, 3.4, and 4.1–3). Three β-subunits of Kv genes (Kvβ1.1, Kvβ1.2, Kvβ1.3) were also identified at the mRNA level. Among five I_K-encoding Kv genes identified at the mRNA level, all Kv proteins, except Kv3.2, were detected in rat tail artery and mesenteric artery tissues using Western blot analysis. More direct evidence from our immunocytochemistry study confirmed the presence of Kv2.1 and Kv1.2 proteins in primarily cultured rat VSMCs.

2.3. K_{Ca} Channels in Vascular SMCs

An increase in intracellular calcium concentration opens calcium-activated potassium channels (K_{Ca}) in many excitable and nonexcitable cells. Biophysical characterization classifies K_{Ca} channels into three subtypes. Big-conductance K_{Ca} channel (BK_{Ca}) has a single-channel conductance of approx 250 pS with symmetrical [K^+] across the cell membrane. Under the same recording conditions, a 20- to 80-pS intermediate-conductance (IK_{Ca}) and a 10- to 15-pS small-conductance (SK_{Ca}) of K_{Ca} channels are also identified *(42)*. Pharmacological sensitivity also set these three K_{Ca} channels apart. Charybdotoxin (ChTX) and iberiotoxin (IbTX) at a nanomolar range specifically block BK_{Ca} channels in VSMCs *(43)*. SK_{Ca} channels are selectively inhibited by apamin (IC_{50}, 0.3 n*M*). The apamin-sensitive SK_{Ca} channels have been found in porcine vascular beds, indicating their possible contributions to the maintenance of intrinsic vascular tone *(44,45)*. The expression of IK_{Ca} channels in VSMCs, or other types of excitable cell, has not been documented. Researchers' understanding of IK_{Ca} channels, both electrophysiological and pharmacological properties, is largely derived from studies on nonexcitable cells *(46–48)*.

Among BK_{Ca}, IK_{Ca}, and SK_{Ca} channels, BK_{Ca} channels are best described and characterized in various tissues including VSMCs. BK_{Ca} channels are composed of two types of noncovalently linked subunits: the pore-forming α-subunit and the accessory β-subunit that affects the electrophysiological and pharmacological properties of K_{Ca} channel complexes *(49)*. K_{Ca} α-subunit shares a great sequence homology with the pore-forming subunits of other types of K channels. The expression of α-subunit of BK_{Ca} channel alone yields the Ca^{2+}-independent current when the intracellular calcium concentration is lower than 100 n*M*, turning the channels into a pure voltage-dependent pore *(50)*. The functional coupling of BK_{Ca} β-subunit with the α-subunit greatly increases the sensitivity to the cytoplasmic calcium and confers the inactivation properties of BK_{Ca} channels *(50–52)*. The single gene origin with a family of alternatively spliced variants can explain the wide difference in unitary conductance, calcium sensitivity, and gating of BK_{Ca} channels in different tissues and even within the same tissue *(53,54)*. The transmembrane segments (S1–S7) near the N-terminus of BK_{Ca} channel α-subunit have amino acid sequence similar to that of the voltage sensor and the pore domain of Kv channels. The charged residues (Arg) in the S4 transmembrane domain move outward when the cell membrane is depolarized and interact with negative residues in S2 and S3 domains. In addition, BK_{Ca} channels have four extrahydrophobic domains (S7–S10) at the C-terminus, which are conservative among species *(55–57)*. S9 and S10 regions are associated with intracellular calcium sensitivity and calcium binding *(58–60)*. A more recent topology model for BK_{Ca} channel α-subunit suggested an additional S0 transmembrane segment that leads to the extracellular location of the N-terminus that is associated with β-subunit regulation in the mammalian BK_{Ca} gene *(61)*. The amino acid sequence of human BK_{Ca} gene (1113 amino acids) shares very high identity with that of rat (1178 amino acids) or mouse (1180 amino acids) (96–97%) but relates less to that of *Drosophila* (1175 amino acids) (49%).

At least four types of β-subunits that couple with BK_{Ca} channel α-subunits have been cloned in human tissues (β1–β4). The β-subunit family of BK_{Ca} channels regulates several critical aspects of channel phenotype such as inactivation and apparent Ca^{2+} sensitivity. Structurally, BK_{Ca} channel β-subunits are about 192–310 amino acid residues in length, having two transmembrane domains (TM1 and TM2) with a long extracellular loop in between and two N-linked glycosylation sites. TM1 and TM2 are similar in amino acid sequence, suggesting a common structure. β_1-Subunit increases the apparent calcium sensitivity at a micromolar range of the BK_{Ca} channel α-subunit, and it is mainly expressed in smooth muscle tissues *(50)*. Intracellular amino acid residues at the N-terminus of β-subunits are thought to have a "ball-like" structure that is vital for BK_{Ca} channel phenotypes *(52)*.

3. K+ CHANNELS AS A TARGET OF CO IN VSMCS

CO-induced vasorelaxation has been documented in many publications from many laboratories worldwide. A comprehensive summary of the cardiovascular effects of CO can be found in the book *Carbon Monoxide and Cardiovascular Functions (62)*. In short, CO is generated in the vascular wall from SMCs and endothelial cells. Both the inducible and constitutive isoforms of heme oxygenase (HO) participate in the catabolism of heme into CO, biliverdin, and iron. Chronic effects of endogenous CO involve the modulated proliferation of both VSMCs and endothelial cells, important processes for vascular remodeling. The acute vascular effect of CO manifests itself as vasorelaxation. Three major mechanisms have been proposed for CO-induced vasorelaxation: activation of the cGMP pathway, stimulation of K^+ channels, and modulation of the cytochrome P450 pathway *(63)*. Although this chapter deals with the stimulation of CO on K^+ channels in VSMCs, the importance of other signaling mechanisms underlying the vascular effects of CO is by no means downplayed.

3.1. Mediation of Vasorelaxant Effects of CO by K+ Channels

As discussed earlier, vascular contractility is closely related to the membrane potential and K^+ channel activities. Researchers appear to have reached a consensus about the involvement of K^+ channels in the vascular effects of CO. The challenges that researchers faced before and encounter now involve what kind of K^+ channels in which types of VSMCs are affected by CO and what mechanisms are responsible.

An earlier study conducted by my colleagues and I showed that CO-induced vasorelaxation was inhibited partially by the blockade of the cGMP pathway and partially by the opening of K^+ channels *(64)*. In the presence of tetraethylammonium (TEA) (30 m*M*), CO-induced relaxation of rat tail artery tissues was significantly reduced from 46 to 19% ($n = 8; p < 0.05$). This result suggests the involvement of K^+ channels in the vascular effect of CO. Given that TEA at high concentration may interfere with various types of K^+ channels, especially K_{Ca} channel and Kv channels, we then tested the interaction of CO with some more specific K^+ channel blockers *(64)*. ChTX and apamin have been widely used as selective inhibitors of high-conductance or small-conductance K_{Ca} channels, respectively. ChTX inhibited the vascular effect of CO as TEA did, whereas apamin had no effect on the CO-induced vasorelaxation. These results strongly suggest that the activation of big-conductance K_{Ca} channels constitutes an important mechanism for the CO-induced vasorelaxation. Finally, the K_{Ca} channel-independent relaxation induced by CO was abolished completely by Rp-8-Br-cGMPS, which blocked the cGMP pathway *(64)*.

Similar observations of the dual regulatory mechanism of CO (i.e., cGMP pathway and K^+ channels) have been made on other types of vascular tissues, such as isolated dog basilar artery segments *(65)*. After the isolated dog cerebral artery segments were contracted with 60 m*M* KCl, subsequent application of CO relaxed this vascular tissue in a dose-dependent fashion. This result appeared against the contribution of K^+ channels to the vasorelaxant effect of CO for 60 m*M* KCl in bath solution already largely reduced the driving force for the outward movement of K^+ ions. However, this high KCl stimulation might already reduce the vasorelaxant effect of CO with a threshold concentration of 57 μ*M* in comparison with a threshold concentration of 1 μ*M* CO to relax the phenylephrine-precontracted rat tail artery tissues *(64)*. It would be more revealing if the investigators were to report the threshold concentration of CO to relax dog cerebral artery tissues precontracted with nondepolarizing agents such as prostaglandin $F_{2\alpha}$ ($PGF_{2\alpha}$) *(65)*. Nevertheless, after dog basilar artery tissues were precontracted with $PGF_{2\alpha}$, a K^+ channel–dependent mechanism of the CO-induced vasorelaxation was demonstrated because the addition of TEA (1 m*M*) partially alleviated the relaxant effect of CO. In this vascular tissue the guanylyl cyclase inhibitor 1H-[1,2,4]oxadiazolo[4,3-a]quinoxalin-1-one (ODQ) (10 μ*M*) also induced a partial blockade of CO's effect. Coapplication of ODQ and TEA did not completely block the effect of CO. This finding left the unanswered question: What is the nature of the cGMP-independent and TEA-resistant mechanism for the relaxant effect of CO on dog cerebral arteries *(65)*? The possibility remains that Kv channels may be modulated by CO since TEA at 1 m*M* may not suffice to block all types of K^+ channels. This would have been confirmed if the inhibition of CO's effects by ODQ had been tested on the vascular tissues precontracted with 60 m*M* KCl.

Vasorelaxation induced by CO may be solely mediated by K^+ channels. Barbé et al. *(66)* exposed Wistar rats for 1 wk to 530 ppm of CO, and the rats were also simultaneously treated daily with either methylene blue to block the cGMP pathway, glibenclamide to block K_{ATP} channels, or apamin to block SK_{Ca} channels. At the end of the treatment period, hearts were isolated and in vitro perfused using the Langendorff method with a constant perfusion pressure. Chronic CO treatment enhanced coronary flow throughout the ischemia-reperfusion protocol compared to the heart preparations without CO treatment. CO treatment also reduced the amplitude of ischemic contraction of cardiac muscles. These protective effects of CO were not altered by methylene blue, which defied the mediating role of the cGMP pathway in CO's effect. Whereas the administration of glibenclamide and apamin both blocked the vasodilatory effects of CO, only apamin suppressed the effect of CO on cardiac contractile recovery. The upregulation of K_{ATP} channels and small conductance K_{Ca} channels in coronary artery SMCs appeared to underline the CO-induced coronary relaxation. The apamin-blocked cardiac contractile recovery in the CO-treated group can be explained by two cellular events. The CO-induced activation of SK_{Ca} channels in coronary artery smooth muscle may help to improve coronary circulation and reduce ischemia/reperfusion damage of cardiac myocytes. Alternatively, the opening of small-conductance K_{Ca} channels in cardiac myocytes from the CO-treated group may cushion ischemia/reperfusion-induced calcium overload by hyperpolarizing membrane potential, which leads to reduced calcium entry through voltage-dependent Ca^{2+} channels in cardiomyocytes. A major challenge for the latter scenario is the lack of evidence for the existence of SK_{Ca} channels in cardiomyocytes.

Direct evidence for the interaction of CO and K_{ATP} channels has not been obtained using the patch-clamp technique on either native SMCs or cell lines that express cloned K_{ATP} channels. Werkstrom et al. *(67)* reported that the CO-evoked relaxations of urethra and the esophagogastric junction inner smooth muscle of pig were not significantly reduced by treatment with glibenclamide, a blocker of K_{ATP} channels. However, K_{ATP} channels unlikely play an important role in these tissues because the addition of glibenclamide had no effect on the spontaneous tension development at resting conditions. The putative effect of CO on ATP production should also be considered because this may indirectly affect the activity of K_{ATP} channels. Also in the study by Werkstrom et al. *(67)*, the presence of 4-aminopyridine-sensitive Kv channels in urethra and the inner smooth muscle at the esophagogastric junction of pig was demonstrated. The CO-induced relaxation was not affected by 4-aminopyridine in these tissues. Hence, CO might not have an effect on Kv channels at least in these tissues.

In pressurized rat gracilis muscle arterioles, CO induced vasorelaxation mainly by opening K^+ channels *(68)*. Inclusion of exogenous CO in the superfusion buffer suppressed the pressure-induced vasoconstriction in chromium mesoporphyrin (CrMP)-treated vessels, a treatment to inhibit endogenous production of CO. On the other hand, the pressure-induced vasoconstriction in vessels treated with both CrMP and TEA (1 m*M*) was not altered by exogenous CO. The study by Zhang et al. suggested that a TEA-sensitive K^+ channel was the target of CO in the pressure-induced arterioles. More and more studies indicate that K^+ channels appear to play a more important role in the effects of CO on small peripheral arteries and cerebral arteries. A similar trend has been associated with the tissue-specific production and effects of another K^+ channel opener, endothelium-derived hyperpolarizing factor. Application of CO to piglet pial arterioles in vivo induced a concentration-dependent vascular dilation *(69)*. The K_{Ca} channel inhibitors TEA and IbTx completely abolished the CO effect. This is another example of the vascular effect of CO being exclusively mediated by K_{Ca} channels.

3.2. Demonstration by Patch-Clamp Technique of Direct Action of CO on K^+ Channels

The direct effect of CO on K_{Ca} channels has been demonstrated in isolated rat tail artery SMCs *(70,71)*. Whole-cell K^+ channel currents in these cells were reversibly increased by CO at concentrations as low as 10 μ*M*. In a control experiment, the effect of CO on whole-cell K^+ channel currents in N1E-115 cells was examined. These cells expressed medium-conductance (98 pS) and SK_{Ca} (5.4 pS) channels, but not the BK_{Ca} channels *(72)*. CO failed to alter whole-cell K_{Ca} channel currents in N1E-115 cells *(71)*. Single-channel studies provide further evidence of the nature of the CO-sensitive K^+ channels. Rat tail artery SMCs possess a BK_{Ca} channel with a single-channel conductance of about 239 pS with symmetric 145 m*M* KCl on both sides of the patch membrane. This channel was calcium sensitive and blocked by externally applied ChTX (100 n*M*), but not by apamin (100 n*M*). Extracellularly or intracellularly applied CO increased the open probability (NPo) of single BK_{Ca} channels in a concentration-dependent fashion without affecting the single-channel conductance. The effect of CO on the NPo was also antagonized by ChTX (100 n*M*). Whether an increased intracellular calcium concentration in the presence of CO occurred was first examined. It was found that CO had no effect on the resting intracellular free calcium concentration in acutely isolated rat tail artery SMCs *(73)*. This conclusion was derived from a fura-2 assay on the average level of intracellular calcium in single cells. Whether calcium concentrations in the microdomain of cytosol, alternatively called "calcium sparks," are increased by CO cannot be concluded from this study *(73)*. What has been confirmed is that calcium sensitivity of K_{Ca} channels was enhanced by CO *(70,71)*. With $[Ca^{2+}]_i$ at 3 μ*M*, the K_{Ca} channels spent about 40% of the time in their open state in the absence of CO, whereas in the presence of CO (10 μ*M*) the NPo increased significantly so that the channels were open about 90% of the time. The increased calcium sensitivity alone, or together with a putative increase in calcium levels adjacent to K_{Ca} channels, would account for the stimulatory effect of CO on BK_{Ca} channels.

In this regard, a recent study on SMCs of porcine cerebral arterioles is especially intriguing *(74)*. Beyond the well-documented stimulatory effect of CO on BK_{Ca} channels in these cells, the Ca^{2+} spark-induced transient K_{Ca} channel was also stimulated by CO. The latter could be attributed to the increased effective coupling of Ca^{2+} sparks to K_{Ca} channels by CO *(74)*. Evidence for this notion includes increased percentage of Ca^{2+} spark frequency in SMCs of intact cerebral arterioles by CO and inhibition of the effect of CO by ryanodine that selectively blocked intracellular calcium release from ryanodine-sensitive pools. However, whether this Ca^{2+} spark-related mechanism also mediated the effect of CO on BK_{Ca} channels in cerebral arterioles is still unknown.

Is the effect of CO on K_{Ca} channels mediated by some known second-messenger systems, such as the cGMP system? The answer may rely on specific cell types and K_{Ca} channel isoforms. In a study by my colleagues and I *(71)* on rat tail artery SMCs, we firmly establish that CO directly opens K_{Ca} channels independent of cGMP or other intracellular second messengers. There are several lines of supporting evidence *(71)*. First, CO increased K_{Ca} channel activity in either inside-out or outside-out cell-free patch-clamp recordings. Second, GTP and cGMP-dependent protein kinase were absent in the patch-clamp recording solutions, which excludes the indirect effect of CO on the K_{Ca} channel via the activation of cGMP-dependent protein kinases. Third, the application of GTP-γ-S, pertussis toxin, or cholera toxin to cell-free membrane patches did not affect K_{Ca} channel currents nor the excitatory effect of CO on K_{Ca} channels. Thus, a membrane-attached G-protein would not explain the effect of CO on K_{Ca} channels.

In single SMCs from rat gracilis muscle arterioles, a 105-pS K_{Ca} channel was recorded from cell-free membrane patches *(68)*. This channel was blocked by IbTx and opened by CO. Although the molecular nature of this K^+ channel remains obscure because of its sensitivity to IbTx and calcium and its small conductance, a direct effect of CO on K^+ channels is indicated because these cell-free studies were carried out in the absence of cytosol. A similar effect of CO on 105-pS K^+ channels has also been demonstrated in SMCs of rat renal interlobar arteries *(75)*.

3.3. Opening of K+ Channels in VSMCs by Endogenous CO

The physiological importance of gasotransmitters cannot be realized based only on the effect of exogenous gases at physiologically relevant concentrations. The effects of these gases generated in vivo have to be shown. This rule also applies to CO *(67,68)*. Leffler et al. *(69)* incubated piglet pial arterioles in vivo with heme-L-lysinate, a substrate of HO, to promote endogenous production of CO. This treatment elicited a dose-dependent dilation that is sensitive to TEA inhibition *(69)*. A similar approach has been adopted in a patch-clamp study. Wu et al. *(76)* showed that the inhibition of HO by CrMP significantly reduced whole-cell K_{Ca} channel currents in rat tail artery SMCs. After an endogenous level of CO was elevated with heme-L-lysinate incubation for 10 min, K_{Ca} channel currents were significantly increased *(76)*. The effect of heme-L-lysinate was not induced by another two end products of HO-catabolized heme metabolism, free iron and biliverdin. Direct application of biliverdin to tail artery SMCs did not alter K_{Ca} currents, and heme-L-lysinate still significantly increased K_{Ca} currents in the presence of the free iron scavenger deferoxamine *(76)*. Together, these results strongly indicate that K_{Ca} channels in rat tail artery SMCs are stimulated by endogenous CO. In SMCs of rat renal interlobar arteries, CrMP treatment significantly reduced the open probability of TEA-sensitive 105-pS K^+ channels *(68)*. Similar results were also reported with SMCs from pig pial arterioles *(74)*. Heme-L-lysinate (100 nM) increased mean NPo of BK_{Ca} channels by 6.9-fold. This effect of heme-L-lysinate developed over a period of 3 to 4 min, much slower than the rapid effect of exogenously applied CO *(74)*.

4. ALTERED EFFECTS OF CO ON ION CHANNELS UNDER PATHOPHYSIOLOGICAL CONDITIONS

Altered CO production and/or changed sensitivity of K^+ channels in VSMCs would result in abnormality in vascular contractility. Reduced expressions of HO-1 and HO-2 proteins have been reported in aortic tissues of young spontaneously hypertensive rats *(77)*. A significantly lower endogenous production of CO has also been detected in women with pregnancy-induced hypertension and preeclampsia *(78)*. The effect of CO on K^+ channels in VSMCs from hypertensive subjects has not been reported.

Abnormal metabolism of CO has been shown in cardiac tissues from streptozotocin-induced diabetic rats *(79)*. In a previous study by my colleagues and I *(80)*, Western blot assay was employed to examine the expression of HO in normal and diabetic rat tissues. The expression of HO-1 proteins was barely detectable in normal rat vascular tissues, but apparently visible in rat spleen. One month after STZ injection, the expression of HO-1 proteins in all vascular tissues tested was significantly upregulated. By contrast, HO-2 levels were not different in all rat tissues examined between normal and diabetic rats *(80)*. It appears that endogenous CO levels in diabetic vascular tissues may actually be increased. We further examined the effects of CO on vascular contractility and K^+ channels in single SMCs from STZ-induced diabetic rats. The CO-induced relaxation of tail artery tissues from diabetic rats was significantly decreased compared with that of nondiabetic control rats. As demonstrated previously using the same tail artery tissue, both the cGMP pathway and K_{Ca} channels are responsible for CO's effect. After the cGMP pathway was blocked using ODQ, the CO-induced relaxation of diabetic tissues was completely abolished. This observation naturally leads to the question as to the existence and function of K_{Ca} channels in tail artery tissues from diabetic rats. A patch-clamp study using the single-channel recording technique did not find abnormality in biophysical properties and pharmacological sensitivity of BK_{Ca} channels in diabetic sSMCs. Interestingly, the sensitivity of K_{Ca} channels in diabetic SMCs to CO was significantly reduced. For instance, CO at 10 μM induced an 81% increase in the mean open probability of single K_{Ca} channels in normal SMCs but had no effect in diabetic SMCs. The reduced sensitivity may be linked to increased glycation of K_{Ca} channel proteins in diabetes. By culturing tail artery SMCs from diabetic rats or normal control rats in hyperglycemic medium containing 25 m*M* glucose for 8 d, the reduced sensitivity of K_{Ca} channels to CO was replicated. This phenomenon was not observed once the 25 m*M* glucose in the medium was replaced with 25 m*M* mannitol. Thus, the role played by the hyperosmolality in the diminished effect of CO on K_{Ca} channels in diabetes would be correspondingly minimized.

Several mechanisms have been proposed to explain the glycation-reduced sensitivity of K_{Ca} channels to CO. First, the interaction of CO with histidine residue of K_{Ca} channels *(70,76)* may be impeded by the glycation of K_{Ca} channels. It has been shown previously that the accessibility to the surface histidines was altered in the glycated RNase A protein *(81)*. Second, glycation of channel proteins has been shown to cause protein thiol oxidation, protein aggregation, and crosslinking *(82)*. It is possible that the glycation of K_{Ca} channels may render the channel gate inflexible, thus decreasing their open probability and responses to CO. Third, the formation of advanced glycation end products may damage the structural integrity of K_{Ca} channel proteins by targeting on hisditine and other side chains *(83)*.

In another group of experiments in our study, normal tail artery SMCs were chronically cultured with 25 m*M* 3-*O*-methylglucose (3-OMG). By the end of 8 d of culture, the sensitivity of K_{Ca} channels in these SMCs to CO was consistently diminished. 3-OMG is a nonmetabolizable glucose analog. The influence of 3-OMG on the effect of CO indicates that the glycation of K_{Ca} channels rather than the metabolism of glucose by cultured VSMCs might be the mechanism for the altered K_{Ca} channel functionality in diabetes. Finally, this "glycation" theory is supported by deglycation experiments. Deglycation of K_{Ca} channels by culturing diabetic SMCs with 5 m*M* glucose for a prolonged period (35 d) regained the sensitivity of channels to CO *(80)*. Whether K_{Ca} channel proteins are heavily glycated in diabetes should be further vigorously tested by isolating these proteins from diabetic SMCs.

5. CONCLUSIONS

Vascular contractility is closely related to the structure and functions of different types of K^+ channels. Three major types of K^+ channels participate in the regulation of excitability of VSMCs. While relying on different sensors (membrane potential for Kv channels, ATP level for K_{ATP} channels, calcium content for K_{Ca} channels), the opening of these three types of K^+ channels leads to the same outcome—membrane hyperpolarization and muscle relaxation. The multiplicity of stimuli, tissue-specific distribution, and variable density of these K^+ channels enable them to differentially react with vasoactive factors and gasotransmitters, including CO. To date, the interaction of CO with different subtypes of K_{Ca} channels in VSMCs has been extensively studied or documented. Especially in some peripheral resistant arteries and cerebral arteries/arterioles, K_{Ca} channels outplay the cGMP pathway to take the full credit for mediating the vasorelaxtion effect of CO. In other types of VSMCs, the cGMP pathway and K^+ channels or even only the cGMP pathway is responsible for the CO effect.

The debate on whether CO interacts with K^+ channels directly or indirectly via stimulation of the cGMP pathway has not been settled. Furthermore, a clear-cut answer to this debate may never be reached. Quite possibly, CO may directly act on K^+ channel proteins, as discussed herein and in Chapter 13, in some cell types but not in others. Missing pieces to this puzzle may be found by examining the existence of physical as well as functional couplings among different subunits of K^+ channel complex, between K^+ channels and various G-proteins or protein kinases, among K^+ channels and cellular metabolites including calcium and ATP, and among K^+ channels and different cytoskeleton structures.

ACKNOWLEDGMENT

This work was supported by a grant-in-aid from the Heart and Stroke Foundation of Saskatchewan, Canada. The author also was supported by an investigator award from the Canadian Institutes of Health Research.

REFERENCES

1. Quayle JM, Nelson MT, Standen NB. ATP-sensitive and inwardly rectifying potassium channels in smooth muscle. Physiol Rev 1997;77:1165–1232.
2. Ablonczy Z, Kono M, Crouch RK, et al. Mass spectrometric analysis of integral membrane proteins at the subnanomolar level: application to recombinant photopigments. Anal Chem 2001;73:4774–4779.

3. Noma A. ATP-regulated K^+ channels in cardiac muscle. Nature 1983;305:147, 148.
4. Murphy ME, Brayden JE. Nitric oxide hyperpolarizes rabbit mesenteric arteries via ATP-sensitive potassium channels. J Physiol (Lond) 1995;486:47–58.
5. Noack T, Edwards G, Deitmer R, et al. Potassium channel modulation in rat portal vein by ATP depletion: a comparison with the effects of levcromakalim (BRL 38227). Br J Pharmacol 1992;107:945–955.
6. Russ U, Metzger F, Kickenweiz E, et al. Binding and effects of K_{ATP} channel openers in the vascular smooth muscle cell line, A10. Br J Pharmacol 1997;122:1119–1126.
7. Xu X-P, Lee KS. Characterization of the ATP-inhibited K^+ current in canine coronary smooth muscle cells. Pflügers Arch 1994;427:110–120.
8. Zhang H-L, Bolton TB. Activation by intracellular GDP, metabolic inhibition and pinacidil of a glibenclamide-sensitive K-channel in smooth muscle cells of rat mesenteric artery. Br J Pharmacol 1995;114:662–672.
9. Beavis AD, Lu Y, Garlid KD. On the regulation of K^+ uniport in intact mitochondria by adenine nucleotides and nucleotide analogs. J Biol Chem 1993;268:997–1004.
10. Inoue I, Nagase H, Kishi K, et al. ATP-sensitive K^+ channel in the mitochondrial inner membrane. Nature 1991;352:244–247.
11. Paucek P, Mironova G, Mahdi F, et al. Reconstitution and partial purification of the glibenclamide-sensitive, ATP-dependent K^+ channel from rat liver and beef heart mitochondria. J Biol Chem 1992;267:26,062–26,069.
12. O'Rourke B. Myocardial K_{ATP} channels in preconditioning. Circ Res 2000;87:845–855.
13. Lorenz E, Alekseev AE, Krapivinsky GB, et al. Evidence for direct physical association between a K^+ channel (Kir6.2) and an ATP-binding cassette protein (SUR1) which affects cellular distribution and kinetic behavior of an ATP-sensitive K^+ channel. Mol Cell Biol 1998;18:1652–1659.
14. Proks P, Gribble FM, Adhikari R, et al. Involvement of the N-terminus of Kir6.2 in the inhibition of the K_{ATP} channel by ATP. J Physiol 1999;514:19–25.
15. Tucker SJ, Gribble FM, Zhao C, et al. Truncation of Kir6.2 produces ATP-sensitive K^+ channels in the absence of the sulphonylurea receptor. Nature 1997;387:179–181.
16. Ashcroft FM, Gribble FM. New windows on the mechanism of action of K_{ATP} channel openers. Trends Pharmacol Sci 2000;21:439–445.
17. Yamada M, Isomoto S, Matsumoto S, et al. Sulphonylurea receptor 2B and Kir6.1 form a sulphonylurea-sensitive but ATP-insensitive K^+ channel. J Physiol (Lond) 1997;499:715–720.
18. Yokoshiki H, Sunagawa M, Seki T, et al. ATP-sensitive K^+ channels in pancreatic, cardiac, and vascular smooth muscle cells. Am J Physiol 1998;274:C25–C37.
19. Cao K, Tang G, Hu D, et al. Molecular basis of ATP-sensitive K^+ channels in rat vascular smooth muscles. Biochem Biophys Res Commun 2002;296:463–469.
20. Miyoshi Y, Nakaya Y, Wakatsuki T, et al. Endothelin blocks ATP-sensitive K^+-channels and depolarises smooth muscle cells of porcine coronary artery. Circ Res 1992;70:612–616.
21. Murphy ME, Brayden JE. Nitric oxide hyperpolarizes rabbit mesenteric arteries via ATP-sensitive potassium channels. J Physiol (Lond) 1995;486:47–58.
22. Wellman GC, Quayle JM, Standen NB. ATP-sensitive K^+ channel activation by calcitonin gene-related peptide and protein kinase A in pig coronary arterial smooth muscle. J Physiol (Lond) 1998;507: 117–129.
23. Kubo M, Quayle JM, Standen NB. Angiotensin II inhibition of ATP-sensitive K^+ currents in rat arterial smooth muscle cells through protein kinase C. J Physiol (Lond) 1997;503:489–496.
24. Standen NB, Quayle JM, Davies NW, et al. Hyperpolarizing vasodilators activate ATP-sensitive K^+ channels in arterial smooth muscle. Science 1989;245:177–180.
25. Zhao W, Zhang J, Lu Y, et al. The vasorelaxant effect of H_2S as a novel endogenous gaseous K_{ATP} channel opener. EMBO J 2001;20:6008–6016.
26. Fleischmann BK, Washabau RJ, Kotlikoff MI. Control of resting membrane potential by delayed rectifier potassium currents in ferret airway smooth muscle cells. J Physiol (Lond) 1993;469:625–638.
27. Gelband CH, Hume JR. $[Ca^{2+}]$i inhibition of K^+ channels in canine renal artery: novel mechanism for agonist-induced membrane depolarization. Circ Res 1995;77:121–130.
28. Lu Y, Zhang J, Tang G, et al. Voltage-dependent K^+ channel current in vascular smooth muscle cells from rat mesenteric arteries. J Membr Biol 2001;180:163–175.
29. Tang G, Wang R. Differential expression of K_V and K_{Ca} channels in vascular smooth muscle cells during one-day culture. Pflügers Arch 2001;442:124–135.
30. Chandy KG, Gutman GA. Voltage-gated K^+ channels. In: North RA, ed. Handbook of Receptors and Channels: Ligand- and Voltage-Gated Ion Channels. CRC Press: Boca Raton, FL, 1995, pp. 1–71.

31. Isom LL, De Jongh KS, Catteral WA. Auxiliary subunits of voltage-gated ion channels. Neuron 1994;12:1183–1194.
32. Kalman K, Nguyen A, Tseng-Crank J, et al. Genomic organization, chromosomal localization, tissue distribution, and biophysical characterization of a novel mammalian shaker-related voltage-gated potassium channel, Kv1.7. J Biol Chem 1998;273:5851–5857.
33. Deal KK, England SK, Tamkun MM. Molecular physiology of cardiac potassium channels. Physiol Rev 1996;76:49–67.
34. Roberds SL, Tamkun MM. Cloning and tissue-specific expression of five voltage-gated potassium channel cDNAs expressed in rat heart. Proc Natl Acad Sci USA 1991;88:1798–1802.
35. Overturf KE, Russell SN, Carl A, et al. Cloning and characterization of a Kv1.5 delayed rectifier K^+ channel from vascular and visceral smooth muscles. Am J Physiol 1994;267:C1231–C1238.
36. Zahradka P, Harris KD, Triggs-Raine B, et al. PCR-based analysis of voltage-gated K^+ channels in vascular smooth muscle. Mol Cell Biochem 1995;145:39–44.
37. Archer SL, Huang JMC, Reeve HL, et al. Differential distribution of electrophysiologically distinct myocytes in conduit and resistance arteries determines their response to nitric oxide and hypoxia. Circ Res 1996;78:431–442.
38. Mays DJ, Foose JM, Philipson LH, et al. Localization of the Kv1.5 K^+ channel protein in explanted cardiac tissue. J Clin Invest 1995;96:282–292.
39. England SK, Uebele VN, Kodali J, et al. A novel K^+ channel β-subunit (hKvβ1.3) is produced via alternative mRNA splicing. J Biol Chem 1995;270:28,531–28,534.
40. Xu C, Tang G, Lu Y, et al. Molecular basis of voltage-dependent delayed rectifier K^+ channels in smooth muscle cells from rat tail artery. Life Sci 2000;66:2023–2033.
41. Xu C, Lu Y, Tang G, et al. Expression of voltage-dependent K^+ channel genes in mesenteric artery smooth muscle cells. Am J Physiol 1999;277:G1055–G1063.
42. Benham CD, Bolton TB, Lang RJ, et al. Calcium-activated potassium channels in single smooth muscle cells of rabbit jejunum and guinea-pig mesenteric artery. J Physiol (Lond) 1986;371:45–67.
43. Brayden JE, Nelson MT. Regulation of arterial tone by activation of calcium-dependent potassium channels. Science 1992;256:532–535.
44. Köhler M, Hirschberg B, Bond CT, et al. Small-conductance, calcium-activated potassium channels from mammalian brain. Science 1996;273:1709–1714.
45. Sokol PT, Hu W, Yi L, et al. Cloning of an apamin binding protein of vascular smooth muscle. J Protein Chem 1994;13:117–128.
46. Jensen BS, Strobaek D, Christophersen P, et al. Characterization of the cloned human intermediate-conductance Ca^{2+}-activated K^+ channel. Am J Physiol 1998;275:C848–C856.
47. Grissmer S, Nguyen AN, Cahalan MD. Calcium-activated potassium channels in resting and activated human T lymphocytes. J Gen Physiol 1993;102:601–630.
48. Partiseti M, Choquet D, Diu A, et al. Differential regulation of voltage- and calcium-activated potassium channels in human B lymphocytes. J Immunol 1992;148:3361–3368.
49. Munujos P, Knaus H, Kaczorowski GJ, et al. Cross-linking of charybdotoxin to high-conductance calcium-activated potassium channels: identification of the covalently modified toxin residue. Biochemistry 1995;34:10,771–10,776.
50. Meera P, Wallner M, Jiang Z, et al. A calcium switch for the functional coupling between α (hslo) and β subunits (K_V, Ca β) of maxi K channels. FEBS Lett 1996;385:127, 128.
51. Wallner M, Meera P, Toro L. Molecular basis of fast inactivation in voltage and Ca^{2+}-activated K^+ channels: a transmembrance β-subunit homolog. Proc Natl Acad Sci USA 1999;96:4137–4142.
52. Xia XM, Ding JP, Lingle CJ. Molecular basis for the inactivation of Ca^{2+}- and voltage-dependent BK channels in adrenal chromaffin cells and rat insulinoma tumor cells. J Neurosci 1999;19: 5255–5264.
53. Ha TS, Jeong SY, Cho SW, et al. Functional characteristics of two BK_{Ca} channel variants differentially expressed in rat brain tissues. Eur J Biochem 2000;267:910–918.
54. Lagrutta A, Shen KZ, North RA, et al. Functional differences among alternatively spliced variants of Slowpoke, a Drosophila calcium-activated potassium channel. J Biol Chem 1994;269:20,347–20,351.
55. Atkinson NS, Robertson GA, Ganetzky B. A component of calcium-activated potassium channels encoded by the Drosophila Slo locus. Science 1991;253:551–555.
56. Butler A, Tsunoda S, McCobb DP, et al. mSlo, a complex mouse gene encoding "maxi" calcium-activated potassium channels. Science 1993;261:221–224.
57. Adelman JP, Shen KZ, Kavanaugh MP, et al. Calcium-activated potassium channels expressed from cloned complementary DNAs. Neuron 1992;9:209–216.

58. Meera P, Wallner M, Song M, et al. Large conductance voltage- and calcium-dependent K^+ channel, a distinct member of voltage-dependent ion channels with seven N-terminal transmembrane segments (S0–S6), an extracellular N terminus, and an intracellular (S9–S10) C terminus. Proc Natl Acad Sci USA 1997;94:14,066–14,071.
59. Knaus HG, Eberhart A, Koch RO, et al. Characterization of tissue-expressed alpha subunits of the high conductance Ca^{2+}-activated K^+ channel. J Biol Chem 1995;270:22,434–22,439.
60. Wei A, Solaro C, Lingle C, et al. Calcium sensitivity of BK-type KCa channels determined by a separable domain. Neuron 1994;13:671–681.
61. Wallner M, Meera P, Toro L. Determinant for β-subunit regulation in high-conductance voltage-activated and Ca^{2+}-sensitive K^+ channels: an additional transmembrance region at the N terminus. Proc Natl Acad Sci USA 1996;93:14,922–14,927.
62. Wang R. Carbon Monoxide and Cardiovascular Functions. CRC Press: Boca Raton, FL, 2002.
63. Wang R. Resurgence of carbon monoxide: an endogenous gaseous vasorelaxing factor. Can J Physiol Pharmacol 1998;76:1–15.
64. Wang R, Wang ZZ, Wu L. Carbon monoxide–induced vasorelaxation and the underlying mechanisms. Br J Pharmacol 1997; 121:927–934.
65. Komuro T, Borsody MK, Ono S, et al. The vasorelaxation of cerebral arteries by carbon monoxide. Exp Biol Med (Maywood) 2001;226:860–865.
66. Barbé C, Rochetaing A, Kreher P. Mechanisms underlying the coronary vasodilation in the isolated perfused hearts of rats submitted to one week of high carbon monoxide exposure in vivo. Inhal Toxicol 2002;14:273–285.
67. Werkstrom V, Ny L, Persson K, et al. Carbon monoxide–induced relaxation and distribution of haem oxygenase isoenzymes in the pig urethra and lower oesophagogastric junction. Br J Pharmacol 1997;120:312–318.
68. Zhang F, Kaide J, Wei Y, et al. Carbon monoxide produced by isolated arterioles attenuates pressure-induced vasoconstriction. Am J Physiol Heart Circ Physiol 2001;281:H350–H358.
69. Leffler CW, Nasjletti A, Yu C, et al. Carbon monoxide and cerebral microvascular tone in newborn pigs. Am J Physiol 1999;276:H1641–H1646.
70. Wang R, Wu L. The chemical modification of K_{Ca} channels by carbon monoxide in vascular smooth muscle cells. J Biol Chem 1997;272:8222–8226.
71. Wang R, Wu L, Wang ZZ. The direct effect of carbon monoxide on K_{Ca} channels in vascular smooth muscle cells. Pflügers Arch 1997;434:285–291.
72. Leinders T, Van Kleef RGDM, Vijverberg HPM. Divalent cations activate small- (SK) and large-conductance (BK) channels in mouse neuroblastoma cells: selective activation of SK channels by cadmium. Pflügers Arch 1992;422:217–222.
73. Wang R. Resurgence of carbon monoxide: an endogenous gaseous vasorelaxing factor. Can J Physiol Pharmacol 1998;76:1–15.
74. Jaggar JH, Leffler CW, Cheranov SY, et al. Carbon monoxide dilates cerebral arterioles by enhancing the coupling of Ca^{2+} sparks to Ca^{2+}-activated K^+ channels. Circ Res 2002;91:610–617.
75. Kaide JI, Zhang F, Wei Y, et al. Carbon monoxide of vascular origin attenuates the sensitivity of renal arterial vessels to vasoconstrictors. J Clin Invest 2001;107:1163–1171.
76. Wu L, Cao K, Lu Y, et al. Different mechanisms underlying the stimulation of K_{Ca} channels by nitric oxide and carbon monoxide. J Clin Invest 2002;110:691–700.
77. Ndisang JF, Wu L, Zhao W, et al. Induction of heme oxygenase-1 and stimulation of cGMP production by hemin in aortic tissues from hypertensive rats. Blood 2003;101:3893–3900.
78. Baum M, Schiff E, Kreiser D, et al. End-tidal carbon monoxide measurements in women with pregnancy-induced hypertension and preeclampsia. Am J Obstet Gynecol 2000;183:900–903.
79. Nishio Y, Kashiwagi A, Taki H, et al. Altered activities of transcription factors and their related gene expression in cardiac tissues of diabetic rats. Diabetes 1998;47:1318–1325.
80. Wang R, Wang ZZ, Wu L, et al. Reduced vasorelaxant effect of carbon monoxide in diabetes and the underlying mechanisms. Diabetes 2001;50:166–174.
81. Baek W-O, Vijayalakshmi. Effect of chemical glycosylation of Rnase A on the protein stability and surface histidines accessibility in immobilized metal ion affinity electrophoresis (IMAGE) system. Biochim Biophys Acta 1997;1336:394–402.
82. Swamy MS, Abraham EC. Glycation of lens MIP26 affects the permeability in reconstituted liposomes. Biochem Biophys Res Commun 1992;186:632–638.
83. Coussons PJ, Jacoby J, McKay A, et al. Glucose modification of human serum albumin: a structural study. Free Radic Biol Med 1997;22:1217–1227.

12 Modulation of Multiple Types of Ion Channels by Carbon Monoxide in Nonvascular Tissues and Cells

Rui Wang

CONTENTS

SUMMARY

Carbon monoxide (CO) is a gasotransmitter. Once generated in cells, CO affects specific cellular functions depending on cell types and specific targets in the cells. Ion channels couple membrane excitability and metabolism to cellular functions. The interaction of CO and ion channels constitutes an important mechanism for the biological effect of CO. CO has been reported to alter the expression or function of K^+ channels, Ca^{2+} channels, Na^+ channels, and other types of nonselective ion channels in different tissues. Different types of K^+ channels are the main target of CO in various tissues including visceral smooth muscle cells. Modulation of Ca^{2+} channel function by CO has been controversial, especially in chemosensitive cells of the carotid body. The diversity of effects of CO on ion channels is best exemplified in neurons. Future studies need to establish more specifically the role of endogenous CO in the regulation of the ion channel, the molecular mechanisms for the CO-induced changes in ion channel function and expression, and the correlation of the effects of CO on ion channels with specific cellular functions. Pathophysiological implications of the effects of CO on ion channels should also be intensively investigated to elucidate the pathological role of abnormal CO metabolism and function.

Key Words: Carbon monoxide; gasotransmitters; ion channels; kidney; neuron; visceral smooth muscle cells.

From: *Signal Transduction and the Gasotransmitters: NO, CO, and H_2S in Biology and Medicine* Edited by: Rui Wang © Humana Press Inc., Totowa, NJ

1. INTRODUCTION

Carbon monoxide (CO) is a member of the gasotransmitter family *(1)*. Both heme oxygenase (HO)-dependent heme metabolism and NADPH-dependent microsomal lipid peroxidation yield CO in cells. These two pathways, HO in particular, leading to endogenous CO production, have been identified in numerous types of cells. This ubiquitous availability of enzymatically produced CO suggests a wide array of biological functions of this gasotransmitter over many different systems.

The respiratory system generally is regarded as the most affected by CO intoxication from environmental exposure. Also in the lung, CO provides potent antiinflammatory protection and antioxidant defense. Patients with inflammatory respiratory diseases have a high level of CO in their exhaled air. The developmental importance of CO should have been realized when the first case of severe newborn jaundice was developed. Accumulated bilirubin in jaundice cannot be isolated from an increased level of CO, both generated as the metabolites of heme in a one-to-one ratio. One example of the effects of CO on development is vessel relaxation and vascular protection in the human placenta. Another is the CO-related expression and function changes of ion channels at different developmental stages. CO-induced vasorelaxation has been demonstrated in numerous types of vascular tissues. Vascular growth, proliferation, and remodeling are also under the influence of endogenous CO. Platelet aggregation is inhibited by CO *(2)*. A protection against cardiac anaphylaxis by CO has been proved. CO affects neuronal functions partially by modulating cerebral circulation. CO itself functions as a retrograde gasotransmitter messenger to participate in the regulation of long-term potentiation. The production and release of some neuronal hormones, such as corticotropin-releasing hormone, gonadotropin-releasing hormone, or hypothalamic hormones, are also subject to the effects of CO. In liver, CO modulates portal perfusion as a relaxant for hepatic arterial circulation, but not as much for hepatic venous circulation. The rate of bile acid uptake by hepatocytes, as well as bile excretion, is affected by CO. Gastrointestinal (GI) smooth muscles relax in response to CO *(3)*. CO may also be an important mediator for reducing the rejection of transplants *(4)*. Readers are referred to two recently published books by Abraham et al. *(5)* and Wang *(6)* for more detailed descriptions of the diverse biological functions of CO. In many of the aforementioned systems, ion channels play critical roles in coupling membrane excitability and metabolism with cellular functions. Therefore, these ion channels are important membrane targets for endogenous substances, including gasotransmitters.

This chapter is devoted to the effects of CO on different ion channels in various cell types. The interaction of CO and ion channels in vascular smooth muscle cells (VSCMs) is not described here; for a discussion of this topic *see* Chapter 11.

2. EFFECTS OF CO ON K^+ CHANNELS

One of the earliest studies examining the effects of CO on ion channels was carried out on cultured urinary bladder monocytes from guinea pig. In this preparation, whole-cell K^+ current was evoked by stepwise depolarization from –65 to +10 mV and the current was believed to be K_{Ca} channels *(7)*. CO superfusion for 2 min induced about 50% inhibition of K_{Ca} currents in these cells. Once CO was removed from the bath solution, the inhibited K_{Ca} currents recovered. This effect of CO was believed to be the result of decreased intracellular calcium concentration in the presence of CO, although the possibility that CO may directly act on K_{Ca} channels independent of intracellular calcium level should also be considered *(8)*. Changes in intracellular calcium concentrations in the presence of CO were not directly measured in this study. Because CO increased cyclic guanosine 5'-monophosphate (cGMP) production from these cells, it was concluded that the inhibition of K_{Ca} channels was mediated by cGMP elevation, which decreased intracellular calcium. High-intensity light exposure also facilitated K_{Ca} channel recovery from CO inhibition, believed to be the result of the dissociation of CO from the CO-Fe^{2+} complex in guanylyl cyclase. The inhibitory effect of CO on K_{Ca} channels was mimicked by sodium nitroprusside (0.01–1 m*M*), a nitric oxide (NO) donor *(7)*. To my knowledge, this is probably the only report showing an inhibitory effect of CO on K_{Ca} channels in any type of cells. One sobering fact about this study is the concentration of CO used. Guinea pig urinary bladder monocytes were superfused with a CO-containing solution made by "bubbling physiological saline in a container of about 120 mL volume with a gas mixture of 80 vol % CO and 20 vol % O_2 (supply rate about 120 mL/min)." Although actual CO concentration was not measured, CO concentration in this study obviously is far beyond the physiological range of CO. Furthermore, application of this solution may also exert a hypoxic effect on K_{Ca} channels on monocytes.

A series of studies has been conducted in jejunal smooth muscle cells (SMCs), in which excitatory effects of exogenous and endogenous CO on K^+ channels have been shown. This series started with human jejunal circular SMCs *(9)*. At a concentration of 1% in Krebs solution, CO induced an initial and transient 175% increase in whole-cell K^+ current in 20 of 22 cells tested. The membrane potential of the perforated cells was also transiently hyperpolarized by about 16 mV. Because the recorded whole-cell K^+ channel currents were sensitive to 1 m*M* quinidine (n = 2), but not to 10 m*M* tetraethylammonium (TEA) (n = 2), these K^+ currents were assigned to be conducted by Kv channels. However, the whole-cell K^+ currents and membrane potential underwent oscillated changes in the presence of 1% CO in four cells (8 cpm in one cell). This observation, together with a small number of cells used in experiments with quinidine and TEA, opens the debate as to whether the CO target is Kv or K_{Ca} channels. Since the expression of HO in this cell preparation was confirmed and since zinc protoporphyrin IX (2 m*M*) decreased whole-cell K^+ channel currents, the physiological importance of endogenous CO on K^+ channels pertinent to functions of these cells was suggested. However, this notion should be approached very cautiously because the effect of zinc protoporphyrin was tested on only one cell.

The results of CO's effect on K^+ channels from human cells were later on repeated by the same group using dog cells *(10)*. The expression of HO-2 was identified in canine jejunum, which constituted the endogenous sources of CO. Exogenous CO (1% in bath solution) increased whole-cell outward current by 285% in canine jejunal SMCs. The effect of exogenous CO was mimicked by 8-bromo-cGMP. The nature of the whole-cell outward K^+ channel, not K_{Ca} channels, was identified to be a Kv channels based on the effectiveness of quinidine (50 μ*M*), and the inability of charybdotoxin (ChTx) (100 n*M*), to inhibit the effect of CO. This pharmacological dissection of Kv and K_{Ca} channels is arguably questionable because ChTx may not be potent enough to block all types of K_{Ca} channels and the calcium sensitivity and single-channel conductance of the K^+ currents have not been determined. CO also hyperpolarized the resting membrane potential from an average of –37 to about –45 mV. Furthermore, the CO-induced oscillation of outward currents, originally observed in human jejunal circular SMCs, was replicated in canine jejunal circular SMCs. During an observation period of more than 15 min, CO evoked cyclic changes (about 20 cpm at +40 mV) in whole-cell outward current.

The issue of the nature of the K^+ channels that were affected by CO surfaced again. Logically, one would attribute this to oscillatory changes in intracellular calcium levels. Fura-2 or Indo-1 measurements were carried out and did not reveal any effect of CO (1%) on intracellular calcium levels in these cells. Does this rule out the linkage between calcium oscillation and K^+ current oscillation in the presence of CO? Not necessarily. In a patch-clamp study *(11)*, CO increased K_{Ca} channel currents in tail artery SMCs but did not increase intracellular calcium levels in these cells measured with a Fura-2 assay *(12)*. This is partially explained by the limitation of fura-2 fluorescence in detecting changes in calcium levels in small microzones.

Jaggar et al. *(13)* reported that CO not only enhanced the activity of big-conductance K_{Ca} (BK_{Ca}) channels in VSMCs but also stimulated Ca^{2+} spark-induced transient K_{Ca} channels. The latter could be attributed to the increased effective coupling of Ca^{2+} sparks to K_{Ca} channels by CO. Evidence for this notion includes an increased percentage of Ca^{2+} spark frequency in SMCs of intact cerebral arterioles by CO, and the inhibition of CO's effect by ryanodine that selectively blocked intracellular calcium release from ryanodine-sensitive pools. Clearly, CO can stimulate BK_{Ca} channels and calcium spark-activated transient K_{Ca} channels in VSMCs. It is still possible that in jejunal SMCs CO actually stimulated one type of K_{Ca} channel, instead of Kv channels *(10)*. In fact, Farrugia et al. *(10)* provide evidence for this latter possibility. After including EGTA (2 m*M*) in the pipet solution to dialyze the cells, CO (1%) induced only a 58% increase in whole-cell K^+ currents, which was significantly lower than the CO-induced increase (285%) in the absence of EGTA. The mediation of CO's effect by the cGMP pathway was further tested in jejunal SMCs. CO induced a marginal increase in cGMP level, merely significant because of large standard errors (from 86 ± 32 to 178 ± 70 pmol/10^6 cells; $n = 5$, $p < 0.05$). 8-Bromo-cGMP (2 m*M*) also slightly increased whole-cell current at a membrane potential more positive than +10 mV (94 ± 37%). Finally, the application of CO in the presence of 8-bromo-cGMP further enhanced K^+ channel currents. These findings taken together demonstrate that CO's effect on K^+ channels in jejunal SMCs was not exclusively, if at all, mediated by the cGMP pathway.

A recent chapter about the jejunal SMC issue described the effect of endogenous CO on membrane excitability. Xue et al. *(14)* reported a 5-mV depolarization of jejunal SMCs isolated from HO-2 knockout mice in comparison with wild-type mice. Once the nitric oxide synthase gene was eliminated from HO-2-deficient mice, the resting membrane potential of jejunal SMCs was further depolarized by about 8 mV. The contractility of jejunal circular smooth muscles in response to electrical field stimulation was also significantly reduced in HO-2-deficient mice. This study demonstrated the crucial importance of endogenous CO in maintaining normal resting membrane potential and relaxant capability of the concerned cells. The endogenous sources of CO in this preparation were believed to be from either enteric neurons or interstitial cells of Cajal. The ion mechanisms for the depolarization of jenunal SMCs and the reduced contractility in the absence of endogenous CO were not provided. Considering the importance of K^+ flux in determining resting membrane potential in general, stimulation of K^+ channels by CO in these SMCs would be a reasonable speculation. Farrugia and Szurszewskt *(15)* depict a scheme of the CO-mediated GI motility control. Interstitial cells of Cajal may serve as the production site of endogenous CO. K^+ channels, being Kv or K_{Ca}, located on intestinal SMCs are the target of endogenous CO. CO as a coupling gasotransmitter coordinates the activities of interstitial cells of Cajal GI SMCs.

The tissue-specific effect of CO and the differential involvement of K^+ channels in the effect of CO can be appreciated from studies on isolated smooth muscle from pig urethra and the esophagogastric junction (GEJ) *(16)*. These muscle preparations developed spontaneous tension in the presence of calcium in the bath solution in vitro. CO (72 μ*M*) relaxed the urethral preparations, which were precontracted with noradrenaline, by about 76%, as well as increased cGMP content. The relaxant effect of CO, however, was not blocked by methylene blue (30 μ*M*), which was used to block the cGMP-mediated pathway. Glibenclamide (1 μ*M*) and 4-aminopyridine (4-AP) (0.1–1 m*M*) did not significantly alter CO's effect, invalidating the involvement of K_{ATP} channels and Kv channels, respectively. Apamin (0.1 μ*M*), iberiotoxin (IbTx) (0.1 μ*M*), or ChTX (0.1 μ*M*) alone or in combination did not change the contraction force in the absence or presence of CO. Blockade of Kv channels with 4-AP alone in urethral tissues induced a contraction. It seems that only Kv channels contribute to the resting membrane potential of SMCs, and, by doing so, affect the resting tension of this tissue. It also seemed that the relaxant effect of CO on this tissue was not mediated by the cGMP pathway, nor by the tested K^+ channels. By contrast, resting membrane potential and resting tension in smooth muscles at the GEJ might be under the control of both Kv and K_{Ca} channels since 4-AP, IbTx, and ChTX application alone or a combination of apamin and ChTX induced contraction force development. Exogenously applied CO also relaxed smooth muscles at the GEJ, which were precontracted with carbachol, by about 86%. Furthermore, CO also increased cGMP content in the GEJ. Different from urethral preparation, CO's effect on the GEJ was reduced by 30% by methylene blue. Methylene blue alone caused spontaneously developed muscle tone in the absence of CO. A combination of ChTx and apamin, but not sole application of 4-AP, IbTx, ChTX, or apamin, reduced CO-induced (24 μ*M*) relaxations of smooth muscles at the GEJ.

These data suggest that the relaxation of muscles at the GEJ in the presence of CO involved a cGMP-related mechanism and a K^+ channel. The nature of this K^+ channel has not been identified, and no electrophysiology or molecular biology experiments have been conducted to address this issue. It has been reported that vasorelaxation induced by endothelium-derived hyperpolarizing factor (EDHF) was mediated by a specific type of K^+ channel that was sensitive only to the coapplication of apamin and ChTX *(17)*. The possibility that CO acts on an EDHF-sensitive K^+ channel remains.

The effects of CO on K^+ channel currents have also been reported in freshly dispersed rabbit corneal epithelial cells *(18)*. An 84% ($n = 14$) increase in K^+ channel currents and a membrane hyperpolarization from –42 to –51 mV were observed in the presence of 1% CO in the bath solution, recorded using the perforated whole-cell voltage-clamp technique. Based on its reversal potential and its inhibition by quinidine or diltiazem, this CO-activated channel was believed to be a K_V channel. Further study using a cell-attached single-channel recording technique demonstrated that CO increased the steady-state open probability (NPo) of the channel depolarized to 0 mV from a holding potential of –40 mV. CO had no effect on single-channel conductance (about 135 pS). The effect of CO on the NPo of K^+ channels appeared to be mediated by cytosolic factors because single-channel NPo in excised inside-out patches ($n = 4$) was not altered by CO. The cytosolic factor critical for the effect of CO could be cGMP because CO at the same concentration (1%) increased cGMP level from 0.41 ± 0.24 to 0.55 ± 0.27 pmol/10^6 cells ($n = 4$; $p < 0.05$).

However, the statistical significance of this marginal increase in cGMP is not really convincing, and thus the role of cGMP in the effect of CO on K^+ channels invites more vigorous tests.

An apical 70-pS K^+ channel is located in the thick ascending limb. This channel contributes to about 80% of the apical K^+ conductance *(19)*. This channel is regulated by NO and cytochrome P450-dependent metabolites, but its calcium sensitivity is not apparent. CO also modulated this 70-pS renal K^+ channel *(20)*. Liu et al. *(20)* first demonstrated the expression of HO-2 at both the mRNA level and protein level in the rat cortex and outer medulla. They then showed that the inhibition of HO-2 by the application of 10 μ*M* chromium mesoporphyrin (CrMP) reversibly reduced the NPo of the apical 70-pS K^+ channel to 26% of the control value. Providing substrate of HO-2, heme-L-lysinate (10 μ*M*), enhanced the single-channel activity in cell-attached patches by 98%. Once HO-2 was inhibited by CrMP, the stimulatory effect of heme-L-lysinate on the channel activity was abolished. The inhibition or enhancement of HO activity not only reduced or increased the production of CO, but also that of biliverdin and iron. However, biliverdin (10 μ*M*) alone had no effect on the channel activity. By contrast, the effect of heme-L-lysinate on the 70-pS K^+ channel activity was mimicked by the effect of exogenous CO. On inside-out membrane patches, CO (100 μ*M*) doubled single-channel activity. The inhibitory effect of CrMP on the channel activity was also antagonized by the exogenously applied CO. These results strongly support the notion that endogenous CO stimulates the apical 70-pS K^+ channel in the rat thick ascending limb *(20)*.

Neuronal K_{Ca} channels may also be modulated by CO. It is known that BK_{Ca} channels of rat chemoreceptor cells are inhibited by hypoxia. Riesco-Fagundo et al. *(21)* showed that the hypoxic inhibition was reversed by CO at the whole-cell and the single-channel levels. Whereas hypoxia decreased the NPo of single K_{Ca} channels by about 50%, application of CO to the hypoxic solution, which was bubbled with a mixture of 80% N_2 and 20% CO to yield a Po_2 of 8.7 mmHg, overcame the inhibitory effect of hypoxia by increasing the NPo of K_{Ca} channels to 150% of the control level. The underlying mechanism was speculated to be the competitive inhibition of O_2 binding to membrane hemoproteins that act as an O_2 sensor to modulate channel activity. Unfortunately, this study did not directly investigate the effect of CO on K_{Ca} channels in the absence of hypoxia. A direct stimulatory effect of CO on K_{Ca} channels may alternatively well explain the reversal of the hypoxic effect without the involvement of an unknown O_2 sensor in the membrane.

3. EFFECTS OF CO ON CA^{2+} CHANNELS

The interaction of CO and Ca^{2+} channels has been studied in a few cases. An earlier study used HO inhibitor to test the effect of endogenous CO on Ca^{2+} channels in AtT-20 pituitary cells *(22)*. Based on their pharmacological sensitivities and electrophysiological properties, the total Ca^{2+} currents in this pituitary cell line were believed to consist of L-, N-, and possibly P-type voltage-dependent Ca^{2+} channels. Extracellular application of zinc-protoporphyrin-IX (ZnPP-IX) irreversibly inhibited the whole-cell Ca^{2+} currents. This inhibitory effect of ZnPP-IX was concentration dependent. However, the effect of ZnPP-IX could not be attributed to the reduced endogenous CO production. There is an argument that the ZnPP-IX-induced attenuation of Ca^{2+} current was abolished by coapplication of superoxide dismutase, suggesting that ZnPP-IX might generate superoxide anion and the latter inhibited Ca^{2+} channels. Another argument was that including an inhibitor of cGMP-dependent protein kinase (KT5823, 1 μ*M*) in the pipet solution did not affect the effect of ZnPP-IX. The counterargument is that endogenous CO might directly interact with Ca^{2+} channel proteins independent of the cGMP pathway. What remained puzzling was that two other HO-2 inhibitors, tin-protoporphyrin-IX and Zn-deuteroporphyrin-bis-glycol, increased Ca^{2+} channel current. Were the stimulatory effects of tin-protoporphyrin-IX and Zn-deuteroporphyrin-bis-glycol on Ca^{2+} channels because of the inhibition of HO-2 and decreased endogenous CO level even if the inhibitory effect of ZnPP-IX on Ca^{2+} channels were not related to HO activity? Three tests should provide evidence to answer this question. First, the existence of HO-2 or HO-1 in these cells should be directly determined. Second, the endogenous level of CO in the presence of these HO inhibitors should be measured. Third, the effect of exogenous CO on calcium channel currents in these cells should be assayed.

Also using HO inhibitors to study the interaction of CO and Ca^{2+} channels, Overholt et al. *(23)* drew a different conclusion about carotid body from that of the aforementioned study on AtT-20 pituitary cells *(22)*. They believed that endogenous CO suppressed the intracellular Ca^{2+} level by inhibiting Ca^{2+} channels *(24)*. Support for this notion included observations that intracellular Ca^{2+} concentrations as well as Ca^{2+} channel currents in glomus cells in carotid body were elevated after ZnPP-IX incubation, and that application of exogenous CO abolished the effect of ZnPP-IX. However, not everyone agreed on this notion. Mokashi et al. *(25)* studied the effect of exogenous CO under normoxic conditions on intracellular Ca^{2+} concentrations in cultured glomus cells of adult rat carotid body. A significant increase in intracellular Ca^{2+} concentration, about sevenfold, was induced by CO. Intracellular calcium levels of carotid body can be elevated because of increased extracellular Ca^{2+} entry, and the latter is critical for chemosensory response to high Pco (partial pressure of CO) stimulation during normoxia and to hypercapnia *(26)*. In a later study by the same research group, the carotid sinus nerves (CSN) with carotid body were isolated from rats *(27)*. It was found that CO perfusion triggered chemosensory discharge of CSN in the dark. Cd^{2+} (200 μ*M*) completely abolished the CO-induced CSN activity. Cd^{2+} is a well-known blocker of voltage-dependent Ca^{2+} channels.

Two hypotheses were proposed correspondingly. The direct interaction of CO with plasma membrane proteins to depolarize membrane and open Ca^{2+} channels was one. The other predicted that CO bound to intracellular protein(s) and released intracellular calcium. The opening of voltage-dependent Ca^{2+} channels was triggered by intracellular calcium release. Both processes would rely on the opening of Ca^{2+} channels and thus could be blocked by Cd^{2+} *(27)*. A key element in the proposed scenarios is the CO-induced membrane depolarization. However, CO has been shown to significantly open whole-cell K^+ currents in isolated rabbit carotid body cells, which had been inhibited by low Po_2. This would lead to membrane hyperpolarization, instead of depolarization *(28)*. One important difference between studies by Rozanov et al. *(27)* and Lopez-Lopez and Gonzalez *(28)* was the concentration of CO used. Pco at 570 mmHg might depolarize carotid body *(27)*, but a hyperpolarization might occur with Pco at 70 mmHg (or 10% CO) *(28)*. Based on this rationale, under physiological conditions endogenous CO may actually open K^+ channels and inhibit the chemosensory response of carotid body.

The CO-induced decrease in intracellular calcium concentration originally was reported in rat aortic tissue segments *(29)*. In rat tail artery SMCs, CO did not affect intracellular calcium concentrations *(12)*. Conversely, CO increased calcium spark in SMCs of porcine cerebral arterioles *(13)*. In visceral SMCs, CO appeared not to affect intracellular calcium concentration *(30)*. CO (about 100 μ*M*) inhibited the 40 m*M* KCl–induced contraction of guinea pig ileum tissues. Increasing concentrations of KCl from 15 to 40 m*M* induced graded increases in intracellular calcium concentrations as well as contraction forces. CO, although inhibiting the contraction force development, only slightly reduced the KCl-induced increase in intracellular calcium. These tissue contractility results did not support a role of calcium channels in the relaxant effect of CO. The effect of CO on voltage-dependent calcium channels in guinea pig ileum SMCs was directly studied using the whole-cell patch-clamp technique. CO (100 μ*M*) had little effect on the peak Ba^{2+} currents (I_{Ba}) when voltage was stepped from –60 to +50 mV (holding potential: –60 mV). It was thus concluded that the relaxant effect of CO on this tissue was not because of the inhibition of L-type Ca^{2+} channels. Because the CO-induced relaxation of ileum tissues was antagonized by 1H-[1,2,4]oxadiazolo[4,3,-a]quinoxalin-1-one (1 μ*M*), a cGMP-dependent mechanism of the effect of CO that may decrease the sensitivity of contractile elements to Ca^{2+} is suggested.

4. EFFECTS OF CO ON NEURONAL ION CHANNELS

One of the earliest discoveries on the physiological importance of CO was neuronal effect of CO, including CO-mediated long-term potentiation *(31,32)* and neuronal hormone release *(33)*. These neuronal effects of CO have been closely related to membrane excitability as well as various types of ion channels in neurons. In addition to the previous discussion of the potential effect of CO on neural K_{Ca} channels *(21)*, several other types of ion channels in neuronal cells might be targets of CO.

4.1. Nonselective Cationic Channels

Pincha et al. *(34)* found that bath application of CO as well as NO increased the firing rate of most locus coeruleus (LC) neurons in rat brain slices. LC neuron firing rate was increased by 80% in 23 of 29 cells by CO within 3–10 min. Similar effects were obtained with bath or intracellular application of selective activators of cGMP-dependent protein kinase. In these neurons, there was a tetrodotoxin (TTX)-insensitive and voltage-independent conductance with an estimated reversal potential of –27 mV. Experimental evidence was provided to show that this conductance did not involve a Na^{+}-Ca^{2+} exchanger, nor a Cl^{-} conductance. Therefore, it was believed to be a nonselective cationic channel. Because NO and CO had the same effect on the firing rate of these neurons and NO stimulated the nonselective cationic channel, Pincha et al. *(34)* concluded that "NO and CO activate noradrenergic neurons of LC via a cGMP-dependent protein kinase and a nonselective cationic channel." Note that no actual experiments have been done to test directly the effect of CO on the nonselective cationic channels in LC neurons. Further caution should be exercised as to the physiological meaning of CO's effect because a saturated CO solution with a CO concentration of about 1 m*M* was used.

4.2. Cyclic Nucleotide-Gated Ion Channels

Many natural odor ligands stimulate olfactory receptor neurons (ORNs) by stimulating the G-protein-coupled cyclic adenosine monophosphate (cAMP) pathway, which then activates the cyclic nucleotide-gated (CNG) channel to generate cationic inward current and membrane depolarization. Leinders-Zufall et al. *(35)* examined the effect of CO on olfactory signal generation in isolated ORNs of the tiger salamander. Exogenously applied CO at a concentration as low as 1 μ*M* consistently generated prominent and reversible inward CNG current, recorded with the whole-cell patch-clamp technique. The EC_{50} of the effect of CO was 2.9 μM. In line with the stimulatory effect of CO on CNG channels, CO (3.1 μM) depolarized the membrane of ORNs by about 22 mV. A cGMP-mediated pathway was suggested to underline this stimulatory effect of CO. In the absence of GTP in the pipet solution or after activation of CNG channels with cGMP, CO lost its stimulatory effect on CNG channels. The blockade of soluble guanylate cyclase activation with LY85383 (10 μM) abolished CO's effect. Thus, CO plays an important role in olfactory signaling by generating cGMP, which stimulates CNG channels. This CO-cGMP pathway and odor-cAMP pathway converge onto the same effector, CNG channels, to coordinate the activity of ORNs.

4.3. Na^+ Channels

Carratu et al. *(36)* investigated the influence of a low level of prenatal CO exposure to the function of rat peripheral nervous system. In 40-d-old rats exposed to CO (75 and 150 ppm) during gestation, the inactivation kinetics of transient Na^+ current recorded from sciatic nerve fibers were significantly slowed. This prolongation of Na^+ currents was accompanied by a right shift in steady-state inactivation of Na^+ channels. Thus, more Na^+ channels in neurons from CO-treated rats were available for opening at a given membrane potential. However, this change disappeared in neurons from 270-d-old CO-exposed rats. Chronic CO exposure during gestation also induced, in neurons from both 40- and 270-d-old rats, a left shift of the voltage-current relationship of Na^+ channels with reversal potential decreased from about +120 to + 100 mV. Carratu et al.'s *(36)* study suggests the importance of CO on the development of the neuronal system at specific developmental stages by regulating the behavior of Na^+ channels in neurons. CO might directly modify Na^+ channel macromolecule to keep the channel in immature status. Alternatively, membrane lipid distribution and density may be altered by CO exposure, affecting the conformation and function of the Na^+ channel proteins embedded within. The protein change and the environmental lipid change could affect the voltage sensor and ion-selective filter of Na^+ channels, influencing voltage-dependent activation and inactivation as well as ion selectivity.

5. CONCLUSIONS

This chapter has summarized the effect of CO on different types of ion channels in different types of cells. Some of the studies reviewed were conducted to explain the toxicological impact of CO on ion channel structure and function. Consequently, a high concentration of CO was used. These studies bear little physiological relevance to the understanding of the importance of endogenous CO. Nevertheless, the information provided would certainly serve as a good reference to the reaction of ion channels to CO under extreme conditions. It is important while evaluating the cited literature to keep in mind the distinction between the effect of CO and that of the secondary hypoxia induced by a high concentration of CO. Finally, comparison of CO concentrations between toxicological studies and physiological experiments is not easy CO-containing solutions were made following different protocols. Therefore, CO concentrations have been expressed as an array of units from parts per million, to percentages, to micromoles.

K^+ channels are the main target of CO in various tissues. Except in cultured bladder monocytes from guinea pig, where CO inhibited K_{Ca} channel currents, CO stimulated various types of K^+ channels in general. Both exogenous and endogenous CO stimulated

K^+ currents in human and canine jejunal SMCs. Although this K^+ conductance was believed to be Kv channel, the possibility that a factual K_{Ca} current was stimulated cannot be excluded yet. Debate is also open as to whether CO's effect on K^+ channels in jejunal SMCs is mediated by the cGMP pathway or not. Evidence has been submitted in rat chemoreceptor cells where CO stimulated K_{Ca} channels. An EDHF-sensitive K^+ channel was proposed to be the target of CO in SMCs from pig GEJ based on the observation that the tissue relaxant effect of CO was blocked by the coapplication of ChTx and apamin. The involvement of cGMP pathway also partially explains the relaxant effect of CO in this tissue. The stimulation of Kv channels by CO was shown in freshly dispersed rabbit corneal epithelial cells, and this effect eventually induced membrane hyperpolarization. In the thick ascending limb of rat kidney, endogenous and exogenous CO stimulated the apical 70-pS K^+ channels.

CO was reported to have no effect on Ca^{2+} channels in pituitary cells or in guinea pig ileum tissues. A major controversy over the effect of CO on Ca^{2+} channels concerns chemosensitive carotid body cells. Some studies showed that endogenous CO inhibited Ca^{2+} channels in carotid body chemosensitive cells. Others showed that exogenous CO increased Ca^{2+} channel currents in the same type of cells. Still another study demonstrated that exogenous CO actually opened K^+ channels and led to membrane hyperpolarization of carotid body, which would inactivate Ca^{2+} channels. Concentration ranges of CO might be the reason for these discrepancies.

Various effects of CO on ion channels in neuronal preparations has been reported. Among the affected ion channels by CO are nonselective cationic channel, CNG ion channels, and Na^+ channels, in addition to K^+ channels.

Despite the progress made in the last few decades, many questions and challenges remain regarding the interaction of CO and ion channels. Future studies need to address more thoroughly whether CO's effects on ion channels are mediated by different second messengers or by a direct interaction between CO and ion channel proteins. The interaction of CO and other gasotransmitters, including NO and hydrogen sulfide, on ion channel functions is very intriguing. The pathophysiological implications of the effects of CO on ion channels have not received due attention. Either the altered endogenous production level of CO or changed sensitivities of ion channels to CO in different diseases would provide pathogenic mechanisms for these diseases. Correspondingly, CO-sensitive ion channels would be good candidates for therapeutic intervention of the related disorders.

ACKNOWLEDGMENTS

This work was supported by the Heart and Stroke Foundation of Saskatchewan, Canada. The author was supported by an investigator award from the Canadian Institutes of Health research.

REFERENCES

1. Wang R. Two's company, three's a crowd—can H_2S be the third endogenous gaseous transmitter? FASEB J 2002;16:1792–1798.
2. Brüne B, Ullrich V. Inhibition of platelet aggregation by carbon monoxide is mediated by activation of guanylate cyclase. Mol Pharmacol 1987;32:497–504.
3. Rattan S, Chakder S. Inhibitory effect of CO on internal anal sphincter: heme oxygenase inhibitor inhibits NANC relaxation. Am J Physiol 1993;265: G799–G804.
4. Sato K, Balla J, Otterbein L, et al. Carbon monoxide generated by heme oxygenase-1 suppresses the rejection of mouse-to-rat cardiac transplants. J Immunol 2001;166:4185–4194.

5. Abraham NG, Alam J, Nath K. Heme Oxygenase in Biology and Medicine. Kluwer Academic/Plenum: New York, 2002.
6. Wang R. Carbon Monoxide and Cardiovascular Functions. CRC Press: Boca Raton, FL, 2001.
7. Trischmann U, Klockner U, Isenberg G, et al. Carbon monoxide inhibits depolarization-induced Ca rise and increases cyclic GMP in visceral smooth muscle cells. Biochem Pharmacol 1991;41:237–241.
8. Wang R, Wu L. The chemical modification of K_{Ca} channels by carbon monoxide in vascular smooth muscle cells. J Biol Chem 1997;272:8222–8226.
9. Farrugia G, Irons WA, Rae JL, et al. Activation of whole cell currents in isolated human jejunal circular smooth muscle cells by carbon monoxide. Am J Physiol 1993;264: G1184–G1189.
10. Farrugia G, Miller SM, Rich A, et al. Distribution of heme oxygenase and effects of exogenous carbon monoxide in canine jejunum. Am J Physiol 1998;274: G350–G358.
11. Wang R, Wang ZZ, Wu L. Carbon monoxide-induced vasorelaxation and the underlying mechanisms. Br J Pharmacol 1997; 121:927–934.
12. Wang R. Resurgence of carbon monoxide: an endogenous gaseous vasorelaxing factor. Can J Physiol Pharmacol 1998;76:1–15.
13. Jaggar JH, Leffler CW, Cheranov SY, et al. Carbon monoxide dilates cerebral arterioles by enhancing the coupling of Ca^{2+} sparks to Ca^{2+}-activated K^+ channels. Circ Res 2002;91:610–617.
14. Xue L, Farrugia G, Miller SM, et al. Carbon monoxide and nitric oxide as coneurotransmitters in the enteric nervous system: evidence from genomic deletion of biosynthetic enzymes. Proc Natl Acad Sci USA 2000;97:1851–1855.
15. Farrugia G, Szurszewski JH. Heme oxygenase, carbon monoxide, and interstitial cells of Cajal. Microsc Res Tech 1999;47:321–324.
16. Werkstrom V, Ny L, Persson K, et al. Carbon monoxide–induced relaxation and distribution of haem oxygenase isoenzymes in the pig urethra and lower oesophagogastric junction. Br J Pharmacol 1997;120:312–318.
17. Zygmunt PM, Hogestatt ED. Role of potassium channels in endothelium-dependent relaxation resistant to nitroarginine in the rat hepatic artery. Br J Pharmacol 1996;117:1600–1606.
18. Rich A, Farrugia G, Rae JL. Carbon monoxide stimulates a potassium-selective current in rabbit corneal epithelial cells. Am J Physiol 1994;267:C435–C442.
19. Lu M, Zhu Y, Balazy M, Reddy KM, et al. Effect of angiotensin II on the apical K^+ channel in the thick ascending limb of the rat kidney. J Gen Physiol 1996;108:537–547.
20. Liu H, Mount DB, Nasjletti A, Wang W. Carbon monoxide stimulates the apical 70–pS K^+ channel of the rat thick ascending limb. J Clin Invest 1999;103:963–970.
21. Riesco-Fagundo AM, Perez-Garcia MT, Gonzalez C, et al. O_2 modulates large-conductance Ca^{2+}-dependent K^+ channels of rat chemoreceptor cells by a membrane-restricted and CO-sensitive mechanism. Circ Res 2001;89:430–436.
22. Linden DJ, Narasimhan K, Gurfel D. Protoporphyrins modulate voltage-gated Ca current in AtT-20 pituitary cells. J Neurophysiol 1993;70:2673–2677.
23. Overholt JL, Bright GR, Prabhakar NR. Carbon monoxide and carotid body chemoreception. Adv Exp Med Biol 1996;410:341–344.
24. Prabhakar NR. NO and CO as second messengers in oxygen sensing in the carotid body. Respir Physiol 1999;115:161–168.
25. Mokashi A, Roy A, Rozanov C, et al. High pCO does not alter pHi, but raises $[Ca^{2+}]i$ in cultured rat carotid body glomus cells in the absence and presence of $CdCl_2$. Brain Res 1998;803:194–197.
26. Roy A, Rozanov C, Iturriaga R, et al. Acid-sensing by carotid body is inhibited by blockers of voltage-sensitive Ca^{2+} channels. Brain Res 1997;769:396–399.
27. Rozanov C, Roy A, Mokashi A, et al. Chemosensory response to high pCO is blocked by cadmium, a voltage-sensitive calcium channel blocker. Brain Res 1999;833:101–107.
28. Lopez-Lopez JR, Gonzalez C. Time course of K^+ current inhibition by low oxygen in chemoreceptor cells of adult rabbit carotid body: effects of carbon monoxide. FEBS Lett 1992;299:251–254.
29. Lin H, McGrath JJ. Carbon monoxide effects on calcium levels in vascular smooth muscle. Life Sci 1988;43:1813–1816.
30. Kwon S, Chung S, Ahn D, et al. Mechanism of carbon monoxide–induced relaxation in the guinea pig ileal smooth muscle. J Vet Med Sci 2001;63:389–393.
31. Verma AD, Hirsch J, Glatt CE, et al. Carbon monoxide: a putative neural messenger. Science 1993;259:381–384.
32. Zhuo MS, Small SA, Kandel ER, et al. Nitric oxide and carbon monoxide produce activity-dependent long-term synaptic enhancement in hippocampus. Science 1993;260:1946–1950.

33. Lamar CA, Mahesh VB, Brann DW. Regulation of gonadotrophin-releasing hormone (GnRH) secretion by heme molecules: a regulatory role for carbon monoxide? Endocrinology 1996;137:790–793.
34. Pineda J, Kogan JH, Aghajanian GK. Nitric oxide and carbon monoxide activate locus coeruleus neurons through a cGMP-dependent protein kinase: involvement of a nonselective cationic channel. J Neurosci 1996;16:1389–1399.
35. Leinders-Zufall T, Shepherd GM, Zufall F. Regulation of cyclic nucleotide–gated channels and membrane excitability in olfactory receptor cells by carbon monoxide. J Neurophysiol 1995;74:1498–1508.
36. Carratu MR, Renna G, Giustino A, De et al. Changes in peripheral nervous system activity produced in rats by prenatal exposure to carbon monoxide. Arch Toxicol 1993;67:297–301.

13 The Molecular Mechanisms Underlying the Effects of Carbon Monoxide on Calcium-Activated K^+ Channels

Lingyun Wu

Contents

SUMMARY

Large-conductance calcium-activated K^+ (BK_{Ca}) channels actively participate in the regulation of membrane potentials. In vascular smooth muscle cells (VSMCs), the opening of BK_{Ca} channels provides a negative feedback in response to membrane depolarization and increased intracellular calcium. Consequent membrane hyperpolarization and closure of voltage-dependent calcium channels leads to relaxation of VSMCs. In this context, a better understanding of the modulatory mechanisms for K_{Ca} channels is critical. Many single-channel studies on the cell-free membrane patches have demonstrated the modulation of BK_{Ca} channels by CO. CO modified histidine residues of BK_{Ca} channel proteins, thus leading to the increased open probability of BK_{Ca} channels. The combination of chemical modification and mutational alteration of BK_{Ca} channels has unmasked a direct effect of CO on α-subunit of BK_{Ca} channels. Although CO increased the activity of heterologously expressed BK_{Ca} channels encoded by BK_{Ca}, α-subunit gene, nitric oxide (NO) failed to do so. Activation of native BK_{Ca} channels in VSMCs by maximally stimulating BK_{Ca}, β-subunit nullified the effect of NO, but not of CO, on BK_{Ca} channels. A better understanding of the pathophysiological impact of the altered interaction of CO and BK_{Ca} channels will provide novel mechanisms for the pathogenesis and maintenance of certain diseases.

Key Words: Carbon monoxide; gasotransmitters; BK_{Ca} channels; vascular smooth muscle cells.

From: *Signal Transduction and the Gasotransmitters: NO, CO, and H_2S in Biology and Medicine* Edited by: Rui Wang © Humana Press Inc., Totowa, NJ

1. INTRODUCTION

The opening of K^+ channels in vascular smooth muscle cells (VSMCs) leads to a K^+ efflux and hyperpolarizes cell membranes. Membrane hyperpolarization not only inactivates voltage-dependent Ca^{2+} channels but also inhibits the agonist-induced intracellular Ca^{2+} release, causing blood vessel relaxation. Conversely, the closure of K^+ channels causes vasoconstriction by depolarizing cell membrane. Many vasorelaxing factors can activate K^+ channels directly or indirectly. Carbon monoxide (CO) generated from the cleavage of the heme ring in hemoproteins *(1,2)* is a potent gasotransmitter *(3–7)*. Nitric oxide (NO) is an endogenous gasotransmitter *(2)* that is synthesized from the terminal guanidino nitrogen atoms of L-arginine by NO synthase (NOS) *(8)*. Growing evidence indicates that direct or indirect modulations of K^+ channels in VSMCs constitute important mechanisms for the vascular effects of CO and NO. This chapter focuses on the molecular basis of K_{Ca} channels in VSMCs and the modulation of K_{Ca} channels by CO.

2. FUNCTION AND MODULATION OF K_{Ca} CHANNELS

K_{Ca} channels are expressed in many excitable and nonexcitable cells and are heterogeneous in their molecular compositions. An increase in intracellular calcium concentration, either globally in the cytosol or locally as "calcium sparks," causes K_{Ca} channels to open, leading to vasorelaxation *(9)*. This negative feedback regulation opposes the vasoconstraction induced by an increase in intracellular calcium. K_{Ca} channels are divided into three subtypes according to their single-channel conductance with symmetrical $[K^+]$ across cell membrane: big-conductance (BK_{Ca}) (approx 250 pS), intermediate-conductance (IK_{Ca}) (20–80 pS), and small-conductance (SK_{Ca}) (10–15 pS) channels *(10)*. BK_{Ca} channels in VSMCs are blocked by micromolar external tetraethylammonium and more specifically by nanomolar charybdotoxin (ChTX) and iberiotoxin (IbTX) *(9)*. SK_{Ca} channels can generate a long-lasting hyperpolarization or the slow after-hyperpolarization after an action potential in most brain neurons. Apamin is a selective SK_{Ca} channel blocker having an IC_{50} of 0.3 n*M*. The apamin-sensitive SK_{Ca} channels have been found in the brain *(11)*, hepatocytes *(12)*, smooth muscle cells (SMCs) from the mouse ileum *(13)*, and SMCs in some porcine vascular beds *(14)*, indicating their possible contributions to the maintenance of intrinsic vascular tone. Human IK_{Ca} channels have recently been detected at high levels in lung, placenta, trachea and salivary gland, liver, bone marrow, and colon *(15)*. IK_{Ca} channels were also cloned from nonexcitable human B- and T-lymphocytes, indicating the possible association of IK_{Ca} channels with immune reactions *(16,17)*. Nevertheless, IK_{Ca} channels are undetectable in human brain, heart muscle, and aortic smooth muscles *(18)*. IK_{Ca} channels can be blocked by ChTX (IC_{50} = 28 n*M*), and clotrimazole (IC_{50} = 153 n*M*), but not by apamin *(18)*.

BK_{Ca} channels sense the changes in both intracellular calcium concentrations and membrane potentials, whereas IK_{Ca} and SK_{Ca} channels are voltage independent and gated only by cytoplasmic calcium. Many endogenous vasoactive substances regulate K_{Ca} channel activities. Vasoconstrictors such as angiotensin II and a thromboxane A_2 agonist (U46619) inhibit the channel opening *(18,19)*. Phosphorylation of channel proteins mediated by a cyclic adenosine monophosphate–dependent protein kinase (protein kinase A [PKA]) *(20)*, G-protein-coupled pathway *(21)*, or cyclic guanosine 5'-monophosphate (cGMP)-dependent protein kinase (PKG) *(22)* also activates K_{Ca} channels in vascular smooth muscles. Endothelium-derived hyperpolarizing factor (EDHF) is defined as a non-NO, nonprostaglandin substance that is released by acetylcholine (ACh) and induces smooth muscle hyperpolarization *(2,23)*. Although the candidacy of K^+ ion *(24)*, epoxyeicosatrienoic acid, or endocannabinoids *(25)* has been proposed, the real nature of EDHF has not yet been identified. EDHF-mediated vasorelaxation in many vascular beds is believed to be effected through the activation of K^+ channels, especially K_{Ca} channels, because EDHF effect can be abolished by ChTX *(26)* and by apamin *(27)* in rat mesenteric arteries. Recently, more and more studies have shown that a combination of ChTX and apamin abolishes the EDHF-induced vasorelaxation in small arteries, such as guinea pig coronary, carotid, and basilar arteries *(24,28,29)*. Whether the combination of ChTX and apamin targets on a novel K_{Ca} channel subtype or multi-K_{Ca} channels in VSMCs needs to be further elucidated.

3. MOLECULAR BASIS OF K_{CA} CHANNELS IN VSMCS

Among different types of K_{Ca} channels, BK_{Ca} channels are best described and characterized in various tissues including VSMCs. To provide more focused background information on BK_{Ca} channels, the molecular basis of IK_{Ca} and SK_{Ca} channels is not discussed in this chapter. BK_{Ca} channels are composed of two types of noncovalently linked subunits: the pore-forming α-subunit and the accessory β-subunit. The latter affects the electrophysiological and pharmacological properties of K_{Ca} channel complexes *(30)*. K_{Ca} α-subunit shares a great sequence homology with the pore-forming subunits of other types of K channels. The expression of α-subunit of BK_{Ca} channel alone yields the Ca^{2+}-independent current when the intracellular calcium concentration is <100 n*M*, turning the channels into a pure voltage-dependent pore *(31)*. The functional coupling of BK_{Ca} β-subunit with the α-subunit greatly increases the sensitivity to the cytoplasmic calcium and confers the inactivation properties of BK_{Ca} channels *(31–33)*. The single gene origin with a family of alternatively spliced variants can explain the wide difference in unitary conductance, calcium sensitivity, and gating of BK_{Ca} channels in different tissues and even within the same tissue *(34,35)*. BK_{Ca} α-subunits are first cloned from *Drosophila* (dSlo) *(36)* and later from mammalian animals: human (hSlo) *(37)*, rat (rSlo) (GenBank accession no. U55995), mouse (mSlo) *(38)*, and canine (cSlo) (GenBank accession no. U41001). BK_{Ca} channels were also cloned from rat brains, adrenal chromaffin cells, or rat insulinoma tumor (RINm5f) cells *(39)*. Human BK_{Ca} α-subunit (hSlo) is mapped to chromosome 10q23.1 *(37)*. The transmembrane segments (S1–S7) near the N-terminus of BK_{Ca} channel α-subunit have amino acid sequence similar to that of the voltage sensor and the pore domain of Kv channels. The charged residues (Arg) in the S4 transmembrane domain move outward when the cell membrane is depolarized and interact with negative residues in the S2 and S3 domains. In addition, BK_{Ca} channels have four extra hydrophobic domains (S7–S10) at the C-terminus, which are conservative among species *(36,38,40)*. The S9 and S10 regions are associated with intracellular calcium sensitivity and calcium binding *(41–43)*. A more recent topology model for BK_{Ca} channel α-subunit suggested an additional S0 transmembrane segment that leads to the extracellular location of the N-terminus and is associated with α-subunit regulation in mammalian BK_{Ca} gene *(44)*. The amino acid sequence of human BK_{Ca} gene (1113 amino acids) shares very high identity with that of rat (1178 amino acids) or mouse (1180 amino acids) (96–97%) but relates less to that of *Drosophila* (1175 amino acids) *(49%)*.

A conventional notion attributes the calcium-binding sites to a defined fragment of the intracellular C-terminus of the α-subunits of BK_{Ca} channels. This fragment of about 400 amino acids contains a conserved group of five consecutive negatively charged aspartic acid residues, called the "calcium bowl" *(43,45,46)*. $^{45}Ca^{2+}$-overlay protein blot assay demonstrated calcium binding to this C-tail *(46,47)*. The second region of the C-terminus, the RCK domain (Regulator of Conductance for K^+), may also account for calcium binding and sensing of BK_{Ca} channels *(48)*. Mutation of the calcium bowl and RCK domains eliminated the calcium sensitivity of BK_{Ca} channels *(48,49)*.

However, recently this conventional notion has been seriously challenged. Piskorowski and Aldrich *(50)* demonstrated that BK_{Ca} channels that lack the whole intracellular C-terminus retain wild-type calcium sensitivity. They concluded that the intracellular C-tail, including the "calcium bowl" and the RCK domain, is not necessary for the calcium-activated opening of these channels. In their opinion, "the intracellular C terminus may function as a domain that mediates or modulates the allosteric coupling between Ca^{2+} binding and opening conformational changes." Thus, the possibility is raised that the Ca^{2+} sensitivity of BK_{Ca} channels is not intrinsic to BK_{Ca} α-subunit. An unknown accessory protein may serve the calcium-binding role. To keep in line with the published cell-free single-channel recording results, a tight association between this soluble protein with BK_{Ca} α-subunit should be remain in cell-free patches.

At least four types of β-subunits that couple with BK_{Ca} channel α-subunits have been cloned in human tissues (β1–β4). The β-subunit family of BK_{Ca} channels regulates several critical aspects of channel phenotype such as inactivation and apparent Ca^{2+} sensitivity. Structurally, BK_{Ca} channel β-subunits are about 192–310 amino acid residues in length, having two transmembrance domains (TM1 and TM2) with a long extracellular loop in between and two N-linked glycosylation sites (**Fig. 1**). TM1 and TM2 are similar in amino acid sequence, suggesting a common structure. The β_1-subunit increases the apparent calcium sensitivity of K_{Ca} complex, and it is mainly expressed in vascular smooth muscle tissues *(31)*. Intracellular amino acid residues at the N-terminus of β-subunits are thought to have a "ball-like" structure that is vital for BK_{Ca} channel inactivation *(33)*. A recent study from MacKinnon's laboratory revealed the crystal structure of K_{Ca} channels from *Methanobacterium thermoautotrophicum (51)*. The investigators proposed that eight RCK domains form a gating ring at the intracellular membrane surface and that this ring uses the free energy of Ca^{2+} binding to change its shape and perform mechanical work to open the pore and permit ion conduction.

Although the molecular structure of K_{Ca} channel complex has been greatly unmasked in recent years, researchers' understanding of the molecular composition of K_{Ca} channels in VSMCs still is very limited. K_{Ca} channel α-subunit has not been cloned from VSMCs, and it is not known whether the different isoforms of BK_{Ca} channels exist in VSMCs. The interaction of α- and β-subunits in VSMCs is also unclear. For example, researchers, knowing that the putative calcium-binding sites are not on the β-subunit, are still puzzled by the mechanisms by which β-subunit alters the calcium sensitivity and voltage dependence of BK_{Ca} channels.

4. MODULATION OF BK_{Ca} CHANNELS BY CO

Heme oxygenase (HO) works in concert with NADPH-cytochrome P450 reductase to cleave the heme ring in hemoproteins into CO, biliverdin, and iron *(3)*. Results from our studies *(1,4)* and others have demonstrated that endogenously generated CO can effectively relax vascular tissues. Like NO, CO acts on at least two cellular targets in VSMCs: soluble guanylyl cyclase and BK_{Ca} channels *(3,52–56)*. Although the debate has not been

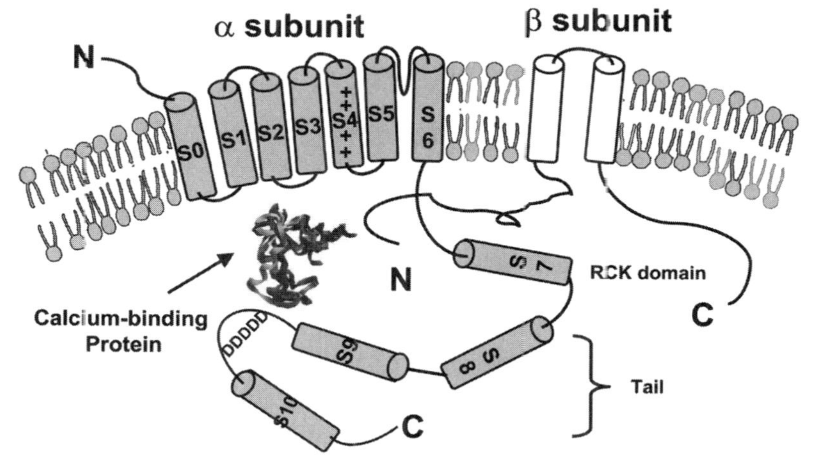

Fig. 1. Schematic topology of K_{Ca} channels. The monomer of BK_{Ca} channel α-subunit has seven putative transmembrane domains (S0–S6) in which the charged residues in S4 are crucial for the responsiveness to voltage changes across the membrane. BK_{Ca} channel α-subunits also have four putative cytoplasmic hydrophobic domains (S7–S10) that are responsible for intracellular Ca^{2+} sensing. BK_{Ca} channel β-subunit has two putative transmembrane domains with a long extracellular loop that may interact with the pore regain of the α-subunits. The N-terminal end of the-subunit has a "ball-like" structure that can regulate the gating property of the channels by pluging or unpluging the cytoplasmic entrance of the channels. The interaction among the N-terminal end, S0, and-subunit is not shown. On the other hand, IK_{Ca}/SK_{Ca} channels interact with cytoplasmic calmodulin instead of specific β-subunits. D, aspartic acid; RCK, Regulator of Conductance for K.

settled on whether CO and NO act on BK_{Ca} channels directly or indirectly by altering intracellular second messengers, many single-channel studies on cell-free membrane patches have been supportive of a direct interaction of CO and NO with K_{Ca} channel proteins. On the other hand, CO and NO stimulate BK_{Ca} channels through quite different molecular mechanisms.

4.1. Interaction of CO and Different Amino Acid Residues of BK_{Ca} Channels

Although the primary structure of K_{Ca} channels in VSMCs is relatively known, the contributions of various amino acids to the gating and conducting of K_{Ca} channels are not yet clear. Modification of one or more amino acid residues may suffice to change the conductance and/or the open probability of K_{Ca} channels *(53,54,57,58)*. As the molecules of gas, CO and NO can freely penetrate plasma membrane and interact with the membrane-spanning K_{Ca} channel proteins. Therefore, the possibility exists that CO and NO may directly interact with certain amino acids of BK_{Ca} channels to alter BK_{Ca} channel behavior. The chemical interaction of CO with cellular proteins has been known. For instance, CO forms hydrogen bonds with the distal histidine residue (His^{64}) in myoglobin *(59)* or histidine 25 in HO *(60)*. Similarly, modification of sulfhydryl groups in cysteine by NO has been shown to inhibit the conduction of an anion-selective channel in oocyte *(61)*.

The following sections discuss the contributions of different amino acid residues to the biophysical properties of BK_{Ca} channels, the interaction of CO with different amino acid residues of BK_{Ca} channels, and the experimental evidence to corroborate these conclusions. For more discussions on the chemical mediation of K_{Ca} channels by NO, readers are referred to Chapter 6.

4.1.1. Role of Histidine Residues of BK_{Ca} Channels

Diethylpyrocarbonate (DEPC) is a specific chemical reagent that reacts with histidine. It is also not membrane permeable. Application of DEPC to different sides of plasma membrane can provide information on the topological difference of histidine residues as evaluated by specific functional assays (such as the patch-clamp technique), if these histidine residues contribute significantly to the protein function (such as the gating or conductance of ion channels). My colleagues and I *(54)* found that exposure of the cytoplasmic surface of cell membranes to DEPC (0.5 m*M*) did not affect the characteristics of BK_{Ca} channels in cell-free membrane patches from rat tail artery SMCs. By contrast, when DEPC was applied to the external surface of cell membranes, the open probability of BK_{Ca} channels, but not single-channel conductance, was reduced. This inhibitory effect of DEPC, if specific for histidine residues, should be a function of pH because DEPC reacts only with the unprotonated imidazole ring. At pH 6.3, a 46% inhibition of the open probability (NPo) of BK_{Ca} channels by DEPC was observed. At pH 5.2, the NPo of K_{Ca} channels only slightly decreased by DEPC treatment (6%). The pH dependence of the effect of DEPC indicated the specific modification of histidine residues. A kinetic analysis of the effect of DEPC on BK_{Ca} channels revealed that the NPo of BK_{Ca} channels began to decrease 1 min after the application of DEPC and decreased by 50% 4 min after DEPC treatment. The decrease in the NPo of BK_{Ca} channels by DEPC was also concentration dependent, following pseudo-first-order kinetics. The reaction order obtained from the slope of the double logarithmic plot was 1.0, indicating that one histidine residue per channel protein might be involved in the modifying effect of DEPC *(54)*.

Knowing that histidine residues may participate in the regulation of open probability of BK_{Ca} channels, my colleagues and I *(54)* continued to investigate the interaction of CO and DEPC on BK_{Ca} channels. We found that DEPC treatment abolished the CO-induced increase in the NPo of single BK_{Ca} channels in outside-out membrane patches, but not in inside-out patches. The inactivated DEPC did not affect the stimulatory effect of CO on K_{Ca} channels, whether DEPC was applied on the internal or outer surface of membrane patches. These results suggest that CO induced a chemical modification of histidine residue located on the extracellular surface of the transmembrane BK_{Ca} channel proteins in VSMCs *(54)*. We also applied hydroxylamine to remove DEPC from imidazoles after treatment of the membrane patch with DEPC. In one case, we found that the inhibitory effect of DEPC on the CO-induced modification of single BK_{Ca} channel currents was removed in the presence of hydroxylamine. An interaction between CO and DEPC on histidine residue was also demonstrated by the CO-induced protection of BK_{Ca} channels from inhibition by DEPC.

To confirm further the involvement of histidine residues in the modifying effect of CO on BK_{Ca} channels, the cells were exposed to illuminated Rose bengal, a treatment specifically for inducing photo-oxidation of histidine residues. Again, the extracellularly applied Rose bengal abolished the CO-induced activation of K_{Ca} channels. However, CO still significantly increased the NPo of BK_{Ca} channels in outside-out patches isolated from cells that were either preincubated with Rose bengal in the absence of light or exposed to illumination in the absence of the dye for 15 min. These results ruled out possible nonspecific damage of K_{Ca} channels induced by nonilluminated dye or by photoinactivation of the BK_{Ca} channels *(54)*.

It appears that the CO-induced increase in the open probability of BK_{Ca} channels in VSMCs may be specifically altered by chemical modification of histidine residues. However, chemical modification of the externally located histidine residue may not be the sole mechanism for the stimulation of BK_{Ca} channels by CO. How does outer side-located histidine residue link to the calcium sensitivity of BK_{Ca} channels? Our previous study *(54)* also showed that CO significantly increased the calcium sensitivity of BK_{Ca} channels. In the presence of 10 μM CO, the $[Ca^{2+}]_i$ open probability of BK_{Ca} curve moved to the left with the Hill coefficients increased from 0.88 to 1.70 *(55)*. As discussed earlier, the Ca^{2+}-binding sites are located on the cytoplasmic side of BK_{Ca} channel proteins. Obviously, external histidine residue is not involved in calcium binding to BK_{Ca} channels. It is possible that the externally located histidine residue is the modulator site. Chemical modification of this residue will induce necessary conformational changes in the channel-forming protein, either changing the apparent calcium affinity of the existed binding sites or unmasking new binding sites. Alternatively, CO may have two acting sites. Modification of the externally located histidine residue changes the gating mechanism, and modification of the cytoplasmically located amino acid residues changes the calcium sensitivity of K_{Ca} channels. Changes in the cytoplasmically located amino acid residues by CO alone may not suffice to induce significant changes in BK_{Ca} channel openings if the external histidine residue has not been modified. In this case, the interaction between CO and the external histidine residue plays a permissive role in the effect of CO on BK_{Ca} channels. Increasing the concentration of CO would simply increase the probability that the modulator sites or Ca^{2+}-binding sites would be modified.

It is also worth noting that there are apparent differences between the effects of CO and DEPC on BK_{Ca} channels. DEPC decreased the NPo of BK_{Ca} channels in a relatively irreversible manner because this reagent is involved in the covalent modification of histidine. On the other hand, CO increased the NPo in a reversible fashion probably because of a relatively weak reaction between CO and the imidazole group of histidine via hydrogen bonds.

4.1.2. Role of Sulfhydryl Groups of BK_{Ca} Channels

The ion permeation and gating of BK_{Ca} channels are controlled by multiple molecular switches. The interaction of CO and amino acid residues other than histidines of BK_{Ca} channels might also play important roles in the regulation of channel function. Sulfhydryl groups of cysteinyl residues of peptides and proteins generally are the most reactive of all amino acid side-chain functionalities under normal physiological conditions. Among the irreversible modifying agents for the sulfydryl groups, *N*-ethylmaleimide (NEM) has been mostly used *(62)*. Exposure of cytoplasmic surface, but not outer surface, of cell membranes to NEM (5 m*M*) significantly decreased the open frequency of single BK_{Ca} channels by 72%. The opening probability of single BK_{Ca} channels and multiple openings of BK_{Ca} channels were significantly reduced *(63)*. However, the single-channel conductance of BK_{Ca} channels in rat tail artery SMCs was not affected by NEM treatment *(63)*. Our study thus provides evidence for the structural or functional involvement of cysteines (sulfydryl group) in BK_{Ca} channels of rat tail artery SMCs, which is consistent with many reports in the literature *(53,64,65)*. For example, NEM included in the pipet solution was found to decrease mean open time by 15-fold and slightly reduced single-channel conductance of single ACh receptor channels of BC3H-11 cells in the "cell-attached" configuration *(65)*.

Lee et al. *(64)* showed that NEM produced an irreversible inhibition of K_{ATP} channel activity when applied to the intracellular surface of excised inside-out patches. Bolotina et al. *(53)* reported that pretreatment of membrane patches with NEM prevented NO-induced activation of BK_{Ca} channels. Interestingly, application of NEM to the extracellular or intracellular side of membrane patches did not alter the effect of CO on the open probability of single BK_{Ca} channels *(63)*. These observations suggest that modification of BK_{Ca} channels by NO or CO may have quite different mechanisms. Although cysteine residues are the target of NO, the effect of CO on BK_{Ca} channels is not mediated by sulfhydryl groups of the channel protein *(63)*.

4.1.3. Role of ε-Amino Groups of BK_{Ca} Channels

An earlier study on GH3 cells showed that chemical modification of the ε-amino groups located on the cytoplasmic surface of membrane using trinitrobenzenesulfonic acid (TNBS) irreversibly increased K_{Ca} channel open probability without affecting single-channel conductance and voltage sensitivity. The increase in open probability is predominantly because of the loss of long-duration closures of the channel, and the increase in the lengths of long-duration openings *(43)*. TNBS was believed to neutralize irreversibly the peptide terminal amino groups and the ε-amino group of lysine *(15,66)*. By contrast, pretreatment of the outer surface of membrane patches of rat tail artery SMCs with TNBS (3 m*M*) for 5 min did not alter the conductance or open probability of BK_{Ca} channels *(63)*. TNBS treatment did not prevent CO (30 μ*M*) from effectively increasing BK_{Ca} channel activity *(63)*. The discrepancy among these reports, regarding the effects of TNBS on BK_{Ca} channels, likely results from differences in the species and cell types, because even similar BK_{Ca} channels from different species may possess different molecular structures. Furthermore, the specificity of the effect of TNBS on lysine residues of BK_{Ca} channels at different concentrations and pH should be determined. Our study based on the interaction of TNBS and CO suggests, but does not conclude, that the modification of BK_{Ca} channels by CO is unlikely because of the modification of internally located lysine residues of the channel protein *(63)*.

Whether extracellularly applied TNBS can affect the biophysical properties of BK_{Ca} channels as well as the effect of CO on BK_{Ca} channels in VSMCs has not been reported. In giant axons of squid, externally applied TNBS has been shown to modify delayed rectifier K channels. After TNBS treatment, the kinetics of macroscopic ionic currents were slowed, the size of currents at large positive voltages were increased, the voltage-dependent probability of channel openings at more positive potentials was shifted, and K channel gating currents altered *(67)*.

4.1.4. Role of Carboxyl Groups of BK_{Ca} Channels

Carboxyl groups located on the cytoplasmic domain of the channel protein may play a role in calcium binding and sensing for BK_{Ca} channels *(68)*. To knock out carboxyl groups or modify them chemically would prove a means to examine the mechanisms for the regulation of calcium sensitivity of BK_{Ca} channels. Trimethyloxonium (TMO) is a highly reactive agent that specifically esterifies carboxyl groups, such as those contained in aspartic or glutamic acid residues *(69)*. A normally negatively charged hydrophilic group will be converted in the presence of TMO to neutral methyl esters, a more hydrophobic residue. It is worth noting that the hydrolyzation of TMO is very rapid and that an accurate concentration of TMO reaching the cell membrane is difficult to obtain. Thus, TMO powder should always be added to the reaction solution immediately before superfusing the membrane patches. The reduced single-channel conductance of K_{Ca} channels *(54)* as well as voltage-gated Na channels (70,71) by TMO has been reported. TMO also decreased the whole-cell ACh-activated channel current by 60% and the cell-attached single-channel conductance by 10% in a muscle cell line *(69)*. In VSMCs, TMO (50 m*M*) applied to the outer surface of membrane patches reduced the unit current amplitude of single BK_{Ca} channels *(63)*. The slope conductance of BK_{Ca} channels was reduced from 256 to 171 pS after TMO treatment. These changes in the biophysical properties of BK_{Ca} channels were not observed after TMO was applied to the cytoplasmic surface of membrane patches *(63)*. These results suggest that the BK_{Ca} channel has carboxyl groups on its extracellular surface and that those carboxyl groups may have a profound influence on channel conductance.

Extracellularly applied TMO had no effect on but ChTX decreased the open probability of single BK_{Ca} channels *(63)*. However, the presence of TMO significantly decreased the blocking effect of ChTX on K_{Ca} channels. MacKinnon and Miller *(58)* reported a similar effect of TMO and also demonstrated that TMO treatment did not change the calcium sensitivity of K_{Ca} channels. ChTX is a highly basic peptide with eight positively charged residues (four lysines, three arginines, and one histidine), but only two negatively charged groups. ChTX blocks K_{Ca} channels by physically occluding the pore and preventing ion conduction *(30)*. Carboxyl groups themselves may not be involved in the gating mechanisms of K_{Ca} channels. However, modification of carboxyl groups by TMO may change the electrostatic forces and thus reduce the affinity of ChTX to K_{Ca} channels.

Different from the interaction of TMO and ChTX on BK_{Ca} channels, TMO pretreatment of outer surface membrane failed to alter CO-induced increases in the open probability of K_{Ca} channels *(63)*. This may be explained if the changes in electrostratic force induced by TMO treatment do not affect the chemical interaction of CO and histidine residues of BK_{Ca} channels.

Our studies provide evidence for the presence of essential sulfhydryl, basic amino acid, and carboxyl groups associated with the normal function of BK_{Ca} channels in rat tail artery SMCs *(63)*. The topography of these amino acid residues is important in terms of the functioning behavior of K_{Ca} channels, including gating mechanism, ion conduction, calcium binding, and voltage and calcium sensitivities. The gating mechanism of BK_{Ca} channels is under the influence of multiple amino acid residues, including histidine and cysteine, although CO may act only on histidine residues *(54)*. The carboxylic acid moieties may actively participate in the control of ion permeation.

The interpretation of data derived from chemical modification of different amino acids has called for caution mainly from the view of specificity. Although modification of BK_{Ca} channels by different reagents is obvious and relatively specific in normal reaction solution, the ionic strength and pH of reaction solutions should be taken into account in the future to better characterize the effect of CO and other chemical reagents on ion channels *(58,69,72)*. Chemical modification of BK_{Ca} channels should also be combined with genomic methods to obtain different mutants of BK_{Ca} channel genes with specifically deleted or replaced target amino acid residues. The altered effect of CO on these mutated BK_{Ca} channels will reveal more specifically the targets of CO on BK_{Ca} channel proteins.

4.2. Different Effects of CO and NO on α- and β-Subunits of BK_{Ca} Channels

Using reverse transcriptase polymerase chain reaction (RT-PCR) analysis, my colleagues and I detected the expression of α- and $β_1$-subunits of K_{Ca} channels in rat vascular tissues (**Fig. 2**). The presence of both α- and β-subunits as well as their functional and

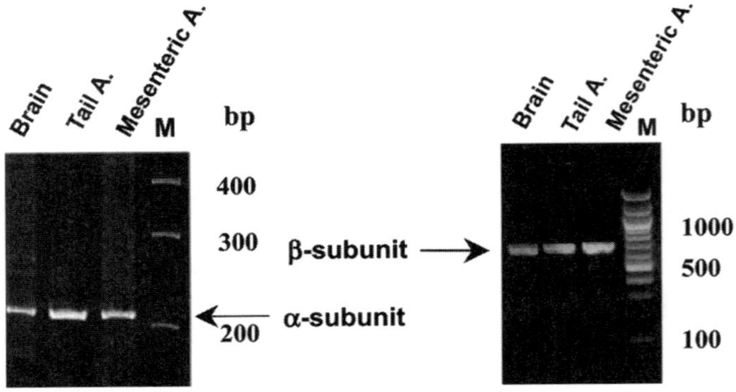

Fig. 2. RT-PCR analysis of expression of α-subunit (212 bp) and β-subunit (691 bp) of BK_{Ca} channels in rat brain and vascular tissues. M, molecular marker; bp, base pairs; A, artery.

structural coupling in VSMCs determine the functionality of native BK_{Ca} channel complex and their response to different modulators. We recently demonstrated that the excitatory effects of CO and NO on BK_{Ca} channels are likely mediated respectively by α- and β-subunits *(73)*. Our results provided a molecular basis for understanding the acute interaction of CO and NO, two endogenous gasotransmitters *(2)*, on the regulation of vascular contractility.

4.2.1. Stimulation of α-Subunit, But Not β-Subunit, of BK_{Ca} Channels in VSMCs by CO

4.2.1.1. Pharmacological Evidence. Dehydrosoyasaponin-I (DHS) is a glycosylated triterpene that acts on BK_{Ca} β-subunit in smooth muscles *(74,75)*. This compound is not membrane permeable. In HEK-293 cells expressing only K_{Ca} α-subunit, DHS failed to inhibit the IbTX binding to plasma membranes *(76)*. However, DHS inhibited IbTX binding to HEK-293 cells expressing both K_{Ca} α- and K_{Ca} β-subunits *(76)*. In a study on VSMCs, DHS concentration dependently increased the NPo of BK_{Ca} channels when applied to the inner membrane surface, but not to the outer surface. After the maximum excitatory effect of DHS was obtained, the same membrane patches were exposed to CO. A further increase in the NPo of BK_{Ca} channels induced by CO was observed *(73)*. The additive effects of DHS and CO on native BK_{Ca} channels indicate that CO might affect BK_{Ca} channels in a β-subunit-independent mechanism.

4.2.1.2. Evidence of β-Subunit Nullification. In a study by my colleagues and I *(73)*, VSMCs were treated with anti-BK_{Ca} β antisense ODN. The specific phosphorothioate antisense oligdeoxynucleotide (ODN) hybridizes the translation initiation codon of BK_{Ca} β-subunit mRNA to eliminate the participation of β-subunit in the assembly of native K_{Ca} channel complex in rat tail artery SMCs. The use of antisense ODN in combination with electrophysiology measurement offers a useful approach for evaluating the contribution of different genes to the native ionic currents. This antisense treatment did not alter single-channel conductance of native BK_{Ca} channels in VSMCs. However, the voltage dependence and calcium sensitivity of native BK_{Ca} channels were significantly reduced. The calcium sensitivity of the native BK_{Ca} channels after BK_{Ca} β antisense ODN treatment is similar to that of BK_{Ca} channels composed of α-subunit alone *(75)*. Moreover, the stimulatory effect of DHS on the native BK_{Ca} channels disappeared after anti-K_{Ca} β antisense ODN treatment. On the other hand, CO effectively stimulated the native BK_{Ca} channel currents after treatment with BK_{Ca} β antisense ODN. This result further shows that the presence of BK_{Ca} β-subunit is not required for the interaction of CO and BK_{Ca} channels *(73)*.

4.2.1.3. Evidence of Heterogous α-Subunit Expression. COS-1 cells transfected with mSlo cDNAs, which encode BK_{Ca} α-subunit, exhibited apparent BK_{Ca} currents. CO significantly increased BK_{Ca} current density especially at membrane potentials positive to +20 mV *(73)*.

4.2.2. Stimulation of β-Subunit of BK_{Ca} Channels in VSMCs by NO

NO has been known for its ability to stimulate BK_{Ca} channels in VSMCs *(53,73)*. A recent study shows that the effects on BK_{Ca} channels of NO rely on the presence of β-subunit *(73)*. This conclusion is supported by three lines of evidence. Pharmacological evidence demonstrated that after the native BK_{Ca} channels in VSMCs were maximally stimulated by DHS, sodium nitroprusside (SNP) (an NO donor) did not further stimulate BK_{Ca} channels. Evidence of β-subunit nullification revealed that the stimulatory effects of SNP and (±)*S*-nitroso-*N*-acetylpenicillamine (another NO donor) on the open probability of BK_{Ca} channels in VSMCs were abolished by treatment with anti-K_{Ca} β-antisense ODN. Evidence of heterologous-subunit expression demonstrated that SNP did not affect BK_{Ca} currents in COS-1 cells expressing only mSlo cDNA *(73)*. Then these data are taken together, it is reasoned that the effect of NO may rely on the presence of K_{Ca} β-subunit.

A higher concentration of CO was needed to stimulate BK_{Ca} channels to achieve a similar excitation induced by lower concentrations of NO, consistent with different vasorelaxant potencies of these two gases. This could be partially explained by the difference in the chemical interactions of CO and NO with K_{Ca} α-subunit and K_{Ca} β-subunit, respectively. A relatively weak reaction between CO and the imidazole group of histidine occurs by the formation of a hydrogen bond, which could be largely responsible for the effect of CO on K_{Ca} channel *(23)*. A much weaker hydrogen bond, compared with the disulfide bond, may result in a greater dissociation rate between CO and its target than between NO and K_{Ca} channel. This speculation merits further investigation.

4.2.3. Interaction of CO and NO on BK_{Ca} Channels in VSMCs

The aforementioned data present a scenario in which CO and NO act on different components of BK_{Ca} channels. Because the functionality of α- and β-subunits of BK_{Ca} channels would affect the overall behavior of the channel complex, it is highly possible that the effects of CO and NO on different subunits may be integrated. It has been shown that the K_{Ca} channel-mediated vasorelaxing effect of CO was significantly decreased by approx 40% in the presence of SNP *(17)*. Is it possible that NO desensitizes BK_{Ca} channels toward CO or vice versa? Recently, my colleagues and I *(73)* demonstrated that the integrated effect of CO and NO on native BK_{Ca} channels in SMCs was not a simple algebraic addition of their individual effects. The pretreatment of cell membranes with SNP abolished the stimulatory effect of CO on the open probability of BK_{Ca} channels *(73)*. When the application order was reversed—i.e., pretreatment of the cell membranes with CO and then applying NO—the stimulatory effect of NO was superimposed on that of CO. Based on these observations, we hypothesize that the CO-sensitive α-subunit of BK_{Ca} channels might also be modified by NO so that the α-subunit became desensitized to CO. If this were the case, not only CO but also DEPC would have no effect on K_{Ca}

channels after NO treatment. In following experiments, we first stimulated the native BK_{Ca} channels in VSMCs with SNP and then applied DEPC to modify outer surface-located histidine residues. Interestingly, the prestimulation with SNP abolished the inhibitory effect of DEPC on BK_{Ca} channels *(73)*. Prestimulation with DHS did not prevent DEPC from inhibiting the open probability of BK_{Ca} channels. It seems that NO has two distinct acting sites. By acting on the inner surface-located DHS-sensitive site (likely on β-subunit), NO increased K_{Ca} channel openings. By acting on the outer surface-located DEPC-sensitive histidine residue, NO reduced K_{Ca} channel sensitivity to the agents that targeted on histidine. Interestingly, after the outer membrane surface was first treated with DEPC, the subsequently applied NO or DHS (inner surface) still increased the NPo of K_{Ca} channels. These results suggested that likely the DEPC-sensitive histidine is located on K_{Ca} α-subunit and that its structural change would not affect the K_{Ca} β-subunit-dependent excitatory effects of NO or DHS.

The molecular nature of the interaction of NO and histidine residue of BK_{Ca} α-subunit is not clear. In reaction with superoxide anions, NO would transform to peroxynitrite, which leads to the oxidation of many proteins. However, this mechanism could not explain the observed NO effect because superoxide dismutase did not change the effect of NO on BK_{Ca} channels, nor did it change the inhibitory effect of NO on the CO-induced increase in NPo. Alternatively, the action of NO on K_{Ca} β-subunit may exert an allosteric effect on the neighboring K_{Ca} α-subunit. The noncovalent association of α- and β-subunits of K_{Ca} channels is tight with a distance in between of <12 Å *(77)*. The conformational changes in K_{Ca} β-subunit would prevent the formation of hydrogen bonds between CO and the imidazole group of histidine of K_{Ca} α-subunit *(54)*. This putative "repulsive" mechanism is reminiscent of the effect of NO on the activation of soluble guanylyl cyclase. On binding to soluble guanylyl cyclase, NO induces the displacement of the iron atom, and this repulsive *trans* effect prevents the access and binding of the proximal histidine of soluble guanylyl cyclase to the heme iron *(78)*.

Vascular tone may be under the tonic influence of CO (ascribed to the long life-span and low vasorelaxant potency of the gas) and regulated by the phasic effect of NO, which has a short half-life and greater vasorelaxant potency. It can be perceived based on studies by my colleagues and I and others that the persistent presence of CO in the surroundings of VSMCs would not affect the sensitivity of BK_{Ca} channels to NO so that the vascular contractility could be acutely and precisely regulated by the NO surge. In cases in which a prolonged NO exposure at normal or higher concentrations is encountered such as in ischemia-reperfusion damage, septicemia, or other pathophysiological situations, the tonic relaxant influence of CO on vascular tone would be reduced because of its decreased effect on BK_{Ca} channels in VSMCs. Thus, the interaction between CO and NO would provide a feedback mechanism to help regain the control of the disturbed vascular contractility. Elucidation of the interaction of CO and NO on the function of BK_{Ca} channels in VSMCs was a very important progress in our understanding of the integrated fine-tuning of cardiovascular functions by gaseous vasoactive factors.

5. CONCLUSION

This chapter has addressed molecular mechanisms for the individual and integrated effects of CO and NO on BK_{Ca} channels in VSMCs. Because the majority of the results reviewed herein are derived from cell-free patch-clamp studies in which the involvement of second messengers was minimized, direct actions of CO and NO on BK_{Ca} are deduced. The direct modification of BK_{Ca} channels by CO is mediated by structural changes of histidine residue in α-subunit, whereas the NO effect on BK_{Ca} channels may closely be linked to the structural changes of cysteine residues on β-subunit. It is worth noting that CO and NO may also affect BK_{Ca} channels indirectly in VSMCs. The NO-induced activation of BK_{Ca} channels mediated by cGMP-dependent mechanisms has been reported *(77,79)*. No study has reported a cGMP-dependent effect of CO on BK_{Ca} channels, to our knowledge. The direct and indirect effects of CO and NO on BK_{Ca} channels may not be necessarily exclusive to each other. The differences in experimental conditions and cell types from different vascular beds may underline different action modes of CO or NO on BK_{Ca} channels *(77)*.

Cell-type-specific effects of CO on BK_{Ca} channels should be further investigated. The complexity and heterogeneity of BK_{Ca} channels in different types of VSMCs determine that CO's effects on BK_{Ca} channels may be cell-type dependent. Different subtypes or isoforms of BK_{Ca} α-subunits may account for large variation in calcium sensitivity *(78)*. Different coupling affinities or ratios between K_{Ca} α- and K_{Ca} β-subunits have also been acknowledged *(80)*. Although the α/β complex form of K_{Ca} channels was predominant, K_{Ca} channels comprising α-subunit alone also existed *(74)*. Furthermore, tissue distribution patterns of α- and β-subunit did not always match, and in certain types of tissues they might not even couple *(79)*. These variances determine that CO and NO may have different effects on BK_{Ca} channels in different types of SMCs, which might provide a feedback through different mechanisms to control or restore the disturbed vascular contractility.

It is expected that in the future the mechanisms for the CO-enhanced calcium sensitivity of BK_{Ca} channels will be unmasked. Elucidation of the role of calcium-binding sites of BK_{Ca} channels in the effect of CO by using point-mutated BK_{Ca} α-subunit is one important approach. Exploration of the possible regulation of BK_{Ca} channels by some cytosolic calcium-binding proteins, such as calmodulin, is another approach. What if CO interacts with this kind of cellular protein and then alters the calcium sensitivity of BK_{Ca} channels? The physical coupling of IK_{Ca} and SK_{Ca} channels and calmodulin has been demonstrated using yeast two-hybrid and coimmunoprecipitation assays *(81,82)*. Even for BK_{Ca} channels, the application of calmodulin-binding peptides has been shown to potently inhibit the channel activity *(83)*. This inhibition is not dependent on the presence of cytosolic free calcium. As per the potential interaction of CO and calmodulin, it has been reported that one calmodulin binds four molecules of heme-CO with an average affinity of 1 μ*M* *(84)*.

The specific interaction of CO and α-subunit of BK_{Ca} channels would be better defined by studying the effects of CO and NO on BK_{Ca} channels of VSMCs from β-subunit knockout mice *(85,86)*. However, caution should be taken in extrapolating the results from these knockout mice to rats or humans because the molecular composition and regulation of BK_{Ca} channels may vary in different species.

ACKNOWLEDGMENTS

This work was supported by a research operating grant from the Canadian Institutes of Health Research (CIHR). The author is a New Investigator of CIHR.

REFERENCES

1. Wang R. Resurgence of carbon monoxide: an endogenous gaseous vasorelaxing factor. Can J Physiol Pharmacol 1998;76:1–15.
2. Wang R. Two's company, three's a crowd: can H_2S be the third endogenous gaseous transmitter? FASEB J 2002;16:1792–1798.

3. Wang R, Wang ZZ, Wu L. Carbon monoxide–induced vasorelaxation and the underlying mechanisms. Br J Pharmacol 1997;121:927–934.
4. Durante W, Peyton KJ, Schafer AI. Platelet-derived growth factor stimulates heme oxygenase-1 gene expression and carbon monoxide production in vascular smooth muscle cells. Arterioscler Thromb Vasc Biol 1999;19:2666–2672.
5. Leffler CW, Nasjletti A, Johnson RA, et al. Contributions of prostacyclin and nitric oxide to carbon monoxide–induced cerebrovascular dilation in piglets. Am J Physiol 2001;280:H1490–H1495.
6. Motterlini R, Gonzales A, Foresti R, et al. Heme oxygenase-1-derived carbon monoxide contributes to the suppression of acute hypertensive responses in vivo. Circ Res 1998;83:568–577.
7. Sabaawy HE, Zhang F, Nguyen X, et al. Human heme oxygenase-1 gene transfer lowers blood pressure and promotes growth in spontaneously hypertensive rats. Hypertension 2001;38:210–215.
8. Furchgott RF, Zawadski JV. The obligatory role of endothelial cells in the relaxation of arterial smooth muscle by acetylcholine. Nature 1980;228:373–376.
9. Brayden JE, Nelson MT. Regulation of arterial tone by activation of calcium-dependent potassium channels. Science 1992;256:532–535.
10. Benham CD, Bolton TB, Lang RJ, et al. Calcium-activated potassium channels in single smooth muscle cells of rabbit jejunum and guinea-pig mesenteric artery. J Physiol (Lond) 1986;371:45–67.
11. Köhler M, Hirschberg B, Bond CT, et al. Small-conductance, calcium-activated potassium channels from mammalian brain. Science 1996;273:1709–1714.
12. Castle NA, Haylett DG, Morgan JM, et al. Dequalinium, a potent inhibitor of apamin-sensitive K+ channels in hepatocytes and of nicotinic responses in skeletal muscle. Eur J Pharmacol 1993;(19)236: 201–207.
13. Vogalis F, Goyal RK. Activation of small conductance Ca(2+)-dependent K+ channels by purinergic agonists in smooth muscle cells of the mouse ileum. J Physiol 1997;1:502(pt 3):497–508.
14. Sokol PT, Hu W, Yi L, et al. Cloning of an apamin binding protein of vascular smooth muscle. J Protein Chem 1994;13:117–128.
15. Jensen BS, Strobaek D, Christophersen P, et al. Characterization of the cloned human intermediate-conductance Ca^{2+}-activated K+ channel. Am J Physiol 1998;275(3 pt 1):C848–C856.
16. Grissmer S, Nguyen AN, Cahalan MD. Calcium-activated potassium channels in resting and activated human T lymphocytes. J Gen Physiol 1993;102:601–630.
17. Partiseti M, Choquet D, Diu A, et al. Differential regulation of voltage- and calcium-activated potassium channels in human B lymphocytes. J Immunol 1992;148:3361–3368.
18. Scornik FS, Toro L. U46619:a thromboxane A_2 agonist, inhibits K_{Ca} channel activity from pig coronary artery. Am J Physiol 1992;262:C708–C713.
19. Toro L, Amador M, Stefani E. ANG II inhibits calcium-activated potassium channels from coronary smooth muscle in lipid bilayers. Am J Physiol 1990;258:H912–H915.
20. Sadoshima J, Akaike N, Kanaide H, et al. Cyclic AMP modulates Ca-activated K channel in cultured smooth muscle cells of rat aortas. Am J Physiol 1988;255(4 pt 2):H754–H759.
21. Kume H, Graziano MP, Kotlikoff MI. Stimulatory and inhibitory regulation of calcium-activated potassium channels by guanine nucleotide–binding proteins. Proc Natl Acad Sci USA 1992;89:11,051–11,055.
22. Taniguchi J, Furukawa KI, Shigekawa M. Maxi K+ channels are stimulated by cyclic guanosine monophosphate–dependent protein kinase in canine coronary artery smooth muscle cells. Pflugers Arch 1993;423:167–172.
23. Garland CJ, Plane F, Kemp BK, et al. Endothelium-dependent hyperpolarization: a role in the control of vascular tone [review]. Trends Pharmacol Sci 1995;16:23–30. (erratum in Trends Pharmacol Sci 1995;16:177).
24. Edwards G, Dora KA, Gardener MJ, et al. K^+ is an endothelium-derived hyperpolarizing factor in rat arteries. Nature 1998;396:269–272.
25. Mombouli JV, Vanhoutte PM. Endothelium-derived hyperpolarizing factor(s): updating the unknown. Trends Pharmacol Sci 1997;18:252–256.
26. Hwa JJ, Ghibaudi L, Williams P, et al. Comparison of acetylcholine-dependent relaxation in large and small arteries of rat mesenteric vascular bed. Am J Physiol 1994;266(3 pt 2):H952–H958.
27. Chen G, Cheung DW. Effect of K^+-channel blockers on ACh-induced hyperpolarization and relaxation in mesenteric arteries. Am J Physiol 1997;272:H2306–H2312.
28. Corriu C, Feletou M, Canet E, et al. Endothelium-derived factors and hyperpolarization of the carotid artery of the guinea-pig. Br J Pharmacol 1996;119:959–964.
29. Yamanaka A, Ishikawa T, Goto K. Characterization of endothelium-dependent relaxation independent of NO and prostaglandins in guinea pig coronary artery. J Pharmacol Exp Ther 1998;285:480–489.

30. Munujos P, Knaus H, Kaczorowski GJ, et al. Cross-linking of charybdotoxin to high-conductance calcium-activated potassium channels: identification of the covalently modified toxin residue. Biochemistry 1995;34:10,771–10,776.
31. Meera P, Wallner M, Jiang Z, et al. A calcium switch for the functional coupling between alpha (hslo) and beta subunits (KV,Ca beta) of maxi K channels. FEBS Lett 1996;382:84–88.
32. Wallner M, Meera P, Toro L. Molecular basis of fast inactivation in voltage and Ca^{2+}-activated K^+ channels: a transmembrance β-subunit homolog. Proc Natl Acad Sci USA 1999;96:4137–4142.
33. Xia XM, Ding JP, Lingle CJ. Molecular basis for the inactivation of Ca^{2+}- and voltage-dependent BK channels in adrenal chromaffin cells and rat insulinoma tumor cells. J Neurosci 1999;19:5255–5264.
34. Ha TS, Jeong SY, Cho SW, et al. Functional characteristics of two BKCa channel variants differentially expressed in rat brain tissues. Eur J Biochem 2000;267:910–918.
35. Lagrutta A, Shen KZ, North RA, et al. Functional differences among alternatively spliced variants of Slowpoke, a Drosophila calcium-activated potassium channel. J Biol Chem 1994;269:20,347–20,351.
36. Atkinson NS, Robertson GA, Ganetzky B. A component of calcium-activated potassium channels encoded by the Drosophila Slo locus. Science 1991;253:551–555.
37. McCobb DP, Fowler NL, Featherstone T, et al. A human calcium-activated potassium channel gene expressed in vascular smooth muscle. Am J Physiol 1995;269:H767–H777.
38. Butler A, Tsunoda S, McCobb DP, et al. mSlo, a complex mouse gene encoding "maxi" calcium-activated potassium channels. Science 1993;261(5118):221–224.
39. Chang CP, Dworetzky SI, Wang J, et al. Differential expression of the alpha and beta subunits of the large-conductance calcium-activated potassium channel: implication for channel diversity. Brain Res Mol Brain Res 1997;45:33–40.
40. Adelman JP, Shen KZ, Kavanaugh MP, et al. Calcium-activated potassium channels expressed from cloned complementary DNAs. Neuron 1992;9:209–216.
41. Meera P, Wallner M, Song M, et al. Large conductance voltage- and calcium-dependent K+ channel, a distinct member of voltage-dependent ion channels with seven N-terminal transmembrane segments (S0–S6), an extracellular N terminus, and an intracellular (S9–S10) C terminus. Proc Natl Acad Sci USA 1997;94(25):14,066–14,071.
42. Knaus HG, Eberhart A, Koch RO, et al. Characterization of tissue-expressed alpha subunits of the high conductance Ca(2+)-activated K+ channel. J Biol Chem 1995;270:22,434–22,439.
43. Wei A, Solaro C, Lingle C, et al. Calcium sensitivity of BK-type K_{Ca} channels determined by a separable domain. Neuron 1994;13:671–681.
44. Wallner M, Meera P, Toro L. Determinant for β-subunit regulation in high-conductance voltage-activated and Ca^{2+}-sensitive K^+ channels: an additional transmembrance region at the N terminus. Proc Natl Acad Sci USA 1996;93:14,922–14,927.
45. Schreiber M, Salkoff L. A novel calcium-sensing domain in the BK channel. Biophys J 1997;73: 1355–1363.
46. Bian S, Favre I, Moczydlowski E. Ca^{2+}-binding activity of a COOH-terminal fragment of the Drosophila BK channel involved in Ca^{2+}-dependent activation. Proc Natl Acad Sci USA 2001;98: 4776–4781.
47. Braun AF, Sy L. Contribution of potential EF hand motifs to the calcium-dependent gating of a mouse brain large conductance, calcium-sensitive K^+ channel. J Physiol 2001;533:681–695.
48. Xia XM, Zeng X, Lingle CJ. Multiple regulatory sites in large-conductance calcium-activated potassium channels. Nature 2002;418:880–884.
49. Bao L, Rapin AM, Holmstrand EC, et al. Elimination of the BKCa channel's high-affinity Ca^{2+} sensitivity. J Gen Physiol 2002;20:173–189.
50. Piskorowski R, Aldrich RW. Calcium activation of BK_{Ca} potassium channels lacking the calcium bowl and RCK domains. Nature 2002;420:499–502.
51. Jiang Y, Lee A, Chen J, et al. Crystal structure and mechanism of a calcium-gated potassium channel. Nature 2002; 417:501, 502.
52. Kelly RA, Smith TW. Nitric oxide and nitrovasodilators: similarities, differences, and interactions. Am J Cardiol 1996;77:2C–7C.
53. Bolotina VM, Najibi S, Palacino JJ, et al. Nitric oxide directly activates calcium-dependent potassium channels in vascular smooth muscle. Nature 1994;368:850–853.
54. Wang R, Wu L. The chemical modification of K_{Ca} channels by carbon monoxide in vascular smooth muscle cells. J Biol Chem 1997;272:8222–8226.
55. Wang R, Wu L, Wang ZZ. The direct effect of carbon monoxide on K_{Ca} channels in vascular smooth muscle cells. Pflügers Arch 1997;434:285–291.

56. Werkstrom V, Ny L, Persson K, et al. Carbon monoxide-induced relaxation and distribution of haem oxygenase isoenzymes in the pig urethra and lower oesophagogastric junction. Br J Pharmacol 1997;120:312–318.
57. Frace AM, Eaton DC. Chemical modification of Ca^{2+}-activated potassium channels of GH3 anterior pituitary cells. Am J Physiol 1992;263:C1081–C1087.
58. MacKinnon R, Miller C. Functional modification of a Ca^{2+}-activated K^+ channel by trimethyloxonium. Biochemistry 1989;28:8087–8092.
59. Yang F, Phillips GN Jr. Crystal structures of CO-, deoxy- and met-myoglobins at various pH values. J Mol Biol 1996;8:762–764.
60. Sun J, Loehr TM, Wilks A, et al. Resonance Raman and EPR spectroscopic studies on heme-heme oxygenase complexes. Biochemistry 1994;33:13,734–13,740.
61. Xu M, Akabas MH. Amino acids lining the channel of the γ-aminobutyric acid type A receptor identified by cysteine substitution. J Biol Chem 1993;268:21,505–21,508.
62. Kenyon GL, Bruice TW. Novel sulfhydryl reagents. In: Colowick SP, Kaplan NO, eds. Methods in Enzymology, Academic: New York, 1997, pp. 407–430.
63. Wang R, Wu L. Interaction of selective amino acid residues of K_{Ca} channels with carbon monoxide in vascular smooth muscle cells. Exp Biol Med 2003;228:474–480.
64. Lee K, Ozanne SE, Hales CN, et al. Effects of chemical modification of amino and sulfhydryl groups on KATP channel function and sulfonylurea binding in CRI-G1 insulin-secreting cells. J Membr Biol 1994;139:167–181.
65. Bouzat CB, Barrantes FJ, Sigworth FJ. Changes in channel properties of acetylcholine receptors during the time course of thiol chemical modifications. Pflügers Arch 1991;418:51–61.
66. Cescatti L, Pederzolli C, Menestrina G. Modification of lysine residues of Staphylococcus aureus alpha-toxin: effects on its channel-forming properties. J Membr Biol 1991;119:53–64.
67. Spires S, Begenisich T. Modification of potassium channel kinetics by amino group reagents. J Gen Physiol 1992;99:109–129.
68. Oberhauser A, Alvarez O, Latorre R. Activation by divalent cations of a Ca^{2+}-activated K^+ channel from skeletal muscle membrane. J Gen Physiol 1988;92:67–86.
69. Pappone PA, Barchfeld GL. Modifications of single acetylcholine-activated channels in BC3H-1 cells: effects of trimethyloxonium and pH. J Gen Physiol 1990;96:1–22.
70. Worley JF, French RJ, Krueger BK. Trimethyloxonium modification of single batrachotoxin-activated sodium channels in planar bilayers: changes in unit conductance and in block by saxitoxin and calcium. J Gen Physiol 1986;87:327–349.
71. Sigworth FJ, Spalding BC. Chemical modification reduces the conductance of sodium channels in nerve. Nature 1980;283:293–295.
72. Christensen BN, Hida E. Protonation of histidine groups inhibits gating of the quisqualate/kainate channel protein in isolated catfish cone horizontal cells. Neuron 1990;5:471–478.
73. Wu L, Cao K, Lu Y, et al. Different mechanisms underlying the stimulation of K_{Ca} channels by nitric oxide and carbon monoxide. J Clin Invest 2002;110:691–700.
74. Tanaka Y, Meera P, Song M, et al. Molecular constituents of maxi K_{Ca} channels in human coronary smooth muscle: predominant alpha + beta subunit complexes. J Physiol (Lond) 1997;502:545–559.
75. McManus OB, Helms LMH, Pallanck L, et al. Functional role of the beta subunit of high conductance calcium-activated potassium channels. Neuron 1995;14:645–650.
76. Wanner SG, Koch RO, Koschak A, et al. High-conductance calcium-activated potassium channels in rat brain: pharmacology, distribution, and subunit composition. Biochemistry 1999;38:5392–5400.
77. Knaus HG, Garcia-Calvo M, Kaczorowski J, et al. Subunit composition of the high conductance calcium-activated potassium channel from smooth muscle, a representative of the mSlo and slowpoke family of potassium channels. J Biol Chem 1994;269:3921–3924.
78. Mistry DK, Garland CJ. Characteristics of single, large-conductance calcium-dependent potassium channels (BK_{Ca}) from smooth muscle cells isolated from the rabbit mesenteric artery. J Membr Biol 1998;164:125–138.
79. Tseng-Crank J, Godinot N, Johansen TE, et al. Cloning, expression, and distribution of a Ca^{2+}-activated K^+ channel beta-subunit from human brain. Proc Natl Acad Sci USA 1996;93:9200–9205.
80. Wu JV, Shuttleworth TJ, Stampe P. Clustered distribution of calcium sensitivities: an indication of hetero-tetrameric gating components in Ca^{2+}-activated K^+ channels reconstituted from avian nasal gland cells. J Membr Biol 1996;154:275–282.
81. Fanger CM, Ghanshani S, Logsdon NJ, et al. Calmodulin mediates calcium-dependent activation of the intermediate conductance K_{Ca} channel, $IK_{Ca}1$. J Biol Chem 1999;274:5746–5754.

82. Schumacher MA, Rivard AF, Bachinger HP, et al. Structure of the gating domain of a Ca^{2+}-activated K^+ channel complexed with Ca^{2+}/calmodulin. Nature 2001;410:1120–1124.
83. Braun AP, Heist EK, Schulman H. Inhibition of a mammalian large conductance, calcium-sensitive K^+ channel by calmodulin-binding peptides. J Physiol 2000;527:479–492.
84. Leclerc L'Hostis E, Leclerc L, Haiech J, et al. Heme-CO binding to tryptophan-containing calmodulin mutants. Biochim Biophys Acta 1996;1313:209–216.
85. Plüger S, Faulhaber J, Fürstenau M, et al. Mice with disrupted BK channel beta1 subunit gene feature abnormal Ca(2+) spark/STOC coupling and elevated blood pressure. Circ Res 2000;87:E53–E60.
86. Brenner R, Peréz G, Bonev A, et al. Vasoregulation by the β1 subunit of the calcium-activated potassium channel. Nature 2000;407:870–876.

14 Carbon Monoxide and Signal Transduction Pathways

Patty J. Lee and Leo E. Otterbein

CONTENTS

SUMMARY

Carbon monoxide (CO) is emerging as an important signaling molecule that exerts a myriad of biological effects that are only recently being uncovered. CO is a diatomic gas that is generated predominantly from heme degradation by the enzyme heme oxygenase. Traditionally considered a biological "waste product" of heme metabolism and, at high doses, lethal, CO clearly has diverse functions including the modulation of neural signals, inflammation, cell proliferation, cell death, and smooth muscle contractility. Interestingly, at concentrations well below those that would otherwise create toxic effects, CO has beneficial effects in various models of injury and inflammation. The precise mechanisms of these CO-mediated effects are yet unknown but are becoming the focus of intense investigations. This chapter reviews the known signal transduction pathways of CO with a special emphasis on the roles of guanylate cyclase, the mitogen-activated protein kinases, and nuclear factor-κB.

Key Words: Carbon monoxide; signal transduction; heme oxygenase; mitogen-activated protein kinases; guanylate cyclase; cyclic guanosine 5'-monophosphate; nuclear factor-κB.

1. INTRODUCTION

Evolutionarily speaking, carbon monoxide (CO) has been present in our environment since the beginning of life as we know it. In fact, evolutionists contend that the creation

From: *Signal Transduction and the Gasotransmitters: NO, CO, and H_2S in Biology and Medicine* Edited by: Rui Wang © Humana Press Inc., Totowa, NJ

of amino acids, believed to be the fundamental backbone of all life, necessitated the presence of CO along with oxygen and nitrogen *(1)*. First discovered and described in the latter part of the 18th century, CO has been rightly labeled as a toxic molecule, one capable of displacing oxygen in the blood and creating tissue hypoxia. This branding has withstood the test of time and is reflected in the widespread use of commercial CO detectors today. To be sure, CO at certain concentrations, like all agents including life-sustaining oxygen, has limits of tolerability. Yet, there remain paradoxes that challenge the widely held belief that CO is merely a toxic gas. The most notable is the presence in every cell of every species the enzyme heme oxygenase (HO), which serves one purpose: to deconstruct heme molecules into biliverdin, thereby resulting in the release of one molecule of CO and one atom of divalent iron. Biliverdin is subsequently converted to bilirubin and excreted as bile. The generation of endogenous CO is seemingly overlooked, but it is fact that on increased stress, whether it be a bacterial infection or an arthritic joint, the expression and activity of HO increases. Concomitantly, there is an obvious increase in CO levels as evidenced by increased levels of the gas in the exhaled breath of sick individuals *(2,3)*. The paradox lies in the fact that increased levels of CO should lead to a decreased level of oxygen delivery to tissues, which should negatively affect outcome. This release of CO generated by HO into the cellular and tissue milieu and transported remotely by the mass transit vehicle hemoglobin is now thought to be more than simply a waste product. It is clear that CO is a potent signaling molecule capable of evoking cellular responses and directing phenotypic changes to dictate cellular responses. It is the ideal chemical messenger in that it is (a) diffusible across lipid bilayers; (b) nonreactive, unlike its sister gas, nitric oxide (NO), which forms the toxic peroxynitrite molecule capable of nitrosylating proteins to exert function; and (c) nonmetabolized and easily excreted. CO is carried in the circulation as carboxyhemoglobin, where partial pressure differences eventually force its release in the lungs and out via exhaled breath.

The field of molecular and cellular signaling is vast, encompassing hundreds if not thousands of cascades, pathways, and cross talk toward the generation of specific gene products, and the knowledge of CO-selective signaling is just beginning to be elucidated. This chapter discusses the current discoveries including the ability of CO to modulate guanylate cyclase/cyclic guanosine 5'-monophosphate (cGMP), mitogen-activated protein kinases (MAPKs), and nuclear factor-κB (NF-κB).

2. GASEOUS SISTER MOLECULES NO AND CO

There are vast amounts of literature on the biology of NO, making an in-depth discussion beyond the scope of this chapter. However, a brief overview is warranted here because it is the first gaseous molecule discovered to behave as a signaling and effector molecule. In 1987, mammalian cells were found to produce NO as a short-lived intracellular gaseous messenger. Since that time, investigations into the functional significance of this phenomenon have been immense. There are two types of NO-producing enzymes, termed *NO synthases* (NOSs)—constitutive and inducible isoforms. Both catalyze the conversion of L-arginine to citrulline with NO gas released as a product. The similarities between NO and CO are staggering, and it is under debate as to whether the response observed with one is truly not dependent on the other as the aforementioned evidence would suggest. CO activates inducible NOS (iNO)S/NO, which in turn leads to increased HO-1 and therefore additional CO. CO, like NO, binds avidly to heme-containing moieties, the most well-described being guanylate cyclase. Additionally, it is known, e.g., that NO, like CO, can chemically react with the heme of cytochrome oxidase (complex IV of the electron transport chain) to decrease the affinity of this enzyme for oxygen. Depending on the cellular milieu, including the continued production and source of NO, the production of low levels of reactive oxygen species (ROS) can lead to either prosurvival or pro-death pathways *(4,5)*.

NO, unlike CO, is highly reactive and thus there are numerous targets within the cell where NO could elicit cellular responses. This is evident in a recent publication by Matsumoto et al. *(6)* in which a modified yeast two-hybrid assay was used to study novel NO-dependent protein-protein interactions that rely on the nitrosylation of proteins. Several recent studies suggest that one essential prosurvival mediator that is altered by NO is the upregulation of HO-1. HO-1, first described in 1968 by Tenhunen et al. *(7)* as the rate-limiting enzyme in the degradation of heme, was, until the early 1990s, labeled as simply a housekeeping gene. The landmark studies by Keyse and Tyrell *(8,9)* showing that nonheme agents such as H_2O_2 and ultraviolet (UV) light could induce this enzyme and lead to subsequent protection from a subsequent stressor changed the view of this molecule. They, and now countless others, have demonstrated that this ubiquitous enzyme is capable of being induced by numerous nonheme stimuli, which results, in most instances, in remarkable cytoprotection both in vitro and in vivo *(10)*.

Most recently, the field has begun to focus on mechanisms of action by which HO-1 imparts its protection. CO has evolved as a potential target because many of the effects observed with HO-1 can be mimicked with low concentrations of CO. In 1995, Kim et al. *(10a)* first demonstrated that NO can upregulate expression of HO-1 in hepatocytes in vitro. Reports by Naughton et al. *(11)* make clear that HO can act as both a sensor to and a target of redox-based mechanisms involving NO. Therefore, CO may serve as a signaling molecule in the modulation of the tissue stress responses and extend our knowledge of the biological function of HO-1 in response to nitrosative stress. Using pharmacological inhibitors of HO-1 enzymatic activity, Choi et al. *(12)* illustrated that NO-mediated protection against cell death induced by glucose deprivation is dependent on HO-1. They further concluded that the protective effects of HO-1 are mediated via the production of CO. We hypothesize from our studies of acute hepatitis that NO-induced upregulation of HO-1 provides antiapoptotic and anti-inflammatory effects to protect hepatocytes against the potential toxicities of NO and other mediators of cell death, i.e., tumor necrosis factor-α (TNF-α). In addition, bursts of NO and production of reactive nitrogen species can potentially provide antimicrobial or antitumor effects. It is interesting that these well-touted toxic gas molecules at appropriate concentrations can impart potent and significant biological effects in a concerted effort to maintain homeostasis.

3. CO AND GUANYLATE CYCLASE

Analogous to NO, CO upregulates cGMP via guanylate cyclase and thereby exerts effects on vasoregulation and neurotransmission. CO-mediated activation of soluble guanylate cyclase (sGC) leads to a severalfold increase in cGMP production, although its potency is 30–100 times lower than that of NO *(13)*. There are certainly complex interrelationships between CO and NO, with the potential for one to substitute for the other depending on the local milieu, but this has been reviewed elsewhere *(14,15)* and is beyond the scope of the current review. In earlier studies, investigators demonstrated that CO inhibits platelet aggregation and relaxes smooth muscle via guanylate cyclase *(15,16)*. Subsequently, evidence emerged that CO arising from heme metabolism through HO activity regulates cGMP production in brain as well as vascular tissues *(17,18)*. The vasodilatory effect of CO may be important for maintaining tissue perfusion during normal and hypoxic states *(19)*. Increased levels of cGMP after hypoxia require HO activity without any dependence on NO *(20)*. Inhibition of guanylate cyclase with methylene blue or 1H-[1,2,4]-oxadiazolo[4,3-a]quinoxalin-1-one (ODQ) abrogated the antiproliferative effect of CO on hypoxic vascular smooth muscle cells (VSMCs) *(19)*. The mechanism whereby CO attenuates cell growth in a cGMP-dependent manner may involve the modulation of cell cycle-specific transcription factors such as E2F *(21)*. CO can also cause airway bronchodilation in guinea pigs administered histamine injections via cGMP *(22)*. Liu et al. *(22)* found that the anti-apoptotic effect of CO on cytokine-treated rat aortic smooth muscle cells was partially dependent on the activation of sGC and was associated with p53 suppression and inhibition of mitochondrial cytochrome-*c* release. They recently extended their studies by demonstrating that p38 MAPK inhibition with the specific chemical inhibitor SB203580 had no effect on the antiapoptotic properties of CO *(24)*. Fujita et al. *(25)* demonstrated that the protective effects of exogenous CO (0.1% or 1000 ppm) prior to lethal lung ischemia-reperfusion (I-R) injury were, at least in part, because of sGC. Mice treated with a guanylate cyclase inhibitor, ODQ, were not rescued from I-R-induced lethality by CO. The survival benefit of CO appeared to be mediated by guanylate cyclase activation and subsequent suppression of plasminogen activator inhibitor-1 in macrophages, which reduced microvascular fibrin deposition.

4. CO AND MAPK

It is clear that CO also uses signaling pathways independent of guanylate cyclase/cGMP, depending on cell type, inducer, and local biologic milieu. MAPKs are a family of serine-threonine protein kinases that are activated in response to various extracellular stimuli *(26)*. Three major MAPK signaling pathways—extracellular signal–regulated protein kinase (ERK), p38 MAPK (p38), and c-Jun NH2-terminal protein kinase (JNK)—have been identified in mammalian cells. Recently, in human airway SMCs, exogenous CO induced growth arrest via the ERK MAPK pathway, whereas guanylate cyclase/cGMP was not involved *(27)*. The CO-induced growth arrest was associated with p21 upregulation and cyclin D1 downregulation. Otterbein et al. *(28)* demonstrated that the anti-inflammatory properties of CO are mediated by p38 MAPK and the upstream MAPK kinase 3 (MKK 3). In RAW 264.7 macrophages and a mouse model of lipopolysaccharide (LPS)-induced sepsis, CO inhibited the expression of LPS-induced proinflammatory cytokines such as TNF-α, interleukin (IL)-1β, and macrophage inflammatory protein-1α while simultaneously increasing expression of the antiinflammatory cytokine IL-10. However, in mice deficient in MKK3 (*mkk3*$^{-/-}$), an upstream activator of p38 MAPK, CO did not affect LPS-induced TNF-α production *(28)*. In addition, CO increased LPS-induced IL-10 levels in *mkk3*$^{+/+}$ mice, compared with *mkk3*$^{+/+}$ mice treated with LPS alone. The alternative MAPKs, ERK and JNK, were not affected by CO, and the ability of CO to modulate cytokines was independent of cGMP and NO.

The ability of CO to suppress inflammation is likely involved in xenograft transplant models in which 400 ppm of CO for 2 d prevented rejection for up to 50 d *(29)*. Presumably, the known ability of CO to decrease platelet aggregation, cause vasodilation, and modulate proinflammatory cytokines contributes to the favorable outcome in xenograft transplantation *(30)*. CO at 250 ppm decreased eosinophil influx and IL-5 production in allergen-challenged mice *(31)*. In a rat model of hyperoxic lung injury, the administration of CO significantly attenuated airway neutrophilia and lung apoptosis *(32)*. There is recent evidence that JNK MAPK and activator protein (AP)-1 mediate the anti-inflammatory properties of CO. Morse et al. *(33)* demonstrated that CO attenuated IL-6 production in LPS-treated macrophages at the transcriptional level. Mutations of the AP-1-binding site in the IL-6 promoter decreased the effect of CO on promoter activity, and CO decreased AP-1 binding in electrophoretic mobility shift assays *(33)*. Mice deficient in the JNK pathway had decreased serum levels of IL-6 and IL-1β in response to LPS compared with control mice.

In addition to the potent anti-inflammatory properties of CO that are mediated by MAPK, CO has antiapoptotic and cytoprotective effects via MAPK. Brouard et al. *(34)* demonstrated that CO could inhibit TNF-α-induced endothelial apoptosis and this involves the activation of the p38 MAPK pathway. Inhibition of p38 MAPK with the selective chemical inhibitor SB203580 or a dominant-negative mutant abrogated the antiapoptotic effects of HO-1. Amersi et al. *(35)* have shown that CO has protective effects against liver I-R injury via p38 MAPK. We recently have shown that low levels of exogenous CO attenuated I-R-induced rat pulmonary endothelial cell and mouse lung apoptosis via p38 MAPK *(36)*. CO differentially modulates the MAPK during I-R injury by decreasing ERK and JNK MAPK activation while increasing p38 MAPK activation in rat pulmonary artery endothelial cells, and this is associated with the attenuation of I-R-induced apoptosis. In the presence of p38 MAPK inhibition, CO lost its ability to attenuate I-R-induced apoptosis in endothelial cells and mouse lung, likely through the inability to decrease caspase 3 activity *(36)*. We extended these studies to demonstrate that CO specifically activates the p38α isoform of p38 MAPK via MKK3 to attenuate I-R-induced apoptosis in endothelial cells and mouse lung *(37)*. There are four known p38 MAPK isoforms—α, β, γ, and δ—of which the α and β are best described. The isoforms are thought to have distinct biological roles and are differentially regulated by upstream MKKs. Generally, MKK6 is thought to regulate α, β, and γ but not δ, whereas MKK3 regulates α, γ, and δ, but not β *(38)*. However, MKK4 has also been shown to regulate p38β in certain cells and is inducer dependent *(39)*, which adds to the potential redundancy of p38 signaling. We have shown that CO activates MKK3 and p38α MAPK to inhibit Fas/Fas ligand expression; activation of caspases 3, 8, and 9; poly (ADP-ribose) polymerase cleavage; and mitochondrial cytochrome-*c* release *(37)*. In addition, CO differentially modulates the pro- and antiapoptotic members of the Bcl-2 family proteins. We correlate our endothelial cell findings in MKK3-deficient ($mkk^{-/-}$) and Fas receptor–deficient ($fas^{-/-}$) mice subjected to lung I-R injury *(37)*.

In rat pulmonary artery endothelial cells treated with TNF-α, exogenous CO similarly enhanced p38 MAPK activation and downregulated ERK activation but had no effect on JNK activation *(40)*. Choi et al. *(12)* demonstrated that in murine embryonic liver cells, the presence of a CO donor significantly decreased glucose deprivation-induced ERK MAPK activation but had no effect on p38 or JNK MAPK activation. They also demonstrated that a CO donor or an ERK MAPK inhibitor blocked glucose deprivation-induced cytotoxicity in liver cells. Therefore, they speculated that CO mediates cytoprotection through ERK MAPK suppression. The use of a p38 MAPK inhibitor had no effects. There is likely signaling specificity depending on the cell type and MAPK inducer utilized. There is also evidence for proapoptotic effects of exogenous CO that are likely concentration dependent. Thom et al. *(41)* showed that bovine pulmonary artery endothelial cells exposed to 100 ppm CO for more than 1 h died via NO- and caspase 1-dependent mechanisms. However, pretreatment of endothelial cells at lower concentrations of CO (10 ppm) conferred resistance to the lethal effects of 100 ppm of CO.

The precise biochemical mechanisms by which CO modulates the MAPK are not clear. Given the avidity of CO for heme-containing moieties, we postulate that CO could be modulating MAPK through unknown or unidentified intermediate heme molecules. This is supported by our observations (unpublished data) that the protective effects of CO require new protein synthesis in TNF-α-induced apoptosis. Further understanding of CO-modulated targets may be achieved by exploring potential downstream targets of p38 MAPK. The p38 MAPK activity assays we used in our I-R apoptosis studies show that CO increases ATF2 phosphorylation, which can then modulate many genes. Potential p38 MAPK targets include known substrates such as the transcription factors CHOP and Elk1, or other kinases such as MSK1 and MAPKAP, which then exert pleiotropic biological effects depending on cell type and stimulus. At this point, however, CO modulation of these proteins, via p38 MAPK, is still speculative and warrants future detailed studies. Brouard et al. *(42)* demonstrated that HO-1/CO cooperates with NF-κB-dependent antiapoptotic genes (*c-IAP2* and *A1*) to protect against TNF-α-mediated endothelial cell apoptosis. However, CO does not appear to activate NF-κB directly but, rather, requires basal NF-κB activity to suppress TNF-α-mediated apoptosis.

5. CO AND NF-κB

NF-κB is a transcription factor composed of a nuclear homo- or heterodimeric complex of a number of different Rel family members *(43)*. The most common form is composed of a 50-kDa (p50) and a 65-kDa (p65/RelA) subunit. In quiescent cells, NF-κB is retained in the cytoplasm by a series of inhibitory IκB proteins: IκBα, IκBβ, IκBε, p100, p105 *(43,44)*. Binding of NF-κB to these IκB proteins prevents the complex from localizing to the nucleus and thereby prevents transcriptional activity. On stimulation of the cell, rapid phosphorylation of IκB results in the release of the NF-κB dimers and subsequent NF-κB translocation and transcriptional activation. IκB is phosphorylated and subsequently degraded via multiple pathways and ultimately proteosome degradation following ubiquitination, which involves upstream kinases including IKKα and β. There are also several upstream MAPK kinases that have been shown to activate IκB such as NF-κB inducing kinase; MEKK1; and, more recently, p38 *(45–47)*. In most instances, this signal transduction pathway is considered the primary means by which NF-κB is activated; however, other means including increases in ROS and UV irradiation also activate NF-κB *(48,49)*. The downstream genes regulated by NF-κB are complex as well as stimulus and cell-type dependent. For instance, in monocytes and endothelial cells, LPS-induced NF-κB activation leads to rapid induction of cytokine expression; however, the same stimulus in endothelial cells can also lead to increased expression of cytoprotective genes including the Bcl family of antiapoptotic proteins *(50,51)*.

Studies by Brouard et al. *(42)* using an endothelial cell model of apoptosis demonstrated that HO-1/CO cooperates with NF-κB-dependent antiapoptotic genes (i.e., *c-IAP2* and *A1*) to protect against TNF-α-mediated apoptosis. This effect is dependent on the ability of HO-1/CO to activate the p38 MAPK signal transduction pathway *(42)*. In an elegant series of experiments, these investigators demonstrated that the ability of HO-1 and CO to provide protection requires NF-κB; IκB overexpression results in loss of protection, but there is no nuclear translocation or activation of the dimer over that observed in control *(52)*. They concluded that although NF-κB is not activated by HO-1/CO, it is required in some manner for the expression of the antiapoptotic battery of genes *(42)*. In a model of endotoxic shock in monocytes, exposure to CO at low concentrations resulted in a marked inhibition of IκBα phosphorylation and NF-κB activation that led to a >50% reduction in granulocyte macrophage colony-stimulating factor production *(52)*. In direct contrast to the aforementioned studies and as a further example of cell specificity, HO-1 and CO in hepatocytes also provides protection; however, in this situation, the activation of NF-κB is critical for cytoprotection. Zuckerbraun et al. *(53)* have shown that the cytoprotection afforded by CO in the liver and in isolated hepatocytes involves a complex cycle where CO activates NF-κB via the generation of low levels of ROS, which leads very rapidly to iNOS and NO expression, which ultimately leads to and requires upregulation of HO-1. Each of these steps in turn is required to prevent fulminant hepatitis induced by TNF/D-galactosamine in mice and from TNF/actinomycin D-induced apoptosis in primary mouse hepatocytes. Clearly, the role of NF-κB remains to be further elucidated, as the preceding summary delineates, but the results regardless of the outcome support the role of this crucial transcription factor as a mechanism by which the HO-1-CO axis operates.

6. CO AND OTHER SIGNALING PATHWAYS

The vascular effects of CO have also been attributed to the inhibition of cytochrome P450 oxygenase and subsequent decreased endothelin-1 production *(54)*. CO also has direct effects on calcium-activated potassium channel activity (K_{Ca}) in VSMCs that were not mediated by cGMP or guanine nucleotide-binding proteins (G_i/G_o or G_s) *(55)*. Kaide et al. *(56)* demonstrated that decreased CO production was accompanied by a decreased number of open potassium channels in SMCs and increased vascular contractility; these effects were reversed with exogenous CO. In a study investigating the mechanisms of CO-mediated dilation of porcine cerebral arterioles, CO was found to dilate arterioles by increasing the effective coupling of calcium sparks to K_{Ca} channels *(57)*. CO can also induce the generation of hydrogen peroxide, which can then act as an intracellular messenger *(58)*. CO-generated ROS is postulated to occur through the induction of manganese superoxide dismutase *(59)* or through the inhibition of catalase *(58)*. Recently, the first eukaryotic transcription factor selectively modulated by CO was discovered. Dioum et al. *(60)* demonstrated that CO inhibits NPAS2, a member of the hypoxia-inducible Factor-1α (HIF-1α) family of proteins. NPAS2 is a homologue of CLOCK, a transcription factor involved in circadian activity in the suprachiasmatic nucleus but also present in cells outside the central nervous system *(60)*. Interestingly, NPAS2 binds to DNA consensus sequences (CACGTG) similar to that of HIF-1 (TACGTG); HIF-1 DNA binding has been shown to decrease in the presence of exogenous CO during hypoxia *(61)*. The potential relationships among CO, NPAS2, and hypoxia-regulated genes remain to be explored and may yield novel insights into CO signaling.

7. CONCLUSION

CO, like NO, is proving to be a unique gaseous signaling molecule that is integral to cellular homeostasis. There are accumulating reports of CO-mediated biological effects in almost all organ systems, but the precise signaling mechanisms and molecular targets remain an evolving area. As reviewed in this chapter, at low doses CO has potent anti-inflammatory,

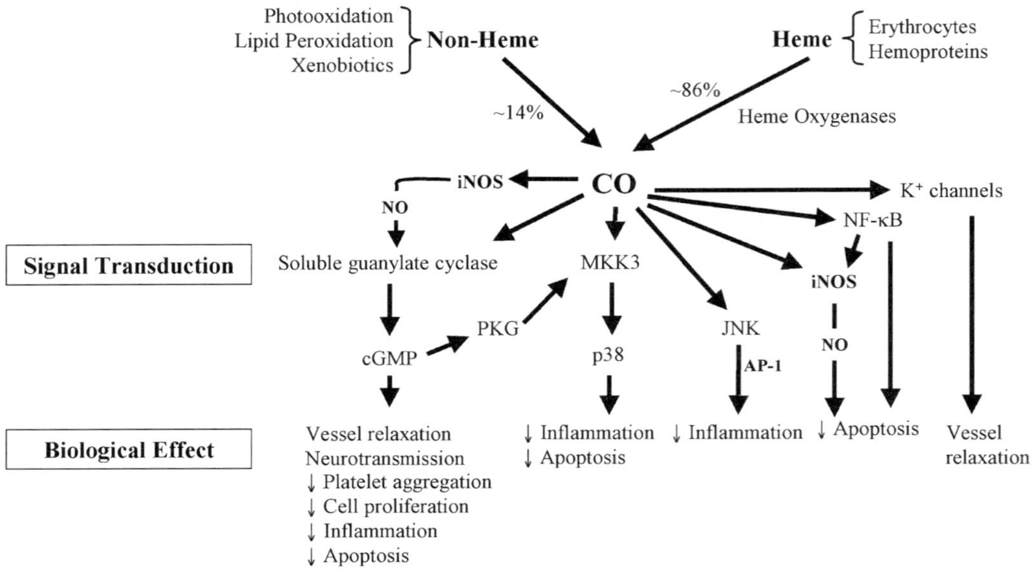

Fig. 1. CO signaling pathways.

antiapoptotic, and antiproliferative effects. Furthermore, the pleiotropic biological effects of CO are mediated by equally complex and varied signaling pathways. As summarized in **Fig. 1**, evidence points to the important roles of guanylate cyclase/cGMP, MAPK, especially p38, and NF-κB, but the list continues to grow. These cumulative observations underscore the importance of further investigations into the precise mechanisms and signaling pathways utilized by CO. There is emerging and convincing evidence for potential therapeutic applications of the HO/CO system. At a time of intense explorations of the genome and proteome as well as the necessary complex bioinformatics, it is the immense simplicity of the diatomic gases that holds potentially new and exciting answers to the intricate machinery of life.

REFERENCES

1. Miyakawa S, Yamanashi H, Kobayashi K, et al. Prebiotic synthesis from CO atmospheres: implications for the origins of life. Proc Natl Acad Sci USA 2002;99:14,628–14,631.
2. Paredi P, Kharitonov SA, Barnes PJ. Exhaled carbon monoxide in lung disease. Eur Respir J 2003; 21:197,198.
3. Zegdi R, Perrin D, Burdin M, et al. Increased endogenous carbon monoxide production in severe sepsis. Intensive Care Med 2002;28:793–796.
4. Moncada S, Erusalimsky JD. Does nitric oxide modulate mitochondrial energy generation and apoptosis? Nat Rev Mol Cell Biol 2002;3:214–220.
5. Beltran B, Quintero M, Garcia-Zaragoza E, et al. Inhibition of mitochondrial respiration by endogenous nitric oxide: a critical step in Fas signaling. Proc Natl Acad Sci USA 2002;99:8892–8897
6. Matsumoto A, Comatas KE, Liu L, et al. Screening for nitric oxide–dependent protein-protein interactions. Science 2003;301:657–661.
7. Tenhunen R, Marver HS, Schmid R. The enzymatic conversion of heme to bilirubin by microsomal heme oxygenase. Proc Natl Acad Sci USA 1968;61:748–755.
8. Keyse SM, Tyrrell RM. Induction of the heme oxygenase gene in human skin fibroblasts by hydrogen peroxide and UVA (365 nm) radiation: evidence for the involvement of the hydroxyl radical. Carcinogenesis 1989;11:787–789.

9. Keyse SM, Tyrrell RM. Heme oxygenase is the major 32–kDa stress protein induced in human skin fibroblasts by UVA radiation, hydrogen peroxide, and sodium arsenite. Proc Natl Acad Sci USA 1989;86:99–103.
10. Otterbein LE, Soares MP, Yamashita K, et al. Heme oxygenase-1:unleashing the protective properties of heme. Trends Immunol 2003;24:449–455.
10a. Kim YM, Bergonia HA, Muller C, et al. Loss and degradation of enzyme-bound heme induced by cellular nitric oxide synthesis. J Biol Chem 1995;270(11):5710–5713.
11. Naughton P, Foresti R, Bains SK, et al. Induction of heme oxygenase 1 by nitrosative stress: a role for nitroxyl anion. J Biol Chem 2002;277:40,666–40,674.
12. Choi BM, Pae HO, Kim YM, et al. Nitric oxide–mediated cytoprotection of hepatocytes from glucose deprivation–induced cytotoxicity: involvement of heme oxygenase-1. Hepatology 2003;37:810–823.
13. Watts RN, Ponka P, Richardson DR. Effects of nitrogen monoxide and carbon monoxide on molecular and cellular iron metabolism: mirror-image effector molecules that target iron. Biochem J 2003; 369:429–440.
14. Maines JD. The heme oxygenase system: a regulator of second messenger gases. Annu Rev Pharmacol Toxicol 1997;37:517–554.
15. Brune B, Ullrich V. Inhibition of platelet aggregation by carbon monoxide is mediated by activation of guanylate cyclase. Mol Pharmacol 1987;32:497–504.
16. Utz J, Ullrich V. Carbon monoxoide relaxes ilial smooth muscle through activation of guanylate cyclase. Biochem Pharmacol 1991;41:1195–2001.
17. Ingi T, Cheng J, Ronnett GV. Carbon monoxide: an endogenous modulator of the nitric oxide–cyclic GMP signaling system. Neuron 1996;16:835–842.
18. Maines MD. Carbon monoxide: an emerging regulator of cGMP in the brain. Mol Cell Neurosci 1993;4:389–397.
19. Morita T, Perrella MS, Lee M, et al. Smooth muscle cell–derived carbon monoxide is a regulator of vascular cGMP. Proc Natl Acad Sci USA 1995;92:1475–1479.
20. Morita T, Mitsialis, SA, Koike H, et al. Carbon monoxide controls the proliferation of hypoxic vascular smooth muscle cells. J Biol Chem 1997;272:32,804–32,809.
21. Durante, W. Carbon monoxide and vascular smooth muscle cell growth. In: Carbon Monoxide and Cardiovascular Functions. (Wang R, ed.) CRC Press: Boca Raton, FL, 2001, pp. 45–65.
22. Cardell LO, Ueki IF, Stjarne P, et al. Bronchodilation in vivo by carbon monoxide, a cyclic GMP related messenger. Br J Pharmacol 1998;124:1065–1068.
23. Liu X, Chapman GB, Peyton KJ, et al. Carbon monoxide inhibits apoptosis in vascular smooth muscle cells. Cardiovasc Res 2002;55:396–405.
24. Liu XM, Chapman GB, Peyton KJ, et al. Antiapoptotic action of carbon monoxide on cultured vascular smooth muscle cells. Exp Biol Med 2003;228:572–575.
25. Fugita T, Toda K, Karimova A, et al. Paradoxical rescue from ischemic lung injury by inhaled carbon monoxide by derepression of fibrinolysis. Nat Med 2001;7:598–604.
26. Cobb MH. MAP kinase pathways. Prog Biophys Mol Biol 1999;71:479–500.
27. Song R, Mahidhara RS, Liu F, et al. Carbon monoxide inhibits human airway smooth muscle cell proliferation via mitogen-activated protein kinase pathway. Am J Respir Cell Mol Biol 2002;27:603–610.
28. Otterbein LE, Bach FH, Alam J, et al. Carbon monoxide has anti-inflammatory effects involving the mitogen-activated protein kinase pathway. Nat Med 2002;6:422–428.
29. Sato K, Balla J, Otterbein L, et al. Carbon monoxide generated by heme oxygenase-1 suppresses the rejection of mouse-to-rat cardiac transplants. J Immunol 2001;166:4185–4194.
30. Otterbein LE. Carbon monoxide: innovative anti-inflammatory properties of an age-old gas molecule. Antioxid Redox Signal 2002;4:309–319.
31. Chapman JT, Otterbein LE, Elias JA, et al. Carbon monoxide attenuates aeroallergen-induced inflammation in mice. Am J Physiol Lung Cell Mol Physiol 2001;281:L209–L216.
32. Otterbein LE, Mantell LL, Choi AMK. Carbon monoxide provides protection against hyperoxic lung injury. Am J Physiol 1999;276:L688–L694.
33. Morse D, Pischke SE, Zhou Z, et al. Suppression of inflammatory cytokine production by carbon monoxide involves the c-Jun NH2 terminal kinase (JNK) pathway and activator protein-1 (AP-1) J Biol Chem 2003;278:36,993–36,998.
34. Brouard S, Otterbein LE, Anrather J, et al. Carbon monoxide generated by heme oxygenase 1 suppresses endothelial cell apoptosis. J Exp Med 2002;192:1015–1026.
35. Amersi F, Shen XD, Anselmo D, et al. Ex vivo exposure to carbon monoxide prevents hepatic ischemia/reperfusion injury through p38 MAP kinase pathway. Hepatology 2002;35:815–823.

36. Zhang X, Shan P, Otterbein LE, et al. Carbon monoxide inhibition of apoptosis during ischemia-reperfusion lung injury is dependent on the p38 mitogen activated protein kinase pathway and involves caspase 3. J Biol Chem 2003;278:1248–1258.
37. Zhang X, Shan P, Alam J, et al. Carbon monoxide modulates Fas/Fas ligand caspases and Bcl-2 family proteins via the p38alpha mitogen-activated protein kinase pathway during ischemia-reperfusion lung injury. J Biol Chem 2003;278:22,061–22,070.
38. Braz JC, Bueno OF, Liang Q, et al. Targeted inhibition of p38 MAPK promotes hypertrophic cardiomyopathy through upregulation of calcineurin-NFAT signaling. J Clin Invest 2003;111:1475–1486.
39. Herlaar E, Brown Z. p38 MAPK signalling cascades in inflammatory disease. Mol Med Today 1999;5:439–447.
40. Sethi JM, Otterbein LE, Choi AM. Differential modulation by exogenous carbon monoxide of TNF-alpha stimulated mitogen-activated protein kinases in rat pulmonary artery endothelial cells. Antioxid Redox Signal 2002;4:241–248.
41. Thom S, Fisher D, Xu YA, et al. Adaptive responses and apoptosis in endothelial cells exposed to carbon monoxide. Proc Natl Acad Sci USA 2000;97:1305–1310.
42. Brouard S, Berberat PO , Tobiasch E, et al. Heme oxygenase-1-derived carbon monoxide requires the activation of transcription factor NF-kappa B to protect endothelial cells from tumor necrosis factor-alpha-mediated apoptosis. J Biol Chem 2002;277:17,950–17,961.
43. Karin M. The beginning of the end: IκB kinase (Ikk) and NFkB activation. J Biol Chem 1999;183: 27,339–27,342.
44. Read MA, Whitley MZ, Williams AJ, et al. NFkB and IkBa: an inducible regulatory system in endothelial activation. J Exp Med 1994;179:503–512.
45. Zechner D, Craig R, Hanford DS, et al. MKK6 activates myocardial cell NF-kB and inhibits apoptosis in a p38 mitogen-activated protein kinase dependent manner. J Biol Chem 1998;273:8232–8239.
46. Vandenberghe W, Plaisance S, Boone E, et al. p38 and extracellular signal regulating kinase mitogen-activated protein kinase pathways are required for nuclear factor-kappaB p65 transactivation mediated by tumor necrosis factor. J Biol Chem 1998;273:3285–3290.
47. Lee FS, Peters RT, Dang LC, et al. MEKK1 activates both IkB kinase alpha and IkB beta. Proc Natl Acad Sci USA 1998;95:9319–9324.
48. Bender K, Gottlicher M, Whiteside S, et al. Sequential DNA damage-independent and -dependent activation of NF-kB by UV. EMBO J 1998;17:5170–5181.
49. Altavilla D, Saitta A, Guarini S, et al. Oxidative stress causes nuclear factor-kappaB activation in acute hypovolemic hemorrhagic shock. Free Radic Biol Med 2001;30:1055–1066.
50. Tamatani M, Che YH, Matsuzaki H, et al. Tumor necrosis factor induces Bcl-2 and Bcl-x expression through NFkappaB activation in primary hippocampal neurons. J Biol Chem 1999;274:8531–8538.
51. Bach FH, Hancock WW, Ferran C. Protective genes expressed in endothelial cells—a regulatory response to injury. Immunol Today 1997;18:483–486.
52. Sarady JK, Otterbein SL, Liu F, et al. Carbon monoxide modulates endotoxin-induced production of granulocyte macrophage colony-stimulating factor in macrophages. Am J Respir Cell Mol Biol 2002;27:739–745.
53. Zuckerbraun BS, Billiar TR, Otterbein SL, et al. Carbon monoxide protects against liver failure through nitric oxide-induced heme oxygenase 1. J Exp Med 2003;198(11):1707–1716.
54. Coceani F, Kelsey L, Seidlitz E. Carbon monoxide-induced relaxation of the ductus arteriosus in the lamb: evidence against the prime role of guanylate cyclase. Br J Pharmacol 1996;118:1689–1696.
55. Wang R, Wu L, Wang Z. The direct effect of carbon monoxide on KCa channels in vascular smooth muscle cells. Eur J Physiol 1997;434:285–291.
56. Kaide JI, Zhang F, Wei Y, et al. Carbon monoxide of vascular origin attenuates the sensitivity of renal arterial vessels to vasoconstrictors. J Clin Invest 2001;107:1163–1171.
57. Jaggar JH, Leffler CW, Cheranov SY, et al. Carbon monoxide dilates cerebral arterioles by enhancing the coupling of Ca^{2+} sparks to Ca^{2+}-activated K^+ channels. Circ Res 2002;91:610–617.
58. Piantadosi CA. Biological chemistry of carbon monoxide. Antioxid Redox Signal 2003;4:259–270.
59. Frankel D, Mehindate K, Schipper HM. Role of heme oxygenase-1 in the regulation of manganese superoxide dismutase gene expression in oxidatively-challenged astroglia. Cell Physiol 2000;185:80–86.
60. Dioum EM, Rutter J, Tuckerman JR, et al. NPAS2: a gas-responsive transcription factor. Science 2002;298:2385–2387.
61. Liu Y, Christou H, Morita T, et al. Carbon monoxide and nitric oxide suppress the hypoxic induction of vascular endothelial growth factor gene via the 5' enhancer. J Biol Chem 1998;273:15,257–15,262.

15 Carbon Monoxide-Induced Alterations in the Expression of K_{Ca} Channels in Pulmonary Artery Smooth Muscle Cells

Eric Dubuis, Prem Kumar, Pierre Bonnet, and Christophe Vandier

CONTENTS

SUMMARY

Carbon monoxide (CO) is a generated gas and can regulate vascular tone. The two major sources of CO both exogenously and endogenously are automotive emissions and cigarette smoke, and the predominant biological source of CO is from degradation of heme by heme oxygenase. Accumulating data demonstrate that exogenous and endogenous CO are vasodilators, acting directly on vascular smooth muscle cells and on Ca^{2+}-activated K^+ channels. Surprisingly, there have been only a few experiments performed on the pulmonary circulation. Indeed, in this singular circulation, lying adjacent to the lung alveoli, CO could be directly delivered from the atmosphere and thus act directly on smooth muscle cells (SMCs) without delivery and by blood hemoglobin. Results from our laboratory have demonstrated that CO exposure can activate Ca^{2+}-activated K^+ channels of pulmonary SMCs. This review focuses on this point and also demonstrates that CO could induce a novel mechanism that we call "CO-induced CO increased sensitivity" in which K^+ channels are the main actor.

Key Words: Chronic carbon monoxide; large-conductance Ca^{2+}-activated K^+ channels; voltage-activated K^+ channels; chronic hypoxia; pulmonary hypertension; gasotransmitter.

From: *Signal Transduction and the Gasotransmitters: NO, CO, and H_2S in Biology and Medicine* Edited by: Rui Wang © Humana Press Inc., Totowa, NJ

1. INTRODUCTION

Carbon monoxide (CO) has often been presented as a "killer" or as a poison *(1)* because of its competition with oxygen for binding to hemoglobin (Hb), locking the compound in the oxy conformation and decreasing oxygen unloading to tissues. This toxicity is perhaps an unfortunate consequence of industrialization and the production of high quantities of exogenous CO. However, this action of CO has also provided a useful tool to support the hypothesis that oxygen sensors are heme proteins in oxygen-sensing cells *(2,3)*. Recently, it was demonstrated that this diatomic gas may constitute the second member of a new class of transmitters called gasotransmitters *(4)*. The physiological resurgence of CO *(5)* resulted in large part because of the demonstration that it could induce relaxation of vascular tissues (independently of hypoxia) and also because it could be endogenously produced by heme oxygenase (HO), which catalyzes the oxidative cleavage of heme to CO, iron, and biliverdin *(6)*. Since the first evidence of a vascular effect of CO in the pulmonary circulation *(7)*, only a few experiments have been performed on this circulation and almost none on the effect of chronic CO on isolated pulmonary myocytes.

This chapter summarizes some of the main findings obtained by ourselves and others on the effect of chronic CO on membrane K^+ channels of pulmonary artery (PA) smooth muscle cells (SMCs).

2. PHARMACOLOGICAL RELEVANCE OF CHRONIC CO EXPOSURE

Global background concentrations of CO are in the range of 50–120 ppb *(8)*. CO is produced by both natural and anthropogenic processes, and recent data suggest that human activities are responsible for about 60% of the CO in the nonurban troposphere *(8)*. The concentration of CO shows a considerable variability depending on where the CO measurement was made. For example, CO concentration was found to be generally <25 ppm inside an automobile under typical driving conditions *(9)*, average concentrations ranged from 2 or 3 ppm to 9 ppm in some US kitchens *(10)*, and levels ranging from 10 to 50 ppm were found for pedestrians and street workers in Toronto *(11)*. The highest peak indoor CO concentration measured was >600 ppm and was associated with emissions from geysers *(12)*. Effects related to the production of hypoxemia for acute exposure to CO have historically been a basis of concern, and in recent years this has grown to include concerns for potential effects from chronic exposure as well. In addition, occupational exposure limits are utilized in numerous countries in which levels differ with the period of exposure. For example, this level ranges from 25 to 50 ppm for a typical 8-h working day and 9 ppm for the recommended multihour ambient air *(8,13)*. The most common source of CO is tobacco smoke. The CO concentration in tobacco smoke is approx 45,000 ppm (4.5%), and a cigarette smoker may be exposed to 400–500 ppm of CO for the time it takes to smoke a cigarette (6 min) with a carboxyhemoglobin (COHb) of 4% (compared with 1% for nonsmokers and 15% for heavy smokers) *(13)*. CO is, of course, not the only compound in cigarette smoke (*see* ref. *14*), and it was demonstrated that 2% CO inhalation (inducing a COHb of 90%) is not a causative factor for cardiopulmonary dysfunction after smoke inhalation *(15)*.

Claude Bernard *(16)* and John Haldane *(17)* described the first scientific studies of the hypoxic effects of CO. The hypoxic and pathological effects of CO result from the formation of COHb, a tight but slowly reversible complex with Hb. This decreases the oxygen-carrying capacity of blood and the availability of oxygen to tissues. Another proposed pathological mechanism of CO, independent of hypoxic stress, is the elevation of the free radical nitric oxide (NO) *(13)*. All the pathological effects occur mainly with high concentrations of CO and/or long exposure periods. The effect of CO inhalation on the pulmonary circulation was not clearly studied, and very high concentrations of CO were used at up to 10,000 ppm *(8)*. Furthermore, when CO is delivered through the alveoli, blood Hb acts more as a "CO buffer" that could prevent (at low concentrations and in normoxia) CO uptake by other tissues. This suggests that if CO has direct effects on SMCs, the more exposed and sensitive cells should be those from the bronchioles and resistive PAs. Thus, in the pulmonary circulation, CO could act as inhaled NO and fit well into the concept of gasotransmitters *(4)*.

3. CHRONIC CO EXPOSURE

3.1. Activation of K_{Ca} Channels in PA Smooth Muscle

3.1.1. Membrane Hyperpolarization of Pressurized PA SMCs

As indicated previously, studying the effect of chronic CO in the pulmonary circulation is pharmacologically relevant. Nevertheless, only a few experiments have been conducted on the PA circulation and almost none on the effect of chronic CO on PA SMCs. Recently, we demonstrated that chronic CO (rats were placed in an exposure chamber containing an air-CO mixture for 3 wk at 530 ppm) increases Ca^{2+}-activated K^+ currents (K_{Ca} currents), which hyperpolarized isolated resistance PA SMCs by approx 10 mV *(18)*. Furthermore, the resting membrane potential of PA SMCs isolated from chronic CO-exposed rats was more negative than for controls, and this was associated with a decrease in membrane resistance in these cells. Chronic CO exposure also hyperpolarizes (by approx 10 mV) the membrane potential of pressurized PA vessels (intraluminal pressure was maintained at 14 mmHg) and iberiotoxin (IbTx), a specific blocker of conductance Ca^{2+}-activated K^+ channels (BK_{Ca}), depolarizes membranes of SMCs from arteries isolated from chronic CO-exposed rats but not those from normal/normoxic rats (**Fig. 1**). By contrast, 4-aminopyridine (4-AP), a more specific blocker of voltage-activated K^+ channels, or Kv channels, still depolarized membranes of all groups of rats (**Fig. 1**). These results suggest that there is a shift in the control of resting membrane potential from a 4-AP-sensitive currents to 4-AP-sensitive currents plus IbTx-sensitive currents. Note that IbTx depolarized the membrane of PA rings to values positive for the threshold for activation of L-type Ca^{2+} currents (–40 mV), which suggests that BK_{Ca} channels act as a feedback mechanism in the regulation of Ca^{2+} entry, as previously suspected in other preparations *(19)*. There was no significant difference between the resting membrane potential of PA rings isolated from rats exposed to chronic CO of 50 or 530 ppm (**Fig. 1B**). This result demonstrates that ambient/low concentrations of CO can induce a functional change in resistive PA SMCs and suggests that K^+ channels are the main sensor/effector of CO, as also observed in other oxygen-sensing cells such as carotid body type I cells *(2,20)*.

3.1.2. Increase in BK_{CA} Currents in Isolated PA SMCs

There is no direct evidence that CO is able to increase K^+ currents in PA SMCs. Furthermore, the physiological and pathological importance of endogenous CO on K^+ channel functions has not been studied directly. Indeed, only studies using blockers of HO to inhibit endogenous CO production were performed and only in systemic vessels *(21–23)*.

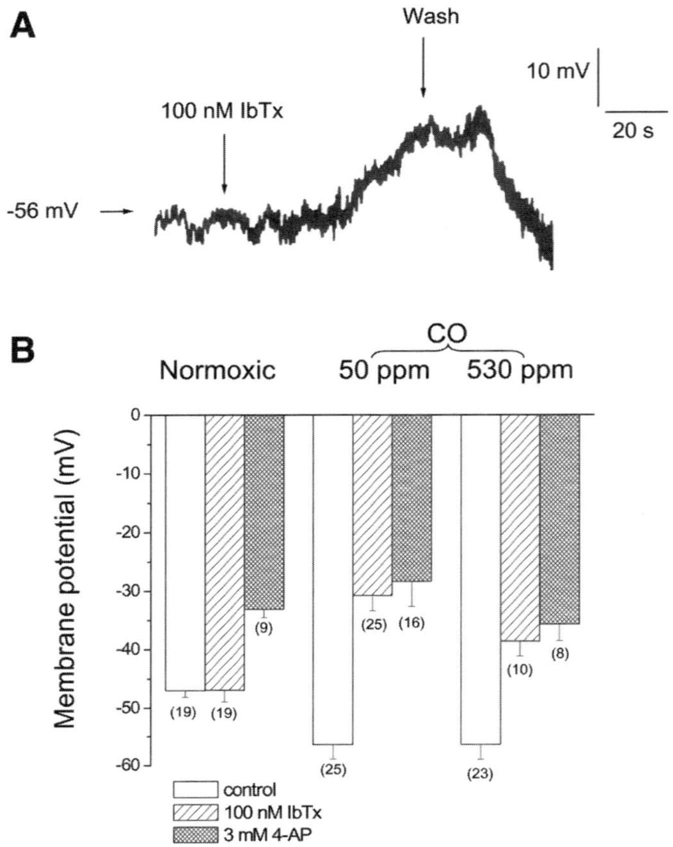

Fig. 1. Effect of IbTx and 4-AP on resting membrane potential of pressurized PA SMCs. **(A)** Example of resting membrane potential recording in pressurized artery obtained in rat exposed to 50 ppm of chronic CO before and after action of IbTx. **(B)** Histogram showing means ± SEM values of resting membrane potential in control condition and after action of IbTx or 4-AP in three populations of rats. The numbers in parentheses indicate the number of cells of pressurized PA from at least three different rats.

Studying the effect of CO on K^+ currents in PA is important because K^+ channel activity is linked to contractile tone, and, therefore, factors that regulate the activity of these channels have a major influence on blood vessel diameter and on PA blood pressure. PA SMCs have a high input resistance in the resting state (reported as 18 GΩ by Evans et al. *[24]* and 4.7 GΩ by Park et al. *[25]* and Dubuis et al. *[18]*) that corresponds to low ionic channel activity near the resting membrane potential. As a consequence, the opening or closure of just a few channels can cause a substantial change in membrane potential and thus modification of arterial tone. On the other hand, PA SMCs have a high density of K_{Ca} channels (mainly BK_{Ca}) and Kv channels. Both of these channels regulate the resting membrane potential of PA SMCs *(26–28)*. Because BK_{Ca} is suspected of acting as a feedback mechanism in the regulation of Ca^{2+} entry *(19)*, a deregulation of these K^+ channels (as observed in chronic hypoxia *(29)*) could change the reactivity of vascular smooth muscle.

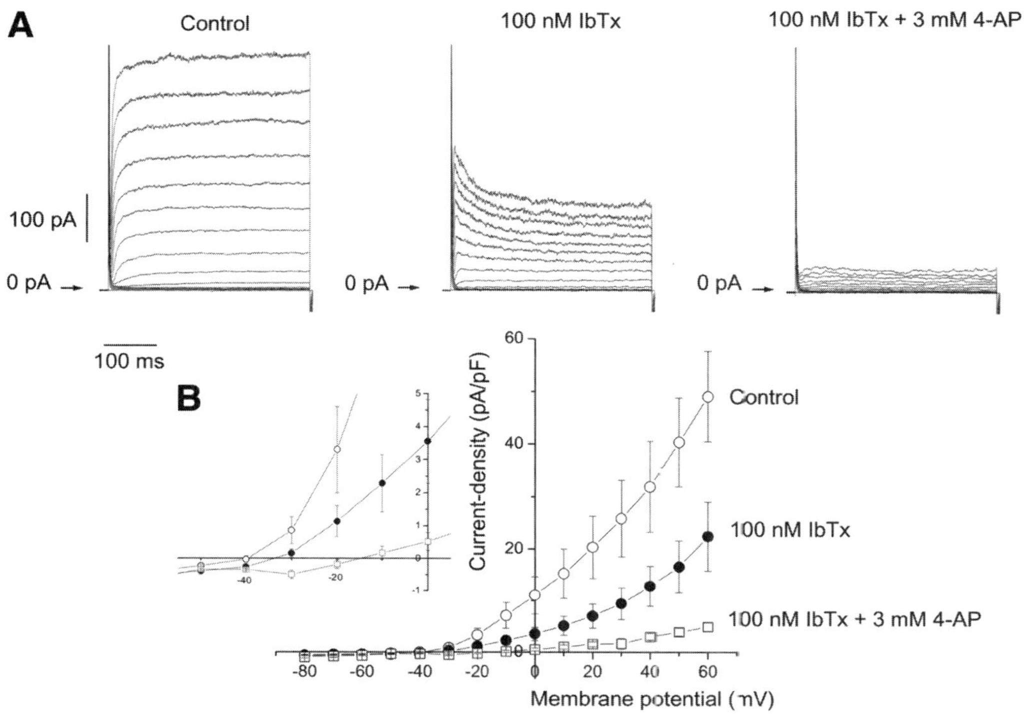

Fig. 2. Effect of IbTx and 4-AP on whole-cell current and current density-voltage relations during voltage step in isolated cells from rat exposed to 50 ppm of chronic CO. **(A)** IbTx (100 n*M*) and 4-AP (3 m*M*) blocked a large component of outward current. **(B)** Current-voltage relations showing effect of IbTx and 4-AP on isolated cells. The inset shows an expanded scale of the same current density-voltage relation between –55 and 5 mV. The results represent the means ± SEM

As we have already stated, pulmonary circulation is a singular circulation close to alveoli by which CO can be delivered. Thus, a small amount of CO arising from alveoli could act directly on the K^+ channels of resistive PA SMCs without delivery by blood Hb; **Figure 2** illustrates this point. We used a small amount of CO (50 ppm—within the range of the occupational exposure limits used in numerous countries) and demonstrated that, at this low concentration, CO could increase K^+ currents.

Figure 2 shows that 50 ppm of chronic CO exposure for 3 wk induced an increase in net outward current of PA SMCs to approx 45 pA/pF at +60 mV (note that in control rats this net current was about 30 pA/pF at +60 *[18]*). Furthermore, IbTx decreased outward currents at potentials close to the resting membrane potential (**Fig. 2B**, inset). This is what we would predict from a current that controls the resting membrane potential. This increase in net outward current results from an increase in IbTx-sensitive current, probably BK_{Ca} current, which represents at least 50% of the net outward current at +60 mV in these chronic CO-exposed rats, compared with 9% in normoxic rats (**Fig. 3**). When we increased the concentration of CO to 530 ppm, there was no significant difference in the effect observed with 50 ppm. Indeed, in rats exposed to 530 ppm of chronic CO, was an increase in the IbTx-sensitive current that represented at +60 mV 45% of the total current (**Fig. 3**). In both chronic CO groups, the amplitude of the 4-AP-sensitive current was not changed, but it represented approx 35% of the net outward current, compared with 55%

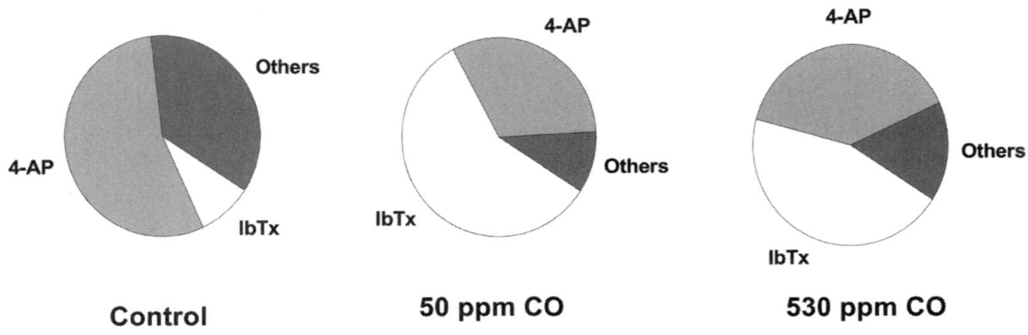

Fig. 3. Representation of contribution (in % of total outward current amplitude) of IbTx and 4-AP-sensitive current at +60 mV obtained in control and in two groups of rats exposed to chronic CO. Chronic CO increased IbTx-sensitive currents but decreased the 4-AP-sensitive current contribution. Because part of the total outward current was not sensitive to IbTx and 4-AP, we named this current "Others," for other current that is not blocked by IbTx and 4-AP.

in control rats (**Fig. 3**). This result suggests that CO treatment could induce a differential expression of K^+ channels in resistance PA SMCs. In this regard, a recent study has already demonstrated a differential expression of Kv and K_{Ca} channels in vascular SMCs (VSMCs) after 1 d of culture *(30)*. Note that in coronary artery SMCs, we observed an increase in 4-AP-sensitive currents with no change in a tetraethylammonium (TEA)-sensitive currents *(31)*. We suspect that this effect results more from a cardiac consequence (hypoxia) than from a direct effect of CO on SMCs *(31)*.

Note that 50 as well as 530 ppm of exogenous CO *(18)* had no effect on mean PA pressure. This can be explained by other and opposite effects of CO that could induce mainly an increase in mean PA pressure. For example, CO could inhibit nitric oxide synthase activity *(32)* favoring a contraction. In addition, this lack of effect of CO on mean PA pressure might be because of an increase in cardiac output induced by chronic CO. Nevertheless, an effect of chronic CO on cardiac output was only observed for concentrations up to those inducing 30–40% COHb *(33)*. For lower COHb and for rats exposed to CO for 3–10 wk at 50 ppm, we observed no change in cardiac output *(34)*.

A part of exogenous CO will be bound, in vivo, to Hb and will have no direct effect on vascular tone. This should induce hypoxia and have in PA an opposite action to CO, inducing a decrease in BK_{Ca} currents and a depolarization of membrane of PA-resistive SMCs. We postulated that at 50 ppm this effect would be minimal and the part of unbound CO could act directly via alveolar on resistance PA to induce vasodilation.

3.1.3. Increase in BK_{Ca} Channels in Isolated PA SMCs

All the results presented suggest that the increase in K^+ currents induced by chronic CO exposure results in an increase in BK_{Ca} channels (mainly because the current is sensitive to IbTx). To further characterize of the channels involved, we recorded single-channel currents from inside-out membrane patches in these resistive chronic CO PA SMCs (**Fig. 4**). In these membrane patches, the activity of a BK_{Ca} predominated because most of the current was blocked by IbTx (pipet tips were briefly loaded with drug-free pipet solution and then back-filled with pipet solution containing 100 n*M* IbTx), and the single-channel conductance was calculated to be 266 ± 5 pS ($n = 4$) in symmetrical 140 m*M* KCl conditions and pCa7. This conductance was obtained from the approxi-

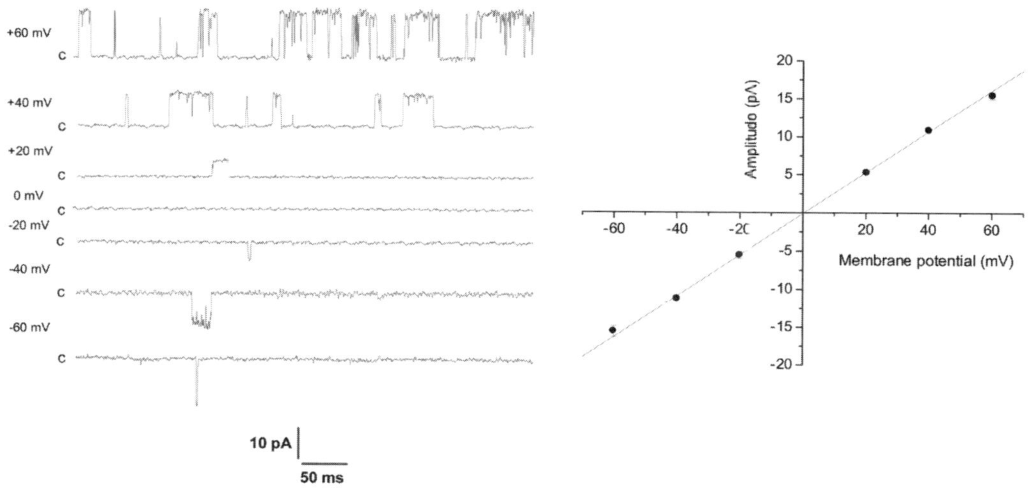

Fig. 4. Typical example of single-channel currents from one inside-out membrane patch (left) and current-voltage relation obtained in four resistive chronic CO PA SMCs (right). The "c" beside each trace indicates the closed channel level for membrane potential held from –60 to +60 mV.

mately linear (between –60 and +60 mV) current-voltage relation (**Fig. 4**) and was in the range of BK_{Ca} single-channel conductances obtained in most SMCs *(35)* but is higher than those found in rat resistive PA SMCs *(36)*. Indeed, we found a conductance that was approx 30 pS less than normoxic resistive rat PA SMCs (unpublished data). The voltage for half-maximal activation ($V_{1/2}$) and the slope factor (*Vs*; obtained from the NPo-voltage relation) at pCa7 were found to be 74 ± 7 mV and 12.9 ± 1.1, respectively in four chronic CO PA SMCs. From these values only the $V_{1/2}$ was significantly lower, by approx 40 mV, than that observed in normoxic resistive rat PA SMCs (unpublished data). All these data clearly demonstrate that chronic CO increases functional BK_{Ca} channel activity by increasing the conductance and open probability for the same amount of intracellular Ca^{2+}. The mechanism involved needs to be elucidated, but a differential coexpression of BK_{Ca} channel α-subunits with the β-subunits induced by chronic CO may explain some of the effects observed *(37)*.

3.2. CO-Induced Increased CO Sensitivity Mechanism

3.2.1. Relaxation of PA Smooth Muscle by Acute CO

Several lines of investigation provide evidence that CO is a general vasodilator acting directly on VSMCs *(5)*. The mechanism of CO-induced vasodilation includes a soluble guanylate cyclase (sGC)-dependent and an sGC-independent mechanism *(5)*. In addition, among the sGC-independent mechanisms, CO could directly activate BK_{Ca} in tail artery SMCs *(38)* and Kv channels in jujenal circular SMCs *(39)*. This direct effect of acute CO on BK_{Ca} could explain the increasing effective coupling of Ca^{2+} sparks to these channels *(40)*.

In the pulmonary circulation, exogenous CO decreases vascular resistance *(7)* and inhibits hypoxic pulmonary vasoconstriction by a cyclic guanosine 5'-monophosphate (cGMP)-independent mechanism *(41)*, probably by a direct vasorelaxation effect of CO on SMCs *(42)*. Recently, we demonstrated that acute CO relaxed PA rings *(43)*. The

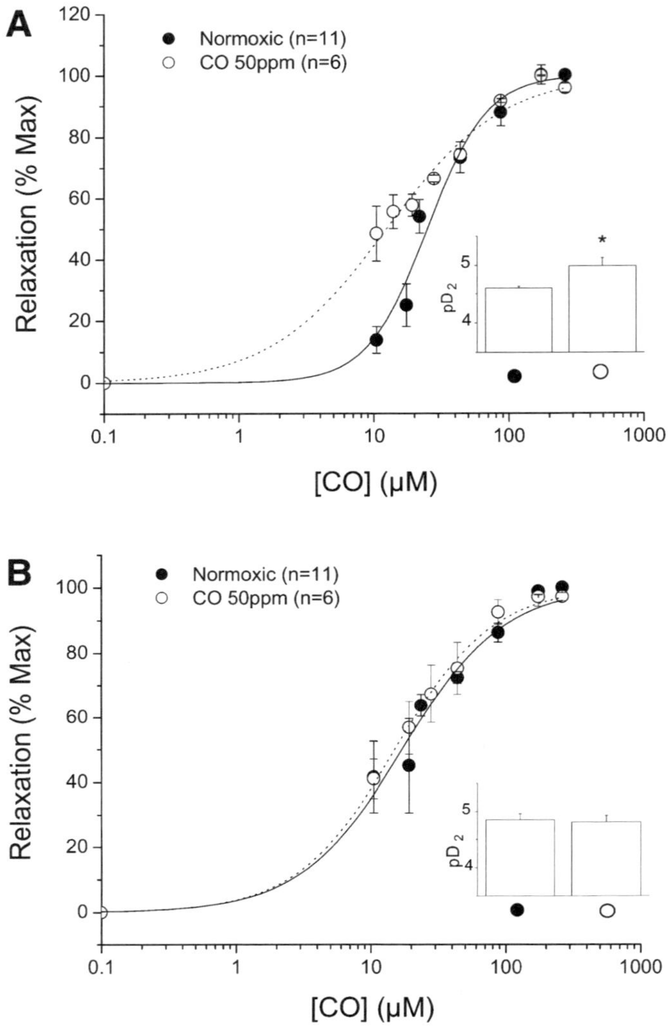

Fig. 5. Concentration-dependent relaxant effect of acute CO in PA (**A**) and thoracic aorta (**B**) rings isolated from normoxic rats and rats exposed to 50 ppm of chronic CO. Changes in tension induced by CO are expressed as a percentage of the maximal relaxation induced by 279 μ*M* CO. The inset in (**B**) shows the means ± SEM pD_2 values (obtained using the Boltzmann function) from the two populations of rings. The lines (solid and dotted) represent fitted lines using the Boltzmann function. $*p < 0.05$ vs normoxic rats.

threshold concentration for the relaxant effect was approx 10 μ*M* with an EC_{50} of 25 μ*M* and a maximal relaxation obtained at 279 μ*M* (**Fig. 5A**). Because we cannot estimate the loss of CO in our open chamber, the actual concentration of CO in the bath might be somewhat lower than the estimated concentration *(44)*. Nevertheless, all these concentrations were in the range of those used in vitro and in other vascular preparations *(5)*. This relaxation was mainly abolished by 1H-[1,2,4]oxadiazolo[4,3,-a]quinoxalin-1-one (ODQ), a selective blocker of sGC *(45)*, and accounted for 60% of sGC activation. Numerous studies with different vascular beds, including PA, support the role of sGC in CO-induced relaxation *(46)* or vasodilation *(47)*. Furthermore, the range of CO used in our experiments is consistent with the in vitro sGC stimulation *(5)*. This effect of CO is linked to an increase in cGMP concentration in the lung *(47)* and in SMCs *(48)*. Like Naik and Walker *(47)*, we demonstrated that CO-induced PA relaxation in vitro is dependent on the activation of sGC but independent of the activation of K^+ channels, suggesting a mechanism different from that observed in tail artery SMCs in which CO induces relaxation by activating both a cGMP signaling pathway and BK_{Ca} channels *(49)*. Nevertheless, we could not totally block the CO relaxation using 10^{-5} *M* ODQ (a concentration high enough to entirely block acute CO relaxation in PA rings) *(46)*, suggesting another mechanism of action of CO.

3.2.2. Enhancement of In Vitro Sensitivity of PA Smooth Muscle by Chronic CO

All the experiments that we performed showed that chronic CO was able to activate BK_{Ca} channel activity by an unknown mechanism. The physiological implication of this effect is not clear, but the consequence is a hyperpolarization of PA SMC membrane, a decrease in PA reactivity, and probably a decrease in sensitivity against agonist or pressure, as was demonstrated with endogenous CO in other preparations *(21–23)*. Because BK_{Ca} channels have a high conductance, a small increase in their activity (mainly by increasing Ca^{2+} sensitivity) could have a large hyperpolarization/repolarization effect on membranes. It has been shown that an increase in expression of these channels could induce compensatory vasodilatory pathways in systemic hypertension *(50)* and that activation of BK_{Ca} could reverse acute hypoxic pulmonary vasoconstriction *(51)*. Because we demonstrated that chronic CO may increase Ca^{2+} sensitivity of BK_{Ca} channels, with the consequence being an increase in functional BK_{Ca}, which regulates the resting membrane potential, we suspected an increase in acute CO sensitivity. Indeed, acute CO sensitivity of 25 μ*M* in normoxic (control) PA rings (without endothelium) was increased to approx 12 μ*M* in rate exposed to 50 ppm of chronic CO exposed rats (**Fig. 5A**). We observed no effect on thoracic aortic rings, suggesting a specific effect of low chronic CO on the PA circulation (**Fig. 5B**). We clearly demonstrated that the main mechanism of this increase in CO sensitivity, which we named "CO-induced increased CO sensitivity" (**Fig. 6**), is an increase in K^+ channel activity regulating PA tone (**Fig. 6**). Indeed, TEA had little effect on the relaxation induced by acute CO on normoxic PA rings *(43)*, but it totally abolished the relaxation in chronic CO-exposed rats (**Fig. 7**). Another effect is an increased the effect of ODQ, suggesting that acute CO activates mainly K^+ channels by increasing sGC activity and/or the cGMP sensitivity of K^+ channels. The mechanism needs to be explored further.

3.3. Increased In Vitro Sensitivity of Hypoxic PA Smooth Muscle

Prolonged exposure to alveolar hypoxia (chronic hypoxia) induces physiological pulmonary vascular remodeling that results in the development of pulmonary hypertension *(52)* with a reduced expression of Kv α-subunit proteins *(53,54)* and of BK_{Ca} channel activity *(29)*. In PA obtained from rats treated with chronic hypoxia (rats were placed in a hypobaric chamber for 3 wk at 0.5 atm), we found that the amplitude of CO relaxation (normalized to the amplitude obtained with 80 m*M* external K^+) as well as CO sensitivity was decreased *(43)*. Compared with normoxic PA rings, the EC_{50} of these hypoxic rings was doubled and increased to 50 μ*M*. By contrast, we found that the amplitude of CO relaxation as well as CO sensitivity was increased in these hypoxic PA rings with 50 ppm of CO was added at the same time as hypoxia (rats were placed in a hypobaric chamber with 50 ppm of CO for 3 wk at 0.5 atm). The acute CO sensitivity was at least doubled

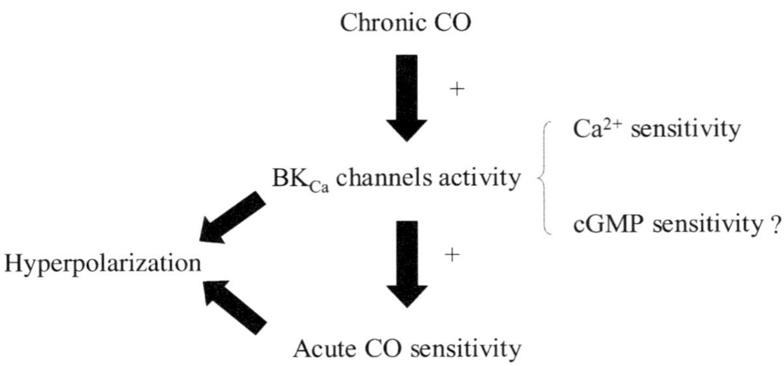

Fig. 6. Mechanism of CO-induced CO-increased sensitivity. Chronic CO increased BK_{Ca} activity, which is responsible for the increase in acute CO sensitivity and for the hyperpolarization.

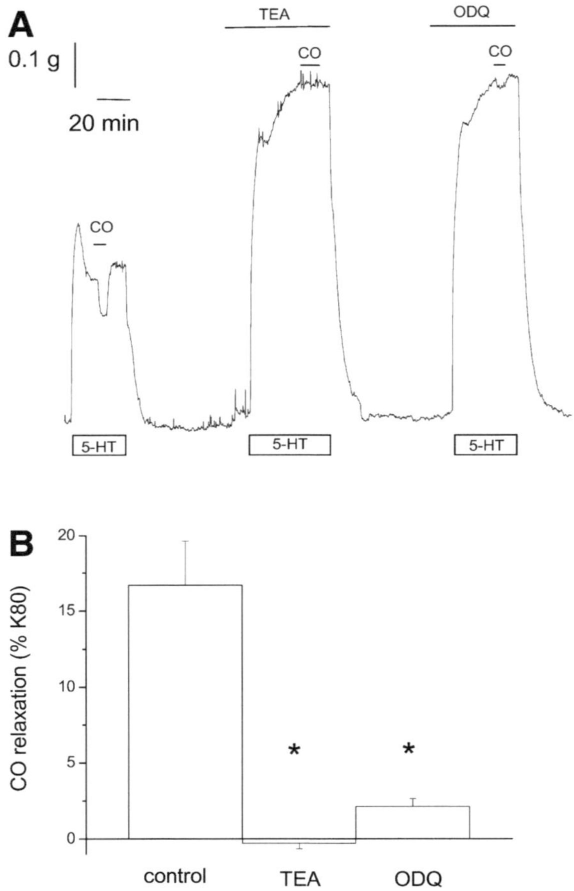

Fig. 7. Vasorelaxant effect of acute CO on PA rings from rat exposed to 50 ppm of chronic CO in presence of soluble guanylate cyclase inhibitor or K^+ channel inhibitor. **(A)** Original representative traces showing the relaxant effect of CO (23 μ*M*) and its decrease by the K^+ channel inhibitor TEA (20 m*M*) and the soluble guanylate cyclase inhibitor ODQ (10^{-5} *M*). **(B)** Relaxant effect of acute CO alone or in presence of 10^{-5} *M* ODQ or 20 m*M* TEA in bath solution. The histogram represents the means ± SEM obtained in six PA rings. *$p < 0.05$ vs control. 5-HT, 5-hydroxytryptamine.

with an EC_{50} of about 10 μM. These results demonstrate that chronic CO can prevent a hypoxia-induced decrease in acute CO sensitivity in PA rings. This may be explained by a different effect of chronic hypoxia and chronic CO on type HO-1 activity. Indeed, Carraway et al. *(55)* showed that chronic exogenous CO at 50 ppm could induce an increase in HO-1 activity in chronic hypoxic rats, and it was demonstrated that HO-1 is transiently increased by chronic hypoxia *(56)*. Furthermore, a sustained activation or an overexpression of HO-1 prevents the development of hypoxic pulmonary hypertension *(57,58)*. Nevertheless, the link between exogenous CO and endogenous CO production needs to be explored.

In chronic hypoxic PA rings, CO relaxation was almost totally abolished by ODQ, suggesting that CO relaxed hypoxic PA rings only through sGC activation *(43)*. The decrease in CO relaxation in chronic hypoxia could thus be explained by an increase in phosphodiesterase type 5 activity and/or a decrease in PA SMC reactivity to cGMP, as demonstrated for NO in PA *(59)*. PA SMCs have a high density of BK_{Ca} channels and these channels (through spontaneous transient outward currents) regulate the resting membrane potential of PA SMCs *(26,28)* and thus PA contraction. Chronic hypoxia is associated with a decrease in KCa channel activity and a reduced ability of cGMP to activate BK_{Ca} probably through inhibition of the cGMP-dependent protein kinase–induced activation of BK_{Ca} *(29)*. This decreased ability of cGMP to activate BK_{Ca}, associated with a decrease in BK_{Ca} activity could explain the decrease in acute CO sensitivity. Then, if chronic CO could prevent the effect of hypoxia on CO relaxation, one should obtain an increase in acute CO sensitivity linked to an increased effect of CO on K^+ channels. This is the case and the CO relaxation was almost totally abolished by TEA, suggesting that CO relaxed hypoxic-CO PA rings only through K^+ channel activation. This increase in sensitivity could be explained, in contrast to chronic hypoxia, by an increase in BK_{Ca} activity, as we already published for chronic CO without chronic hypoxia *(18)*. Indeed, the range of CO used and its EC_{50} in our experiments was consistent with the in vitro BK_{Ca} stimulation *(5)*. Furthermore, in systemic arteries it was demonstrated that endogenously generated CO affected renal tone and attenuated vasoconstrictor reactivity by increasing BK_{Ca} activity *(21)*, a phenomenon also observed in chronic hypoxia for renal *(60)* and mesenteric arteries *(61)*. Nevertheless, the regulation of BK_{Ca} by cGMP needs to be explored further. The decreased effect of ODQ on acute CO-induced relaxation could be explained in hypoxic-CO PA rings by a decrease in sGC activity. Indeed, a recent study demonstrated that prolonged overexpression of HO-1 (observed in hypoxic-CO PA rings) decreased sGC activity by limiting the availability of cellular heme *(62)*.

4. CONCLUSION

These results demonstrate that chronic CO as well as acute CO can activate K^+ channels and, in particular, BK_{Ca} channels in PA SMCs. The mechanism by which chronic CO increases functional expression of BK_{Ca} channels needs to be explored and several heme proteins may be involved. This effect of chronic CO is pharmacologically and physiologically relevant because the concentration of CO used are in the range of the observed environmental concentrations. This activation of K^+ channels by chronic CO in PA SMCs may have therapeutic implications because it has been suggested that drug therapies acting on K^+ channel expression may have potential value for the treatment of vascular diseases *(63)*. More specifically, this increase in BK_{Ca} currents induced by chronic CO (exogenously and/or endogenously) may provide a novel target for therapy during pulmonary hypertension in a fashion similar to NO *(64)*. In this regard, a recent study has already demonstrated a protective effect of inhaled CO (0.1%) in the lung ischemia/reperfusion model *(65)*.

ACKNOWLEDGMENTS

The authors thank Marie-Astrid Coolen for secretarial assistance, and Catherine Girardin and Manuel Rebocho for technical assistance. This work was supported by a grant from Agence de l'Environnement et de la Maîtrise d'Energie (ADEME) (98 93 029), la Fondation Simone et Cino Del Duca, and the Wellcome Trust. E. Dubuis holds a doctoral fellowship from ADEME and Conseil Regional du Centre.

REFERENCES

1. Weaver LK. Carbon monoxide poisoning. Crit Care Clin 1999;15:297–317.
2. Barbe C, Al-Hashem F, Conway AF, et al. A possible dual site of action for carbon monoxide-mediated chemoexcitation in the rat carotid body. J Physiol 2002;543:933–945.
3. Kumar P, Dubuis E, Vandier C. Regulation of K^+ currents by CO in carotid body type I cells and pulmonary artery smooth muscle cells. Adv Exp Med Biol 2003;536:147–154.
4. Wang R. Two's company, three's a crowd: can H_2S be the third endogenous gaseous transmitter? FASEB J 2002;16:1792–1798.
5. Wang R. Resurgence of carbon monoxide: an endogenous gaseous vasorelaxing factor. Can J Physiol Pharmacol 1998;76:1–15.
6. Tenhunen R, Marver HS, Schmid R. The enzymatic conversion of heme to bilirubin by microsomal heme oxygenase. Proc Natl Acad Sci USA 1968;61:748–755.
7. Duke HN, Killick EM. Pulmonary vasomotor responses of isolated perfused cat lung to anoxia. J Physiol 1952;117:303–316.
8. World Health Organization. Carbon Monoxide: Environmental Health Criteria 213. 2nd ed. Geneva, Switzerland, World Health Organization: 1999.
9. Petersen GA, Sabersky RH. Measurements of pollutants inside an automobile. J Air Pollut Control Assoc 1975;25:1028–1032.
10. Wade WA, Cote WA, Yocom JE. A study of indoor air quality. J Air Pollut Control Assoc 1975;25:933–939.
11. Wright GR, Jewczyk S, Onrot J, et al. Carbon monoxide in the urban atmosphere: hazards to the pedestrian and the street-worker. Arch Environ Health 1975;30:123–129.
12. Brunekreef B, Smit J, Biersteker K, et al. Indoor carbon monoxide pollution in The Netherlands. Environ Int 1982;8:193–196.
13. Raub JA, Mathieu-Nolf M, Hampson NB, et al. Carbon monoxide poisoning—a public health perspective. Toxicology 2000;145:1–14.
14. Traber DL, Bradford DW. Pulmonary changes induced by the administration of carbon monoxide and other compounds in smoke. In: Penney DG, ed. Carbon Monoxide. 1st ed. CRC Press: New York, 1996, pp. 87–108.
15. Sugi K, Theissen JL, Traber LD, et al. Impact of carbon monoxide on cardiopulmonary dysfunction after smoke inhalation injury. Circ Res 1990;66:69–75.
16. Bernard C. Leçons sur les effets des substances toxiques et médicamenteuses. Bailliere: Paris, 1857.
17. Haldane JS. The relation of the action of carbonic oxide to oxygen tension. J Physiol 1895;18:201–217.
18. Dubuis E, Gautier M, Melin A, et al. Chronic carbon monoxide enhanced IbTx-sensitive currents in rat resistance pulmonary artery smooth muscle cells. Am J Physiol Lung Cell Mol Physiol 2002;283: L120–L129.
19. Brayden JE, Nelson MT. Regulation of arterial tone by activation of calcium-dependent potassium channels. Science 1992;256:532–535.
20. Riesco-Fagundo AM, Perez-Garcia MT, Gonzalez C, et al. O(2) modulates large-conductance Ca(2+)-dependent K(+) channels of rat chemoreceptor cells by a membrane-restricted and CO-sensitive mechanism. Circ Res 2001;89:430–436.
21. Kaide JI, Zhang F, Wei Y, et al. Carbon monoxide of vascular origin attenuates the sensitivity of renal arterial vessels to vasoconstrictors. J Clin Invest 2001;107:1163–1171.

22. Liu H, Mount DB, Nasjletti A, et al. Carbon monoxide stimulates the apical 70-pS K^+ channel of the rat thick ascending limb. J Clin Invest 1999;103:963–970.
23. Zhang F, Kaide J, Wei Y, et al. Carbon monoxide produced by isolated arterioles attenuates pressure-induced vasoconstriction. Am J Physiol Heart Circ Physiol 2001;281:H350–H358.
24. Evans AM, Osipenko ON, Gurney AM. Properties of a novel K^+ current that is active at resting potential in rabbit pulmonary artery smooth muscle cells. J Physiol 1996; 496:407–420.
25. Park MK, Bae YM, Lee SH, et al. Modulation of voltage-dependent K^+ channel by redox potential in pulmonary and ear arterial smooth muscle cells of the rabbit. Pflügers Arch 1997;434:764–771.
26. Bae YM, Park MK, Lee SH, et al. Contribution of Ca^{2+}-activated K^+ channels and non-selective cation channels to membrane potential of pulmonary arterial smooth muscle cells of the rabbit. J Physiol 1999;514:747–758.
27. Osipenko ON, Alexander D, MacLean MR, et al. Influence of chronic hypoxia on the contributions of non-inactivating and delayed rectifier K currents to the resting potential and tone of rat pulmonary artery smooth muscle. Br J Pharmacol 1998; 124:1335–1337.
28. Vandier C, Delpech M, Bonnet P. Spontaneous transient outward currents and delayed rectifier K^+ current: effects of hypoxia. Am J Physiol Lung Cell Mol Physiol 1998; 275:L145–L154.
29. Peng W, Hoidal JR, Karwande SV, et al. Effect of chronic hypoxia on K^+ channels: regulation in human pulmonary vascular smooth muscle cells. Am J Physiol Cell Physiol 1997;272:C1271–C1278.
30. Tang G, Wang R. Differential expression of K_V and KCa channels in vascular smooth muscle cells during 1-day culture. Pflügers Arch 2001;442:124–135.
31. Barbe C, Dubuis E, Rochetaing A, et al. A 4-AP-sensitive current is enhanced by chronic carbon monoxide exposure in coronary artery myocytes. Am J Physiol Heart Circ Physiol 2002;282: H2031–H208.
32. Durante W, Schafer AI. Carbon monoxide and vascular cell function (review). Int J Mol Med 1998;2: 255–262.
33. Penney DG. Hemodynamic response to carbon monoxide. Environ Health Perspect 1988;77:121–130.
34. Melin A, Obert P, Rebocho M, et al. Cardiac morphology and function following long-term exposure to carbon monoxide at high altitude in rats. J Toxicol Environ Health A 2002;65:1981–1998.
35. Carl A, Lee HK, Sanders KM. Regulation of ion channels in smooth muscles by calcium. Am J Physiol Cell Physiol 1996;271:C9–C34.
36. Albarwani S, Heinert G, Turner JL, et al. Differential K^+ channel distribution in smooth muscle cells isolated from the pulmonary arterial tree of the rat. Biochem Biophys Res Commun 1995;208:183–189.
37. Toro L, Wallner M, Meera P, et al. Maxi-K(Ca), a unique member of the voltage-gated K channel superfamily. News Physiol Sci 1998;13:112–117.
38. Wang R, Wu L, Wang Z. The direct effect of carbon monoxide on KCa channels in vascular smooth muscle cells. Pflügers Arch 1997;434:285–291.
39. Farrugia G, Irons WA, Rae JL, et al. Activation of whole cell currents in isolated human jejunal circular smooth muscle cells by carbon monoxide. Am J Physiol Gastrointest Liver Physiol 1993;264: G1184–G1189.
40. Jaggar JH, Leffler CW, Cheranov SY, et al. Carbon monoxide dilates cerebral arterioles by enhancing the coupling of Ca^{2+} sparks to Ca^{2+}-activated K^+ channels. Circ Res 2002;91:610–617.
41. Tamayo L, Lopez-Lopez JR, Castaneda J, et al. Carbon monoxide inhibits hypoxic pulmonary vasoconstriction in rats by a cGMP-independent mechanism. Pflügers Arch 1997;434:698–704.
42. Steinhorn RH, Morin FC, Russell JA. The adventitia may be a barrier specific to nitric oxide in rabbit pulmonary artery. J Clin Invest 1994;94:1883–1888.
43. Dubuis E, Gautier M, Melin A, et al. Chronic carbon monoxide exposure of hypoxic rats increases in vitro sensitivity of pulmonary artery smooth muscle. Can J Physiol Pharmacol 2003;81:711–719.
44. McLaughlin BE, Chretien ML, Choi C, et al. Potentiation of carbon monoxide-induced relaxation of rat aorta by YC-1 [3-(5'-hydroxymethyl-2'-furyl)-1-benzylindazole]. Can J Physiol Pharmacol 2000;78: 343–349.
45. Garthwaite J, Southam E, Boulton CL, et al. Potent and selective inhibition of nitric oxide-sensitive guanylyl cyclase by 1H-[1,2,4]oxadiazolo[4,3-a]quinoxalin-1-one. Mol Pharmacol 1995;48:184–188.
46. Villamor E, Perez-Vizcaino F, Cogolludo AL, et al. Relaxant effects of carbon monoxide compared with nitric oxide in pulmonary and systemic vessels of newborn piglets. Pediatr Res 2000;48:546–553.
47. Naik JS, Walker BR. Homogeneous segmental profile of carbon monoxide–mediated pulmonary vasodilation in rats. Am J Physiol Lung Cell Mol Physiol 2001; 281:L1436–L1443.
48. Morita T, Perrella MA, Lee ME, et al. Smooth muscle cell–derived carbon monoxide is a regulator of vascular cGMP. Proc Natl Acad Sci USA 1995;92:1475–1479.

49. Wang R, Wang Z, Wu L. Carbon monoxide–induced vasorelaxation and the underlying mechanisms. Br J Pharmacol 1997;121:927–934.
50. Liu Y, Hudetz AG, Knaus HG, et al. Increased expression of Ca^{2+}-sensitive K^+ channels in the cerebral microcirculation of genetically hypertensive rats: evidence for their protection against cerebral vasospasm. Circ Res 1998;82:729–737.
51. Farrukh IS, Peng W, Orlinska U, et al. Effect of dehydroepiandrosterone on hypoxic pulmonary vasoconstriction: a Ca(2+)-activated K(+)-channel opener. Am J Physiol Lung Cell Mol Physiol 1998;274:L186–L195.
52. Reeves JT, Wagner WW Jr., McMurtry IF, et al. Physiological effects of high altitude on the pulmonary circulation. Int Rev Physiol 1979;20:289–310.
53. Platoshyn O, Yu Y, Golovina VA, et al. Chronic hypoxia decreases K(V) channel expression and function in pulmonary artery myocytes. Am J Physiol Lung Cell Mol Physiol 2001;280:L801–L812.
54. Wang J, Juhaszova M, Rubin LJ, et al. Hypoxia inhibits gene expression of voltage-gated K^+ channel alpha subunits in pulmonary artery smooth muscle cells. J Clin Invest 1997;100:2347–2353.
55. Carraway MS, Ghio AJ, Suliman HB, et al. Carbon monoxide promotes hypoxic pulmonary vascular remodeling. Am J Physiol Lung Cell Mol Physiol 2002;282:L693–702.
56. Carraway MS, Ghio AJ, Carter JD, et al. Expression of heme oxygenase-1 in the lung in chronic hypoxia. Am J Physiol Lung Cell Mol Physiol 2000;278:L806–L812.
57. Christou H, Morita T, Hsieh CM, et al. Prevention of hypoxia-induced pulmonary hypertension by enhancement of endogenous heme oxygenase-1 in the rat. Circ Res 2000;86:1224–1229.
58. Minamino T, Christou H, Hsieh CM, et al. Targeted expression of heme oxygenase-1 prevents the pulmonary inflammatory and vascular responses to hypoxia. Proc Natl Acad Sci USA 2001;98 :8798–8803.
59. Jernigan NL, Resta TC. Chronic hypoxia attenuates cGMP-dependent pulmonary vasodilation. Am J Physiol Lung Cell Mol Physiol 2002;282:L1366–L1375.
60. O'Donaughy TL, Walker BR. Renal vasodilatory influence of endogenous carbon monoxide in chronically hypoxic rats. Am J Physiol Heart Circ Physiol 2000;279:H2908–H2915.
61. Gonzales RJ, Walker BR. Role of CO in attenuated vasoconstrictor reactivity of mesenteric resistance arteries after chronic hypoxia. Am J Physiol Heart Circ Physiol 2002;282:H30–H37.
62. Abraham NG, Quan S, Mieyal PA, et al. Modulation of cGMP by human HO-1 retrovirus gene transfer in pulmonary microvessel endothelial cells. Am J Physiol Lung Cell Mol Physiol 2002;283: L1117–L1124.
63. Berger MG, Rusch NJ. Voltage and calcium-gated potassium channels: functional expression and therapeutic potential in the vasculature. In: Sabatier JM, ed. Perspectives in Drug Discovery and Design. Kluwer Academic Publishers: Netherlands, 1999, pp. 313–332.
64. Adnot S, Raffestin B, Eddahibi S. NO in the lung. Respir Physiol 1995;101:109–120.
65. Fujita T, Toda K, Karimova A, et al. Paradoxical rescue from ischemic lung injury by inhaled carbon monoxide driven by derepression of fibrinolysis. Nat Med 2001;7:598–604.

IV

Gas of the Rotten Egg: Hydrogen Sulfide as the Third Gasotransmitter

16 Hydrogen Sulfide Production and Metabolism in Mammalian Tissues

Kenneth N. Maclean and Jan P. Kraus

CONTENTS

SUMMARY

Evidence for a physiological role for H_2S rests primarily on observations that this compound is endogenously produced in specific mammalian tissues relevant to its proposed role as a neuromodulator and vasorelaxant. Multiple mammalian enzymes have the potential to catalyze the desulfuration of cysteine to form H_2S including cystathionine β-synthase (CBS; EC 4.2.1.22), cystathionine γ-lyase (CSE; 4.4.1.1), cysteine aminotransferase (EC 2.6.1.3), mercaptopyruvate sulfurtransferase (MST; EC 2.8.1.2), rhodanese (thiosulfate: cyanide sulfurtransferase) (EC 2.8.1.1), and cysteine lyase (EC 4.2.1.10). Presently, there is insufficient knowledge regarding the H_2S forming activities of cysteine lyase, MST, and rhodanese to assess their possible role in the physiological production of this compound. There is some evidence that CSE plays a role in mammalian H_2S production, but a recent biochemical study of human CSE casts doubt on the ability of this enzyme to perform this function. The regulatory and catalytic characteristics of CBS share a striking number of common features with mammalian neuronal and endothelial nitric oxide synthases and thus appears to be eminently suited to the task of producing H_2S as a cell messenger. Our current knowledge of human sulfur metabolism is insufficient to exclude the possibility of other enzymes playing a role in the synthesis of H_2S.

From: *Signal Transduction and the Gasotransmitters: NO, CO, and H_2S in Biology and Medicine* Edited by: Rui Wang © Humana Press Inc., Totowa, NJ

Key Words: Cystathionine β-synthase; cystathionine γ-lyase; mercaptopyruvate sulfurtransferase; cysteine aminotransferase; serine sulfhydrase; hemeprotein; autoinhibitory domain; vasorelaxant; neuromodulator.

1. INTRODUCTION

Hydrogen sulfide (H_2S) is a highly toxic compound that has a median lethal concentration of the same order of magnitude as that of cyanide. Because of the perceived toxicity of H_2S, the possible physiological role of reactions generating H_2S in vivo has not received much attention until recently. Despite its toxicological effects, there is increasing evidence that H_2S is endogenously produced in a range of mammalian cells and that this compound functions as a gasotransmitter *(1)* in a membrane receptor-independent manner analogous to that of nitric oxide (NO) or carbon monoxide (CO). Consequently, interest in the possible mechanism by which H_2S could be produced in a manner consistent with its proposed physiological roles has grown. The purpose of this chapter is to critically assess the methods used for determination of H_2S and to summarize what is currently known about the biochemistry and regulation of candidate enzymes for the production of this compound in mammals.

2. DETERMINATION OF H_2S IN ANIMAL TISSUES

Much of the evidence for a physiological role for H_2S rests on observations that this compound is endogenously produced in specific mammalian tissues that are relevant to its proposed role as a neuromodulator and vasorelaxant *(1–3)*. However, it should be appreciated that accurate determinations of very small concentrations of labile gaseous compounds such as NO or H_2S can be problematic, and it is important to rigorously assess the methodologies used to avoid possible artifactual results.

Mammalian tissues contain sulfur compounds in a labile form, which can be liberated as H_2S by treatment with dithiothreitol (DTT) or by acidification. These compounds include acid-labile sulfur, sulfane sulfur, and protein-associated sulfur. Acid-labile sulfur is the sulfur liberated as H_2S when tissues are treated under acidic conditions and is contained in iron-sulfur clusters. Other labile sulfur species include sulfane sulfur compounds such as thiosulfates, persulfides of cysteine, cysteamine, glutathione and protein, thiotaurine, thiocystine, protein-associated sulfur, and other compounds. These sulfur species often are referred to as "bound sulfur" and are defined as divalent sulfur that is easily liberated by reduction with excess thiols *(4)*. An excellent review of this subject was recently published *(5)*. Most labile sulfur is liberated as inorganic sulfide, and, therefore, the determination of inorganic sulfide is the basis of labile sulfur determination in tissues. Inorganic sulfide can be present as H_2S, HS^-, or S_2^- in the aqueous phase and as H_2S in the gas phase. Thus, it appears that it is difficult to distinguish any free preexisting sulfide present in the tissues from that derived from labile sulfur compounds during analysis. However, Ogasawara et al. *(4)*, were unable to detect any free sulfide ions in various tissues following deproteinization of the samples. Similarly, no free sulfide was detected in the liver and heart by gas chromatographic analysis *(6)*. Labile sulfur can be determined spectrophotometrically with or without derivatization of the resulting sulfide with or without a subsequent high-performance liquid chromatography (HPLC) step. Ubuka *(5)* has recently reviewed these and other methods.

3. PHYSIOLOGICAL CONCENTRATIONS OF H_2S IN THE CIRCULATION AND SPECIFIC TISSUES

The concentration of H_2S that can be released from human serum with DTT is 1.16 ± 0.1 μ*M* in humans and only slightly higher in rabbits, mice, and guinea pigs, and up to 1.88 ± 0.65 μ*M* in rats. Surprisingly, it was found to be significantly higher (6.18 ± 0.96 μ*M*) in bovine samples *(7)*. Tissue concentrations of labile sulfur tend to be much higher, and using the same method in five different rat tissues, it was determined that the lowest concentration of labile sulfur is found in the brain (31.05 ± 6.24 nmol/g), whereas in the kidney the concentration was as high as 363.9 ± 104.8 nmol/g. The relative proportions of bound and acid-labile sulfur vary significantly among different mammalian tissues, and although nearly 90% of the total releasable sulfur in the kidney was found to be in the bound form, 100% of the sulfur in the heart was of the acid-labile type present primarily in the mitochondrial fraction *(4)*. Similar results were obtained using a different method of H_2S release and measuring it by sulfide-derived methylene blue determination with ion interaction reverse-phase HPLC *(8)*. Three mouse tissues were compared: the brain was the lowest, at 68.9 ± 11.3 nmol/g, whereas the kidney was the highest, at 200.1 ± 46.4 nmol/g; the liver had an intermediate value of 144.5 ± 12.4 nmol/g *(9)*. The consistency of these results using two different methods strongly indicates that these results are credible. Other investigators have also investigated sulfide concentrations in rat tissues and expressed the concentrations of H_2S as micrograms per gram tissue *(10,11)*. The concentrations that they determined in different brain subsections and various rat organs are of the same order of magnitude as those expressed as nanomoles per milligram of tissue (see values just cited). The observation that specific tissues appear to contain relatively higher concentrations of H_2S is consistent with a possible physiological role for this compound.

4. ENZYMATIC PRODUCTION OF ENDOGENOUS H_2S

The major source of H_2S in mammals is cysteine. Although bacterial cysteine desulfhydrases are well characterized, no such enzyme has ever been isolated from mammalian tissues. Several enzymes present in mammalian tissues may catalyze the desulfuration of cysteine. These include cystathionine β-synthase (CBS) (EC 4.2.1.22), cystathionine γ-lyase (CSE) (4.4.1.1), and cysteine aminotransferase (EC 2.6.1.3) along with mercaptopyruvate sulfurtransferase (MST) (EC 2.8.1.2), rhodanese (thiosulfate: cyanide sulfurtransferase) (EC 2.8.1.1), and cysteine lyase (EC 4.2.1.10).

Several laboratories investigated the capacity of different tissues to produce H_2S. Stipanuk and Beck *(12)* showed that CBS and CSE were active when rat liver or kidney homogenates were incubated with very high concentrations of cysteine (280 m*M*) and pyridoxal 5'-phosphate (PLP) at pH 7.8. Cysteine aminotransferase in combination with 3-mercaptopyruvate sulfurtransferase catalyzed essentially all of the sulfide production at pH 9.7 with 160 m*M* cysteine. At a more physiological concentration of cysteine (2 m*M*), CBS and CSE both appeared to be active in cysteine desulfuration, whereas the aminotransferase pathway did not. A more recent study by Ogasawara et al. *(4)* examined the tissue-specific and subcellular distribution of enzyme capacity for sulfide production in rats. Bound sulfur was found largely in the cytosolic fraction primarily in the kidney and liver. Sulfide production capacity from cysteine was greatest in liver cytosol and correlated well with the distribution of CSE in tissues and subcellular fractions. CBS, for some reason, was not studied.

4.1. Cystathionine β-Synthase

CBS is a rather promiscuous enzyme that will catalyze numerous β-replacement reactions based on the following general reaction scheme:

$$XCH_2CH\ (NH_2)\ COOH + YH \rightarrow XH + YCH_2CH\ (NH_2)\ COOH$$

in which X = OH; SH; *O*-acyl, *S*-alkyl, Cl, or CN; and Y = SH, *S*-alkyl, *S*-EtOH, *S*-Et-NH_2, or *S*-CH_2 CH (NH_2) COOH. The reaction for which this enzyme is best known is the condensation of homocysteine and serine to form cystathionine. This reaction represents the first committed step in the biosynthesis of cysteine from methionine by transsulfuration *(13)*. CBS forms homotetramers of 63-kDa subunits *(14)* that may associate in solution into large multimers *(15,16)*. In addition to binding PLP, mammalian CBS binds heme *(17)*. The function of this heme group is unknown, but recent work has indicated that it is not directly involved in the catalytic mechanism *(18–21)*.

Mammalian CBS is activated 2.5- to 5-fold by *S*-adenosyl-L-methionine (AdoMet) *(18,19)*. The possible relevance of the CBS-AdoMet response and heme ligand to the production of H_2S as a cell messenger compound are discussed in detail later. Deficiency of CBS in humans is a recessively inherited disorder of metabolism and is the major cause of homocystinuria *(13)*. Braunstein and colleagues *(24,25)* proposed more than 30 yr ago that CBS also has a serine sulfhydrase activity and that it is capable of catalyzing numerous β-replacement reactions including the ones in which H_2S is a substrate or a product according to the general scheme shown previously. **Table 1** summarizes some of the reactions that CBS can catalyze to produce H_2S from cysteine or cysteine from H_2S and serine. Although these alternative reactions are theoretically possible and have been demonstrated in vitro, the question arises, do they occur in vivo? For instance, does CBS in addition to its well-documented activity in cystathionine production also produce cysteine directly from serine or is it also involved in generating H_2S? Braunstein and Goryachenkova *(24)* stated that reaction rates are highest and nearly equal with serine and with cysteine when homocysteine is the replacing agent. Our preliminary data (unpublished) comparing the catalytic efficiencies of the various reactions suggest that L-serine is a significantly better substrate than cysteine when homocysteine is the cosubstrate ($K_{cat}/K_m = 2.64$ vs 0.44). This finding is consistent with another of our observations that serine inhibits the formation of H_2S from cysteine and 2-mercaptoethanol by 50% even in the presence of a sixfold higher concentration of cysteine, suggesting that the enzyme strongly prefers serine to cysteine. Conversely, our preliminary experiments suggest that the catalytic efficiency of the serine sulfhydrase reaction (formation of cysteine from serine and H_2S) is the highest of all the CBS activities investigated ($K_{cat}/K_m = 8.31$ [m$M^{-1}\cdot s^{-1}$]), raising the possibility that CBS catalyzes the formation both of cystathionine and cysteine from serine in vivo. However, lower reaction rates for alternative reactions do not preclude a dual role for CBS in both cysteine biosynthesis and H_2S production. Logically, the cellular requirements for a highly toxic labile gaseous messenger compound will be much lower than those of a conditionally essential amino acid required for both protein synthesis and redox homeostasis. Consequently, the variance in catalytic efficiency seen in the various CBS activities may actually represent differential tailoring of different CBS activities for specific functions in vivo.

Table 1
CBS-Catalyzed Reactions Producing or Utilizing H_2S

Substrate	*Cosubstrate*	*Product*	*References*
L-Cysteine	Homocysteine	L-Cystathionine + **H_2S**	*25,43,105,106*
L-Cysteine	β-Mercaptoethanol	*S*-(2-Hydroxyethyl)-L-cysteine + **H_2S**	*24,25,106*
L-Cysteine	1-Mercapto-2-propanol	*S*-(2-Hydroxypropyl)-L-cysteine + **H_2S**	*105,106*
L-Cysteine	Cysteamine	*S*-Aminoethyl-L-cysteine + **H_2S**	*105*
L-Cysteine	Methanethiol	*S*-Methyl-L-cysteine + **H_2S**	*25*
L-Cysteine	Thioglycollic acid	*S*-Carboxymethyl-L-cysteine + **H_2S**	*25*
L-Cysteine	1-Butanethiol	*S*-Butyl-L-cysteine + **H_2S**	*105*
L-Cysteine	1-Pentanethiol	*S*-Pentyl-L-cysteine + **H_2S**	*105*
L-Cysteine	3-Mercapto-1,2-propanediol	S-(2,3-Dihydroxy) propyl-L-cysteine + **H_2S**	*105,106*
L-Cysteine	DTT	*S*-(2, 3- Dihydroxy, 4-mercapto) butyl-L-cysteine + **H_2S**	*105*
L-Cysteine	H_2O	L-serine + **H_2S**	*24,25*
L-Serine	**H_2S**	L-cysteine + **H_2S**	*25,106, 107*

4.2. Cystathionine γ-Lyase

CSE, also known as γ-cystathionase or cyst(e)ine desulfhydrase, catalyzes the second step of transsulfuration: cleavage of the cystathionine C-γ-S bond yielding L-cysteine, α-ketobutyrate, and ammonia. The rat enzyme was also reported to catalyze the formation of two gases, H_2S and NH_3, and pyruvate from cysteine *(24,26)*. The enzyme is a homotetramer composed of approx 40- to 4kDa subunits and carries one PLP coenzyme per monomer. Mammalian CSE is structurally similar to the bacterial transsulfuration enzymes, CSE and cystathionine β-lyase *(19,26–30)*. Deficiency of CSE in humans causes an essentially benign condition named cystathioninuria *(13)*. Steegborn et al. *(31)* recently conducted biochemical investigations of purified recombinant human CSE that brought into question the ability of this enzyme to produce H_2S from cysteine in vivo. They found that human CSE cleaves cystathionine almost exclusively at the C-γ-S bond of cystathionine rather than at the C-γ-S bond. L-Cysteine was degraded at least two orders of magnitude slower than L-cystathionine at respective concentrations of 0.55 m*M*, whereas degradation of L-cystine was almost undetectable. Both reactions involve the splitting of C-β-S bonds. These investigators concluded that the human enzyme shows high reaction specificity toward C-γ-S bonds.

4.3. Rhodanese and MST

Sulfurtransferases (EC 2.8.1.1–5) are widely distributed enzymes that catalyze the transfer of sulfane sulfur from a donor molecule, such as thiosulfate or 3-mercaptopyruvate, to a nucleophilic acceptor, such as cyanide or 3-mercaptoethanol. If the acceptor molecule is a thiol, a disulfide (RSSR) and H_2S will be formed. Although the enzyme was discovered in 1953 *(32)* its physiological role is poorly understood. The enzyme was recently purified *(33)* and its corresponding cDNA cloned *(34)*. It was concluded from these studies that MST is a member of the rhodanese enzyme family, which is localized in mitochondria. Rhodanese catalyzes transfer of the sulfur atom from thiosulfate to sulfur acceptors such as cyanide or thiol compounds and also has MST activity, indicating that this enzyme is also capable of generating H_2S. Nagahara and Nishino *(35)* also found that MST possesses both MST and rhodanese activities. Immunocytochemical and biochemical studies showed that the enzyme is widely distributed in rat tissues and present in both mitochondria and cytoplasm *(36)*.

4.4. Cysteine Lyase

Cysteine lyase is a PLP-dependent enzyme catalyzing the formation of cysteic acid and H_2S from L-cysteine and sulfite. The enzyme was originally identified in chicken embryo yolk sac *(37)*. Cysteine lyase was redefined later by Braunstein's group *(24)* as an exclusively β-replacing lyase with a strict specificity for the primary substrate L-cysteine, and for several sulfur-containing cosubstrates, as shown here:

$$HSCH_2\text{-}CHNH_3^+\text{-}COO\text{-} + RSH \rightarrow H_2S + RSCH_2\text{–} CHNH_3^+\text{-}COO^-$$

$$(R = H_3C;\ H_5C_2,\ HO \cdot Et,\ NH_2 \cdot Et,\ \text{-}OOC\text{-}CH(NH_3^+)\text{-}CH_2,\ etc.)$$

Although maximum reaction rates with 2-mercaptoethanol were much higher than with sulfite, the only physiologically abundant replacing agent in yolk sac is sulfite. The natural function of cysteine lyase appears to be the production of cysteic acid—the precursor of the fetal metabolite taurine. Most of the studies on this enzyme were performed in chicken; evidence that this enzyme exists in mammalian eggs or tissues awaits future research.

5. EVOLUTIONARY RELATIONSHIPS AMONG H_2S-PRODUCING ENZYMES

A common feature of enzymes capable of catalyzing H_2S production appears to be their relative promiscuity in terms of substrate range and the ability to catalyze multiple related, yet functionally distinct, reactions. In some cases, consideration of the evolutionary relationships of these enzymes offers some clues to the likelihood of their playing a role in the endogenous production of H_2S as a signaling compound in mammals.

5.1. Cystathionine β-Synthase

The evolutionary relationships of human CBS offer insight into the possible origin of its multiple activities including the ability both to produce and to consume H_2S. Human CBS has a modular structure with a catalytic domain flanked by N- and C-terminal domains (**Fig. 1**). The enzyme is a member of the β family of PLP-dependent enzymes sharing a common type II fold structure with *O*-acetylserine sulfhydrylase (OASS), threonine deaminase, 1-aminocyclopropane-1-carboxylate deaminase, and the β-subunit of tryptophan synthase *(38–40)*. Sequence conservation of the 45-kDa central catalytic domain of human CBS with OASS is particularly pronounced in the active center of CBS, and the architecture of the active site surrounding the PLP-binding site at Lys 119 is almost identical *(16,40,41)*. The common origin of CBS with OASS is further confirmed by the observation that many CBS enzymes in bacteria and protozoa appear to possess varying degrees of both CBS and OASS activities *(42)*. All CBS enzymes that have been studied to date appear to be capable of multiple activities, but the relative efficiencies of these activities seem to vary significantly in different organisms. CBS from the nematode *Panagrellus redivivus* catalyzes the replacement of the β-SH group of cysteine with 2-mercaptoethanol to yield a thioether, *S*-(2-hydroxyethyl) cysteine, and H_2S with significantly greater efficiency than it catalyzes the formation of cystathionine from serine

Fig. 1. Domain organization of CBS and cysteine synthase (OASS) enzymes from *Homo sapiens* and other organisms. Filled boxes represent conserved domains; narrow lines denote nonconserved regions. The area designated "CBS/CS" represents a highly conserved region present in both CBS and CS enzymes; this region also displays significant structural conservation with several other members of the β family of PLP-dependent enzymes such as serine/threonine deaminases and the β-subunit of tryptophan synthase *(38,39)*. In human CBS, this region comprises residues 84–382. The CBS domains are hydrophobic sequence motifs of approx 55 amino acids that appear to be involved in the autoinhibitory function of the C-terminal domain *(20,66)*. Two of these motifs are present in the C-terminal regions of both human and yeast CBS. The star indicates the approximate position of heme binding in the human CBS enzyme (Cys-52 and His-65). CBS HS, *H. sapiens* (Swiss-Prot accession no. P35520); CBS SC, *S. cerevisiae* (Swiss-Prot P32582); CBS TC, *T. cruzi* (Swiss-Prot Q9BH24); CS ST, *Salmonella typhimurium* (Swiss-Prot P12674).

and homocysteine *(43)*. The variance of the relative efficiencies of these different reactions catalyzed by CBS from different organisms suggests a somewhat fluid mosaic of mechanistically related CBS reactions that are evolutionarily tailored to the particular requirements of the organism in question.

Mammalian CBS differs from CBS of more primitive organisms in several aspects that may pertain to its possible role in the production of H_2S as a cell messenger. First, mammalian CBS is a hemeprotein with the heme ligand covalently bound by residues Cys 52 and His 65 *(40)*. The possible function of this ligand is discussed in detail in the following section, but it is clear that it does not play a direct role in catalysis, which can be explained solely by participation of PLP in the reaction mechanism *(41,44)*. Second, the heme-binding residues are conserved only in mammals and fish and the absence of heme in CBS purified from *Saccharomyces cerevisiae* and *Trypanasoma cruzi* indicates that this ligand is not evolutionarily conserved in CBS in lower eukaryotes *(18,45,46)*. The implication of these findings is that the function of heme is unique to higher organisms. Given the possible role of the CBS heme group in signal transduction from messenger compounds such as NO, CO, and H_2S, and the role that these compounds appear to play in modulating vascular tone and brain functions such as long-term potentiation, it is interesting to note that the heme-binding residues (and, by extension, the heme group) are restricted to organisms that possess closed circulations and higher neurological functions. However, the absence of the conserved heme-binding residues does not unequivocally prove the absence of heme, and investigation of the possible role of heme in human CBS would benefit from further assessment of the heme status of CBS from other lower organisms.

The kinetics of the regulation of CBS by AdoMet are eminently suitable for the regulation of H_2S as a cellular messenger compound and are discussed in detail in the following section. Again, like heme, this regulation has only been observed in mammalian CBS, whereas CBS from yeast *(18,45)*, *T. cruzi (46)*, and *Drosophila melanogaster* (unpublished results) is not regulated by this compound. Identification of the AdoMet-binding site and subsequent analysis of its evolutionary conservation will provide useful information in assessing the possibility that the mammalian CBS AdoMet response is a recent functional adaptation in higher organisms of functional relevance to H_2S metabolism.

5.2. Cystathionine γ-Lyase

Because of the similarity between CSE and other related enzymes with different functions, the evolutionary relationships of CSE are difficult to interpret with the limitations of our current knowledge. What has become clear from studies in a wide range of organisms is that these enzymes are capable of a broad range of activities, and evidence has accrued regarding the significant overlap in their range of functions. CSE belongs to the γ family of PLP-dependent enzymes that incorporates *O*-succinylhomoserine (thiol)-lyase and *O*-acetylhomoserine (thiol)-lyase, which catalyze γ-replacement or γ-elimination reactions *(39)*. Determination of the CSE crystal structures has recently revealed that both yeast and human CSE are virtually identical at their active sites to cystathionine γ-synthase (CGS) from *Escherichia coli*. Both CSEs and bacterial CGSs exhibit γ-synthase and γ-lyase activities depending on their position in the metabolic pathway and substrate availability. Plant CGSs use homoserine phosphate instead of *O*-succinylhomoserine as one substrate. This is reflected by a partially different active site structure in plant CGSs. This group of enzymes has a glutamate (E333 in yeast CSE), which binds to the distal group of cystathionine or the amino group of cysteine *(30,31)*. The high degree of similarity in these enzymes serves to emphasize the potential for CSE to have multiple roles in mammalian tissues.

5.3. Rhodanese and MST

The evolutionary relationship between rhodanese and MST is quite clear because these enzymes share a very high degree of sequence conservation (up to 50% identity at the amino acid level), particularly around the active site and have essentially overlapping ranges of activity. In an elegant series of experiments, Nagahara and colleagues *(33–35)*, very clearly demonstrated the interrelationship between these enzymes. They found that conversion of key residues in the active site of recombinant rat liver rhodanese decreased rhodanese activity and increased MST activity. Sequence elements in rhodanese designated "rhodanese domains" are ubiquitous structural modules occurring in the three major evolutionary phyla. They are found as tandem repeats, with the C-terminal domain hosting the properly structured active-site cysteine residue, as single domain proteins or in combination with other distinct protein domains. An increasing number of reports indicate that these rhodanese modules serve as versatile sulfur carriers that have adapted their function to fulfill the need for reactive sulfane sulfur in distinct metabolic and regulatory pathways *(47)*.

The biological role of the rhodanese family of sulfurtransferases is still the subject of considerable debate *(48)*. Proposed functions include detoxification of cyanide compounds *(49)*, maintenance of sulfane levels *(50)*, thiamin biosynthesis *(51)* and formation of prosthetic groups in iron-sulfur clusters *(52)*. The involvement of different rhodanese-like sulfurtransferases in different biological functions is suggested by the wide variability in the residues of the active center loop *(47)*. The major point from consideration of the evolutionary relationships of these enzymes is that they have the capacity to surprise us with the breadth of their functions and that other related transferases that could play a significant role in the endogenous production of H_2S may remain to be discovered.

6. REGULATION OF H_2S PRODUCTION

All of the candidate enzymes for the production of H_2S in mammalian tissues are capable of multiple reactions using a relatively wide range of substrates that have the potential to act in mutually exclusive competition. Consequently, rational assessment of the likely role of these enzymes in the production of H_2S in vivo depends, to a certain extent, on our knowledge of how these enzymes are regulated. The purpose of this section is to summarize the current knowledge regarding the regulation of these enzymes in human tissues with a view to assessing their suitability for generating H_2S as a gasotransmitter in vivo.

6.1. Cystathionine β-Synthase

Although the regulation of CBS has only recently begun to receive detailed attention, it is by far the best understood of all the enzymes that are likely to play a role in the endogenous production of H_2S in mammals. Early characterizations focused on the direct modulation of CBS activity by AdoMet and limited proteolysis *(22,23)*. The advent of molecular biological techniques has subsequently facilitated the elucidation of the mechanisms and regulatory principles that underlie transcriptional regulation of the gene. To date, limited experimental evidence has indicated that both the cystathionine- and H_2S-forming activities of CBS are increased in equal measure when CBS activity is stimulated/repressed. Clearly there is considerable scope for further investigation, and the possible differential regulation of the multiple reactions of CBS remains an intriguing possibility.

6.1.1. Transcriptional Regulation

The complete genomic sequence of human CBS has been determined *(53)*, and the transcriptional start sites of five human CBS mRNA isoforms, designated CBS –1a, –1b, –1c, –1d, and –1e, respectively, have been mapped *(54)*. Two promoter regions upstream of exons –1a and –1b have been identified *(53)*. Recently, we have shown that both human and yeast CBS are coordinately regulated with proliferation and that the human CBS –1b promoter is serum and fibroblast growth factor-responsive and is downregulated by oxidative stress, growth arrest owing to nutrient depletion or the action of differentiation-inducing reagents *(9,55)*.

The CBS –1b promoter is activated by a synergistic interaction between the transcription factors NF-Y and either Sp1 or Sp3 *(56)*. Recent work in our laboratory using site-directed mutagenesis, DNase I footprinting, and deletion analysis of 5' proximal sequence revealed that 210 bp of proximal sequence is sufficient for maximal promoter activity. As little as 32 bp of the –1b proximal promoter region is capable of driving transcription in HepG2 cells, and this activity is entirely dependent on the presence of a single overlapping Sp1/Egr1-binding site. We have recently found that both human CBS promoters are transactivated by Sp1 and Sp3 but only the –1b promoter is subject to a previously unreported site-specific synergistic activation between Sp1 and Sp3. Experiments in Sp1-deficient fibroblasts have indicated that the role of Sp1 in driving CBS transcription is both dominant and indispensable (unpublished results). In the context of gasotransmitter production, it is interesting to note that Sp1 also plays a crucial role in the regulation of neuronal and endothelial nitric oxide synthase (NOS) *(57,58)* as well as soluble guanyl synthase *(59)*.

The role of Sp1 in regulation of the production of NO and H_2S is intriguing because NO itself acts to impair the ability of Sp1 to activate transcription from a range of Sp1-driven promoters including CBS *(60,61)* (unpublished results) and may thus serve as a site of regulatory cross talk between these two messenger systems.

6.1.2. Tissue Distribution

Any physiological role for H_2S in terms of regulation of vascular tone and/or neurotransmitter function is likely to be compartmentalized into specific relevant tissues. Consequently, it is unlikely that the enzyme(s) involved in its production will be ubiquitously expressed in all tissues. In terms of tissue distribution, the CBS promoter regions constitute an interesting paradox in that the multiple sites of transcriptional initiation, high GC content, Sp1 regulation, inverted CAAT box, and absence of a classic TATA box sequence are all typical characteristics of an ubiquitously expressed "housekeeping gene." Yet, CBS is expressed in a highly tissue-specific manner and appears to be developmentally regulated *(62–65)*. The precise tissue distribution of CBS awaits a complete analysis, but, to date, significant levels of CBS activity have been located in the liver, brain (particularly in the cerebellum and hippocampus), kidney (proximal tubule), pancreas, and intestine. The mechanism by which the CBS promoters have restricted expression is not presently understood. We have recently shown that the CBS 5'-flanking sequence does not confer tissue-specific expression and that several KLF members of the Sp1 family of transcription factors can act to repress both CBS promoters by competitively displacing Sp1 binding (unpublished observations). These KLF transcription factors are not ubiquitously expressed and have a relatively limited tissue distribution and suggest a novel mechanism by which the Sp1-dependent expression of both CBS promoters can be regulated in a tissue-specific fashion. The limited tissue distribution of CBS is thus consistent with a role for this enzyme in the production of H_2S as a physiological messenger compound.

6.1.3. Direct Modulation of CBS Activity By C-Terminal Autoinhibitory Domain

Although transcriptional mechanisms are important for controlling relative expression levels of enzymes in different tissues, it should be appreciated that the time scale of transcriptional responses from initial stimulus through to final translation of protein is relatively long. A key requisite of any enzyme involved in the biosynthesis of potentially toxic labile gaseous messenger compounds is the ability to respond to regulatory signals by modulating activity in a rapid "real-time" fashion. This kind of response inevitably requires direct (and preferably reversible) regulation of the activity of preexisting pools of enzyme in response to appropriate stimuli. Mammalian CBS is subject to such regulation and is activated 2.5- to 5-fold by AdoMet *(22,23,66)* with an apparent activation constant of 15 μM *(67,68)*. Recent work has focused on the mechanisms by which the activity of the CBS protein is regulated by AdoMet. It has been demonstrated that binding of AdoMet to human CBS is accompanied by a change in fluorescence spectra that is indicative of an induced conformational change *(16)*. Mammalian CBS can also be activated to a level comparable with that induced by AdoMet by partial thermal denaturation *(69)* or limited proteolysis. This latter form of activation results in the removal of the N- and C-terminal regions of CBS to leave an activated "active core" consisting of residues 37–413 *(14,16,69)*. Recently, it has been suggested that this proteolytic activation of CBS may be induced by tumor necrosis factor-α in vivo *(70)*. All three of these forms of CBS activation are mutually exclusive, suggesting that they proceed through a common mechanism. A role for the C-terminal region of CBS in this activation was further suggested by the observation that specific mutations in this region constitutively activate CBS and render it insensitive to further activation by AdoMet *(66,71,72)*. The C-terminal regulatory region of CBS acts as an autoinhibitory domain and a conformational change induced by either heat, AdoMet binding, specific C-terminal point mutations, or complete removal by limited proteolysis act to displace this domain from its zone of inhibition *(20,69,71)*.

6.1.4. Role of Heme in Regulating CBS Activity

As already described, CBS is unique among PLP-dependent enzymes in that it has a heme group bound by residues Cys 52 and His *(17,40)*, 37. The exact function of this heme group is currently unknown. Previously, one group presented data indicating that the heme group is directly involved in the CBS catalytic mechanism *(73)*. However, subsequent work has definitively disproved this, and a regulatory signal transduction or structural role is now considered to be the likely function of this ligand *(18,20,45,70)*. The heme group and the PLP-binding site are juxtaposed in the human CBS structure *(40)*, and recent ^{31}P nuclear magnetic resonance studies of the PLP group in human CBS suggested regulatory communication involving these ligands as a consequence of changes in the heme oxidation state *(75)*. The idea of the heme group acting as a modulator of enzyme activity in concert with changes in cellular requirements is attractive because this ligand is an obvious candidate site for regulatory interactions with compounds such as NO, CO, or possibly H_2S. Although some data have been accrued showing decreased CBS activity as a consequence of interaction of the CBS heme ligand with both NO and CO, the NO concentrations required to observe this effect are too high to be physiologically relevant *(21)*. Interestingly, the CBS heme group resembles the heme group of the bacterial CO sensor CooA *(76)* in that they both have axial cysteinate and neutral nitrogen ligands in the ferric state that resist ligand displacement by common ferric and ferrous ligands. The affinity of the CBS heme group for CO is much higher than that observed for NO and is comparable with that seen for CooA, suggesting that in vivo heme-mediated regulatory interactions between CBS and CO are feasible and merit further investigation *(21)*.

Other lines of inquiry have indicated that the CBS heme group plays a structural role and is important in ensuring correct folding of the protein *(20)* and in preventing aggregation of the enzyme into insoluble complexes *(77)*. One possibility that would reconcile regulatory and structural functions for heme is suggested by recent studies of inducible NOS (iNOS), which also has a regulatory heme ligand in humans. NO appears to limit intracellular assembly of dimeric iNOS by preventing heme insertion, decreasing cellular heme availability, and inhibiting dimerization of heme containing iNOS monomers *(78,79)*. It is conceivable that NO could inhibit the correct assembly of functional CBS by decreasing the insertion and/or availability of heme.

6.1.5. Regulation of CBS Shares Many Common Features With NOS

Although numerous different enzymes can conceivably contribute to the endogenous production of H_2S in humans, CBS is a particularly attractive candidate because of the numerous regulatory features it appears to share with other enzymes involved in the biosynthesis of gaseous messenger compounds. Although previous reports that CBS is regulated by calcium efflux, sodium nitroprusside, or calmodulin binding *(80,81)* have not been found to be reproducible (our unpublished results; W. D. Kruger, personal communication, July, 2003), there is striking conservation in many aspects of the regulation of CBS and the NOSs. In common with human CBS, neuronal NOS (nNOS) has a GC-rich TATA-less promoter with multiple sites of transcriptional initiation that are regulated by complex interactions among members of the Sp1/KLF family of transcription factors. Additionally, the activity of nNOS is also modulated both by NO interactions with a covalently linked heme group *(82)* and by regulatory ligand-binding-induced conformational changes mediated by an autoinhibitory domain *(83,84)*. Many of the regulatory features that are conserved between CBS and the NOSs are eminently suitable for rapid regulation of the biosynthesis of labile messenger compounds and thus are consistent with a possible role for CBS in the synthesis of H_2S as a cell-signaling compound.

6.1.6. Evidence for Alternative CBS Functions In Vivo

It has been postulated that the role of CBS in neural tissue may differ from that in other tissues because CSE, the next enzyme in the transsulfuration pathway, has been reported to be absent in the brain. Although recent immunohistochemical studies in our laboratory have found CSE expression in the brain, it remains possible that the level of expression could be significantly lower than that of CBS, which would explain previous reports of cystathionine accumulation in the mammalian nervous system *(85,86)*. Less controversial is the observation that CBS is expressed in the absence of CSE during early development *(87)*. This observation clearly puts into question the role of CBS during early development and would be consistent with a function for the enzyme independent of cysteine biosynthesis. In this context, expression levels of CBS are central to our understanding of the possible role of this enzyme in the production of H_2S as either a neuromodulator or regulator of vascular tone. As an example, we have found that CBS activity is consistently present at low levels in cultured aortic and colonic smooth muscle cells (SMCs) (3–5 mU/mg of protein) (unpublished results). This level of activity is very low compared to other cell lines such as the hepatoma cell line HepG2 (30–60 mU/mg of protein) or normal fibroblasts (10–40 mU/mg). CBS activity was so low that neither aortic nor colonic SMCs were able to grow in cysteine-free medium, implying that CBS is expressed in these cells for a function other than cysteine biosynthesis. An obvious candidate function for low-level CBS activity in aortic/colonic SMCs is the production of H_2S as regulator of vascular/colonic tone. It is readily conceivable that such a function would require relatively low levels of CBS because of the toxic nature of this signal compound. This possibility offers a potential explanation for the previously reported incidence of lower blood pressure in people with Down's syndrome in whom CBS is reported to be overexpressed *(88–90)*.

6.2. Regulation of CSE

To date, there has been no investigation of any CSE promoter region, and nothing is known about which transcription factors regulate this gene. A relatively limited number of investigations have been performed to examine changes in CSE activity and message levels in animal tissues, and it is clear that our understanding of the regulation of this enzyme is very limited. In common with its partner in transsulfuration, CBS, expression of the CSE gene has been reported to be repressed by insulin and to be upregulated during liver regeneration *(91)*. Some work regarding CSE regulation has not been investigated for CBS, so it is currently unclear whether these enzymes are coordinately regulated in these scenarios. Observations in this category include the apparent repression of CSE as a consequence of human immunodeficiency virus infection *(92)* and significant upregulation during lactation in rats *(93)*. One group has reported that CSE transcription and protein turnover rates are affected by B6 availability *(94)*. There are currently two known instances when CBS and CSE are differentially regulated. First, CSE expression in both yeast and mammals appears to be induced by oxidative stress *(95–97)*, whereas transcription of the human CBS gene is very clearly repressed by reactive oxygen species *(65)*. Second, as described in the previous section, the temporal and spatial expression patterns of CSE differ significantly from those of CBS during embryogenesis through to early neonatal life prompting the question, what is the function of CBS during this developmental period?

6.3. MST and Rhodanese

There is a distinct paucity of information regarding the regulation of the related sulfurtransferases MST and rhodanese, which clearly limits the ability to assess their possible contribution to the synthesis of H_2S in mammalian tissues. Nothing is known about their transcriptional control in humans, possible posttranslational regulation, or how many of their multiple enzymatic functions actually occur in vivo. There have been significant investigations into the distribution of both of these enzymes in mammalian tissues. Both MST and rhodanese appear to be widely expressed in animals, but the organ distribution of rhodanese in animals appears to vary significantly from one species to another. In rats, MST is predominantly localized in proximal tubular epithelium of the kidney, pericentral hepatocytes in the liver, cardiac cells in the heart, and neuroglial cells in the brain *(36)*. In all mammals examined to date, rhodanese activity is highest in liver and kidney cortex *(98)*. In the rat nervous system, MST was found to have largely negligible activity compared to rhodanese, making it an unlikely candidate for the production of H_2S as a neuromodulator. Conversely, rhodanese activity was widely distributed in mammalian neural tissues. In rats, the olfactory bulb showed the highest rhodanese activity, and high activity was also observed in the thalamus, septum, hippocampus, and dorsal part of the midbrain. Rhodanese activity was low in various parts of the cerebral cortex *(99,100)*.

Rhodanese has more typically been studied in the context of heme metabolism and detoxification of cyanide, but it is clear that it may have multiple functions in vivo. In this context, it is interesting to note that there is some evidence that this enzyme is activated by protein kinase C phosphorylation *(101)* and may be regulated at least in part by AdoMet *(102)*. It was observed that AdoMet is unable to interact with the sulfur-substituted rhodanese complex suggesting that AdoMet binding acts to blockade the thiosulfate-binding sites *(103)*. The idea that intracellular AdoMet concentrations could act to portion rhodanese into alternative functions including perhaps H_2S production is an intriguing possibility that merits future investigation. Evidence for a direct role for rhodanase in H_2S metabolism has recently emerged *(104)*. In the colon, rhodanese is located primarily in the submucosa and crypts of the colon, and its inhibition dramatically decreases the ability of the colon to actively remove H_2S indicating that this enzyme is the principal enzyme for detoxification of H_2S in this particular milieu.

7. CONCLUSION

Although there are problems with the various methodologies used to detect H_2S, there is currently good evidence to support the conclusion that this compound is produced in mammals and that this finding is consistent with its putative role as a neuromodulator and vasorelaxant. There are several candidate proteins capable of producing H_2S, but more work remains to be done to assess the likelihood of these reactions occurring in vivo in the presence of alternative substrates and competing activities. The regulatory and catalytic characteristics of CBS appear to be the most suited to the task of producing H_2S as a gasotransmitter, but our current knowledge of sulfur metabolism in humans is insufficiently complete to rule out the possibility of other enzymes playing a role in the synthesis of H_2S.

ACKNOWLEDGMENTS

We wish to express our gratitude to Dr. Jana Oliveriusova for assistance with preparation of the figure. This work was supported by National Institutes of Health grants PO1HD08315 and HL65217 to J.P. Kraus.

REFERENCES

1. Wang R. Two's company, three's a crowd: can H_2S be the third endogenous gaseous transmitter? FASEB J 2002;16:1792–1798.
2. Kimura H. Hydrogen sulfide as a neuromodulator. Mol Neurobiol 2002;26:13–19.
3. Hosoki R, Matsuki N, Kimura H. The possible role of hydrogen sulfide as an endogenous smooth muscle relaxant in synergy with nitric oxide. Biochem Biophys Res Commun 1997;237:527–531.
4. Ogasawara Y, Isoda S, Tanabe S. Tissue and subcellular distribution of bound and acid-labile sulfur, and the enzymic capacity for sulfide production in the rat. Biol Pharm Bull 1994;17:1535–1542.
5. Ubuka T. Assay methods and biological roles of labile sulfur in animal tissues. J Chromatogr B Analyt Technol Biomed Life Sci 2002;781:227–249.
6. Ubuka T, Abe T, Kajikawa R, et al. Determination of hydrogen sulfide and acid-labile sulfur in animal tissues by gas chromatography and ion chromatography. J Chromatogr B Biomed Sci Appl 2001;757: 31–37.
7. Ogasawara Y, Ishii K, Togawa T, et al. Determination of bound sulfur in serum by gas dialysis/high-performance liquid chromatography. Anal Biochem 1993;215:73–81.
8. Savage JC, Gould DH. Determination of sulfide in brain tissue and rumen fluid by ion-interaction reversed-phase high-performance liquid chromatography. J Chromatogr 1990;526:540–545.
9. Mitchell TW, Savage JC, Gould DH. High-performance liquid chromatography detection of sulfide in tissues from sulfide-treated mice. J Appl Toxicol 1993;13:389–394.
10. Warenycia MW, Goodwin LR, Benishin CG, et al. Acute hydrogen sulfide poisoning: demonstration of selective uptake of sulfide by the brainstem by measurement of brain sulfide levels. Biochem Pharmacol 1989;38:973–981.
11. Dorman D, Moulin F, McManus B, et al. Cytochrome oxidase inhibition induced by acute hydrogen sulfide inhalation: correlation with tissue sulfide concentrations in the rat brain, liver, lung, and nasal epithelium. Toxicol Sci 2002;65:18–25.
12. Stipanuk MH, Beck PW. Characterization of the enzymic capacity for cysteine desulphhyration in liver and kidney of rat. Biochem J 1982;206:267–277.
13. Mudd SH, Levy HL, Kraus JP. Disorders of transsulfuration. In: Scriver CR, Beaudet AL, Sly WS, et al., eds. The Metabolic and Molecular Bases of Inherited Disease. McGraw-Hill: New York, 2001, pp. 2007–2056.
14. Skovby F, Kraus JP, Rosenberg LE. Biosynthesis and proteolytic activation of cystathionine β-synthase in rat liver. J Biolog Chem 1984;259:588–593.
15. Kraus JP, Rosenberg LE. Cystathionine β-synthase from human liver: improved purification scheme and additional characterization of the enzyme in crude and pure form. Arch Biochem Biophys 1983;222:44–52.
16. Kery V, Poneleit L, Kraus JP. Trypsin cleavage of human cystathionine β-synthase into an evolutionary conserved active core: structural and functional consequences. Arch Biochem Biophys 1998;355:222–232.
17. Kery V, Bukovska G, Kraus JP. Transsulfuration depends on heme in addition to pyridoxal 5'-phosphate: cystathionine β-synthase is a heme protein. J Biol Chem 1994;269:25,283–25,288.
18. Jhee KH, McPhie P, Miles EW. Yeast cystathionine beta-synthase is a pyridoxal phosphate enzyme but, unlike the human enzyme, is not a heme protein. J Biol Chem 2000;275:11,541–11,544.

19. Nagasawa T, Kanzaki H, Yamada H. Cystathionine γ-lyase from Streptomyces phaeochromogenes. Methods Enzymol 1987;143:486–492.
20. Oliveriusova J, Kery V, Maclean KN, et al. Deletion mutagenesis of human cystathionine beta-synthase: impact on activity, oligomeric status, and S-adenosylmethionine regulation. J Biol Chem 2002;277:48,386–48,394.
21. Taoka S, Banerjee R. Characterization of NO binding to human cystathionine beta-synthase: possible implications of the effects of CO and NO binding to the human enzyme. J Inorg Biochem 2001;87: 245–251.
22. Finkelstein JD, Kyle WE, Martin JJ, et al. Activation of cystathionine synthase by adenosylmethionine and adenosylethionine. Biochem Biophys Res Commun 1975;66:81–87.
23. Koracevic D, Djordjevic V. Effect of trypsin, S-adenosylmethionine and ethionine on serine sulfhydrase activity. Experientia 1977;33:1010, 1011.
24. Braunstein AE, Goryachenkova EV. The β-replacement-specific pyridoxal-*P*-dependent lyases. Adv Enzymol 1984;56:1–89.
25. Braunstein AE, Goryachenkova EV, Tolosa EA, et al. Specificity and some other properties of liver serine sulphhydrase: Evidence for its identity with cystathionine β-synthase. Biochim Biophys Acta 1971;242:247–260.
26. Matsuo Y, Greenberg DM. A crystalline enzyme that cleaves homoserine and cystathionine. I. Isolation procedure and some physicochemical properties. J Biol Chem 1958;230:545–560.
27. Nagasawa T, Kanzaki H, Yamada H. Cystathionine γ-lyase of Streptomyces phaeochromogenes: the occurrence of cystathionine γ-lyase in filamentous bacteria and its purification and characterization. J Biol Chem 1984;259:10,393–10,403.
28. Kanzaki H, Kobayashi M, Nagasawa T, et al. Purification and characterization of cystathionine γ-synthase type II from Bacillus sphaericus. Eur J Biochem 1987;163:105–112.
29. Kanzaki H, Nagasawa T, Yamada H. Insight into the active site of Streptomyces cystathionine γ-lyase based on the results of studies on its substrate specificity. Biochim Biophys Acta 1987;913:45–50.
30. Messerschmidt A, Worbs M, Steegborn C, et al. Determinants of enzymatic specificity in the Cys-Met-metabolism PLP-dependent enzymes family: crystal structure of cystathionine γ-lyase from yeast and intrafamiliar structure comparison. Biol Chem 2003;384:373–386.
31. Steegborn C, Clausen T, Sondermann P, et al. Kinetics and inhibition of recombinant human cystathionine γ-lyase: toward the rational control of transsulfuration. J Biol Chem 1999;274:12,675–12,684.
32. Meister A. Conversion of the alpha-keto analog of cysteine to pyruvate and sulfur. Fed Proc 1953; 12:245.
33. Nagahara N, Okazaki T, Nishino T. Cytosolic mercaptopyruvate sulfurtransferase is evolutionarily related to mitochondrial rhodanese: striking similarity in active site amino acid sequence and the increase in the mercaptopyruvate sulfurtransferase activity of rhodanese by site-directed mutagenesis. J Biol Chem 1995;270:16,230–16,235.
34. Nagahara N, Nishino T. Role of amino acid residues in the active site of rat liver mercaptopyruvate sulfurtransferase: cDNA cloning, overexpression, and site-directed mutagenesis. J Biol Chem 1996;271:27,395–27,401.
35. Nagahara N, Nishino T. Mercaptopyruvate sulfurtransferase is a rhodanese family. Seikagaku 1996;68:197–201.
36. Nagahara N, Ito T, Kitamura H, et al. Tissue and subcellular distribution of mercaptopyruvate sulfurtransferase in the rat: confocal laser fluorescence and immunoelectron microscopic studies combined with biochemical analysis. Histochem Cell Biol 1998;110:243–250.
37. Chapeville F, Fromageot P. Formation of sulfite, cysteic acid and tourine from the sulfate of the embryonic egg. Biochim Biophys Acta 1957;26:538–558.
38. Swaroop M, Bradley K, Ohura T, et al. Rat cystathionine β-synthase: gene organization and alternative splicing. J Biol Chem 1992;267:11,455–11,461.
39. Alexander FW, Sandmeier E, Mehta PK, et al. Evolutionary relationships among pyridoxal-5'-phosphate-dependent enzymes: regio-specific alpha, beta and gamma families. Eur J Biochem 1994;219: 953–960.
40. Meier M, Janosik M, Kery V, et al. Structure of human cystathionine beta-synthase: a unique pyridoxal 5'- phosphate-dependent heme protein. EMBO J 2001;20:3910–3916.
41. Kery V, Poneleit L, Meyer JD, et al. Binding of pyridoxal 5'-phosphate to the heme protein human cystathionine beta-synthase. Biochemistry 1999;38:2716–2724.
42. Mino K, Ishikawa K. Characterization of a novel thermostable O-acetylserine sulfhydrylase from Aeropyrum pernix K1. J Bacteriol 2003;185:2277–2284.

43. Papadopoulos AI, Walker J, Barrett J. A novel cystathionine β-synthase from *Panagrellus redivivus* (Nematoda). Int J Biochem Cell Biol 1996;28:543–549.
44. Jhee KH, Niks D, McPhie P, et al. The reaction of yeast cystathionine beta-synthase is rate-limited by the conversion of aminoacrylate to cystathionine. Biochemistry 2001;40:10,873–10,880.
45. Maclean KN, Janosik M, Oliveriusova J, et al. Transsulfuration in Saccharomyces cerevisiae is not dependent on heme: purification and characterization of recombinant yeast cystathionine beta-synthase. J Inorg Biochem 2000;81:161–171.
46. Nozaki T, Shigeta Y, Saito-Nakano Y, et al. Characterization of transsulfuration and cysteine biosynthetic pathways in the protozoan hemoflagellate, Trypanosoma cruzi: isolation and molecular characterization of cystathionine beta-synthase and serine acetyltransferase from Trypanosoma. J Biol Chem 2001;276:6516–6523.
47. Bordo D, Bork P. The rhodanese/Cdc25 phosphatase superfamily: sequence-structure-function relations. EMBO Rep 2002;3:741–746.
48. Nandi DL, Horowitz PM, Westley J. Rhodanese as a thioredoxin oxidase. Int J Biochem Cell Biol 2000;32:465–473.
49. Baumeister RG, Schievelbein H, Zickgraf-Rudel G. Toxicological and clinical aspects of cyanide metabolism. Arzneimittelforschung 1975;25:1056–1064.
50. Westley J. Mammalian cyanide detoxification with sulphane sulphur. Ciba Found Symp 1988;140: 201–218.
51. Palenchar PM, Buck CJ, Cheng H, et al. Evidence that ThiI, an enzyme shared between thiamin and 4-thiouridine biosynthesis, may be a sulfurtransferase that proceeds through a persulfide intermediate. J Biol Chem 2000;275:8283–8286.
52. Pagani S, Bonomi F, Cerletti P. Enzymic synthesis of the iron-sulfur cluster of spinach ferredoxin. Eur J Biochem 1984;142:361–366.
53. Kraus JP, Oliveriusova J, Sokolova J, et al. The human cystathionine β-synthase (CBS) gene: complete sequence, alternative splicing and polymorphisms. Genomics 1998;52:312–324.
54. Bao L, Vlcek C, Paces V, et al. Identification and tissue distribution of human cystathionine beta-synthase messenger-RNA isoforms. Arch Biochem Biophys 1998;350:95–103.
55. Ratnam S, Maclean KN, Jacobs RL, et al. Hormonal regulation of cystathionine beta-synthase expression in liver. J Biol Chem 2002;277:42,912–42,918.
56. Ge Y, Konrad MA, Matherly LH, et al. Transcriptional regulation of the human cystathionine beta-synthase-1b basal promoter: synergistic transactivation by transcription factors NF-Y and Sp1/Sp3. Biochem J 2001;357:97–105.
57. Saur D, Seidler B, Paehge H, et al. Complex regulation of human neuronal nitric-oxide synthase exon 1c gene transcription: essential role of Sp and ZNF family members of transcription factors. J Biol Chem 2002;277:25,798–25,814.
58. Zhang R, Min W, Sessa WC. Functional analysis of the human endothelial nitric oxide synthase promoter: Sp1 and GATA factors are necessary for basal transcription in endothelial cells. J Biol Chem 1995;270:15,320–15,326.
59. Liang F, Schaufele F, Gardner DG. Functional interaction of NF-Y and Sp1 is required for type a natriuretic peptide receptor gene transcription. J Biol Chem 2001;276:1516–1522.
60. Wang S, Wang W, Wesley RA, et al. A Sp1 binding site of the tumor necrosis factor alpha promoter functions as a nitric oxide response element. J Biol Chem 1999;274:33,190–33,193.
61. Zhang J, Wang S, Wesley RA, Danner RL. Adjacent sequence controls the response polarity of nitric oxide–sensitive Sp factor binding sites. J Biol Chem 2003;278:29,192–29,200.
62. Mudd SH, Finkelstein JD, Irreverre F, et al. Transsulfuration in mammals: microassays and tissue distributions of three enyzmes of the pathway. J Biol Chem 1965;240:4382–4392.
63. Quere I, Paul V, Rouillac C, et al. Spatial and temporal expression of the cystathionine beta-synthase gene during early human development. Biochem Biophys Res Commun 1999;254:127–137.
64. Robert K, Vialard F, Thiery E, et al. Expression of the cystathionine beta synthase (CBS) gene during mouse development and immunolocalization in adult brain. J Histochem Cytochem 2003; 51:363–371.
65. Maclean KN, Janosik M, Kraus E, et al. Cystathionine beta-synthase is coordinately regulated with proliferation through a redox-sensitive mechanism in cultured human cells and Saccharomyces cerevisiae. J Cell Physiol 2002;192:81–92.
66. Janosik M, Kery V, Gaustadnes M, et al. Regulation of human cystathionine beta-synthase by S-adenosyl-L-methionine: evidence for two catalytically active conformations involving an autoinhibitory domain in the C-terminal region. Biochemistry 2001;40:10,625–10,633.

67. Kozich V, Kraus JP. Screening for mutations by expressing patient cDNA segments in *E. coli*—homocystinuria due to cystathionine β-synthase deficiency. Hum Mutat 1992;1:113–123.
68. Roper MD, Kraus JP. Rat cystathionine β-synthase: expression of four alternatively spliced isoforms in transfected cultured cells. Arch Biochem Biophys 1992;298:514–521.
69. Janosik M, Kery V, Gaustadnes M, et al. Regulation of human cystathionine beta-synthase by S-adenosyl-L-methionine: evidence for two catalytically active conformations involving an autoinhibitory domain in the C-terminal region. Biochemistry 2001;40:10,625–10,633.
70. Zou CG, Banerjee R. Tumor necrosis factor-alpha-induced targeted proteolysis of cystathionine beta-synthase modulates redox homeostasis. J Biol Chem 2003;278:16,802–16,808.
71. Shan X, Dunbrack RL Jr., Christopher SA, et al. Mutations in the regulatory domain of cystathionine beta-synthase can functionally suppress patient-derived mutations in cis. Hum Mol Genet 2001;10:635–643.
72. Maclean KN, Gaustadnes M, Oliveriusova J, et al. High homocysteine and thrombosis without connective tissue disorders are associated with a novel class of cystathionine beta-synthase (CBS) mutations. Hum Mutat 2002;19:641–655.
73. Taoka S, Ojha S, Shan X, et al. Evidence for heme-mediated redox regulation of human cystathionine β-synthase activity. J Biol Chem 1998;273:25,179–25,184.
74. Bruno S, Schiaretti F, Burkhard P, et al. Functional properties of the active core of human cystathionine beta-synthase crystals. J Biol Chem 2001;276:16–19.
75. Taoka S, Lepore BW, Kabil O, Ojha S, Ringe D, Banerjee R. Human cystathionine beta-synthase is a heme sensor protein: evidence that the redox sensor is heme and not the vicinal cysteines in the CXXC motif seen in the crystal structure of the truncated enzyme. Biochemistry 2002;41:10,454–10,461.
76. Reynolds MF, Parks RB, Burstyn JN, et al. Electronic absorption, EPR, and resonance raman spectroscopy of CooA, a CO-sensing transcription activator from R. rubrum, reveals a five-coordinate NO-heme. Biochemistry 2000;39:388–396.
77. Janosik M, Oliveriusova J, Janosikova B, et al. Impaired heme binding and aggregation of mutant cystathionine beta-synthase subunits in homocystinuria. Am J Hum Genet 2001;68:1506–1513.
78. Albakri QA, Stuehr DJ. Intracellular assembly of inducible NO synthase is limited by nitric oxide–mediated changes in heme insertion and availability. J Biol Chem 1996;271:5414–5421.
79. Chen Y, Panda K, Stuehr DJ. Control of nitric oxide synthase dimer assembly by a heme-NO-dependent mechanism. Biochemistry 2002;41:4618–4625.
80. Eto K, Ogasawara M, Umemura K, et al. Hydrogen sulfide is produced in response to neuronal excitation. J Neurosci 2002;22:3386–3391.
81. Eto K, Kimura H. The production of hydrogen sulfide is regulated by testosterone and S-adenosyl-L-methionine in mouse brain. J Neurochem 2002;83:80–86.
82. Griscavage JM, Fukuto JM, Komori Y, et al. Nitric oxide inhibits neuronal nitric oxide synthase by interacting with the heme prosthetic group: role of tetrahydrobiopterin in modulating the inhibitory action of nitric oxide. J Biol Chem 1994;269:21,644–21,649.
83. Nishida CR, Ortiz de Montellano PR. Autoinhibition of endothelial nitric-oxide synthase: identification of an electron transfer control element. J Biol Chem 1999;274:14,692–14,698.
84. Lane P, Gross SS. The autoinhibitory control element and calmodulin conspire to provide physiological modulation of endothelial and neuronal nitric oxide synthase activity. Acta Physiol Scand 2000;168:53–63.
85. Sturman JA, Rassin DK, Gaull GE. Relation of three enzymes of transsulphuration to the concentration of cystathionine in various regions of monkey brain. J Neurochem 1970;17:1117–1119.
86. Sturman JA, Gaull GE, Niemann WH. Cystathionine synthesis and degradation in brain, liver and kidney of the developing monkey. J Neurochem 1976;26:457–463.
87. Levonen AL, Lapatto R, Saksela M, et al. Human cystathionine γ-lyase: developmental and in vitro expression of two isoforms. Biochem J 2000;347(pt 1):291–295.
88. Murdoch JC, Rodger JC, Rao SS, et al. Down's syndrome: an atheroma-free model? Br Med J 1977;2:226–228.
89. Richards BW, Enver F. Blood pressure in Down's syndrome. J Ment Defic Res 1979;23:123–135.
90. Morrison RA, McGrath A, Davidson G, et al. Low blood pressure in Down's syndrome: a link with Alzheimer's disease? Hypertension 1996;28:569–575.
91. Teshigawara M, Matsumoto S, Tsuboi S, et al. Changes in levels of glutathione and related compounds and activities of glutathione-related enzymes during rat liver regeneration. Res Exp Med (Berl) 1995;195:55–60.
92. Martin JA, Sastre J, de la Asuncion JG, et al. Hepatic γ-cystathionase deficiency in patients with AIDS. JAMA 2001;285:1444–1445.

93. Awata S, Nakayama K, Sato A, et al. Changes in cystathionine t-lyase levels in rat during lactation. Biochem Mol Biol Int 1993;31:185–191.
94. She QB, Hayakawa T, Tsuge H. Alteration in the phosphatidylcholine biosynthesis of rat liver microsomes caused by vitamin B6 deficiency. Biosci Biotechnol Biochem 1995;59:163–167.
95. Godon C, Lagniel G, Lee J, et al. The H_2O_2 stimulon in Saccharomyces cerevisiae. J Biol Chem 1998;273:22,480–22,489.
96. Habib GM, Shi ZZ, Ou CN, et al. Altered gene expression in the liver of γ-glutamyl transpeptidase-deficient mice. Hepatology 2000;32:556–562.
97. Gross C, Kelleher M, Iyer VR, et al. Identification of the copper regulon in Saccharomyces cerevisiae by DNA microarrays. J Biol Chem 2000;275:32,310–32,316.
98. Aminlari M, Li A, Kunanithy V, et al. Rhodanese distribution in porcine (Sus scrofa) tissues. Comp Biochem Physiol B Biochem Mol Biol 2002;132:309–313.
99. Mimori Y, Nakamura S, Kameyama M. Regional and subcellular distribution of cyanide metabolizing enzymes in the central nervous system. J Neurochem 1984;43:540–545.
100. Wrobel M, Wlodek L, Srebro Z. Sulfurtransferases activity and the level of low-molecular-weight thiols and sulfane sulfur compounds in cortex and brain stem of mouse. Neurobiology (Bp) 1996;4: 217–222.
101. Maduh EU, Baskin SI. Protein kinase C modulation of rhodanese-catalyzed conversion of cyanide to thiocyanate. Res Commun Mol Pathol Pharmacol 1994;86:155–173.
102. Buzaleh AM, Vazquez ES, Batlle AM. The effect of cyanide intoxication on hepatic rhodanese kinetics. Gen Pharmacol 1990;21:219–222.
103. Buzaleh AM, Vazquez ES, Del Carmen Batlle AM. In vitro effect of cyanide, thiosulphate and S-adenosyl-L-methionine on the activity of rhodanese and other enzymes. Gen Pharmacol 1991;22:281–286.
104. Picton R, Eggo MC, Merrill GA, et al. Mucosal protection against sulphide: importance of the enzyme rhodanese. Gut 2002;50:201–205.
105. Walker J, Barrett J. Biochemical characterisation of the enzyme responsible for "activated L-serine sulphydrase" activity in nematodes. Exp Parasitol 1992;74:205–215.
106. Walker J, Barrett J, Thong K-W. The identification of a variant form of cystathionine β-synthase in nematodes. Exp Parasitol 1992;75:415–424.
107. Pieniazek NJ, Stepien PP, Paszewski A. An *Aspergillus nidulans* mutant lacking cystathionine β-synthase: identity of L-serine sulfhydrylase with cystathionine β-synthase and its distinctness from O-acetyl-L-serine sulfhydrylase. Biochim Biophys Acta 1973;297:37–47.

17 Toxicological and Environmental Impacts of Hydrogen Sulfide

Sheldon H. Roth

CONTENTS

SUMMARY

Hydrogen sulfide (H_2S) is a very toxic gas at high concentrations. Because it occurs in nature and is produced by numerous industrial activities, it is regarded as both an environmental and industrial pollutant. It is colorless, is heavier than air, and has a characteristic odor of rotten eggs at low concentrations; however, at higher concentrations the olfactory response is lost. Because H_2S can affect many different tissues and organs, it has been termed a broad spectrum toxicant. This chapter focuses on the toxicological effects on the respiratory tract, eye, brain, and olfactory system. An overview on dose-response relationships, sensitive populations, persistent effects, and mechanisms of action is also presented. In addition, a brief synopsis of the environmental effects of H_2S is included. Although the toxicological and environmental impacts of H_2S have been studied for many decades, there are still many concerns about the potential effects of low levels on humans and the ecosystem in general.

Key Words: Environmental toxicant; physiological responses; toxicokinetics; health effects; sensitive populations; dose response.

From: *Signal Transduction and the Gasotransmitters: NO, CO, and H_2S in Biology and Medicine* Edited by: Rui Wang © Humana Press Inc., Totowa, NJ

1. INTRODUCTION

Sulfur is a nonmetallic, greenish-yellow substance that is abundantly distributed throughout the universe. It was known in ancient times as burning stone or brimstone *(1)*. Sulfur forms many compounds when combined with metals and other substances; one of the most familiar is hydrogen sulfide (H_2S). This gaseous substance is one of the major compounds involved in the natural sulfur cycle *(2)*. Many reviews and articles refer to this substance as an environmental and industrial pollutant, toxic to both animals and people and having a characteristic odor of rotten eggs. Because the gas is colorless, heavier than air, and undetected by the sense of smell at higher concentrations, there is the potential for toxic exposures at high levels. It has been well documented that acute exposure to high concentrations can produce various adverse or lethal effects *(3–6)*. The potential risks associated with chronic or repeated low-level exposure have not been well established and perhaps underestimated. Although the toxicological and environmental impacts of H_2S have been studied for many decades, there are still many concerns about the potential effects of low levels on humans and the ecosystem in general.

Because H_2S can affect many organ systems, it has been termed a broad spectrum toxicant *(6)*. The threshold responsiveness is variable, depending on concentration, rate of exposure, and tissue type *(2)*. Tissues with a high oxygen demand are the most susceptible to the toxicant; thus, the brain is considered to be a primary and critical target organ *(5,7)*. H_2S is described as one of the leading causes of sudden death in the workplace *(8,9)*. Over the years, many reviews and documents describing the toxicity of H_2S have been published *(2–7,10)*.

1.1. Historical Account Toxicity of H_2S

In the classic review by Beauchamp and colleagues *(2)*, reference is made to one of the earliest writings of H_2S by Zosimus of Panopolis. In the third century, Zosimus described "Water of Sulfur" or "Divine Water" capable of changing substances to gold *(2)*. In 1713, the Italian physician Bernardino Ramazzini reported in his text titled "De Morbis Artificum" the first account of H_2S poisoning *(11)*. Ramazzini described the irritation and inflammation of the eyes of sewer workers and suggested that these symptoms were caused by a volatile acid that was also capable of blackening coins made of copper and silver. Over a century later, Victor Hugo used this theme for his novel *Les Miserables*. In 1775, Carl Wilhelm Scheele discovered that combining acid with metal (poly) sulfides or heating sulfur in hydrogen gas could produce H_2S; the gas was later analyzed by Berthollet in 1796. In 1934, Sayers *(12)* reported the acute and subacute effects of H_2S in sewers and sewage treatment plants. For many years, H_2S has been regarded as an irritant gas *(13)*, and it is interesting that the initial reports of eye irritation resulting from H_2S became the basis for establishing a threshold limit value for exposure in the workplace.

1.2. Chemical and Physical Properties

H_2S is a flammable and colorless gas that is heavier than air (*see* **Table 1**). At low concentrations it has the characteristic offensive odor of rotten eggs. It is soluble in both polar and nonpolar solvents (e.g., 1 g will dissolve in 242 mL of water, 94.3 mL of alcohol, and 48.5 mL of diethyl ether at 20°C). The gas is readily oxidized to form sulfur dioxide, sulfates, or elemental sulfur (*see* **Eqs. 1** and **2**). The undissociated form is lipid soluble and, therefore, will easily penetrate biological membranes. In aqueous solution, H_2S

Table 1
Chemical and Physical Properties of H_2S

Properties	Gas, colorless, flammable, characteristic offensive odor
Synonyms	Dihydrogen monosulfide, dihydrogen sulfide, hydrogen sulfuric acid, sulfur hydride, sulfureted hydrogen, sewer gas, stink damp
Molecular formula	H_2S
Molecular weight	34.08
Melting point	–85.49°C
Boiling point	–60.33°C
Density (gas)	1.192 g/L (air = 1.00 g/L)
Vapor pressure	18.5 atm (18.75 × 10^5 Pa) at 20°C 23.9 atm (23.9 × 10^5 Pa) at 30°C
Explosive limits in air	4.3–46% by volume
Soluble in	Water, alcohol, ether, methanol, acetone, gasoline, kerosene, crude oil, CS_2, glycerol, aqueous solutions of amines, alkali, carbonates, bicarbonates, hydrosulfides
Odor threshold	0.0008–0.20 mg/m^3 (0.0005–0.13 ppm)
pK_a values	7.04 for HS^- ; 11.96 for S^{2-}
Conversion factors	1% by volume = 10,000 ppm 1 ppm = 1.4 mg/m^3 1 mg/L = 717 ppm standard temperature pressure

dissociates into a hydrosulfide ion (HS^-) and sulfide ion (S^{2-}), as shown in **Eq. 3**. This reaction is dependent on the pH of the solution and pK_a of the gas. At physiological pH (7.4), approximately one-third of the total sulfide will be in the undissociated form and two-thirds as the hydrosulfide ion. A portion of the gas will also evaporate from solution because of the low vapor pressure (*see* **Table 1**). Aqueous solutions containing bromine, chlorine, or iodine may react with H_2S to form elemental sulfur.

$$2H_2S + 3\ O_2 \rightarrow 2H_2O + SO_2 \quad (1)$$

$$2H_2S + O_2 \rightarrow 2H_20 + 2S \quad (2)$$

$$H_2S \rightarrow H^+ + HS^- \rightarrow H^+ + S^{2-} \quad (3)$$

1.3. Sources and Uses

H_2S occurs in nature and is produced by numerous industrial activities; therefore, it is termed both an environmental and industrial pollutant. It is used by various industries and is also a byproduct of many processes *(3)*.

1.3.1. Natural Sources

Ubiquitous in nature, H_2S occurs as a constituent of natural gas, petroleum, sulfur deposits, volcanic gases, and sulfur springs. It is produced in many areas containing bacterial or proteinaceous materials, such as swamps, lagoons, tidal flats, marshes, bogs, mudflats, and near-surface waters, as a result of decomposition by bacteria, fungi, and actinomycetes. Consequently it appears in sewage treatment facilities and is associated with fish aquacultures and livestock production. Oxidation to sulfate and elemental sulfur balances this production and forms the basis of the environmental sulfur cycle *(2)*. It is estimated that 90% of atmospheric H_2S is the result of natural sources *(3,4,6)*.

1.3.2. Anthropogenic Sources

A substantial number of human activities are responsible for the release of H_2S from natural sources. These include the extraction, transport, and processing of natural gas and other petroleum deposits; coal mining; and development of geothermal generators. Natural gas that contains sulfur compounds such as H_2S is termed *sour gas* and, depending on the deposit (geographical formation), may contain levels ranging from <1 to >90% by volume *(6)*. Other anthropogenic sources are livestock production facilities, animal-processing plants, wastewater treatment plants, sewers, sanitary landfills, and some water treatment plants.

1.3.3. Industrial Sources

Many industrial operations potentially use or produce H_2S including tanneries, kraft paper mills, petroleum refineries, natural gas plants, coke ovens, foundries, food-processing plants, producers of heavy water and rayon, and some chemical plants. H_2S is used primarily for the production of elemental sulfur and sulfuric acid, heavy water, metallurgy, analytical agents, and disinfectants in agriculture.

1.4. Fate and Transport

H_2S released into the atmosphere undergoes chemical and photochemical oxidation reactions that produce sulfuric acid and/or sulfate, contributing to acid rain *(3)*. Photo-oxidation can occur in the presence of oxygen, ozone, and SO_2. In a relatively clean atmosphere, H_2S can exist unchanged for approx 1–2 d compared to a mere few hours in a polluted environment. The lifetime of the gas in the atmosphere is also dependent on dispersion patterns, geographical features, temperature, humidity, light, wind, and pressure *(6)*. It has been estimated that H_2S can last for approx 18 h on average *(3)*.

1.5. Ambient Levels and Guidelines

Ambient levels can vary greatly, depending on location and various other factors. Data are limited or lack specifics about sampling times and methods. It has been estimated that ambient levels in the United States are between 0.11 and 0.33 ppb (0.15–0.46 $\mu g/m^3$) *(3)*. In undeveloped areas, concentrations range between 0.02 and 0.07 ppb (0.028–0.09 $\mu g/m^3$) *(3)*. It has been suggested that the general population in urban centers usually is not exposed to concentrations exceeding 1 $\mu g/m^3$ (0.7 ppb). Time-average values may not reflect concentrations that occur over brief periods of time. If these values are much greater than the average by several orders of magnitude, there is potential for health impacts.

2. TOXICOKINETICS

2.1. Absorption

There is a lack of quantitative information regarding the toxicokinetics of H_2S in humans. Most studies have been conducted in animals *(2)*. Because H_2S is a gaseous substance, the primary route of exposure is the respiratory system, and absorption occurs rapidly from the lungs. Absorption through the skin is minimal *(2)*.

2.2. Distribution

H_2S is widely distributed throughout the body primarily as the undissociated gas or HS^- ion *(2,14)*. Following exposure, H_2S can be found in numerous body compartments including blood, brain, lung, liver, kidney, and spleen *(15–17)*. Although the values may not be reliable, the results provide an indication of wide distribution.

2.3. Metabolism (Detoxification)

Animal studies have demonstrated three primary metabolic pathways involved in the detoxification of H_2S: oxidation to sulfate, methylation, and reaction with metalloproteins and disulfide-containing proteins *(2)*. Other possible pathways include conjugation with GSH *(4)*. H_2S can also bind to methemoglobin and cytochrome oxidase *(18)*. Although the interaction with methemoglobin may be a detoxification pathway, the combination with cytochrome oxidase is more likely involved in the toxicity of H_2S and is discussed further in **Subheading 8.** The primary detoxification pathway (oxidation) occurs in the liver and, to a smaller degree, in blood, kidney, and intestinal mucosa *(4,14,19)*. The sulfide anion is converted by hepatic and kidney oxidases to thiosulfate and sulfate *(2)*. One study in volunteers exposed to low concentrations (8–30 ppm) of H_2S reported increased urinary levels of thiosulfate, in support of oxidation as the primary metabolic pathway in humans *(3)*. Elevated blood and urinary levels of thiosulfate were also found in rabbits exposed for 1 h to 100–200 ppm of H_2S *(20)*. Rapid metabolism to sulfate has often been interpreted as an explanation for H_2S being unlikely to accumulate; however, reactions with essential proteins may be responsible for toxicity *(2)* as well as residual effects *(6)*.

2.4. Elimination

Observations of thiosulfate in urine of both animals and humans *(20)* suggest that the primary route of elimination is the kidney.

3. HEALTH EFFECTS

Many previous reviews have described the adverse effects of H_2S on various tissues and organs *(2–7,10)*. Because there have been very few controlled studies in human subjects, most of the data have been derived from case reports or studies using experimental animals or model systems *(3–5)*. It is well recognized that there is a significant lack of scientific data on chronic exposure to low concentrations of H_2S *(3,6)*. A draft of the recent EPA Toxicological Review of Hydrogen Sulfide is available at www.epa.gov/iris.

3.1. Major Target Organs

Because H_2S is able to affect many different organs, it has been referred to as a broad spectrum toxicant. Tissues with a high oxygen demand or exposed mucous membranes such as the lung, nervous system, heart, and eyes are most susceptible to the effects of H_2S at low concentrations *(6)* and can be considered the major target organs. The nervous system appears to be the primary target organ in relation to overall toxicity, and this has been reviewed in detail *(7)*. The toxic effects of H_2S on the major target organs are discussed in detail in this chapter, but the effects of H_2S on other tissues and organs are presented in **Table 2** for brevity. The effects of H_2S on the cardiovascular system are reviewed in Chapter 19.

3.2. Respiratory Tract and Lung

3.2.1. Human Studies

The major route of exposure to H_2S in humans is the respiratory tract. There is considerable literature describing the effects of H_2S on the respiratory system *(3)*. Following

Table 2
Effects of Hydrogen Sulfide

Target	*Symptoms*
Carcinogenesis	No significant data
Cardiovascular system	No significant effects at low levels (<20 ppm); transient changes in electrocardiograms at higher concentrations; decrease in blood pressure, ventricular extrasystoles, arrhythmias, migraine
Development	No significant effects in growth parameters; significant, subtle alterations in architecture; growth characteristics; and neurochemistry of developing (rat) brain
Endocrine	Decrease in milk production (cows); increase in plasma cortisol levels; dose-dependent lesions of thyroid gland; reduction or blockage of oxytoxin receptor
Eye	Irritation and inflammation of conjunctival and corneal tissue (keratoconjunctivitis): blurred vision, halo around lights, "gritty" sensation or pain
Gastrointestinal tract	Nausea, vomiting, diarrhea, stomach cramps
Hepatic tissue (liver)	No significant data
Hematopoietic (blood-forming) system	Variable results; may inhibit heme synthesis and disturb iron metabolism
Immunological system	Depression of macrophage function
Kidney	No significant data
Mutagenesis	No significant data; may be a weak mutagen (Ames' test and in *Drosophila melanogaster*)
Nervous system	Fatigue, vertigo, anxiety, olfactory paralysis, convulsions, collapse, unconsciousness, memory loss (amnesia), depression, headache, morphological and neurochemical alterations in brain tissue
Neuropsychological	Impairment of cognitive function, changes in emotion
Olfactory system	Odor detection, anosmia (loss of smell), olfactory fatigue, olfactory nerve paralysis, decreased sensitivity with age
Reproductive system	Ambiguous reports; high levels may result in dystocia (increase in delivery time)
Respiratory system	Dyspnea, pulmonary tract irritation, bronchitis, rhinitis, pharyngitis, laryngitis, sore throat, cough, chest pain, pulmonary edema, cyanosis, hemoptysis, pneumonia, increased mucosal permeability, breathlessness, wheezing
Skin	Reversible discoloration, spots, rash

exposure to acute and high (often accidental) concentrations of H_2S, numerous respiratory effects are observed including increased respiratory symptoms *(21–24)* and objective evidence of airway obstruction *(25)*; reduction in residual lung volume *(26)*; pulmonary edema *(27)*; bronchial mucosal ulceration *(24,25,27)*; bronchial hyperreactivity *(21)*; and reactive airways dysfunction syndrome *(21,28)* including sore throat, cough, and dyspnea *(3)*.

The mechanism of action has also been studied. H_2S is considered an irritant; the effects are associated with direct damage to the upper respiratory tract epithelium *(14)*. There may be an initial increase in respiratory frequency on exposure as a result of stimulation of the carotid bodies and chemosensors *(29)*. The rate and depth of ventilation may progress to a rapid, deep breathing (hyperpnea) pattern *(3)*; however, if the duration and/or concentration are sufficient, death may result because of respiratory failure or arrest *(2,3)*. The effect of H_2S on breathing control was examined in animals *(30)*. It was observed that NaHS was able to induce apnea more effectively if injected peripherally, suggesting that the transmission between lung and brain was essential and the lung should be considered the primary site of action of H_2S. Bicarbonate is able to prevent apnea, providing evidence of inhibition of cytochrome-*c* oxidase *(31)*. It was also demonstrated that sensory neurons may play a major role in pulmonary defense against the effects of H_2S *(32)*. Although respiratory symptoms usually diminish over time, some may persist for several months *(33)*.

Controlled studies in volunteer subjects have been conducted with low concentrations of H_2S and of relatively short duration *(34–38)*. Exposure to 5 ppm for more than 16 min following exercise did not reveal effects on ventilation mechanics, but there was a significant increase in maximum oxygen uptake *(34)*. At exposures to 2–5 ppm, the respiratory exchange ratio was significantly decreased, but this was attributed to a nonsignificant trend toward increased oxygen uptake and decreased carbon dioxide output *(34)*. No significant changes in respiratory parameters were reported in a study with 5 ppm for 30 min *(35)*. In addition, pulmonary function tests in both males and females were not altered during inhalation of 10 ppm of H_2S for 15 min at elevated metabolic and ventilation rates *(36)*.

A group of 10 patients with asthma (those with severe asthma were excluded) were exposed to 2 ppm of H_2S for 30 min *(39)*. In general, there were no significant changes in airway resistance or specific airway conductance; however, two of the subjects appeared sensitive and exhibited bronchial obstruction *(39)*.

Numerous epidemiological studies have been conducted to evaluate whether chronic or long-term exposure to low levels of H_2S results in respiratory effects *(16,39–42)*. Although some have reported an increase in symptoms *(43)*, it is still difficult to resolve whether significant effects can be attributed directly to H_2S. For many of the studies, concentrations of H_2S were not determined and subjects may have been exposed to other substances. Richardson *(40)*, in a cross-sectional study of sewer workers, reported a statistically significant decrease in spirometric values (FEV_1/FVC). In addition, there was a higher odds ratio (OR) for obstructive lung disease in these workers compared to with control subjects. A questionnaire-based study of 175 oil and gas workers *(22)* was conducted to assess sour-gas exposure that caused symptoms or loss of consciousness. Workers *(34%)* who stated that they had experienced exposures serious enough to cause symptoms showed no decrease in spirometric values *(22)*, and workers who had lost consciousness (8%) experienced shortness of breath with physical activity, wheezing, and tightness in the chest. It was concluded that the symptoms were consistent with bronchial hyperresponsiveness *(22,24)*.

A series of studies *(44–49)* examined subjects in the South Karelia, Finland, area. Pulmonary function of pulp mill workers exposed previously to H_2S at concentrations of 10 ppm or less were not significantly changed *(39)*. Numerous studies have been conducted on communities living near pulp mills. Residents of these communities are exposed to "pollutant mixtures" including particulates, sulfur dioxide, and mercaptans, and, therefore, the role of H_2S is unclear *(3)*. Subjects living in polluted communities experienced significant increases in nasal symptoms; cough and breathlessness or wheezing were also increased but not significantly *(44)*. The effects appeared to be dose related *(44)*. A follow-up study found a dose-related increase in the probability of nasal and pharyngeal irritation of subjects living in the vicinity of a pulp mill *(48)*. Increased ORs were revealed for nasal irritation, cough, breathlessness or wheezing, as well as headaches or migraine *(49)*; however, this study was complicated by the time periods and high levels of other reduced sulfur compounds including sulfur dioxide *(3)*. Because subjects were exposed to mixtures of pollutants (although H_2S appears to be the dominant sulfur contaminant), it was concluded that these studies demonstrate that low levels of H_2S in combination with other sulfur-containing pollutants can have an adverse effect on respiratory health *(3)*. Pulmonary function of workers in viscose-rayon plants exposed to H_2S for an average 12.3 yr was found not to be significantly different from that of a control group *(50)*. The investigators concluded that chronic exposure to low levels of H_2S would not cause adverse effects on pulmonary health *(50)*. Investigations of the health effects of H_2S from geothermal sources in Rotorua, New Zealand by Bates et al. *(42)* revealed elevated mortality risks for diseases of the respiratory system particularly among Maori women. There was also an elevated rate for nasal cancers, although this involved only four cases *(51)*. The limitations of previous studies *(51)* were apparently addressed in the more recent study by Bates et al. *(42)*, stating that the health effects could be associated with the exposure *(42)*. A study of communities exposed to intensive livestock operations found higher rates of stuffy, runny, and burning nose; stuffy sinuses; sore or scratchy throat; and excessive coughing *(52)*. Increased respiratory symptoms were also found in a community near an intensive swine operation *(53)*.

3.2.2. Animal Studies

In controlled laboratory studies using animal models, respiratory effects such as nasal inflammation have been reported following exposure to H_2S *(32,54,55)*. Earlier studies by Lopez et al. *(56)* revealed that higher concentrations of H_2S could produce a transient increase in cellular content and biochemical properties of nasal lavage from rats exposed to 10–400 ppm of H_2S. They concluded that the observations were consistent with cytotoxicity of the nasal epithelium, and that the olfactory epithelium was more sensitive to toxic insult than the respiratory epithelium *(56)*. Lopez et al. *(57)* reported transient histological alterations in a later study, mainly in rats exposed to higher concentrations (440 ppm). They also noted nasal lesions in rats exposed to 400 ppm of H_2S *(58)*. The effects of H_2S on surface properties of pulmonary surfactant have also been examined *(59)*. Bronchoalveolar lavage fluid from exposed animals (200 and 300 ppm of H_2S) contained elevated concentrations of protein and showed marked increases in surface tension. The lungs also showed areas of atelectasis, patchy alveolar edema and perivascular edema *(59)*. Exposure to H_2S may also compromise bacterial inactivation in animals *(60,61)*.

Recent studies by Dorman and colleagues *(54,55)* have extended the previous studies of H_2S on nasal properties. Mild to marked sensory neuron loss and basal cell hyperplasia of the olfactory mucosa were observed in rats following subchronic exposure to concentrations of 10–80 ppm of H_2S *(55)*. On examination of the nasal mucosa, no effects were observed at 10 ppm of H_2S; however, nasal lesions of the olfactory mucosa were found following exposures of 30 and 80 ppm *(54)*. Similar nasal lesions were observed in exposed mice in previous studies *(62)*, although only minimal lesions were observed in rats (63,64).

3.3. Neurological Effects

Many of the symptoms following exposure to H_2S can be attributed to direct actions on the central nervous system *(5)*. Although it has been indicated that repeated exposure is necessary for neurological effects *(2)*, it is evident that acute exposures of sufficient concentration can elicit significant responses *(7)*. Case reports following industrial and/or accidental exposures have provided the majority of data, whereas studies on animal systems have yielded more information at the cellular and molecular levels and potential mechanisms of action. It is accepted that symptoms that occur rapidly following exposure such as dizziness, lack of coordination, and headache are more likely because of direct action on the brain rather than anoxia *(16)*. Multiple neurological effects can occur following exposure to relatively high concentrations *(2,3)*, and it is not surprising that permanent nerve damage can occur with very high levels. The effects at very high concentrations have been compared with those of anesthetics, suggesting a nonspecific mechanism because of the lipid solubility of H_2S *(65)*.

Several studies have provided evidence that H_2S can directly alter the electrical properties of nervous tissue *(2,7)*. The effects are concentration dependent and complex including biphasic responses *(2,66)*. Biphasic effects of H_2S have been reported previously *(67)*. High concentrations of sulfide are capable of blocking spontaneous activity, whereas low concentrations increase the firing activity of rat hippocampal neurons *(2,4)*. Exposure of the volatile salt NaHS to in vitro hippocampal neurons resulted in a rapid initial membrane hyperpolarization and a decrease in membrane resistance *(68)*. Further hyperpolarization was observed immediately after washout *(2,68)*. Similar actions were observed on dorsal raphe serotonergic neurons *(66)*. Recently, it has been demonstrated that low concentrations of gaseous H_2S can also produce a biphasic response in hippocampal neurons: initial depression on exposure and long-lasting exhancement on washout *(69)*. Studies on rat hippocampal preparations suggest that the effects of H_2S may involve $GABA_A$ receptors *(70)*. The variety of concentration- and time-dependent effects observed at the cellular levels may provide an explanation for the diversity of neurophysiological symptoms.

3.3.1. Neurochemistry

Changes in brain neurotransmitter levels have been reported to occur in the mature mammalian brain following H_2S exposure *(2,7)*. Neurotransmitter levels could be increased or decreased, varying with neurotransmitter, brain region, dose of H_2S, and even technique. The changes in neurotransmitter levels of mature brain are similar to those that occur in the developing system (described in **Subheading 5.2.** and *see* ref. *7*). The measured increase in brain catecholamines (5-hydroxytryptamine [5-HT], dopamine, epinephrine, and norepinephrine) was suggested to be a result of a sulfide-induced inhibition of monoamine oxidase (MAO) *(2)*. Sulfide-induced accumulation of extracellular synaptosomal glutamate suggests that higher concentrations of glutamate found in the hippocampus of exposed rats may result from a reduced capacity of nerve endings to take up neurotransmitter *(71)*.

3.3.2. Behavior and Memory

The effects of H_2S on brain physiology, neurochemistry, and morphology suggest that H_2S is capable of inducing alterations in behavior and memory. The literature describing the effects of H_2S on cognitive and/or motor deficits in humans and animals *(72)* consists of primarily epidemiological or case reports. There is some degree of controversy as to whether or not low-level exposure, acute or prolonged, produces behavioral and/or memory changes, and what concentrations are effective. A recent study *(72)* demonstrated that repeated exposure of H_2S at a relatively high concentration (125 ppm) could impair learning of newer and more difficult tasks (as tested by a radial arm maze task), whereas the animals were able to retain previously learned material. Struve et al. *(73)* observed significant reductions in motor activity and water maze performance following short-term exposures to 80 ppm H_2S, although there appeared to be a lack of neuropathology *(55)*. By comparison, a recent study reported no effects on motor activity, passive avoidance, or acoustic startle response in exposed rat pups *(55)*. The persistent effects on cognitive and motor functions are discussed in **Subheading 6.**

3.4. The Eye

Since the early writings of Ramazzini describing eye irritation of sewer workers, there have been numerous studies demonstrating that exposure can result in ophthalmic effects including irritation or alteration of corneal epithelium *(33,44,48,51,52)*. The corneal epithelium appears to be quite sensitive to low levels of H_2S. Exposure to H_2S has also been associated with keratoconjunctivitis, punctate corneal erosion, blepharospasm, lacrimation, and photophobia *(3,74)*. Several investigators have reported significant increases in the incidence of eye disorders even in cases when ambient levels of H_2S occur in the parts-per-billion range *(33,41,44)*. However, it is feasible that these ocular effects may result from peak concentrations of H_2S or to coexposure to mercaptans and/or other sulfides *(3)*. Many of the earlier studies involved locations where multiple substances were emitted. It has been well established that viscose-rayon workers can exhibit ophthalmic effects known as "gas eye," but this may be a result of the combined exposure to H_2S and carbon disulfide *(75)*. A recent discussion of gas eye *(14)* concludes that "overall the balance of evidence suggests that eye irritation is a phenomenon only associated with high exposures to H_2S, in the hundreds of ppm," p. 10.

Animal studies have provided evidence that high acute (often lethal) concentrations of H_2S can produce eye irritation. Effects were noted in rats exposed to 400 ppm of H_2S for 4 h, but there was no effect at 200 ppm *(3,58)*. By comparison, no treatment-related histopathological changes were detected in rats exposed long term (90 d at 6 h/d, 5 d/wk) to 10–80 ppm of H_2S *(62–64)*.

4. DOSE–RESPONSE RELATIONSHIPS

The majority of data on dose/concentration–response relationships has been obtained from animal studies using different routes of administration/exposure and, to some degree, compounds (e.g., gaseous H_2S gas, Na_2S, NaHS). Observations of humans exposed to different levels of H_2S indicate that there is a progressive increase in severity of adverse effects as the concentration increases (**Fig. 1**). The odor response appears to be the most sensitive *(76)*. However, there is considerable variation—for example, 83% of the population can detect H_2S between 0.5 and 30 ppb. By comparison, the threshold for irritant properties is 2.5–20 ppm *(76)*. Tables listing dose/concentration and corresponding health impacts have been published in several reviews and reports *(6,76,77)*. At high concentrations, H_2S is absorbed immediately by the lungs, as evidenced by rapid responses at concentrations above 500 ppm. The severity of effect and, to some degree, the site of action are dependent on the concentration and duration of exposure. The effects at high concentrations appear to be similar for both humans and animals *(6)*. Recently, Brown and Strickland *(78)* have attempted to derive a chemical-specific dose-duration response curve to help identify toxicity markers from rat and mice data using meta-analysis techniques. They showed that the curves for rat fit the data exceedingly well and exhibit a threshold-like response followed by a steep incline as concentration is increased. This

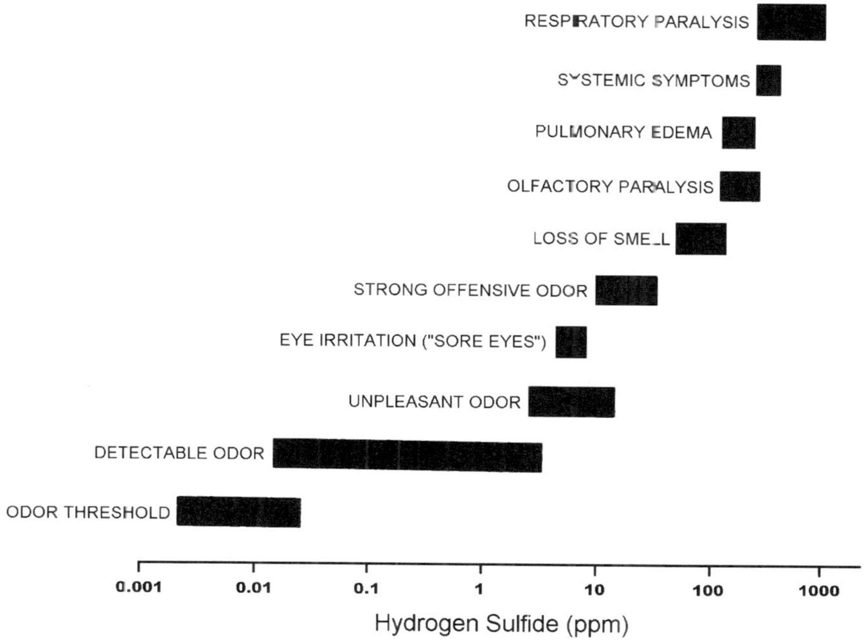

Fig. 1. The effects of H_2S appear to be concentration dependent.

method may prove to be very useful for generating no observed adverse effect level (NOAELs) for H_2S.

5. SENSITIVE POPULATIONS

5.1. Reproduction

There are limited data on the effects of H_2S on reproduction. An increased rate of abortion has been reported in women who worked in the rayon textile or paper products industries or whose husbands were employed in rayon textile or chemical-processing jobs *(79)*. Rats exposed to 20–75 ppm of H_2S 7 h/d during gestation exhibited a dose-dependent, but variable, increase in partition time *(80)*. A study conducted by the Chemical Industry Institute of Toxicology *(62–64)* found no treatment-related histological changes in female or male rat reproductive organs. More recently, an extensive study was conducted to assess whether perinatal exposure to H_2S (10–80 ppm) alters pregnancy outcomes *(55)*. The results indicated no significant changes in reproductive performance including mating and fertility indices, postimplantation loss/litter, number of late resorptions or stillbirths, number of live pups, litter size, average length of gestation, average number of implants in the female, and sperm number and quality in the male *(55)*. There were single incidences of adverse effects noted at the 80-ppm exposure level, such as abnormal sperm, ovarian cysts, and squamous metaplasia of the endometrium *(55)*. An earlier study also reported no evidence of embryotoxicity at exposure levels of 220 ppm for 3 h/d for 7 d *(81)*.

5.2. Development

The study by Dorman et al. *(55)* also evaluated developmental parameters of newborn pups. Although they reported no significant structural malformations, they did note malformations including kinked tail, agnesis of the tail, anophthalmia, small rear legs and body, frontal bone defects, hypognathia, and skin lesions; however, none of these were dose dependent. There were no significant differences in pup weight gain, pinnae detachment, surface righting, incisor eruption and negative geotaxis, vaginal patency, preputial separation, and eyelid separation *(55)*. In addition, no changes were observed in surface righting, motor activity, acoustic startle response, or passive avoidance tests. These results are similar to those of Saillenfait et al. *(82)*, who reported that exposure of pregnant rats to 50–150 ppm of H_2S did not produce any signs of maternal toxicity or adverse effects on the developing fetus, but differ from those reported previously by Hayden et al. *(80)*. Although significant elevations in maternal blood glucose levels were observed in rats exposed to H_2S (20–75 ppm) during gestation and postnatal periods *(83)*, there were no changes in serum protein, lactate dehydrogenase (LDH), serum glutamic oxaloacetic transaminase (SGOT), or alkaline phosphatase in either treated dams or pups *(83)*.

The effects of H_2S on the developing brain have been previously reviewed *(7)*. Reversible encephalopathy observed in a 20-mo-old child was attributed to exposure to H_2S (0.6 ppm) for nearly a year *(84)*. Infants breast-fed by mothers working in rayon factories showed signs of retarded development and listlessness, which was thought to be because of H_2S *(85)*. Following repeated exposure to H_2S (20–75 ppm) during gestation and lactation, newborn rats were found to have alterated developmental characteristics *(80)*, altered architectural and growth characteristics *(86)*, and significant changes in growth patterns of the cerebellum *(87)*. In a series of similar studies, changes in neurotransmitter content of developing rat brain were observed *(88)*. Levels of aspartate, glutamate, and GABA in the cerebrum and cerebellum were significantly depressed, but not glycine. The observed increased levels of taurine may reflect maternal sources *(89)* and the return to control values appeared to coincide with the maturation of the blood–brain barrier *(89)* or to reflect a change in the capacity to metabolize taurine. Changes in neurotransmitter levels could reflect alteration in synthesis and/or release of neurotransmitter or a loss of neurons *(2)*. In later studies, it was revealed that levels of serotonin (5-HT) and norepinephrine were altered in the cerebellum and frontal cortex of developing rats following repeated exposure to similar concentrations of H_2S *(90)*. The effects on 5-HT levels appeared to be concentration dependent *(91)*. Using similar protocols but determining the monoamine levels for up to 60 d revealed that alterations in these neurotransmitters gradually returned monoamine to control levels by d 45 *(86)*. These results provided evidence that repeated or chronic exposure to low levels of H_2S may affect the development of the mammalian brain and contribute to long-term abnormalities in motor and cognitive function *(2)*. Dorman et al. *(55)* found no evidence of gross or histological brain pathology in H_2S-exposed pups. However, they were not able to confirm the previous findings of brain cell growth characteristics *(87)* because they used light microscopy rather than detailed morphological techniques *(55)*. Behavioral testing has not demonstrated that alterations in brain neurotransmitters have a functional impact *(55)*, but further examination of the neurochemistry and functional properties of the developing brain in H_2S-exposed animals is required *(73)*.

5.3. Children

It is often stated that "children are not small adults" and that they differ in their exposures, toxicokinetics and susceptibility to toxins *(3)*. In humans, synaptogenesis extends from 6 mo of gestation to several years following birth *(92)*. Although there is limited evidence that children may be more susceptible to H_2S, there are some data that support the concept that children are more vulnerable. In a study conducted in Alberta, Canada assessing the health of a population living downwind from natural gas refineries, children, but not adults, reported significantly more persistent cough, phlegm, and wheeze compared with unexposed control subjects *(93)*. The children were exposed to a mixture of H_2S and SO_2. Children living near a sulfate cellulose plant in Sweden were found to have significantly more symptoms related to bronchial asthma and increased bronchial hyperreactivity *(48)*. A more recent study documented a considerable increase in the risk of nasal symptoms, cough, eye symptoms, and headache in children exposed long-term to ambient air malodorous sulfur compounds released from pulp mills *(47)*. Clearly there is a need for further research addressing the impact of H_2S on children.

6. PERSISTENT EFFECTS

There are numerous reports that describe persistent or permanent effects in humans following acute high-dose exposure to H_2S *(3)*. Some individuals appear to be resistant and recover completely without any symptoms *(18)*. A report of delayed lung injury described a male experiencing dyspnea, chest tightness, and hemoptysis 3 wk after exposure; 5 mo later, he still showed symptoms of dyspnea on exertion and decreased lung volume and CO diffusion capacity *(33)*. However, the majority of reports describing long-term effects have been related to neurological and/or neuropsychological symptoms *(7)*. Exposed individuals have developed various cognitive, motor and behavioral deficits that seem to persist for months or years *(94–99)*. Several of these case reports have been described in detail *(3)*. One study followed several patients for 5–10 yr and documented "permanent" neurological symptoms including vision and memory impairment; rigid movements; reduced motor function; slight tremor; ataxia; psychosis; abnormal learning, retention, and motor function; and slight cerebral atrophy *(94,95)*.

There is concern that persistent neurobehavioral dysfunction may occur following chronic low-dose exposure to H_2S. Over the years, Kilburn *(100–106)* has published several articles on this subject. He has studied people exposed to H_2S from accidental releases and living downwind from oil refineries *(104,105)*. According to him, these studies have provided justification for and added strength to his hypothesis that H_2S causes chronic brain damage, as demonstrated by neurobehavioral dysfunction *(104)*. In recent years, he has provided further explanations of and justification for his procedures and conclusions *(106)*. Kilburn *(105)* has also recognized that the effects on the brain may be generalized and testing difficult. The association of long-term or repeated exposures and impaired motor and cognitive function suggests that H_2S is a cumulative toxicant *(2,6,7)*. This generally is not accepted because it is assumed that H_2S is rapidly metabolized and excreted *(2)*. Animals studies have demonstrated that repeated exposure to H_2S (25–75 ppm at 3 h/d for 5 d) produced cumulative dose-dependent changes in rat hippocampal electroencephalogram activity, which were reversed in approximately 2 wk *(107)*. The issue of persistent effects in humans as a result of H_2S exposure continues to be controversial and requires testing under controlled conditions. In addition to studies addressing delayed, persistent, or permanent effects, it is necessary to quantitatively determine the toxicokinetics of this gaseous toxicant in humans.

7. EFFECTS ON OLFACTORY SYSTEM

The majority of organisms respond to novel odors that enter their environment. Humans and animals are able to detect and discriminate a variety of chemicals that result in instinctive or learned behaviors associated with food selection, mating, prey–predator interactions, and reproduction *(108–110)*. The response to an odor is highly subjective *(111)*. In humans, the odor threshold of H_2S has been estimated to be in the range of 0.003–0.02 ppm *(4)*. There is considerable individual variation, but the threshold to H_2S was surprisingly similar for people from Japan and the Netherlands *(112)*. It has been proposed that odor can affect certain areas of the brain including the cortex, subcortex, cranial nerves, as well as the peripheral nervous system *(113)*.

The physiology of odor perception and health symptoms has been reviewed recently *(114)*. In summary, health symptoms from odors can result from the sensation (odor) or the chemical (odorant), and irritant compounds such as H_2S can produce various systemic responses. The link between irritant sensations and odor sensations is, in part, a result of the central projections of the olfactory and trigeminal systems *(114)*. The response times for unpleasant odors are significantly shorter than for pleasant ones *(115)*. It has been suggested that health symptoms in sensitive individuals may persist for longer periods of time and exacerbate existing medical conditions (e.g., airway responsiveness) *(114)*.

Release of H_2S from various sources has resulted in numerous complaints about odor and associations with health effects. The strong and distinctive odor of H_2S is readily detected. Symptoms associated with odor have been documented for various sources including oil batteries, petroleum refineries, kraft paper mills, petrochemical plants, waste treatment facilities, commercial mushroom composting operations, and water systems that receive pollutant discharge *(116–119)*.

There has been concern that H_2S can produce chronic olfactory deficits, and neurotoxic effects of low concentrations may be a result of combined exposure with other agents *(113)*. As discussed earlier, 83% of the population can detect H_2S between 0.5 and 30 ppb *(114)*, but at higher concentrations the olfactory mechanisms are impaired. This impairment, often referred to as olfactory paralysis, has been estimated to occur at levels between 100 and 200 ppm *(120,121)*. Whereas some studies conclude that there is no convincing evidence of health effects as a result of low-level exposure (odor) to H_2S, others have stated that exposure to odor is associated with health symptoms, although the mechanisms are unknown. It is evident that the health impact of H_2S odor still needs to be resolved *(114,122)*.

8. MECHANISMS OF ACTION

Inhibition of cytochrome oxidase generally is considered the primary mechanism for H_2S toxicity *(71,121)*. Cytochrome oxidase, the terminal enzyme of oxidation, is important in the regulation of cellular energy production, and inhibition of these enzyme systems would lead to termination of oxidative phosphorylation, a major source of adenosine triphosphate (ATP) synthesis *(123)*. Disruption of oxidative metabolism would affect the target organs with high oxygen requirements, such as brain and heart *(29)*. The inhibition of cytochrome oxidase is similar to that of cyanide *(71)*, but the mechanisms may differ *(3)*. However, it has been suggested that the mechanism of H_2S toxicity involves numerous complex interactions with various enzyme systems *(124)*. This theory would support the concept that H_2S is a broad-spectrum toxicant *(6)*. It has been demonstrated that sulfides are also capable of inhibiting carbonic anhydrase, another major respiratory enzyme *(71)*. Other studies have reported inhibition of MAO, cholinesterase, and Na^+/K^+ adenosine triphosphatase *(2)*; reduction of synaptosomal oxygen consumption and ATP concentration inhibited uptake of glutamate *(71)*.

The mechanisms of neural toxicity have been recently reviewed in detail *(7)*. The effects on excitable membranes include selective activation of K^+ channel conductance and alteration in Ca^{2+} influx *(2,66,68)*. A reduction of sodium influx was also proposed *(2)* but has not been verified. It has been suggested that the effects of H_2S may involve $GABA_A$ receptors *(70)* but not serotonergic (5-HT) pathways *(125)*. Changes in brain neurochemistry may contribute to the toxicity of H_2S on excitable tissues *(7)*.

The effects on isolated brain preparations appear to be complex. Biphasic responses have been observed *(2,66,67)*; low concentrations increased firing activity whereas higher concentrations blocked spontaneous activity *(2,7)*.

It has been well documented that H_2S functions as a neuromodulator in rat brain *(126,127)*. The role of H_2S as a neuromodulator is discussed in greater detail in other chapters (e.g., *see* Chapter 18). It has been proposed that physiological concentrations selectively enhance *N*-methyl-D-aspartate (NMDA) receptor-mediated responses and facilitate the induction of hippocampal long-term potentiation *(126)*. Following exposure of gaseous H_2S to preparations of rat hippocampal slices, it has been shown that long-lasting synaptic enhancement occurs during washout *(69)*. The reduced effects of H_2S in the presence of excess Mg or in the absence of electrical stimulation *(69)* support the proposed mechanism of action involving the NMDA receptor. These actions of H_2S may account for symptoms such as anxiety and cognitive impairment reported in humans following exposure to low levels of H_2S.

9. ENVIRONMENTAL IMPACTS

There is ample evidence that H_2S can affect the health of domestic and wild animals, natural vegetation and agriculture, buildings and equipment. It is beyond the scope of this chapter to review the environmental impacts of H_2S. In addition to the release of H_2S from natural and anthropogenic sources described in **Subheadings 1.3.1.** and **1.3.2.**, animals themselves, particularly those housed in intensive livestock operations, can produce large amounts of H_2S as well as other gases *(52)*. In addition, agitation or mixing of animal slurries can release high levels of the toxic gas *(128)*. These emissions may be responsible for adverse health effects of both animals and humans *(129)*

Sulfide is widely distributed in the aquatic environment *(130,131)*. In his review, Bagarinao *(131)* states: "Sulfide is more than just a disagreeable odor from a stagnant marsh: it is a serious menace to all aerobic organisms as a toxicant," p. 22. He adds that H_2S may influence the health, survival, productivity, and distribution of aquatic organisms. There are many sources of aquatic H_2S including stagnant waters *(132)*, sea floor sediment *(133,134)*, and eruptions primarily from hydrothermal vents *(135)*. Sulfide concentrations can vary considerably in marine habitats *(131)*, from barely detectable to several hundred micromolar at hydrothermal vents and several millimolar in sediment pore water in salt marshes and sewage outfalls. Although many marine organisms have adapted to these high concentrations of sulfide *(130)*, sudden increases in concentration may have adverse effects on marine ecology and valuable coastal fisheries *(131,134,135)*. The mechanisms of action in aquatic species may be similar to those for mammal and humans.

Sulfur is essential for plant growth, function, and essential synthesis of metabolites including amino acids and proteins *(136)*. Thus, plants can serve as an active sink for atmospheric H_2S *(137)*. Plants may also be a source of reduced sulfur gases released to the atmosphere. As a result of excess release, high levels of H_2S may adversely affect plant biomass production and produce visible injury and defoliation *(137)*. It appears that H_2S is more toxic to plants than to humans, and plants may be more susceptible to H_2S than SO_2 *(137)*.

10. CONCLUSION

Although the toxicological and environmental impacts of H_2S have been studied for many decades, there are still many concerns about the potential effects of low levels on humans and the ecosystem in general. It is well established that high levels can produce significant effects on many organisms. There is some evidence that repeated exposure may result in cumulative effects on target systems. The challenge remains to determine the threshold levels, and whether persistent effects can occur following chronic exposure. Because H_2S is an odoriferous agent, it is possible that the olfactory system plays a major role in the response, and may provide some explanation for the reported health impacts that occur at very low levels. More research is required on susceptible populations such as the developing organisms, children, the elderly, and those with cardiovascular and respiratory symptoms. Clearly, there is much to be revealed concerning the toxicological and environmental impacts of this gaseous pollutant.

REFERENCES

1. Morris J. The Brimestone Battles. Houston Chronicle, November 8, 1997:97.
2. Beauchamp RO, Bus JS, Popp JA, et al. A critical review of the literature on hydrogen sulfide toxicity. Crit Rev Toxicol 1984;13:25–97.
3. Agency for Toxic Substances and Disease Registry (ATSDR). Toxicological Profile for Hydrogen Sulfide. US Department of Health and Human Services: Atlanta, 1999, p. 179.
4. US Environmental Protection Agency (US EPA). Health Assessment Document for Hydrogen Sulfide. EPA/600/8–86/026F. Environmental Criteria and Assessment Office: Research Triangle Park, NC, 1992.
5. Reiffenstein RJ, Hulbert W, Roth SH. Toxicology of hydrogen sulphide. Annu Rev Pharmacol Toxicol 1992;32:109–134.
6. Roth SH. Hydrogen sulfide. In: Corn M, ed. Handbook of Hazardous Materials. Academic: San Diego, 1993, pp. 367–376.
7. Roth SH. Toxicological effects of hydrogen sulfide in mammalian brain. Environ Epidemiol Toxicol 1999;1:241–248.
8. Ellenhorn MJ. Diagnosis and treatment of human poisoning. In: Schurwald S, Ordog G, and Wasserberger J, eds. Ellenhorn's Medical Toxicology. Williams & Wilkins: Baltimore, 1997, pp. 1489–1493.
9. Fuller DC, Suruda AJ. Occupationally related hydrogen sulfide deaths in the United States from 1984 to 1994. J Occup Environ Med 2000;42:939–942.
10. Au WW, Legator MS. Special issue on health hazards from exposure to hydrogen sulfide. Environ Epidemiol Toxicol 1999;1:201–269.
11. Smith RP. Hydrogen sulphide: overview, history, theories and unknowns. In: Prior MG, Roth SH, Green FHY, et al., eds. Proceedings of International Conference on Hydrogen Sulphide Toxicity. Sulphide Research Network, Edmontory, Alberta, Canada, 1989, pp.1–13.
12. Sayers RR. Gas hazards in sewers and sewage-treatment plants. Public Health Rep.1934;49:145 (cited in reference 2).
13. Haggard HW. The toxicology of hydrogen sulphide. J Ind Hyg 1925;7:113–121.
14. Costigan MG. Hydrogen sulfide: UK occupational exposure limits. Occup Environ Med 2003;60: 308–312.

15. Goodwin LR, Francom D, Dieken FP, et al. Determination of sulfide in brain tissue by gas dialysis/ion chromatography: post-mortem studies and two case reports. J Anal Toxiol 1989;13:105–109.
16. Milby TH, Baselt RC. Health hazards of hydrogen sulfide: current status and future directions. Environ Epidemiol Toxicol 1999;1:262–269.
17. Jappinen P, Tenhunen R. Hydrogen sulfide poisoning: blood sulfide concentration and changes in haem metabolism. Br J Indust Med 1990;47:283–285.
18. Guidotti TL. Hydrogen sulphide. Occup Med 1996;46:367–371.
19. Suarez FL, Levitt MD. Hydrogen sulfide production and detoxification in the colon. Environ Epidemiol Toxicol 1999;1:256–261.
20. Kage S, Nagata T, Kimura K, et al. Usefulness of thiosulfate as an indicator of hydrogen sulfide poisoning in forensic toxicological examinations: a study with animal experiments. Jpn J Forensic Toxicol 1992;10:223–227.
21. Bomans P, Rappoort G, Malbrain M, et al. Acute hydrogen sulfide intoxication. Clinical presentation and sequelae in five subjects. Eur Respir J 1997;10(suppl. 25):232s.
22. Hessel PA, Herbert FA, Melenka LS, et al. Lung health in relation to hydrogen sulfide exposure in oil and gas workers in Alberta, Canada. Am J Ind Med 1997;31:554–557
23. Salano R, Copello F. An epidemiologic study of a group of workers employed in the maintenance of a sewer network and of urban waste water treatment plants. Med Lav 1998;89:393–403.
24. Hessel PA, Melenka LS. Health effects of acute hydrogen sulfide exposures in oil and gas workers. Environ Epidemiol Toxicol 1999;1:201–206.
25. Van Aalst JA, Isakov R, Polk JD, et al. Hydrogen sulfide inhalation injury. J Burn Care Rehabil 2000;21:248–253.
26. Buick JB, Lowry RC, Magge TRA. Is a reduction in residual volume a sub-clinical manifestation of hydrogen sulfide intoxication. Am J Ind Med 2000;37:296–299.
27. Tanaka S, Fijimoto S, Tamagaki Y, et al. Bronchial injury and pulmonary edema caused by hydrogen sulfide poisoning. Am J Emerg Med 1999;17:427–429.
28. Cormier Y, Coll B, Laviolette M, et al. Reactive airways dysfunction syndrome (rads) following exposure to toxic gases of a swine confinement building. Eur Respir J 1996;9:1090, 1091.
29. Ammann HM. A new look at physiologic respiratory response to H_2S poisoning. J Hazard Mater 1986;13:369–374.
30. Almeida AF, Guidotti TL. Differential sensitivity of lung and brain to sulfide exposure: a peripheral mechanism for apnea. Toxicol Sci 1999;50;287–293.
31. Huckstorf C, Streller T, Mewes H, et al. Cytochrome a species as putative oxygen sensor for carotid body chemoreceptor discharge. Eur J Physiol 2000;(suppl 439):R400 (abstract P24-5).
32. Prior M, Green F, Lopez A, et al. Capsaicin pretreatment modifies hydrogen sulphide–induced pulmonary injury in rats. Toxicol Pathol 1990;18:279–288.
33. Parra O, Monso E, Gallego M, et al. Inhalation of hydrogen sulphide: a case of subacute manifestations and long term sequelae. Br J Ind Med 1991;48:286, 287.
34. Bhambhani Y, Singh M. Physiological effects of hydrogen sulfide inhalation during exercise in healthy men. J Appl Physiol 1991;71:1872–1877.
35. Bhambhani Y, Burnham R, Snydmiller G, et al. (1994) Comparative physiological responses of exercising men and women to 5 ppm hydrogen sulfide exposure. Am Ind Hyg Assoc J 1994;55: 1030–1035.
36. Bhambhani Y, Burnham R, Snydmiller G, et al. Effects of 10-ppm hydrogen sulfide inhalation on biochemical properties of skeletal muscle in exercising men and women. J Occup Environ Med 1996;38:1012–1017.
37. Bhambhani Y, Burnham R, Snydmiller G, et al. Effects of 5 ppm hydrogen sulfide inhalation of biochemical properties of skeletal muscle in exercising men and women. Am Ind Hyg Assoc J 1996;57:464–468.
38. Bhambhani Y, Burnham R, Snydmiller G, et al. Effects of 10-ppm hydrogen sulfide inhalation in exercising men and women. J Occup Environ Med 1997;39:122–129.
39. Jappinen P, Vilkka V, Marttila O, et al. Exposure to hydrogen sulphide and respiratory function. Br J Ind Med 1990;47:824–828.
40. Richardson DB. Respiratory effects of chronic hydrogen sulfide exposure. Am J Ind Med 1995;28:99–108.
41. Bates MN, Garrett N, Graham B, et al. Air pollution and mortality in the rotorua geothermal area. Aust NZ J Public Health 1997;21:581–586.
42. Bates MN, Garrett N, Shoemack P. Investigation of health effects of hydrogen sulfide from a geothermal source. Arch Environ Health 2002;57:405–411.

43. Legator MS, Singleton CR, Morris DL, et al. Health effects from chronic low-level exposure to hydrogen sulfide. Arch Environ Health 2001;56:123–131.
44. Jaakkola JJK, Vilkka V, Marttila O, et al. The South Karelia air pollution study: The effects of malodorous sulfur compounds from pulp mills on respiratory and other symptoms. Am Rev Respir Dis 1990;142:1344–1350.
45. Haahtela T, Marttila O, Vikka V, et al. The South Karelia air pollution study: acute health effects of malodorous sulfur air pollutants released by a pulp mill. Am J Public Health 1992;82:603–605.
46. Martilla O, Haahtela T, Silakoski I, et al. The South Karelia air pollution study: relationship of outdoor and indoor concentrations of malodorous sulfur compounds released by pulp mills. J Air Waste Manage Assoc 1994;44:1093–1096.
47. Martilla O, Jaakkola JJK, Vilkka V, et al. The South Karelia air pollution study: the effects of malodorous sulfur compounds from pulp mills on respiratory and other symptoms in children. Environ Res 1994;66:152–159.
48. Martilla O, Jaakkola JJK, Partti-Pellinen K, et al. The South Karelia air pollution study: daily symptom intensity in relation to exposure levels of malodorous sulfur compounds from pulp mills. Environ Res 1995;71:122–127.
49. Partti-Pellinen K, Marttila O, Vilkka V, et al. The South Karelia air pollution study: Effects of low-level exposure to malodorous sulfur compounds on symptoms. Arch Environ Health 1996;51: 315–319.
50. Higashi T, Toyama T, Sakurai H, et al. Cross sectional study of respiratory symptoms and pulmonary functions in rayon textile workers with special reference to hydrogen sulfide exposure. Ind Health 1983;21:281–292.
51. Bates MN, Garrett N, Graham B, Read D. Cancer incidence, morbidity and geothermal air pollution in Rotorua. Int J Epidemiol 1998;27:10–14.
52. Wing S, Wolf S. Intensive livestock operations, health, and quality of life among eastern North Carolina residents. Environ Health Perspect 2000;108:233–238.
53. Thu K, Donham K, Ziegenforn R, et al. A control study of the physical and mental health of residents living near a large-scale swine operation. J Agric Safety Health 1997;3:13–26.
54. Brenneman KA, James RA, Gross EA, et al. Olfactory neuron loss in adult male cd rats following subchronic inhalation exposure to hydrogen sulfide. Toxicol Pathol 2000;28:326–333.
55. Dorman DC, Brenneman KA, Struve MF, et al. Fertility and developmental neurotoxicity effects of inhaled hydrogen sulfide in sprague-dawley rats. Neurotoxicol Teratol 2000;22:71–84.
56. Lopez A, Prior M, Yong S, et al. Biochemical and cytological alterations in the respiratory tract of rats exposed for 4 hours to hydrogen sulfide. Fundam Appl Toxicol 1987;9:753–762.
57. Lopez A, Prior M, Lillie LE, et al. Histologic and ultrastructural alterations in lungs of rats exposed to sub-lethal concentrations of hydrogen sulfide. Vet Pathol 1988;25:376–384.
58. Lopez A, Prior M, Yong S, et al. Nasal lesions in rats exposed to hydrogen sulfide for four hours. Am J Vet Res 1988;49:1107–1111.
59. Green FHY, Schurch S, De Sanctis GT, et al. Effects of hydrogen sulfide exposure on surface properties of lung surfactant. J Appl Physiol 1991;70:1943–1949.
60. Rogers RE, Ferin J. effect of hydrogen sulfide on bacterial inactivation in the rat lung. Arch Environ Health 1981;36(5): 261–263.
61. Robinson AV. Effect of in vitro exposure to hydrogen sulfide on rabbit alveolar macrophages cultured on gas-permeable membranes. Environ Res 1982;27:491–500.
62. Chemical Industry Institute of Toxicology (CIIT). 90-Day Vapor Inhalation Toxicity Study of Hydrogen Sulfide in B6C3F1Mice. Report to the Chemical Industry Institute of Toxicology, Research Triangle Park, NC, by ToxiGenics, Inc., CIIT Docket #42063, 1983.
63. Chemical Industry Institute of Toxicology (CIIT). 90-Day Vapor Inhalation Toxicity Study of Hydrogen Sulfide in Fischer 344 Rats. Report to the Chemical Industry Institute of Toxicology, Research Triangle Park, NC, by ToxiGenics, Inc., CIIT Docket #22063, 1983.
64. Chemical Industry Institute of Toxicology (CIIT). 90-Day Vapor Inhalation Toxicity Study of Hydrogen Sulfide in Sprague-Dawley Rats. Report to the Chemical Industry Institute of Toxicology, Research Triangle Park, NC, by ToxiGenics, Inc., CIIT Docket #32063, 1983.
65. Beck JF, Donini JC, Maneckjee A. The influence of sulfide and cyanide on axonal function. Toxicology 1983;26:37–45.
66. Kombian SB, Reiffenstein RJ, Colmers WF. The actions of hydrogen sulfide on dorsal raphe serotonergic neurons in vitro. J Neurophysiol 1993;70:81–96.
67. Evans CL. The toxicity of hydrogen sulphide and other sulphides. Q J Exp Physiol 1967;52:231–248.

68. Baldelli RJ, Reiffenstein RJ, Colmers WF. Hydrogen sulfide inhibits rat hippocampal CA1 pyramidal cells and synaptic potentials in vitro: mechanism of toxicity? Proc Soc Neurosci 1989;15:1348.
69. Roth SH, Strosher L, Bennington RB, et al. Synaptic enhancement following exposure to low level hydrogen sulphide. Toxicology 2001;164:217.
70. Roth SH, Skrajny B, Boguszewicz J, et al. Inhibition of hippocampal activity in vitro by low concentrations of hydrogen sulphide: possible involvement of $GABA_A$ receptors. In: The International Toxicologist, Horizons in Toxicology: Preparing for the 21st Century, Abstracts of the VII International Congress of Toxicology, The Wellington Group, International, Seattle, WA, 1995, abstract 99-P-66.
71. Nicholson RA, Roth SH, Zhang A, et al. Inhibition of respiratory and bioenergetic mechanisms by hydrogen sulfide in mammalian brain. J Toxicol Environ Health 1998;54:491–507.
72. Partlo LA, Sainsbury RS, Roth SH. Do repeat exposures to hydrogen sulfide impair radial arm maze performance? Proc Soc Neurosci 1998;28:92–99.
73. Struve MF, Brisbois JN, James RA, et al. Neurotoxicological effects associated with short-term exposure of sprague-dawley rats to hydrogen sulfide. Neurotoxicol 2001;22:375–385.
74. Luck J, Kaye SB. An unrecognised form of hydrogen sulphide keratoconjunctivitis. Br J Ind Med 1989;46:748, 749.
75. Vanhoorne M, de Rouck A, de Bacquer D. Epidemiological study of eye irritation by hydrogen sulphide and/or carbon disulphide exposure in viscose rayon workers. Ann Occup Hyg 1995;39: 307–315.
76. Schiffman SS, Auvermann BW, Bottcher RW. Health effects of aerial emissions from animal production waste management systems. In: Proceedings of International Symposium: Addressing Animal Production and Environmental Issues. Raleigh, NC State University Raleigh, 2001.
77. Svendsen K. The Nordic expert group for criteria documentation of health risks from chemicals and the Dutch expert committee on occupational standards. 127. Hydrogen sulphide. 2001:Stockholm, Sweden. www.niwl.se/.
78. Brown KG, Strickland JA. Utilizing data from multiple studies (meta-analysis) to determine effective dose-duration levels. Example: rats and mice exposed to hydrogen sulfide. Regul Toxicol Pharmacol 2003;37:305–317.
79. Hemminki K, Niemi M-L. Community study of spontaneous abortions: relation to occupation and air pollution by sulfur dioxide, hydrogen sulfide, and carbon disulfide. Int Arch Occup Environ Health 1982;51:55–63.
80. Hayden LJ, Goeden H, Roth SH. Growth and development in the rat during sub-chronic exposure to low levels of hydrogen sulfide. Toxicol Ind Health 1990;6:389–401
81. Andrew RD, Renne RA, Cannon WC. Reproductive toxicity testing for effects of H_2S in rats. In: Pacific Northwest Laboratory Annual Report for 1979 to the DOE Assistant Secretary for Environment: Part I. Biomedical Sciences. Report No. PNL-3300 PTI. US Department of Energy, Pacific Northwest Laboratory: Richland, WA, 1980 pp. 276–278.
82. Saillenfait AM, Bonnet P, De Ceaurriz J. Effects of inhalation exposure to carbon disulfide and its combination with hydrogen sulfide on embryonal and fetal development in rats. Toxicol Lett 1989;48:57–66.
83. Hayden LJ, Goeden H, Roth SH. Exposure to low levels of hydrogen sulfide elevates circulating glucose in maternal rats. J Toxicol Environ Health 1990;31:45–52.
84. Gaitonde UB, Sellar RJ, O'Hare AE. Long-term exposure to hydrogen sulfide producing subacute encephalopathy in a child. Br Med J 1987;294:614.
85. Glebova L. Permissible tolerance of hydrogen sulfide in the atmosphere. Gig Sanit 1950;4:19–21 (cited in reference 4).
86. Roth SH, Skranjy B, Reiffenstein RJ. Alteration of the morphology and neurochemistry of the developing mammalian nervous system by hydrogen sulfide. Clin Exp Pharmacol Physiol 1995;22:379, 380.
87. Hannah RS, Roth SH. Chronic exposure to low concentrations of hydrogen sulfide produces abnormal growth in developing cerebellar purkinje cells. Neurosci Lett 1991;122:225–228.
88. Hannah RS, Hayden LJ, Roth SH. Hydrogen sulfide exposure alters the amino acid content in developing rat CNS. Neurosci Lett 1989;99:323–327.
89. Hannah RS, Bennington R, Roth SH. A relationship between hydrogen sulfide exposure and taurine levels in maternal rats. Proc West Pharmacol Soc 1990;33:177–179.
90. Skrajny B, Hannah RS, Roth SH. Low concentrations of hydrogen sulfide alter monoamine oxidase levels in the developing rat central nervous system. Can J Physiol Pharmacol 1992;70:1515–1518.
91. Skrajny B, Hannah RS, Roth SH. Monoamine levels in the developing rat cns are altered by low concentrations of hydrogen sulfide. Can J Physiol Pharmacol 1993;70:1515–1518.

92. Ikonomidou C, Bittigau P, Ishimaru MJ, et al. Ethanol-induced apoptotic neurodegenration and fetal alcohol syndrome. Science 2000;287:1056–1060.
93. Dales RE, Spitzer WO, Suissa S, et al. Respiratory health of a population living downwind from natural gas refineries. Am Rev Respir Dis 1989;139:595–600.
94. Tvedt B, Skyberg K, Aeserud O, et al. Brain damage caused by hydrogen sulfide: a follow-up study of six patients. Am J Ind Med 1991;20:91–101.
95. Tvedt B, Edland A, Skyberg K, et al. Delayed neuropsychiatric sequelae after acute hydrogen sulfide poisoning: affection of motor function, memory, vision, and hearing. Acta Neurol Scand 1991;84: 348–351.
96. Wasch HH, Estrin WJ, Yip P, et al. Prolongation of the P-300 latency associated with hydrogen sulfide exposure. Arch Neurol 1989;46:902–904.
97. Callender TJ, Morrow L, Subramanian K, et al. Three-dimensional brain metabolic imaging in patients with toxic encephalopathy. Environ Res 1993;60(2):295–319.
98. Snyder JW, Safir EF, Summerville GP, et al. Occupational fatality and persistent neurological sequelae after mass exposure to hydrogen sulfide. Am J Emerg Med 1995;3:199–203.
99. Schneider JS, Tobe EH, Mozley PD Jr, et al. Persistent cognitive and motor deficits following acute hydrogen sulfide poisoning. Occup Med 1998;48(4):255–260.
100. Kilburn KH, Warshaw RH. Epidemiology of adverse health effects from environmental chemicals. Int J Occup Med Toxicol 1994;3(3): 329–349.
101. Kilburn KH, Warshaw RH. Neurotoxic effects from residential exposure to chemicals from an oil reprocessing facility and superfund site. Neurotoxicol Teratol 1995;17:89–102.
102. Kilburn KH, Warshaw RH. Hydrogen sulfide and reduced-sulfur gases adversely affect neurophysiological functions. Toxicol Indust Health 1995;11:185–197.
103. Kilburn KH. Exposure to reduced sulfur gases impairs neurobehavioral function. South Med J 1997;90:997–1006.
104. Kilburn KH. Hydrogen sulfide exposure from refineries in cities. In: Kilburn KH, ed. Chemical Brain Injury. John Wiley: New York, 1998, pp. 92–128.
105. Kilburn KH. Evaluating health effects from exposure to hydrogen sulfide: central nervous system dysfunction. Environ Epidemiol Toxicol 1999;1:207–216.
106. Kilburn KH. Function testing for chemical brain damage: a review. Arch Environ Health 2001;56: 132–137.
107. Skrajny B, Reiffenstein RJ, Sainsbury RS, et al. Effects of repeated exposures of hydrogen sulfide on rat hippocampal EEG. Toxicol Lett 1996;84:43–53.
108. Leonardos, G, Kendall D, Barnard N. Odor threshold determinations of 53 odorant chemicals. J Air Pollut Control Assoc 1969;19:91–100.
109. Dorman DC, Brenneman KA, Struve MF. Experimental investigations into the neurotoxicity and nasal toxicity of hydrogen sulfide in rats. Environ Epidemiol Toxicol 1999;1:249–255.
110. Smith DV, Shepherd GM. Chemical senses: taste and olfaction. In: Zigmon MJ, Bloom FE, Landis SC, et al., eds. Fundamental Neuroscience. Academic: San Diego, 1999, pp. 9–37.
111. Serjak ET, Hibbard CS. Beyond the stench: odors and health risk. Paper presented at the 89th Annual Meeting of the Air & Waste Management Association, San Diego, California, October 24–25, 1996. 96–TP6.01.
112. Hoshika Y, Imamura T, Muto G, et al. International comparison of odor threshold values of several odorants in Japan and The Netherlands. Environ Res 1993;61(1): 78–83.
113. Hirsch AR, Zavala G. Long term exposure on the olfactory system of exposure to hydrogen sulfide. Occup Environ Med 1999;56:284–287.
114. Schiffman SS, Walker JM, Dalton P, et al. Potential health effects of odor from animal operations, wastewater treatment, and recycling of byproducts. J Agromed 2000;7:7–81.
115. Bensafi M, Rouby C, Farget V, et al. Asymmetry of pleasant vs. unpleasant odor processing during affective judgment in humans. Neurosci Lett 2002;328:309–313.
116. White MC, Inserra SG, Berger SA, et al. Health concerns for communities exposed to hydrogen sulfide—a perspective from two communities. Environ Epidemiol Toxicol 1999;1:236–240.
117. Muezzinoglu A. A study of volatile organic sulfur emissions causing urban odors. Chemosphere 2003;51:245–252.
118. Logue JN, Ramaswamy K, Hersh JH. Investigation of illness associated with exposure to hydrogen sulfide among Pennsylvania school students. J Environ Health 2001;63:9–13.
119. Bordado CM, Gomes FP. Emission and odour control in Kraft pulp mills. J Cleaner Prod 2003; 11:797–801.

120. Brenneman KA, James RA, Gross EA, et al. Olfactory neuron loss in adult male cd rats following subchronic inhalation exposure to hydrogen sulfide. Toxicol Pathol 2000;28:326–333.
121. Dorman DC, Moulin FJ-M, McManus BE, et al. Cytochrome oxidase inhibition induced by acute hydrogen sulfide inhalation: correlation with tissue sulfide concentrations in the rat brain, liver, lung and nasal epithelium. Toxicol Sci 2002;65:18–25.
122. Shusterman, D. Critical review: the health significance of environmental odor pollution. Arch Environ Health 1992;47(1):76–87.
123. Khan AA. Biochemical effects of hydrogen sulfide toxicity. In: Prior MG, Roth SH, Green FHY, et al., eds. Proceedings of International Conference on Hydrogen Sulphide Toxicity. Sulphide Research Network, Edmonton, Alberta, Canada, 1989, pp. 79–89.
124. Thompson RW, Valentine HL, Valentine WM. Cytotoxic mechanisms of hydrosulfide anion and cyanide anion in primary rat hepatocyte cultures. Toxicology 2003;188:149–159.
125. Roth SH, Skranjy B. The inhibition of in vitro hippocampal rhythmical activity by hydrogen sulfide does not involve serotonergic transmission. Proc Can Biol Soc 1996:39.
126. Abe K, Kimura H. (1996) The possible role of hydrogen sulfide as an endogenous neuromodulator. J Neurosci 1996;16:1066–1071.
127. Wang R. Two's company, three's a crowd: can H_2S be the third endogenous gaseous transmitter? FASEB J 2002;16:1792–1798.
128. Groves JA, Ellwood PA. Gases in agricultural slurry stores. Ann Occup Hyg 1991;35:139–151.
129. Newman JR, Schreiber RK. Air pollution and wildlife toxicology: an overlooked problem. Environ Toxicol Chem 1988;7:381–390.
130. Julian D, Dalia WE, Arp AJ. Neuromuscular sensitivity to hydrogen sulfide in the marine invertebrate *urechis caupo.* J Exp Biol 1998;201:1393–1403.
131. Bagarinao T. Sulfide as an environmental factor and toxicant: tolerance and adaptations in aquatic organisms. Aquatic Toxicol 1992;24:21–62.
132. Kularatne KIA, Dissanayake DP, Mahanama KRR. Contribution of dissolved sulfates and sulfites in hydrogen sulfide emission from stagnant water bodies in Sri Lanka. Chemosphere 2003;52:901–907.
133. Christensen JK, Hoyer B, Kryger L, et al. Sulfides in the anaerobic environment: the determination of hydrogen sulfide and acid-soluble metallic sulfides in sea-floor sediment. J Chem Educ 1998;75:1605.
134. Gopakuman G, Kullyamma VJ. Effect of hydrogen sulphide on two species of penaeid prawns *Penaeus indicus* (H. Milne Edwards) and *Metapenaeus dobsoni* (Miers). Bull Environ Contam Toxicol 1996;57:824–828.
135. Weeks SJ, Currie B, Bakun A. Massive emissions of toxic gas in the Atlantic. Nature 2002;415: 493–494.
136. De Kok LJ, Castro A, Durenkamp M, et al. Sulphur in plant physiology. In: Proceedings No. 500, The International Fertiliser Society, York, UK, 2002, pp. 3–26.
137. De Kok LJ, Stuiver CEE, Stulen I. Impact of atmospheric H_2S on plants. In: De Kok LJ, Stulen I, eds. Responses of Plant Metabolism to Air Pollution and Global Change. Backhuys Publishres: Leiden, The Netherlands, 1998, pp. 51–63.

18 Hydrogen Sulfide and the Regulation of Neuronal Activities

Hideo Kimura

CONTENTS

SUMMARY

Hydrogen sulfide (H_2S) is a well-known toxic gas, and most studies about H_2S have been devoted to its toxic effects. Recently, however, three groups discovered that the brain contains relatively high concentrations of endogenous H_2S. This discovery accelerated the identification of an H_2S-producing enzyme, cystathionine β-synthase (CBS), in the brain. In addition to the well-known regulators for CBS, S-adenosyl-L-methionine and pyridoxal-5'-phosphate, it was recently found that Ca^{2+}/calmodulin-mediated pathways are involved in the regulation of CBS activity. H_2S is produced in response to neuronal excitation and alters hippocampal long-term potentiation, a synaptic model for memory. The production of H_2S in the brain is also regulated by testosterone. We describe herein recent progress in the study of H_2S as a novel gasotransmitter in the brain.

Key Words: Hydrogen sulfide; gasotransmitter; cystathionine β-synthase; cystathionine γ-lyase; NMDA receptor; calmodulin; testosterone; Alzheimer disease.

1. INTRODUCTION

Since the first description of H_2S toxicity in 1713 *(1)*, most studies about H_2S have been devoted to its toxic effects with little attention paid to its physiological function *(2)*. Warenycia et al. found that the rat brain contains endogenous H_2S *(3)*, and endogenous concentrations of H_2S have also been measured in human and bovine brain *(4,5)*. The

From: *Signal Transduction and the Gasotransmitters: NO, CO, and H_2S in Biology and Medicine* Edited by: Rui Wang © Humana Press Inc., Totowa, NJ

relatively high concentrations of H_2S in the brain (50 to 160 μ*M*) suggest that it has a physiological function.

Endogenous H_2S in the brain is formed from L-cysteine by a pyridoxal-5'-phosphate-dependent enzyme, cystathionine β-synthase (CBS) *(6–11)*. The CBS inhibitors hydroxylamine and aminooxyacetate suppress H_2S production, whereas a CBS activator, *S*-adenosyl-L-methionine (SAM), enhances it. These observations suggest that CBS is a major enzyme that produces H_2S in the brain.

Two other gasotransmitters, nitric oxide (NO) and carbon monoxide (CO), are endogenously produced by enzymes localized to the brain. NO is synthesized by NO synthase (NOS) via the metabolism of arginine to citrulline *(12,13)*, and CO is produced by heme oxygenase via the metabolism of heme to biliverdin *(14,15)*. Both NO and CO enhance the induction of hippocampal long-term potentiation (LTP), a synaptic model of learning and memory *(16–22)*. The activities of NOS are regulated by Ca^{2+}/calmodulin, and NO is released when *N*-methyl-D-aspartate (NMDA) receptors are activated by L-glutamate *(23,24)*. By contrast, the regulation of CO production by neuronal excitation is not understood *(22)*.

H_2S production in the brain is regulated by the Ca^{2+}-and calmodulin-mediated pathways *(11)*. In addition, physiological concentrations of H_2S specifically potentiate the activity of NMDA receptors, and hippocampal LTP is altered in CBS knockout mice *(10,11)*. H_2S can also regulate the release of corticotropin-releasing hormone (CRH) from the hypothalamus *(25)*. Based on these observations, it has been proposed that H_2S may function as a neuromodulator in the brain *(10,11)*. This chapter outlines the identification of an H_2S-producing enzyme, CBS, in the brain and its regulation. It also discusses the physiological function of H_2S in the brain and the possible involvement of H_2S in disease.

2. PRODUCTION OF H_2S

The discovery of endogenous H_2S in the brain prompted us to identify the enzyme that produces H_2S. H_2S can be formed from cysteine by pyridoxal-5'-phosphate-dependent enzymes, including CBS and cystathione γ-lyase (CSE) *(6–9)*. Both CBS and CSE have been studied intensively in the liver and kidney, but little is known about them in the brain. CBS mRNA is highly expressed in the brain, especially in the hippocampus, whereas CSE mRNA is not detectable *(10)*. The production of H_2S from brain homogenates is suppressed by the CBS-specific inhibitors hydroxylamine and aminooxyacetate, whereas it is not suppressed by CSE-specific inhibitors D,L-propargylglycine and β-cyano-L-alanine. A CBS activator, SAM, enhances the production of brain H_2S *(10)*. These observations suggest that CBS is a major H_2S-producing enzyme in the brain.

CBS is dependent on pyridoxal 5'-phosphate and heme, and its activity is enhanced by SAM *(26,27)*. No other regulators for this enzyme had been found. However, we have recently shown that CBS activity is mediated by Ca^{2+} and calmodulin *(11)*. CBS activity is suppressed by calmodulin-specific inhibitors, W13 and trifluoroperazine, and CBS and calmodulin coimmunoprecipitate. The calmodulin binding consensus sequence has also been identified in CBS *(11,28)*. The enzymatic activity of CBS has two metabolic outcomes *(6,29)*. Most studies have been devoted to a pathway in which CBS catalyzes the reaction with substrate homocysteine to produce cystathionine *(29)*, but little attention has been paid to another pathway in which CBS produces H_2S from L-cysteine as a substrate *(6,10,11)*. SAM regulates CBS activity in both metabolic pathways *(10,11,26)*, and a model for CBS regulation by SAM has been proposed *(30)*. A similar mechanism may also function in the regulation of CBS by Ca^{2+}/calmodulin *(11)*. In the absence of Ca^{2+}/calmodulin, the carboxy-terminal domain may cover the catalytic domain, and CBS activity remains at a basal level. When Ca^{2+}/calmodulin binds to the 19 amino acid calmodulin-binding consensus sequence, the catalytic domain is exposed by opening of the carboxy-terminal domain and CBS becomes active. This model is supported by the observation that the CBS mutant (1-396), which is deficient in the 19 amino acid Ca^{2+}/calmodulin-binding sequence, is constantly active even in the absence of Ca^{2+}/calmodulin *(11)*.

We also found that the endogenous level of H_2S in the brain changes as a function of age and sex, and that H_2S production in the brain is partly regulated by testosterone and SAM. Male brains contain more H_2S than female brains at each age, suggesting the involvement of testosterone in the regulation of the H_2S level, as may occur in the liver *(31,32)*. The application of testosterone to female mice increases H_2S and SAM in the brain, almost reaching the levels of males. By contrast, castration of male mice decreases the levels of testosterone, H_2S, and SAM. These observations suggest that both testosterone and SAM are involved in the regulation of brain H_2S. Because SAM is an activator of CBS *(6,10,11,26)*, it is likely that SAM is a downstream regulator of H_2S production induced by testosterone.

The difference in endogenous testosterone levels between male and female brains is much greater than that of H_2S, and age-dependent changes in the testosterone level do not correlate with those of H_2S. The age-dependent changes in SAM are less great than those of H_2S, and the ages that give peak values are also different between SAM and H_2S *(32)*. These observations, in conjunction with another finding—that H_2S production is regulated by the Ca^{2+}/calmodulin-mediated pathway—suggest that there are at least three ways in which endogenous H_2S can be regulated. The first is the fast regulation via the Ca^{2+}- and calmodulin-mediated pathway observed when neurons are electrically excited *(11)*. The second is the slow regulation via the testosterone- and SAM-mediated pathway. Finally, there must be a regulation of the basal H_2S level by another pathway to account for the discrepancy in the sex- and age-dependent changes in testosterone levels and the levels of SAM. Glucocorticoids are a candidate, for they regulate SAM synthesis in the liver *(33)*.

3. REGULATION OF SYNAPTIC ACTIVITY BY H_2S

What is the function of H_2S in the brain? Physiological concentrations of H_2S modify LTP, and LTP is altered in the brains of CBS knockout mice *(10,11)*. By contrast, concentrations of H_2S greater than the physiological basal level specifically suppress excitatory postsynaptic potentials (EPSPs) *(10)*. This suppression initially was thought to result from the toxic effect of H_2S. However, H_2S production can be locally and transiently increased in response to neuronal excitation, and the suppression of EPSPs still occurs *(10,11)*.

Because H_2S regulates the release of CRH from the hypothalamus *(25)*, it is possible that H_2S may modify the release of neurotransmitters. Therefore, the effect of H_2S on glutamate receptor activation was examined, and it was shown that physiological concentrations of H_2S specifically enhance NMDA receptor-mediated responses *(10)*. This modification of NMDA receptor activity by H_2S may not be the direct effect of H_2S but may partly be because of the activation of cyclic adenosine monophosphate pathways by H_2S *(34)*. The NMDA receptor subunits have specific sites directly phosphorylated by protein kinase A, and H_2S may activate this pathway *(35,36)*. At present, it is not clear whether or not there is an effect of H_2S on glutamate release.

H_2S modifies hippocampal LTP by selectively enhancing NMDA receptor-mediated responses *(10,11)*, and testosterone and estradiol also alter hippocampal LTP *(37,38)*. These observations, together with our results showing that testosterone controls brain H_2S levels, suggest that H_2S can be one of the final effectors of the hormones that modify LTP. Another possible effector, NO, which modifies hippocampal LTP, is also regulated by testosterone *(39–41)*. Because the episodic memory is different between sexes and memory for object identity is age dependent *(42,43)*, sex- and age-dependent changes in the level of H_2S may influence these cognitive functions.

4. INVOLVEMENT OF H_2S IN DISEASES OF THE NERVOUS SYSTEM

There is a good amount of data suggesting that defects in H_2S metabolism may be involved in central nervous system disease. The CBS gene is encoded on chromosome 21q22.3 *(44,45)*, a region associated with Down's syndrome *(46,47)*, and it has been proposed that H_2S may be involved in the cognitive dysfunction associated with Down's syndrome *(48)*. Loss of CBS activity causes homocysteinuria, an autosomal recessive disease characterized, in part, by mental retardation *(29)*. CBS interacts with Huntingtin, mutants of which cause Huntingtins disease *(49)*. Finally, polymorphisms of CBS gene is significantly underrepresented in children with high IQ compared with those with average IQ, suggesting that CBS activity may be involved in the cognitive function *(50)*. These observations in conjunction with the findings described earlier suggest that CBS and its product H_2S may regulate some aspects of synaptic activity and modify cognitive function.

Recent studies have shown that abnormalities in the cerebral microvasculature are relevant to the cause of dementia, including Alzheimer's disease (AD) *(51,52)*. Whereas CBS is the major, if not exclusive, enzyme-producing H_2S in the brain, another H_2S-producing enzyme, CSE, was identified as the major H_2S-producing enzyme in the smooth muscle *(53)*. Although exogenously applied H_2S alone relaxes smooth muscle, much lower concentrations of H_2S greatly enhance the smooth muscle relaxation induced by NO *(53)*. H_2S also hyperpolarizes smooth muscle by activating K_{ATP} channels *(54)*. Based on these observations, it is likely that H_2S may also regulate cerebral blood flow.

Because the levels of SAM, an activator of CBS, are lower in the brains of AD patients than in the brains of healthy individuals *(55)*, the endogenous levels of H_2S in AD brains could be lower than those of control brains. To examine this possibility the endogenous H_2S levels in the brains of AD patients were measured and compared with the brains of age-matched healthy individuals. The endogenous H_2S levels of the brains of the AD patients were significantly lower than those of the brains of age-matched healthy individuals *(56)*.

Because CBS is the major enzyme that produces H_2S in the brain, there are three possibilities that may cause the changes in the endogenous H_2S levels. These are differences in the levels of the substrate for CBS, the amount of CBS, or the activity of CBS. There was no significant difference in the amount of endogenous free cysteine or that of CBS between AD and control brains.

Because Morrison et al. *(55)* reported that the amounts of SAM, a CBS activator *(10,11)*, in AD brains were lower than in control brains *(55)*, we attempted to confirm their data by measuring the amounts of SAM in our AD brain samples. We found that the endogenous levels of SAM in AD brains were significantly less than those of control brains. The data obtained by Morrison et al. *(55)* suggest that the low endogenous level of H_2S in AD brains may be caused by the decreased activity of CBS because of the lack of SAM.

The activity of CBS is regulated by SAM *(10,11,26)* and the serum homocysteine level *(57)* is higher in patients with AD than in healthy individuals, suggesting that the activity of CBS must be reduced in AD brains. To test this possibility, the best experiment may be to measure CBS activity. However, because the CBS activity in frozen rat brain samples is decreased to <10% of that in fresh samples (unpublished observation), it is impossible to accurately measure H_2S-producing activity of CBS extracted from frozen AD brains. Alternatively, the amounts of homocysteine in the brains of AD and dementia of Alzheimer type (DAT) patients were measured by the high-performance liquid chromatography and compared with those of control brains. The endogenous homocysteine levels in AD brains were indeed significantly greater than in control brains. These observations confirm that CBS activity is decreased in AD brains, and that this decline in CBS activity may be because of the lack of SAM. The reduction of H_2S and the CBS activity may be involved in some aspects of the cognitive decline associated with AD.

5. CONCLUSION

The physiological relevance of two H_2S-producing enzymes, CBS and CSE, has been identified, and the regulation of their activities has also been determined. A novel finding that H_2S production by CBS is regulated by Ca^{2+} and calmodulin led to the observation that H_2S is produced in response to neuronal excitation. Exogenously applied H_2S modifies LTP, and LTP is altered in the brain of CBS knockout mice. H_2S can regulate some aspects of synaptic activity and modify cognitive function. In smooth muscle, H_2S enhances NO-induced relaxation and can regulate the activity of CSE. Both gaseous smooth muscle relaxants strongly interact with each other.

A few candidates for molecular targets for H_2S have been identified. These are NMDA receptors in the brain and the K_{ATP} channel in smooth muscle. The mechanism of the activation of these targets has not been solved, and it is not known if it is a direct or indirect effect. Because H_2S is a very active molecule, more targets are expected to be found. After H_2S stimulates its targets, it has to be cleared from its site of action. The mechanism of clearance is not understood. The study of H_2S as a physiologically active molecule, a gasotransmitter *(58)*, has just begun, but understanding the mechanisms underlying its physiological function may provide new insight into neurotransmission.

ACKNOWLEDGMENT

This work was supported by a grant from the National Institute of Neuroscience, National Center of Neurology and Psychiatry, Japan.

REFERENCES

1. Ramazzini B. Diseases of Workers 1713 (translated from the Latin text De Morbis Artificum by WC Wright, 1940). University of Chicago Press: Chicago. Reprinted 1964 in History Med., vol. 23.
2. Reiffenstein RJ, Hulbert WC, Roth SH. Toxicology of hydrogen sulfide. Annu Rev Pharmacol Toxicol 1992;32:109–134.
3. Warenycia MW, Goodwin LR, Benishin CG, et al. Acute hydrogen sulfide poisoning: demonstration of selective uptake of sulfide by the brainstem by measurement of brain sulfide levels. Biochem Pharmacol 1989;38:973–981.

4. Goodwin LR, Francom D, Dieken FP, et al. Determination of sulfide in brain tissue by gas dialysis/ion chromatography: postmortem studies and two case reports. J Anal Toxicol 1989;13:105–109.
5. Savage JC, Gould DH. Determination of sulfides in brain tissue and rumen fluid by ion-interaction reversed-phase high-performance liquid chromatography. J Chromatogr 1990;526:540–545.
6. Stipanuk MH, Beck PW. Characterization of the enzymic capacity for cysteine desulphhydration in liver and kidney of the rat. Biochem J 1982;206:267–277.
7. Griffith OW. Mammalian sulfur amino acid metabolism: an overview. In: Jakoby WB and Griffith OW, ed. Methods in Enzymology. Academic: New York, 1987, pp. 366–376.
8. Erickson PF, Maxwell IH, Su LJ, et al. Sequence of cDNA for rat cystathionine γ-lyase and comparison of deduced amino acid sequence with related *Escherichia coli* enzymes. Biochem J 1990;269:335–340.
9. Swaroop M, Bradley K, Ohura T, et al. Rat cystathionine β-synthase. J Biol Chem 1992;267: 11,455–11,461.
10. Abe K, Kimura H. The possible role of hydrogen sulfide as an endogenous neuromodulator. J Neurosci 1996;16:1066–1071.
11. Eto K, Ogasawara M, Umemura K, et al. Hydrogen sulfide is produced in response to neuronal excitation. J Neurosci 2002;22:3386–3391.
12. Palmer RMJ, Ashton DS, Moncada S. Vascular endothelial cells synthesize nitric oxide from L-arginine. Nature 1988;333:664–666.
13. Bredt DS, Snyder SH. Nitric oxide, a novel neuronal messenger. Neuron 1992;8:3–11.
14. Maines MD. Heme oxygenase: function, multiplicity, regulatory mechanisms, and clinical applications. FASEB J 1988;2:2557–2568.
15. Verma A, Hirsch DJ, Glatt CE, et al. Carbon monoxide: a putative neural messenger. Science 1993;259:381–384.
16. O'Dell TJ, Hawkins RD, Kandel ER, Arancio O. Tests of the roles of two diffusible substances in long-term potentiation: evidence for nitric oxide as a possible early retrograde messenger. Proc Natl Acad Sci USA 1991;88:11,285–11,289.
17. Schuman EM, Madison DV. A requirement for the intercellular messenger nitric oxide in long-term potentiation. Science 1991;254:1503–1506.
18. Haley JE, Wilcox GL, Chapman PF. The role of nitric oxide in hippocampal long-term potentiation. Neuron 1992;8:211–216.
19. Stevens CF, Wang Y. Reversal of long-term potentiation by inhibitors of haem oxygenase. Nature 1993;364:147–149.
20. Zhuo M, Small SA, Kandel ER, et al. Nitric oxide and carbon monoxide produce activity-dependent long-term synaptic enhancement in hippocampus. Science 1993;260:1946–1950.
21. Bliss TVP, Collingridge GL. A synaptic model of memory: long-term potentiation in the hippocampus. Nature 1993;361:31–39.
22. Snyder SH, Ferris CD. Novel neurotransmitters and their neuropsychiatric relevance. Am J Psychiatry 2000;157:1738–1751.
23. Garthwaite J, Charles SL, Chess-Williams R. Endothelium-derived relaxing factor release on activation of NMDA receptors suggests role as intercellular messenger in the brain. Nature 1988;336:385–388.
24. Bredt DS, Snyder SH. Isolation of nitric oxide synthetase, a calmodulin-requiring enzyme. Proc Natl Acad Sci USA 1990;87:682–685.
25. Russo CD, Tringali G, Ragazzoni E, et al. Evidence that hydrogen sulphide can modulate hypothalamo-pituitary-adrenal axis function: *in vitro* and *in vivo* studies in the rat. J Neuroend 2000;12:225–233.
26. Finkelstein JD, Kyle WE, Martin JJ, et al. Activation of cystathionine synthase by adenosylmethionine and adenosylethionine. Biochem Biophys Res Commun 1975;66:81–87.
27. Kery V, Bukovska G, Kraus JP. Transsulfuration depends on heme in addition to pyridoxal 5'-phosphate. J Biol Chem 1994;269:25,283–25,288.
28. Rhoads AR, Friedberg F. Sequence motifs for calmodulin recognition. FASEB J 1997;11:331–340.
29. Mudd SH, Levy HL, Skovby F. Disorders of transsulfuration. In: Scriver CR, Beaudet AL, Sly WS, et al., eds. The Metabolic Basis of Inherited Disease. McGraw-Hill: New York, 1989 ,pp. 693–734.
30. Shan X, Dunbrack RLJ, Christopher SA, et al. Mutation in the regulatory domain of cystathionine β-synthase can functionally suppress patient-derived mutations in cis. Hum Mol Genet 2001;10:635–643.
31. Manteuffel-Cymborowska M, Chmurzynska W, Grzelakowska-Sztabert B. Tissue-specific effects of testosterone on *S*-adenosymethionine formation and utilization in the mouse. Biochim Biophys Acta 1992;1116:166–172.
32. Eto K, Kimura H. The production of hydrogen sulfide is regulated by testosterone and S-adenosyl-L-methionine in mouse brain. J Neurochem 2002;83:80–86.

33. Gil B, Pajares MA, Mato JM, et al. Glucocorticoid regulation of hepatic *S*-adenosylmethionine synthetase gene expression. Endocrinology 1997;138:1251–1258.
34. Kimura H. Hydrogen sulfide induces cyclic AMP and modulates the NMDA receptor. Biochem Biophys Res Commun 2000;267:129–133.
35. Leonard AS, Hell JW. Cyclic AMP-dependent protein kinase and protein kinase C phosphorylate N-methyl-D-aspartate receptors at different sites. J Biol Chem 1997;272:12,107–12,115.
36. Tingley WG, Ehlers MD, Kameyama K, et al. Characterization of protein kinase A and protein kinase C phosphorylation of the N-methyl-D-aspartate receptor NR1 subunit using phosphorylation site-specific antibodies. J Biol Chem 1997;272:5157–5166.
37. Harley CW, Malsbury CW, Squires A, et al. Testosterone decreases CA1 plasticity in vivo in gonadectomized male rats. Hippocampus 2000;10:693–697.
38. Foy MR. 17 beta-estradiol: effect on CA1 hippocampal synaptic plasticity. Neurobiol Learn Mem 2001;76:239–252
39. O'Dell TJ, Hawkins RD, Kandel ER, et al. Tests of the roles of two diffusible substances in long-term potentiation: evidence for nitric oxide as a possible early retrograde messenger. Proc Natl Acad Sci USA 1991;88:11,285–11,289.
40. Schuman EM, Madison DV. A requirement for the intercellular messenger nitric oxide in long-term potentiation. Science 1991;254:1503–1506.
41. Singh R, Pervin S, Shryne J, et al. Castration increases and androgens decrease nitric oxide synthase activity in the brain: physiologic implications. Proc Natl Acad Sci USA 2000;97:3672–3677.
42. Herlitz A, Yonker JE. Sex differences in episodic memory: the influence of intelligence. J Clin Exp Neuropsychol 2002;24:107–114.
43. Schiavetto A, Kohler S, Grady CL, et al. Neural correlates of memory for object identity and object location: effects of aging. Neuropsychologia 2002;40:1428–1442.
44. Skovby F, Krassikoff N, Francke U. Assignment of the gene for cystathionine beta-synthase to human chromosome 21 in somatic cell hybrids. Hum Genet 1984;65:291–294.
45. Munke M, Kraus JP, Ohura T, et al. The gene for cystathionine beta-synthase (CBS) maps to the subtelomeric region on human chromosome 21q and to proximal mouse chromosome 17. Am J Hum Genet 1988;42:550–559.
46. Korenberg JR, Kawashima H, Pulst SM, et al. Molecular definition of a region of chromosome 21 that causes features of the Down Syndrome phenotype. Am J Hum Genet 1990;47:236–246.
47. Kraus JP. Molecular analysis of cystathionine β-synthase-a gene on chromosome 21. In: Molecular Genetics of Chromosome 21 and Down Syndrome. Prog Clin Biol Res 1990;360:201–214.
48. Kamoun P. Mental retardation in Down syndrome: a hydrogen sulfide hypothesis. Med Hypotheses 2001;57:389–392.
49. Boutell JM, Wood JD, Harper PS, et al. Huntingtin interacts with cystathionine β-synthase. Hum Mol Genet 1998;7:371–378.
50. Barbaux S, Plomin R, Whitehead AS. Polymorphisms of genes controlling homocysteine/folate metabolism and cognitive function. NeuroReport 2000;11:1133–1136.
51. Kalaria RN. The role of cerebral ischemia in Alzheimer's disease. Neurobiol Aging 2000;21:321–330.
52. De la Torre JC, Mussivand T. Can disturbed brain microcirculation cause Alzheimer's disease? Neurol Res 1993;15:146–153.
53. Hosoki R, Matsuki N, Kimura H. The possible role of hydrogen sulfide as an endogenous smooth muscle relaxant in synergy with nitric oxide. Biochem Biophys Res Commun 1997;237:527–531.
54. Zhao W, Zhang J, Lu Y, et al. The vasorelaxant effect of H_2S as a novel endogenous gaseous KATP channel opener. EMBO J 2001;20:6008–6016.
55. Morrison LD, Smith DD, Kish SJ. Brain S-adenosylmethionine levels are severely decreased in Alzheimer's disease. J Neurochem 1996;67:1328–1331.
56. Eto K, Asada T, Arima K, et al. Brain hydrogen sulfide is severely decreased in Alzheimer's disease. Biochem Biophys Res Commun 2002;293:1485–1488.
57. Clarke R, Smith D, Jobst KA, et al. Folate, vitamin B12, and serum total homocysteine levels in confirmed Alzheimer disease. Arch Neurol 1998;55:1449–1455.
58. Wang R. Two's company, three's a crowd—can H_2S be the third endogenous gaseous transmitter? FASEB J 2002;16:1792–1798.

19 The Role of Hydrogen Sulfide as an Endogenous Vasorelaxant Factor

Rui Wang, Youqin Cheng, and Lingyun Wu

CONTENTS

SUMMARY

Hydrogen sulfide (H_2S) is a small molecule of gas with important physiological functions. The endogenous production of H_2S has been demonstrated in vascular smooth muscle cells (VSMCs), catalyzed by cystathionine γ-lyase (CSE). An elevated endogenous H_2S level in animals leads to decreased blood pressure without altering heart rate. At physiologically relevant concentrations, H_2S relaxes vascular tissues by directly acting on VSMCs. Inhibition of CSE induces a slowly developed hypertension. Increased endogenous production of H_2S, on the other hand, reduces the contraction of isolated vascular tissues. The vasorelaxant effect of H_2S is partially mediated by endothelium, but K_{ATP} channels in VSMCs are the major target of this gas. Dependent on the type of vascular tissue, the production of H_2S and the mechanisms for the vasorelaxant effects of H_2S vary. Vasorelaxation of resistant arteries induced by H_2S is much greater than that of conduit arteries. Although the regulation of endogenous production of H_2S in vascular tissues has been unclear, nitric oxide has been shown to increase H_2S production in vascular tissues. Taking these novel observations together, the importance of H_2S as a gasotransmitter in homeostatic control of cardiovascular function has been greatly appreciated. Accurate measurement of the endogenous level, in addition to the production rate,

From: *Signal Transduction and the Gasotransmitters: NO, CO, and H_2S in Biology and Medicine* Edited by: Rui Wang © Humana Press Inc., Totowa, NJ

of H_2S in cardiovascular tissues has not been achieved to date. Alterations in the production and function of endogenous H_2S under different pathophysiological conditions have not been examined. Specific endogenous stimulators and inhibitors for H_2S metabolism are still largely unknown. These encumbrances are not viewed as insurmountable obstacles but challenges and prospects through which future breakthroughs in the understanding of gasotransmitter biology and medicine will be made.

Key Words: Contractility; cystathionine β-synthase; cystathionine γ-lyase, gasotransmitter; hydrogen sulfide; smooth muscle cells.

1. INTRODUCTION

Endogenous production of hydrogen sulfide (H_2S) has been demonstrated in many mammalian tissues, including vascular tissues. Nonenzymatic production of H_2S requires the participation of elemental sulfur and the oxidation of glucose *(1)*. Increased oxidative stress and hyperglycemia will promote H_2S production. More important to the regulation of physiological functions of mammalian cells is H_2S produced from the transsulfuration pathway *(2)*. L-Cysteine serves as the precursor of H_2S and the enzymatic reaction is accomplished by two pyridoxal-5'-phosphate-dependent enzymes: cystathionine β-synthase (CBS) (EC 4.2.1.22) and cystathionine γ-lyase (CSE) (EC 4.4.1.1) *(1,3–6)*. CSE is also named cysteine desulfhydrase *(7)*. CBS is critical for the transsulfuration of homocysteine to generate cystathionine. L-Cysteine is generated either from cystathionine in the presence of CSE or directly from cystine. Either CBS or CSE catalyzes the final transformation of L-cysteine to H_2S, ammonium, and pyruvate. It has been shown that ammonia, H_2S, and pyruvate cannot inhibit CSE activity *(8)*. Whether these L-cysteine products affect the activity of CBS is still unknown.

Although our understanding of the role of H_2S in the homeostatic control of cardiovascular functions is still in its infancy, research progress over the last 6 yr has seriously challenged the traditional view of H_2S as a mere toxic waste in the cardiovascular system. For example, the H_2S-producing enzymes were located in vascular tissues. In addition, endogenous production of H_2S was directly measured in vascular tissues as well as in the circulation. Hemodynamic changes in cardiovascular functions were also examined using both exogenously applied and endogenously generated H_2S. Furthermore, vascular relaxant effects of H_2S were demonstrated, and the cellular targets and signal transduction pathways of H_2S were explored. Finally, the interaction of H_2S and other gasotransmitters on vascular functions was investigated. All these developments, which are reviewed in this chapter, signal the beginning of a new wave of enthusiasm for exploring the physiological and pathophysiological functions of H_2S in the cardiovascular system.

2. SELECTIVE EXPRESSION OF H_2S-PRODUCING ENZYMES IN CARDIOVASCULAR TISSUES

2.1. Expression of CSE in Cardiovascular System

The expression of CSE in mammalian tissues is both tissue-type specific and developmental stage related. Although absent in brain and lungs *(7,9)*, high expression levels of CSE are reported in liver and ileum *(5,10)*. CSE activity has been demonstrated in adult human liver tissues, but it is not detectable in fetal, premature, and full-term neonatal liver tissue *(11)*. This developmental stage-dependent change in CSE activity indicates an important implication for this enzyme as well as its enzymatic products including H_2S in the developmental regulation.

Zhao et al. *(12)*, for the first time, cloned CSE from rat mesenteric artery tissues containing an opening reading frame of 1197 bp to encode a 398 amino acid peptide. The full sequence of this clone (GenBank accession no. AB052882) was identical to that of the CSE clone derived from rat liver from the same laboratory (GenBank accession no. AY032875). The expression of CSE in portal vein, thoracic aorta *(5)*, mesenteric artery, tail artery, and pulmonary arteries of rats *(12)* has been demonstrated at the mRNA level. Mostly because of the lack of specific antibodies against CSE, CSE protein expression in cardiovascular tissues has not been confirmed yet. Quantitative studies using RNase protection assay showed that the transcriptional expression levels of CSE were different among different vascular tissues *(12)*. The highest CSE mRNA level was found in pulmonary artery, followed by aorta, tail artery, and mesenteric artery *(12)*. An *in situ* hybridization study located CSE mRNA in the vascular smooth muscle layer, not the endothelial layer, of rat aortic wall *(12)*. Zhao et al. *(12)* further demonstrated the presence of CSE mRNA in purified and cultured vascular smooth muscle cells (VSMCs), but not in cultured vascular endothelial cells. The distribution pattern of H_2S-generating enzymes in vascular tissues is very different from that of nitric oxide synthases (NOSs) and heme oxygenase (HO). NOS and HO, enzymes generating nitric oxide (NO) and carbon monoxide, respectively, are expressed in both VSMCs and endothelial cells. Furthermore, whether the expression of CSE in cardiovascular tissues is age dependent has not been studied.

D,L-Propargylglycine (PPG) and β-cyano-L-alanine (CLA) selectively inhibit CSE *(5,12–14)*. H_2S production in both rat liver and ileum tissues was inhibited in vitro by PPG and CLA in a concentration-dependent manner *(10)*. Intraperitoneal injection of PPG significantly suppressed H_2S production in liver, aorta, and ileum tissues *(10)*. These pharmacological studies indirectly demonstrate the expression of CSE in liver, ileum, and vascular tissues.

2.2. Expression of CBS in Cardiovascular System

In comparison to the expression of CSE, the expression of CBS is much less tissue-type selective. The enzymatic activity of CBS has been demonstrated in rat liver, pancreas, and kidney; human brain and liver *(15,16)*; and mouse pancreas, liver, kidney, and brain *(15)*. Cardiovascular tissues are among very few tissues that may not possess CBS. In human atrium and ventricle tissues, the expression of CBS was undetectable using an enzyme assay and Western blot analysis *(17)*. Limited studies on human vascular tissues, including internal mammary arteries, saphenous veins, coronary arteries, or aortic arteries, also reported the absence of the activity and/or expression of CBS *(15,17)*. CBS activity was negligible in extracts of cultured human aortic endothelial cells *(18)*. Wang et al. *(19)* reported the activity of CBS as reflected by the production of cystathionine in cultured human umbilical venous endothelial cells. These cells had been cultured for 14 d in Dulbecco's modified Eagle's medium with the addition of 100 μM L-homocysteine. Unfortunately, their study did not actually measure the protein or mRNA levels of CBS. One hypothesis remains that CBS is not detectable in vascular endothelial cells under normal conditions. Either a high level of homocysteine in the culture medium upregulates the expression of CBS or the expression of CBS in vascular endothelial cells is highly dependent on developmental stages so that the expression of CBS in adult endothelial cells becomes undetectable. Recently, Zhao et al. *(12)* also tried to detect CBS mRNA in adult rat vascular tissues, including endothelial cells. Their results demonstrated the lack of CBS mRNA in these vascular tissues.

Taken together, the generation of H_2S from cardiovascular tissues seems to involve an enzymatic system excluding CBS. This assumption leads to an open debate on the metabolism of homocysteine in the cardiovascular system. As described before, CBS is the rate-limiting enzyme for the transformation of homocysteine to cystathionine. Without CBS expression in the cardiovascular system, homocysteine has to be metabolized in other systems, and this makes the cardiovascular system especially fragile to hyperhomocysteinemic damage. Another hypothesis speculates that CBS may function as an inducible H_2S-generating enzyme in vascular endothelial cells. In response to an elevation of circulating homocysteine level, CBS in vascular endothelial cells would be upregulated. This process would not only affect homocysteine metabolism but also influence vascular contractility and remodeling under different physiological and pathophysiological conditions. Alternatively, homocysteine metabolism in the cardiovascular system may be realized by the remethylation pathway, instead of the transsulfuration pathway, catalyzed by B12-dependent methionine synthase (B12MS) (EC 2.1.1.13) *(17)*.

CBS activity can be specifically inhibited by aminooxyacetate (AOA) and hydroxylamine. A recent study demonstrated that AOA at 1 m*M* abolished H_2S production in liver tissues, and partially inhibited H_2S production in the ileum *(10)*. In these in vitro assays, H_2S production in liver was only slightly inhibited by PPG and CLA at concentrations of about 1 m*M*, indicating that in liver CBS is the predominant enzyme catalyzing the H_2S-producing reaction. Zhao et al. *(12)* also injected AOA (17 mg/kg) intraperitoneally into rats. Different rat tissues were isolated 3 h after the injection and tissue H_2S production was measured. It was found that in vivo AOA treatment did not induce any significant inhibition of H_2S production in liver, ileum, and aortic tissues *(10)*. The difference between in vivo and in vitro effects of AOA on H_2S production from rat liver and ileum tissues indicates a poor membrane permeability of AOA in comparison to PPG. Because AOA was directly added to homogenized tissues in in vitro assay, membrane permeability would not be a limiting factor for AOA to inhibit CBS activity.

3. ENDOGENOUS PRODUCTION OF H_2S IN VASCULAR TISSUES AND ITS MODULATION

Both the endogenous level and endogenous production rate of H_2S have been determined in different tissues. The latter is determined by both the activity of H_2S-generating enzymes and the sufficiency of enzyme substrates, whereas the former is the actual level of H_2S under physiological conditions. The H_2S level in the circulation was reported to be about 10 μ*M* in Wistar rats *(20)*, 46 μ*M* in Sprague-Dawley rats *(12)*, and 10–100 μ*M* in humans *(21)*. The tissue level of H_2S is known to be higher than its circulating level. The endogenous concentration of H_2S in rat, human, and bovine brain tissues is in the range of 50–160 μ*M* *(22–24)*. The endogenous level of H_2S in cardiovascular tissues has not been determined. The endogenous production rate of H_2S, however, has been determined in many different types of vascular tissues. A wide spectrum of production rates of H_2S has been reported depending on the type of blood vessels. The homogenates of rat thoracic aortae have a higher production rate than those of portal vein of rats *(5)*. The production rate of H_2S in rat tail artery tissues is higher than that of rat aorta and mesenteric artery tissues *(12)*.

4. BLOOD PRESSURE CHANGE INDUCED BY H_2S

To test the potential role of H_2S in blood pressure (BP) regulation, Zhao et al. treated normotensive Sprague-Dawley rats daily with PPG (34 mg/[kg·day] intraperitoneally) to inhibit CSE activities *(10)*. With a daily injection of PPG, systolic BP started to increase for the first week of injection and became significantly higher after 2 wk whereas BP of the control rat group remained normotensive. This observation speaks for the importance of endogenous H_2S in maintaining BP at normotensive range. The inhibition of CSE by PPG would reduce the endogenous H_2S production from vascular tissues. Consequently, the vasorelaxant influence of H_2S on the basal tone of vascular tissues would be reduced, which may lead to an elevated peripheral vascular resistance and an increased BP. While PPG treatment continued, the elevated BP gradually subsided and returned to a normotensive level 4 wk after drug injection. This short-lived hypertensive effect of PPG treatment may stem from the accumulation of substrates of CSE after CSE inhibition. An elevated L-cysteine level may in turn stimulate the synthesis of H_2S even with residual CSE activities. It is also possible that with a prolonged inhibition of the CSE/H_2S system other compensatory mechanisms may be upregulated to lower BP toward a normal level. Finally, chronic treatment with PPG may lead to desensitization of CSE to the inhibitor, which can be confirmed by directly measuring CSE activities, and endogenous levels of H_2S in circulation and in vascular tissues following different post-PPG treatment periods. Additionally, whether PPG affects cellular functions distantly related to the inhibition of CSE should also be considered.

One of the criteria for defining gasotransmitters is that "functions of endogenous gases can be mimicked by their exogenously applied counterparts" (*see* Chapter 1). BP changes induced by reducing endogenous H_2S production have been confirmed by introducing exogenous H_2S. A transient (about 30 s) but significant decrease in mean arterial BP (13–30 mmHg) of anesthetized rats was observed after an intravenous bolus injection of H_2S at 2.8 or 14 μmol/kg of body wt provoked *(12)*. The short duration of the hypotensive effect of H_2S could be attributed to the scavenging of H_2S by metalloproteins, disulfide-containing proteins, thio-*S*-methyl-transferase, and heme compounds. The administration of H_2S as a bolus injection also partially explains the transient effect. This hypotensive effect is not because of inhibition of the respiratory system by H_2S, which would have led to a reflective change in heartbeat. The same experiment showed that H_2S injection in these rats did not alter heart rate *(12)*. Pretreatment of rats for 20 min with glibenclamide to specifically block K_{ATP} channels significantly antagonized the hypotensive effect of injected H_2S. On the other hand, pinacidil application mimicked the H_2S-induced transient decrease in BP *(12)*. These in vivo results indicated that the hypotensive effect of H_2S was likely provoked by the relaxation of resistance blood vessels through the opening of K_{ATP} channels.

5. VASORELAXATION INDUCED BY H_2S

Hosoki et al. *(5)* reported the relaxation of rat portal vein and aortic tissues induced by NaHS, an H_2S donor. These vascular tissues were precontracted with 1 μ*M* norepinephrine. A potent relaxation of portal vein tissues was induced by NaHS with an EC_{50} of 160 μ*M*. However, aortic tissues were much less sensitive to NaHS with an EC_{50} >1 m*M* *(5)*. Zhao and colleagues *(12,25)* further investigated the H_2S-induced relaxation of rat aortic tissues. In their studies, aortic tissues were precontracted with either phenylephrine or KCl.

The threshold concentrations of H_2S to initiate relaxation were 18 and 60 μ*M* for 20 m*M* KCl- and phenylephrine-precontracted tissues, respectively. It seems that the potency of vasorelaxant effect of H_2S is affected by the stimulus used to precontract the tissues. H_2S relaxed phenylephrine (0.3 μ*M*)-precontracted aortic tissues *(12)* much more potently (EC_{50} of 125 μ*M*) than did NaHS on norepinephrine-precontracted aortic tissues *(5)*. An even greater vasorelaxation effect of H_2S on small mesenteric arteries has been observed in isolated and perfused mesenteric artery bed (MAB) *(26)*. Although rat aortic and mesenteric artery tissues produce comparable levels of H_2S *(12)*, H_2S is about sixfold more potent in relaxing rat MAB (EC_{50} of 22 μ*M*) than rat aortic tissues. The higher sensitivity of MAB to H_2S emphasizes the importance of H_2S in regulating peripheral resistance and hence BP. Differences in distributions of H_2S targets, sensitivities of voltage-dependent calcium channels to membrane potential change, and sensitivities of contractile proteins to intracellular calcium levels between conduit and resistant arteries may be the mechanisms to be explored further.

The involvement of various signal transduction pathways in the vascular effect of H_2S has been studied. The vascular effect of H_2S is dependent on extracellular calcium entry but independent of the activation of the cyclic guanosine 5'-monophosphate (cGMP) pathway in aortic tissues *(25)*. Treatment of vascular tissues with indomethacin, staurosporine, or SQ22536 did not change the effect of H_2S. Thus, the vasorelaxant effects of H_2S are unlikely mediated by prostaglandin, protein kinase C, or cyclic adenosine monophosphate pathways, respectively. The inclusion of superoxide dismutase and catalase in the bath solution did not to alter the vasorelaxant effect of H_2S *(12,25)*. Therefore, the participation of superoxide anion or hydrogen peroxide in the H_2S-induced vasorelaxation cannot be approved. Studies in our laboratory have demonstrated that H_2S is an endogenous opener of adenosine triphosphate-sensitive K^+ (K_{ATP}) channels in VSMCs. For details of the effect of H_2S on K_{ATP} channels, readers are referred to Chapter 21.

The extent to which the vasorelaxant effects of H_2S depend on a functional endothelium has been examined. In one study *(5)*, the relaxation of rat aortic tissues induced by H_2S was not altered by the removal of endothelium. Unfortunately, the concentration of H_2S used to study the endothelium-dependent vasorelaxation was not specified. In two other studies, the endothelium dependency of the vasorelaxant effect of H_2S was shown. Interestingly, this endothelium dependence was closely related to the concentrations of H_2S. The removal of endothelium attenuated the relaxation of rat aortic tissues induced by H_2S at a single dose (180 μ*M*) *(12)*, but the maximum relaxation induced by H_2S at concentrations ≥1 m*M* was irrelevant to the presence of endothelium *(12)*. The absence of an intact endothelium shifted the H_2S concentration–response curve to the right with the IC_{50} changed from 136 to 273 μ*M*. Recently, we found that the endothelium dependence of H_2S effect was more pronounced in isolated and perfused rat MAB *(26)*. Removal of the functional endothelium significantly reduced the H_2S-induced relaxation of MAB by about sevenfold. The EC_{50} of H_2S changed from 22 μ*M* in the presence of endothelium to 161 μ*M* in the absence of endothelium ($p < 0.05$).

This tissue-type-selective endothelium-dependent effect of H_2S is similar to the tissue-type-selective release of endothelium-derived hyperpolarizing factor (EDHF). The smaller the size of the arteries, the more contribution of EDHF to the endothelium-dependent vasorelaxation *(27)*. EDHF rarely has a role to play in regulating the tone of conduit arteries, but it is important in coronary artery, mesenteric artery, and carotid artery *(28)*. Is it possible that H_2S released from VSMCs stimulates endothelium of small peripheral resistant arteries to release EDHF? In fact, coapplication of charybdotoxin (ChTx) and apamin to the endothelium-intact tissues reduced the H_2S-induced vasorelaxation *(12)*, suggesting the inhibited release or effect of EDHF. In this scenario, H_2S may have two targets: K_{ATP} channels in VSMCs and ChTx/apamin-sensitive K_{Ca} channels in vascular endothelial cells, the target of EDHF. The activation of these two types of channels by H_2S would compound hyperpolarization of smooth muscle cells, leading to vasorelaxation.

A low concentration of H_2S (0.1 mg/L of air or 72 ppm) exposure has been reported to change electrocardiogram features of rabbit heart, with flattened and inverted T waves *(29)*. Beyond this earlier report, our understanding of the cardiac effects of H_2S is very poor. Especially considering the importance of K_{ATP} channels in ischemia/reperfusion heart damage and the demonstrated stimulation of K_{ATP} channels by H_2S in VSMCs, it is highly expected that H_2S is an important gasotransmitter involved in the regulation of heart functions in specific situations.

6. INTEGRATED VASCULAR EFFECTS OF H_2S AND NO

The upregulation of CSE by NO has been reported *(12)*. Incubating the cultured VSMCs with (±)*S*-nitro-*N*-acctylpenicillamine, an NO donor, for 6 h significantly increased the transcriptional level of CSE. This transcriptional regulation of H_2S-generating enzyme may provide a chronic mechanism by which H_2S production can be modulated in a prolonged time scale. Zhao et al. *(12)* first reported the acute production/release of H_2S by NO in vascular tissues. Incubating the homogenized rat vascular tissues with different concentrations of sodium nitroprusside (SNP), an NO donor, increased the accumulated H_2S production during a 90-min period. Putatively, NO increases the activity of cGMP-dependent protein kinases, which in turn stimulates CSE. This mechanism is supported by the finding that the blockade of cGMP-dependent protein kinase abolished the NO-induced increase in H_2S level in vascular tissues *(10)*. It is also possible that NO directly acts on CSE protein. Rat mesenteric artery CSE protein contains 12 cysteines that are the potential substrate of *S*-nitrosylation. Currently, the three-dimensional structure of CSE is unknown and which cysteine contains the free –SH group cannot yet be determined with certainty. However, nitrosylation of a certain free –SH group of CSE in the presence of NO does represent a possibility *(13)*. The correlation of SNP and NO release has been questioned as SNP can exert its effects independently of NO *(30,31)*. To this end, a recent study demonstrated that 1,1-diethyl-2-hydroxy-2-nitrosohydrazine, another NO donor, also stimulated H_2S production from rat aortic tissues *(10)*. It is worth mentioning that NO is also able to destroy iron-sulfur clusters, especially under acidic conditions, when considering the interaction between NO and H_2S.

Although H_2S or NO alone relaxed vascular tissues, the integrated vascular effects of the two gasotransmitters are more complicated than a simple algebraic summation of individual actions. Hosoki et al. *(5)* observed that the vasorelaxant effect of SNP was enhanced by incubating rat aortic tissues with 30 μ*M* NaHS. By contrast, Zhao and Wang *(25)* found that pretreating aortic tissues with 60 μ*M* H_2S inhibited the vasorelaxant effect of SNP. This discrepancy may be partially explained by the experimental conditions of these studies. Hosoki et al. *(5)* used norepinephrine (1 μ*M*)-precontracted helical tissue strips of aorta from Wistar rats, and Zhao and Wang *(25)* used phenylephrine (0.3 μ*M*)-precontracted aortic rings from Sprague-Dawley rats. The tissue damage of helical strips is certainly greater than that of ring preparations. Moreover, the maximal contraction could be induced by 1 μM norepinephrine while 0.3 μM phenylephrine (a submaximal concentration) only induced about 90% of the maximal contraction of rat aortic tissues. The advantage of using a submaximal concentration of phenylephrine is that the tissue can react with the relaxant agent in a more sensitive way. Should sulfide interact with numerous enzymes and other macromolecules, H_2S may inhibit cGMP accumulation. The inhibition of cGMP pathway by H_2S *per se* may not suffice to alter the vascular tone, i.e., to evoke contraction. However, it might exert an inhibitory influence on the NO-induced vasodilation as the latter is mainly mediated by the cGMP pathway. An earlier study showed that the NO-induced relaxation of rabbit aorta and the increase in cGMP level were inhibited by L-cysteine and L-homocysteine *(32)*. Because both L-cysteine and L-homocysteine are endogenous precursors of H_2S and L-cysteine shares a relaxant effect similar to that of H_2S *(33)*, this study also lends support to a possible inhibitory effect of H_2S on the vascular effect of NO.

7. CONCLUSION

The cardiovascular effects of H_2S have gained appreciation in recent years. Depending on the types of vascular tissues, the endogenous production rate of H_2S is different and the vasorelaxant potency of H_2S also differs. Moreover, the endothelium dependency of vascular effect of H_2S may be related to the location and size of blood vessels. Similar to EDHF, the vascular effects and endothelium dependency of H_2S effect are more prominent in small resistant arteries. The direct interaction of H_2S and VSMCs is the major mechanism for the vasorelaxation and hypotension effects of this gasotransmitter. Abnormal metabolism of H_2S may have a significant impact on the cardiovascular functions. Patients with inherited abnormalities of the methionine metabolism exhibit significantly elevated concentrations of homocysteine measured as homocysteine or cysteine-homocysteine mixed disulfide. These patients are prone to arteriosclerotic vascular complications during childhood. Homocysteine causes endothelial cell injury and cell detachment that initiates the development of arteriosclerosis. Hyperhomocysteinemia is potentially accompanied by a reduced circulating level of H_2S. Therefore, arteriosclerotic vascular complications of these subjects should not be simply explained by the homocysteine-induced endothelial cell injury and cell detachment *(34)*. A lower level of circulating H_2S may also affect the structure and function of VSMCs, thus joining homocysteine as compounding pathogenetic factors for arteriosclerotic cerebrovascular disease.

Alterations in the production and function of endogenous H_2S under different pathophysiological conditions have not been examined. Accurate measurement of the endogenous level in addition to the production rate of H_2S in cardiovascular tissues has not been achieved to date. Specific endogenous stimulators and inhibitors for H_2S metabolism are still largely unknown. The altered vascular contractility in the presence of H_2S has been extensively studied, but the effect of H_2S on vascular proliferation and apoptosis still needs to be examined. These encumbrances are not viewed as insurmountable obstacles but challenges and prospects through which future breakthroughs in the understanding of H_2S biology and medicine will be made.

ACKNOWLEDGMENTS

This work was supported by a research grant from the Natural Sciences and Engineering Research Council of Canada. R. Wang is supported by an investigator award from the Canadian Institute of Health Research and regional partnership program of Saskatchewan.

REFERENCES

1. Wang R. Two's company, three's a crowd: can H_2S be the third endogenous gaseous transmitter? FASEB J 2002;16:1792–1798.
2. Mudd SH, Finkelstein JD, Irreverre F, et al. Transsulfuration in mammals: microassays and tissue distributions of three enzymes of the pathway. J Biol Chem 1965;240:4382–4392.
3. Bukovska G, Kery V, Kraus JP. Expression of human cystathionine beta-synthase in Escherichia coli: purification and characterization. Protein Express Purif 1994;5:442–448.
4. Erickson PF, Maxwell IH, Su LJ, et al. Sequence of cDNA for rat cystathionine γ-lyase and comparison of deduced amino acid sequence with related Escherichia coli enzymes. Biochem J 1990;269:335–340.
5. Hosoki R, Matsuki N, Kimura H. The possible role of hydrogen sulfide as an endogenous smooth muscle relaxant in synergy with nitric oxide. Biochem Biophys Res Commun 1997;237:527–531.
6. Stipanuk MH, Beck PW. Characterization of the enzymic capacity for cysteine desulphhydration in liver and kidney of the rat. Biochem J 1982;206:267–277.
7. Simpson RC, Freedland RA. Fractors affecting the rate of gluconeogenesis from L-cysteine in the perfused rat liver. J Nutr 1976;106:1272–1278.
8. Kurzban GP, Chu L, Ebersole JL, et al. Sulfhemoglobin formation in human erythrocytes by cystalysin, an L-cysteine desulfhydrase from Treponema denticola. Oral Microbiol Immunol 1999;14:153–164.
9. Abe K, Kimura H. The possible role of hydrogen sulfide as an endogenous neuromodulator. J Neurosci 1996;16:1066–1071.
10. Zhao W, Ndisang JF, Wang R. The modulation of endogenous production of H_2S in rat tissues. Can J Physiol Pharmacol 2003;81:848–853.
11. Levonen AL, Lapatto R, Saksela M, et al. Human cystathionine gamma-lyase: developmental and in vitro expression of two isoforms. Biochem J 2000;347(pt 1):291–295.
12. Zhao W, Zhang J, Lu Y, et al. H_2S is an endogenous K_{ATP} channel opener in vascular smooth muscle cells. EMBO J 2001;20:6008–6016.
13. Stamler JS, Toone EJ, Lipton SA, et al. (S)NO signals: translocation, regulation, and a consensus motif. Neuron 1997;18:691–696.
14. Uren JR, Ragin R, Chaykovsky M. Modulation of cysteine metabolism in mice—effects of propargylglycine and L-cyst(e)ine-degrading enzymes. Biochem Pharmacol 1978;27:2807–2814.
15. Bao L, Vlcek C, Paces V, et al. Identification and tissue distribution of human cystathionine beta-synthase mRNA isoforms. Arch Biochem Biophys 1998;350:95–103.
16. Mudd SH, Finkelstein JD, Irreverre F, et al. Transsulfuration in mammals: microassays and tissue distributions of three enzymes of the pathway. J Biol Chem 1965;240:4382–4392.
17. Chen P, Poddar R, Tipa EV, et al. Homocysteine metabolism in cardiovascular cells and tissues: implications for hyperhomocysteinemia and cardiovascular disease. Adv Enzyme Regul 1999;39:93–109.
18. Jacobsen DW, Savon SR, Stewart RW, et al. Limited capacity for homocysteine catabolism in vascular cells and tissues: a pathophysiologic mechanism for arterial damage in hyperhomocysteinemia? Circulation 1995;91:29 (abstract).
19. Wang J, Dudman NP, Wilcken DE, et al. Homocysteine catabolism: levels of 3 enzymes in cultured human vascular endothelium and their relevance to vascular disease. Atherosclerosis 1992;97:97–106.
20. Mason J, Cardin CJ, Dennehy A. The role of sulphide and suphide oxidation in the copper molybdenum antagonism in rats and guinea pigs. Res Vet Sci 1978;24:104–108.
21. Richardson CJ, Magee EA, Cummings JH. A new method for the determination of sulphide in gastrointestinal contents and whole blood by microdistillation and ion chromatography. Clin Chim Acta 2000;293:115–125.
22. Goodwin LR, Francom D, Dieken FP, et al. Determination of sulfide in brain tissue by gas dialysis/ion chromatography: postmortem studies and two case reports. J Anal Toxicol 1989;13:105–109.
23. Savage JC, Gould DH. Determination of sulfide in brain tissue and rumen fluid by ion-interaction reversed-phase high-performance liquid chromatography. J Chromatogr 1990;526:540–545.
24. Warenycia MW, Goodwin LR, Benishin CG, et al. Acute hydrogen sulfide poisoning: demonstration of selective uptake of sulfide by the brainstem by measurement of brain sulfide levels. Biochem Pharmacol 1989;38:973–981.
25. Zhao W, Wang R. The H_2S-induced vasorelaxation and the underlying cellular and molecular mechanisms. Am J Physiol 2002;283:H474–H480.
26. Wang R, Cheng YQ, Ndisang JF. Potent vasorelaxant effect of hydrogen sulfide on rat mesenteric artery bed. Circulation 2003;108:IV–165 (abstract).

27. Busse R, Edwards G, Félétou M, et al. EDHF: bringing the concepts together. Trends Pharmacol Sci. 2002;23:374–380.
28. Triggle CR, Dong H, Waldron GJ, et al. Endothelium-derived hyperpolarizing factor(s): species and tissue heterogeneity. Clin Exp Pharmacol Physiol 1999;26:176–179.
29. Kosmider S, Rogala E, Pacholek A. Electrocardiographic and histochemical studies of the heart muscle in acute experimental hydrogen sulfide poisoning. Arch Immun Ther Exp 1967;15:731–740.
30. Garry MG, Richardson JD, Hargreaves KM. Sodium nitroprusside evokes the release of immunoreactive calcitonin gene-related peptide and substance P from dorsal horn slices via nitric oxide-dependent and nitric oxide-independent mechanisms. J Neurosci 1994;14:4329–4337.
31. Ogita K, Shuto M, Yoneda Y. Nitric oxide-independent inhibition by sodium nitroprusside of the native N-methyl-D-aspartate recognition domain in a manner different from that by potassium ferrocyanide. Neurochem Int 1998;33:1–9.
32. Li J, Liu XJ, Furchgott RF. Blockade of nitric oxide-induced relaxation of rabbit aorta by cysteine and homocysteine. Zhongguo Yao Li Xue Bao 1997;18:11–20.
33. Sidhu R, Singh M, Samir G, et al. L-cysteine and sodium hydrosulphide inhibit spontaneous contractility of isolated pregnant rat uterine strips *in vitro*. Pharmacol Toxicol 2001;88:198–203.
34. Brattstrom LE, Hardebo JE, Hultberg BL. Moderate homocysteinemia—a possible risk factor for arteriosclerotic cerebrovascular disease. Stroke 1984;15:1012–1016.

20 Hydrogen Sulfide and Visceral Smooth Muscle Contractility

Philip K. Moore

Contents

SUMMARY

Whether hydrogen sulfide (H_2S) plays a part in the regulation of visceral smooth muscle contractility is not yet known. However, a growing body of evidence suggests this to be the case. For example, cystathione γ-lyase (CSE) occurs in the ileum whereas exogenous H_2S relaxes gastrointestinal, urogenital, and uterine smooth muscle preparations at concentrations similar to those found naturally occurring in rat and human blood. The mechanism of the smooth muscle relaxant effect of H_2S is also unclear. Although an effect of this gasotransmitter on visceral smooth muscle K^+-adenosine triphosphate channels (i.e., similar to what occurs in vascular smooth muscle) cannot be excluded at this stage, the inability of glibenclamide pretreatment to inhibit relaxation of guinea pig ileum to H_2S suggests that alternative mechanisms may occur at the cellular level. Recent work suggests that H_2S synergizes with nitric oxide in relaxing guinea pig ileum. Interestingly, the possibility that endogenous H_2S may regulate ileal smooth muscle contractility, at least in vitro, is suggested by the use of propargylglycine and β-cyanoalanine, which

From: *Signal Transduction and the Gasotransmitters: NO, CO, and H_2S in Biology and Medicine* Edited by: Rui Wang © Humana Press Inc., Totowa, NJ

inhibit CSE enzyme activity and thereby effectively remove the endogenous H_2S-mediated smooth muscle relaxant effect. Both drugs cause a time- and dose-dependent augmentation of the response of guinea pig ileum to electrical (field) stimulation. Evaluation of the effect of H_2S on visceral smooth muscle is still very much in its infancy, and many more experiments are required before the physiological, pathophysiological, and perhaps even the clinical significance of this gas can be fully determined.

Key Words: Hydrogen sulfide; ileum; uterus; colon; propargylglycine; β-cyanoalanine.

1. INTRODUCTION

Gaseous mediators (gasotransmitters) such as nitric oxide (NO) and carbon monoxide (CO) play important roles in many physiological and pathological processes. Over the last few years, numerous publications have appeared in the literature strongly suggesting that hydrogen sulfide (H_2S) may act as an additional, physiologically important gaseous mediator possibly working in concert with NO (for a review, *see* ref. *1*).

For many years H_2S has been recognized as an environmental toxicant (about 1 μg/m^3) *(2)* that has permeated the food chain to appear in detectable amounts in dairy and meat products *(3)*. That H_2S may also be synthesized by living cells has been known for well over 100 yr. For example, numerous bacteria and helminths synthesize H_2S. These include the marine worm *Riftia pachyptila* found in deep-sea hydrothermal vents *(4)*, and *Desulfovibrio* in the gut *(5)*. In mammals, H_2S occurs in human flatus *(6)* and feces *(7)*.

It is less well known that H_2S is also a natural synthetic product of mammalian cells. For example, both rat (50 μ*M*); *(8)* and human (10–100 μ*M*); *(9)* blood contains H_2S, and higher concentrations (60–150 μ*M*) occur in rat brain homogenates *(10)*. In addition, renal cortical tubules also synthesize H_2S *(11)* although the biological significance of H_2S in the kidney is completely unknown.

The biosynthesis of H_2S by mammalian cells is most probably largely accounted for by the activity of two enzymes: cystathionine γ-lyase (CSE) and cystathionine β-synthase (CBS). Both of these enzymes have been extensively characterized and are known to occur in mammals as well as nonmammalian species such as helminths *(12)*. Human CSE exists as two separate isoforms *(13)*. For many years, the sole function of CSE and CBS was believed to be the interchange of methionine and cysteine by the so-called transsulfuration pathway *(14)*. However, as a result of recent research, it is now becoming increasingly clear that this is an altogether too simplistic view of the biological role of these enzymes, both of which are now recognized to use alternative amino acids (apart from cystathionine) as substrates. In this context, it should be noted that Stipanuk and Beck *(15)* were the first to note that rat liver CSE converts cysteine and homocysteine into thiocysteine and H_2S.

Like both NO and CO, when administered acutely and at high concentrations, H_2S is toxic by virtue of complexing with the Fe^{3+} of mitochondrial cytochrome oxidase and thereby blocking cellular oxidative metabolism *(16)*. Indeed, fatalities (80 in the 10-yr period from 1984 to 1994) resulting from overexposure to H_2S, usually in industrial settings, are not uncommon *(17)*.

However, recent results raise the intriguing possibility that "physiologically relevant" concentrations of H_2S (i.e., similar to those attained in blood or tissue homogenates, as already discussed) are biologically active with the major (but by no means not the only) "targets" so far identified being, first, the control of smooth muscle reactivity and, second, the regulation of numerous different central nervous system (CNS) functions.

The role played by H_2S in the control of blood vessel caliber and in the CNS is described in detail in Chapters 18 and 19. Therefore, this chapter concentrates on the intriguing, although still speculative, possibility that H_2S, possibly working in concert with NO, plays a physiological role in the regulation of visceral smooth muscle contractility and perhaps a pathophysiological role in diseases of the gastrointestinal (GI) tract.

2. BIOSYNTHESIS AND BREAKDOWN OF H_2S IN VISCERAL SMOOTH MUSCLE

Regarding the GI tract, most of the attention to date has been focused on H_2S production by the colon. The majority of reports suggest that colonic bacteria account for the bulk of the H_2S formed in this particular region of the GI tract. These are anaerobic bacteria that produce H_2S either from sulfate-containing mucin sulfate or from sulfur-containing amino acids such as cysteine *(18)*. Certainly, the relatively high concentrations of H_2S that are found in human flatus *(6)* as well as human *(5,7,19)* and rat *(20,21)* faces are most likely to be of bacterial origin.

The possibility that at least a minor part of the H_2S found in faces may originate from host cells (vs bacteria) is suggested by the observation that rat enterocytes convert cysteine (and cystine) into H_2S in a CSE-dependent manner *(22)*. Hosoki et al. *(23)* demonstrated H_2S production by homogenates of rat ileum. They used a spectrophotometric assay (based on acidic conversion of methylene blue) to demonstrate substantial (approximately equivalent to that in rat aorta and portal vein) ileal H_2S-forming capacity. They also noted the presence of both CSE and CBS mRNA in rat ileal homogenates. My colleagues and I have recently detected quantitatively similar H_2S-synthesizing ability in homogenates prepared from guinea pig and rabbit ileum (unpublished findings). In general terms, CBS seems to occur in most tissues, but CSE is less widespread, being absent from heart, lung, spleen, adrenal gland, and brain.

In addition to the ability to synthesize H_2S, several mechanisms exist for the rapid removal of H_2S once formed at intestinal sites. Like NO, H_2S binds to and can thus be "quenched" by hemoglobin (Hb). The product is sulfhemoglobin *(24)*. However, unlike NO, numerous enzymes that break down H_2S have also been identified in the GI tract. These include thiol methyltransferase in rat stomach, intestine, and colon *(20,25)* and rhodanese (transfers sulfhydryl groups to cyanide, forming thiocyanate and sulfate) in human colon *(26)*. Rhodanese is a mitochondrial enzyme that is located in colonic epithelial crypt cells as well as in the submucosa in humans. In tissue homogenates, rhodanese avidly catabolizes added H_2S even at the very high concentrations (1 to 2 m*M*) generated locally by bacteria in the colon *(26)*.

Whether enzymatic breakdown, Hb binding, or some type of chemical reaction (perhaps with other gaseous mediators) or a combination of all three is involved, it is clear that breakdown of H_2S, at least in the colon, is very rapid and efficient, with >90% of H_2S and related gases (e.g., methanethiol) being removed on passage from the cecum to the rectum in the intact rat *(27)*.

Thus, mechanisms for the biosynthesis (by either CSE, CBS, or possibly both enzymes) and the rapid catabolism of H_2S in the GI tract seem to be established. However, many important questions still remain to be answered. For example, which cell type(s) forms this mediator? Is endogenous H_2S synthesized/released in the GI tract and, if so, how are these processes controlled? What function(s) does H_2S serve, once released, and how does H_2S interact with other mediators and transmitters to bring about these functions? Finally, is H_2S produced in the GI tract of humans? As will become apparent from the remainder of this chapter, answers to the majority of these questions are not yet forthcoming.

3. RELAXATION OF VISCERAL SMOOTH MUSCLE IN VITRO BY H_2S

Hosoki et al. *(23)* provided the first report of the smooth muscle relaxant effect of H_2S in the GI tract. They noted that graded application of NaHS (stable H_2S donor) caused a concentration-dependent relaxation (EC_{50} of 180 μ*M*) of acetylcholine (ACh)-precontracted guinea pig ileum.

Subsequently, we confirmed this seminal observation using isolated guinea pig ileum *(28)*. In our hands, H_2S (again provided by NaHS) proved to be somewhat more potent, with an EC_{50} of 84 μ*M*. The disparity in potency between the two studies may relate to our use of a lower concentration (88 n*M*, the approximate EC_{70}, vs 1 μ*M*) of ACh. In similar experiments, NaHS also caused dose-related inhibition of the response of the isolated guinea pig ileum to endogenous ACh released as a result of continuous application (0.1 Hz) of electrical (field) stimulation. In this case, the EC_{50} for NaHS was 80 μ*M*, very similar to its potency against the bath applied (i.e., exogenous ACh).

In the course of this work, we also noted that the smooth muscle relaxant effect of H_2S with respect to bath exogenous and endogenous ACh was time dependent. Using bath-applied ACh, the inhibitory effect of NaHS was apparent (34% inhibition) using a dose interval time of 5 s, increased (80%) at 60 s, and declined thereafter (43% at 120 s) (**Fig. 1A**). A very similar time course of action was apparent in experiments using electrically stimulated tissues.

These observations suggest that H_2S is rapidly released from NaHS in the organ bath and thereafter rapidly disposed of, either by catabolism (ileal enzymes, as discussed earlier), or by uptake into smooth muscle cells, or by perhaps an as yet undefined chemical interaction. In this latter context, it may be of interest that aqueous solutions of H_2S have been estimated to comprise approx 30% "free" H_2S with the remainder present as hydrosulfide anion (HS^-). Whether the active (i.e., smooth muscle relaxant) moiety is H_2S or HS^- is not known. It has been known for many years that H_2S is inactivated by bivalent cations such as zinc *(27)* and bismuth *(19)*. Indeed, use is made of such ions in the manufacture of H_2S filters to purge the gas from industrial sites. However, the concentration of these metals in the organ bath is almost certainly insufficient to react with H_2S in the conditions of the experiments. Whether H_2S (or indeed HS^- anions) are destabilized by the presumably high Po_2 of the organ bath environment is not known but would seem to warrant further evaluation.

Within different visceral smooth muscles, the relaxant effect of H_2S is not confined to guinea pig ileum. Thus, we observed that NaHS also produces graded relaxation of the spontaneous, pendular, rhythmic, contractile activity of the rabbit ileum preparation (EC_{50} of 76 μ*M*), whose effect was similar (in potency terms) to that observed in the phenylephrine-precontracted rabbit aorta (EC_{50} of 61 μ*M*) (**Fig. 1B**). Furthermore, NaHS reduced the contractile response of the isolated rat ileum (108 μ*M*) and vas deferens (EC_{50} of 65 μ*M*) in response to electrical stimulation, indicating a similar inhibitory effect toward both endogenous ACh (ileum) and noradrenaline (vas deferens) (**Fig. 1C**).

Apart from vascular smooth muscle (dealt with elsewhere in this volume), other types of smooth muscle that relax in response to applied NaHS have also been identified. For example, Sidhu et al. *(29)* noted that NaHS reduced both the spontaneous contractility

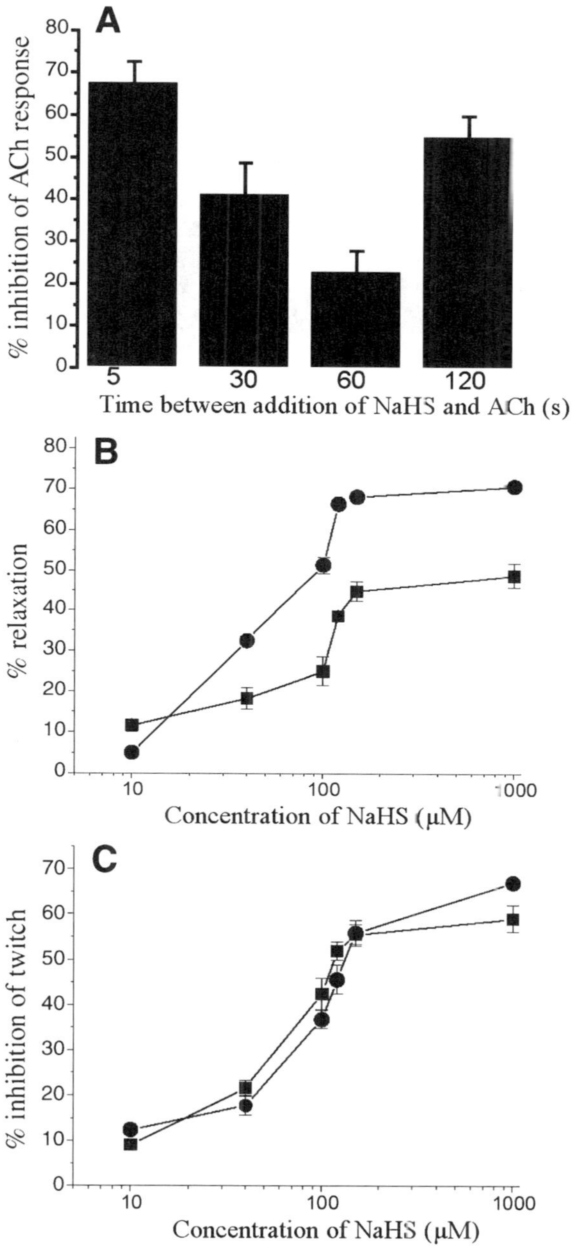

Fig. 1. **(A)** Time-dependent smooth muscle relaxant effect of NaHS in isolated guinea pig leum. NaHS (80 µ*M*) was added to the bath at timed intervals (5–120 s, as indicated) prior to injection of an approximate EC_{70} of ACh (88 n*M*). Results show the percentage of inhibition of the response to ACh and are the mean ± SE ($n = 10$). **(B)** Dose-related smooth muscle relaxant effect of NaHS in spontaneously relaxing rabbit ileum (circles) and phenylephrine-precontracted rabbit aorta (squares). Results show the percentage of muscle relaxation and are the mean ± SE ($n = 6$). **(C)** Dose-related smooth muscle relaxant effect of NaHS in the electrically stimulated rat vas deferens (squares, 0.1 Hz) and ileum (circles, 5 Hz) preparations. Results show the percentage of muscle relaxation and are the mean ± SE ($n = 10$).

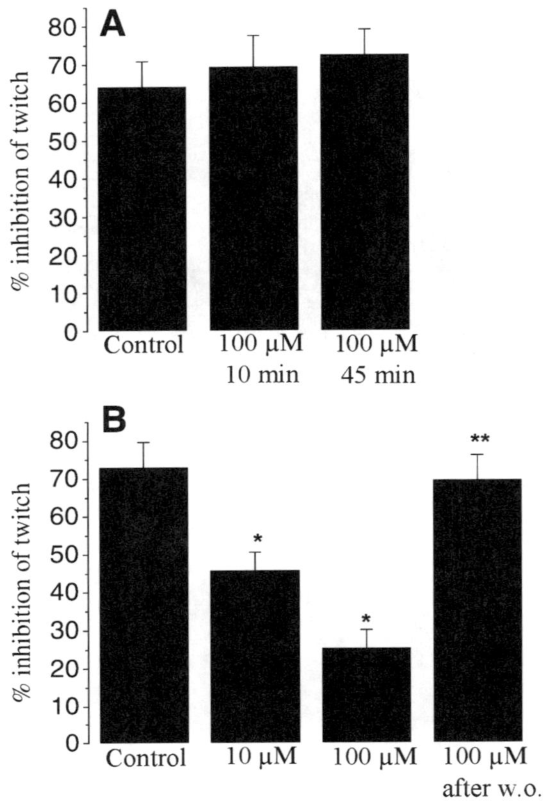

Fig. 2. Effect of **(A)** glibenclamide (100 µ*M*) and **(B)** IBMX (10 µ*M*) on smooth muscle relaxant effect of NaHS (80 µ*M*) in electrically stimulated (0.1 Hz) guinea pig ileum preparation. Results show the percentage of inhibition of the twitch response to electrical stimulation and are the mean ± SE ($n = 8$). $^{*}p < 0.05$ vs control, $^{**}p < 0.05$ vs exposure to IBMX (100 µ*M*) before washout (w.o.).

and the response to oxytocin of pregnant rat myometrial strips. Similar results had been noted previously *(30)*. These studies suggest a novel tocolytic effect of H_2S. Whether this basic pharmacological observation can be translated into a clinical application for drugs affecting H_2S synthesis or activity in the uterus remains to be determined.

4. MECHANISM OF RELAXANT EFFECT OF H_2S IN VISCERAL SMOOTH MUSCLE

One of the principal unresolved issues concerning the effect of H_2S on visceral smooth muscle that has yet to be properly addressed is the precise mechanism of action. A considerable body of evidence supports the possibility that H_2S relaxes vascular smooth muscle both in vitro and in vivo by an effect on smooth muscle K_{ATP} channels *(8,31)*.

However, in our experiments, NaHS did not affect response of the guinea pig ileum to added KCl and neither were contractions of the same preparation to electrical stimulation affected by the K_{ATP} channel blocker glibenclamide (**Fig. 2A**). K_{ATP} channels are found in guinea pig ileum and their activation does bring about glibenclamide-sensitive relaxation responses *(32)*. Thus, it would appear from the present observations that visceral and vascular smooth muscle may differ in the manner in which they relax to H_2S.

In further experiments, we demonstrated that the relaxant effect of NaHS in the electrically driven ileum was unaffected by prior exposure of tissues to agents that either inhibit nitric oxide synthase (L-N^G nitroarginine methylester) or cyclooxygenase (indomethacin) enzyme activity or block opioid receptors (naloxone), thus ruling out the possibility that H_2S releases endogenous NO, prostanoids, or opioids to bring about its smooth muscle relaxant effect in this tissue.

However, we did observe that prior exposure of ileal segments to increasing bath concentrations of isobutylmethylxanthine (IBMX), a potent inhibitor of cyclic nucleotide (cyclic adenosine monophosphate [cAMP] and cyclic guanosine 5'-monophosphate [cGMP]) phosphodiesterase and adenosine receptor antagonist, did result in a dose-related and easily reversible (by washing) inhibition of the smooth muscle relaxant effect of NaHS in the electrically stimulated guinea pig ileum preparation (**Fig. 2B**). At the concentrations used, IBMX had only a weak and readily reversed inhibitory effect on contractile activity in its own right. A role for either adenosine or cAMP/cGMP in the vasorelaxant effect of H_2S seems unlikely *(31)*. However, numerous reports do suggest that H_2S promotes cAMP accumulation in central neurons. Interestingly, the inhibitory effect of IBMX on the relaxant response of the ileum to H_2S implies that accumulation of cAMP and/or cGMP counters (not contributes to) the effect of NaHS on ileal smooth muscle relaxation. Clearly, these experiments are difficult to interpret at the present time.

Additional work to probe the cellular mechanisms underlying the effect of H_2S in nonvascular tissue is needed. Such experiments should perhaps concentrate first on the part played by cAMP/cGMP and perhaps adenosine in this response. The limited amount of information gleaned to date suggests that H_2S may act on different cellular targets in different types of muscle. If this is indeed the case, then H_2S quite clearly differs from other gasotransmitters such as NO and CO, which relax smooth muscle by a single, seemingly ubiquitous mechanism, namely, activation of muscle soluble guanylate cyclase activity.

5. WHETHER OR NOT H_2S RELAXES GI SMOOTH MUSCLE IN CONCERT WITH NO

Whether H_2S acts alone on visceral smooth muscle, or in concert with other gaseous mediators such as NO and/or CO, is controversial. For example, reports in the literature indicate that the vasorelaxant effect of sodium nitroprusside (SNP) (an NO donor) in rat aorta can be either *increased (23,33)* or *decreased (8)* by H_2S.

We *(28)* have noted previously that NaHS and SNP act together in the electrically stimulated guinea pig ileum preparation to bring about an inhibition of smooth muscle contractility that is considerably greater than what might be expected from a simple additive effect. Therefore, these data imply but do not prove, a synergistic interaction between the two gasotransmitters in this tissue.

The mechanism(s) by which H_2S and NO may interact together to cause enhanced visceral (and perhaps vascular) smooth muscle relaxation is likely to be complex. Numerous potential "crossover" points between the two mediators have already been identified. For example, H_2S stabilizes NO in solution *(34)* and augments NO release from *S*-nitrosothiols *(35)*. Furthermore, NO acts on cultured smooth muscle cells to increase both H_2S production and CSE expression *(8)*. Finally, as mentioned previously, both NO and H_2S bind to, and are consequently quenched by, Hb *(24)*.

Thus, the overall biological effect of H_2S on smooth muscle (and other systems) may well be inextricably linked to the synthesis and activity of NO. Whether a similar interaction exists between H_2S and CO (or indeed other gaseous mediators) is not known and warrants study.

6. POSSIBLE ROLE OF H_2S AS A NEUROMODULATOR IN GI SMOOTH MUSCLE

Experiments described to date have largely concentrated on the pharmacological effects of exogenous H_2S (usually generated *in situ* from NaHS). Although of value, such work does not provide any clues about the potential physiological and/or pathophysiological roles of endogenous H_2S in the control of GI contractility.

One approach to this very desirable research goal (previously been found to be extremely useful in "unraveling" the principal physiological functions of NO) is the use of drugs that interfere with the synthesis or activity of H_2S. A range of reasonably potent CSE and CBS inhibitors is available. For example, propargylglycine (PAG) is an irreversible inhibitor of CSE in vitro *(36)*, and when administered to rats, it causes an almost complete inhibition of liver CSE enzyme activity in vivo *(37,38)*. PAG is well absorbed and readily crosses biological membranes *(39,40)*. In addition, β-cyanoalanine (β-CA) is a potent and reversible inhibitor of CSE *(38,41)*, and CBS is inhibited by aminooxyacetic acid (AOAA) *(42)*. Finally, as noted previously, H_2S avidly binds to bismuth subsalicylate *(6)*.

Until recently, none of these various drugs had been exploited as "tools" to investigate the physiological effects of H_2S. We have now reported the effect of some of these agents on contractility of guinea pig ileum *(28)*. Intriguingly, exposure to both PAG and β-CA (CSE inhibitors) but not to AOAA (CBS inhibitor) resulted in a slowly developing (first apparent at about 15 min) but long-lasting (until at least 60 min) increase in the contractile response to electrical stimulation (**Fig. 3**). These data suggest that CSE (but not CBS) enzyme activity is responsible for the biosynthesis of H_2S that causes smooth muscle relaxation in guinea pig ileum.

In separate experiments, we demonstrated that administration of L-cysteine base (substrate for H_2S formation) rapidly reversed the procontractile effect of β-CA (but not PAG) in the electrically stimulated guinea pig ileum preparation. The disparity in the effect of L-cysteine on the response to the two CSE inhibitors may reflect the nature of the enzyme binding of PAG (irreversible *[36]*) as opposed to β-CA (reversible, competitive *[38]*). A non-specific effect of L-cysteine on ileal smooth muscle contractility can be ruled out because at the same concentration and over the same time course, addition of L-cysteine did not affect the response of the electrically driven guinea pig ileum in the absence of either PAG or β-CA.

At this stage, we cannot exclude the possibility that both PAG and β-CA (but not AOAA) bring about a nonspecific facilitation of contractions of the ileum. However, this seems unlikely given the ability of L-cysteine to reverse the effect of β-CA as well as the lack of effect of this compound on the response to applied ACh.

Accordingly, we propose that inhibition of CSE (but not CBS) removes a source of endogenous H_2S from the ileum that normally acts as a smooth muscle relaxant in this tissue under these experimental conditions. The cellular site of H_2S production within the ileum is not yet known. The finding that β-CA exposure did not similarly augment the contractile response to applied ACh in this preparation suggests that H_2S is generated as a *direct* consequence of electrical activity within the intramural nerves rather than as an *indirect* result of the subsequent contraction of the smooth muscle. This would imply a neuronal localization for H_2S production (and, hence, for CSE) in this tissue. Interest-

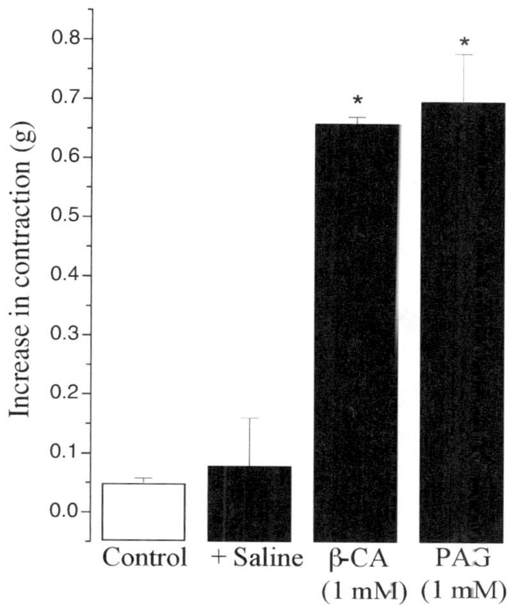

Fig. 3. Increase (g) in contractile response to electrical stimulation (0.1 Hz) of isolated guinea pig ileum in the presence of PAG, β-CA (both at 1 m*M*), or appropriate volume of saline vehicle. "Control" indicates response of the treatment to continued electrical stimulation in the absence of any addition. Results are the mean ± SE ($n = 6$).

ingly, in the CNS, H_2S formation (most probably generated by CBS in this instance), also occurs in neurons *(43)*, and a strong case has been made that H_2S functions as a neuromodulator in the brain possibly linked to the activity of the excitatory neurotransmitter glutamate *(43,44)*.

Accordingly, we propose that H_2S acts as a neuromodulator in the peripheral nervous system as well as in the CNS. Although further experiments are clearly necessary to probe this possibility in greater detail, our current "working hypothesis" is illustrated diagrammatically in **Fig. 4**.

7. PATHOPHYSIOLOGICAL SIGNIFICANCE OF H_2S IN THE GI TRACT

It is probably premature to speculate on the pathophysiological roles of H_2S in the GI tract when the precise physiological functions of this gasotransmitter remain uncertain. Nevertheless, some reference to this intriguing possibility is warranted here.

We have already noted the putative role of H_2S as a neuromodulator of the enteric nervous system. Whether disordered H_2S synthesis, activity, or catabolism plays any part in GI disease states such as colic or ileus has not been directly evaluated. Certainly, no gross disturbance in GI function has been noted in CBS knockout mice *(43,45)*. However, this may not be entirely surprising when one bears in mind the seemingly exclusive role for CSE in the ileum (as outlined earlier), and to the best of our knowledge, CSE knockout animals have yet to be developed. Moreover, no GI problems were identified in animals chronically administered the CSE inhibitor PAG *(46)*. Nevertheless, it should be pointed out that none of the studies quoted here specifically set out to evaluate GI function in these

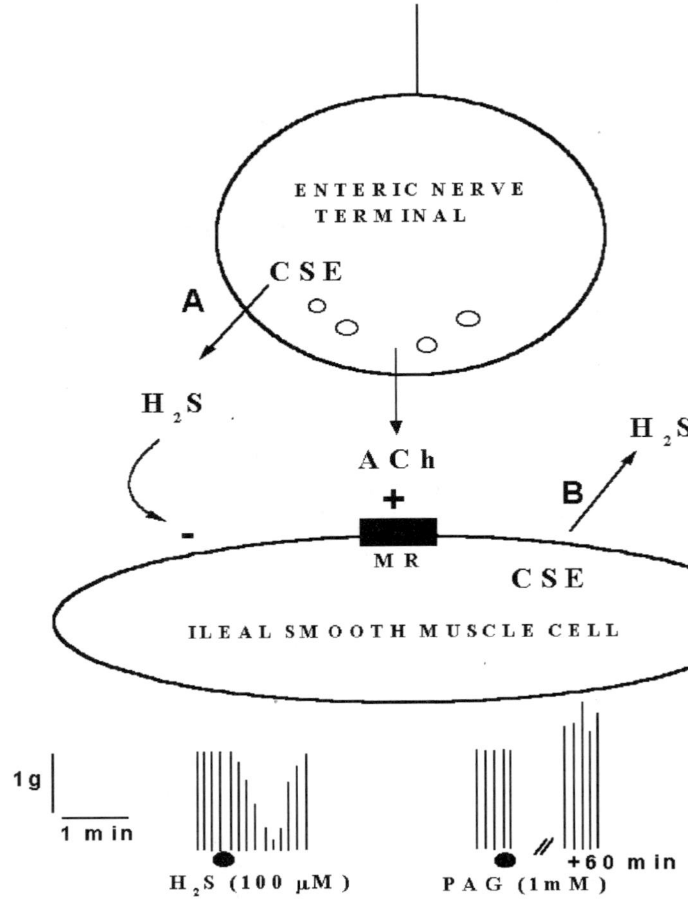

Fig. 4. Diagrammatic representation of possible role of H_2S in regulation of parasympathetic neurotransmission in visceral smooth muscle. Typical responses of the electrically stimulated guinea pig ileum preparation (0.1 Hz) to NaHS and to PAG are shown in the illustration at the bottom. +, Contractile response of guinea pig ileum to released ACh; –, relaxant response to released H_2S. In A, H_2S biosynthesis occurs presynaptically, and in B, the production of H_2S is a postsynaptic event. MR, muscarinic receptor.

animals and, as such, a role for H_2S in gut muscular disease should not be discounted. Finally, whether the formation of H_2S, either by or within the vicinity of mucosal blood vessels, may regulate mucosal blood flow (perhaps in concert with NO and other mediators such as prostacyclin) and in this way protect against ulcer formation also requires further attention.

8. CONCLUSION

To date, very little information is available about either the biosynthesis of H_2S by nonvascular smooth muscle or the effect of H_2S thereon. This is surprising when one bears in mind that all of the reports on this subject to date have identified a smooth muscle relaxant effect of H_2S (usually as NaHS) at concentrations well within what might be considered the "physiological" range. Furthermore, it is likely that parts of the GI tract (e.g., colon) are routinely exposed to concentrations of H_2S that are at least an order of magnitude higher than this range. In addition, H_2S, like NO, may occur at high concentrations (i.e., above the physiological range) in small "pockets" at or close to its target cells. Further work to characterize the effect of both exogenous and endogenous H_2S on smooth muscle of the GI and genitourinary tracts is needed. Furthermore, experiments to evaluate the role of H_2S in regulating GI transit in vivo as well as human studies would be of great value. We believe that the judicious and carefully controlled use of CSE and/or CBS inhibitors as well as the application of H_2S "quenching agents" may prove useful in this particular quest.

REFERENCES

1. Wang R. Two's company, three's a crowd: can H_2S be the third endogenous gaseous transmitter? FASEB J 2002;16:1792–1798.
2. World Health Organization. Air Quality Guidelines for Europe. WHO Regional Office for Europe: Copenhagen, 1987.
3. Kraft AA, Brant AW, Ayres JC. Detection of hydrogen sulphide in packaged meats and in broken-out shell eggs. Food Technol 1956;10:443, 444.
4. Goffredi SK, Childress JJ, DeSaulniers NT, et al. Sulfide acquisition by the vent worm Riftia Pachyptila appears to be via uptake of HS^- rather than H_2S. J Exp Biol 1997;200:2609–2616.
5. Picher MCL, Beatty ER, Cummings JH . The contribution of sulphate reducing bacteria and 5-aminosalicylic acid to faecal sulphide in patients with ulcerative colitis. Gut 2000;46:64–72.
6. Suarez FL, Springfield J, Levitt MD. Identification of gases responsible for the odour of human flatus and evaluation of a device purported to reduce this odour. Gut 1998;43:100–104.
7. Florin TH. Hydrogen sulfide and total acid-volatile sulfides in faeces, determined with a direct spectrophotometric method. Clin Chim Acta 1991;196:127–134.
8. Zhao W, Zhang J, Lu Y, et al. The vasorelaxant effect of H_2S as a novel endogenous gaseous K_{ATP} channel opener. EMBO J 2001;20:6008–6016.
9. Richardson CJ, Magee EA, Cummings JH. A new method for the determination of sulphide in gastrointestinal contents and whole blood by microdistillation and ion chromatography. Clin Chim Acta 2000;293:115–125.
10. Warenycz MW, Steele JA, Karpinski E, et al. Hydrogen sulfide in combination with taurine or cysteic acid reversibly abolishes sodium currents in neuroblastoma cells. Neurotoxicology 1989;10:191–199.
11. Stipanuk MH, De la Rosa J, Hirschberger LL. Catabolism of cyst(e)ine by rat renal cortical tubules. J Nutr 1990;120:450–458.
12. Walker J, Barrett J. Cystathionine beta-synthase and gamma cystathionase in helminths. Parasitol Res 1991;77:709–713.
13. Levonen A-L, Lapatto R, Saksela M, et al. Human cystathionine-γ-lyase: development and in vitro expression of two isoforms. Biochem J 2000;347:291–295.
14. Finkelstein JD. Methionine metabolism in mammals. J Nutr Biochem 1990;1:228–237.
15. Stipanuk MH, Beck PW. Characterisation of the enzymatic capacity for cysteine desulphydration in liver and kidney of the rat. Biochem J 1982;206:267–277.
16. Gosselin RE, Smith RP, Hodge HC. Hydrogen sulfide. In: Clinical Toxicology of Commercial Products. Williams & Wilkins: Baltimore, 1991.
17. Fuller DC, Suruda AJ. Occupationally related hydrogen sulfide deaths in the United States from 1984 to 1994. J Occup Environ Med 2000;42:939–942.
18. Kadota H, Ishida Y. Production of volatile sulfur compounds by microorganisms. Appl Microbiol 1971;22:522–529.
19. Suarez FL, Furne JK. Springfield J, et al. Bismith subsalicylate decreases hydrogen sulfide release in the human colon. Gastroenterology 1998;114:923–929.
20. Furne J, Springfield J, Koenig T, et al. Oxidation of hydrogen sulfide and methanothiol to thiosulfate by rat tissues: a specialised function of the colonic mucosa. Biochem Pharmacol 2000;62:255–259.
21. Levitt MD, Springfield J, Furne J, et al. Physiology of sulfide in the rat colon: use of bismuth to assess colonic sulfide production. J Appl Physiol 2002;92:1655–1660.
22. Coloso RM, Stipanuk MH. Metabolism of cyst(e)ine in rat enterocytes. J Nutr 1998;119:1914–1924.
23. Hosoki R, Matsuki N, Kimura H. The possible role of hydrogen sulfide as an endogenous smooth muscle relaxant in synergy with nitric oxide. Biochem Biophys Res Commun 1997;237:527–531.

24. Searcy DG, Lee SH Sulfur. reduction by human erythrocytes. J Exp Zool 1998;282:310–322.
25. Weisiger RA, Pinkus LM, Jakoby WB. Thiol S-methyltransferase: suggested role in detoxication of intestinal hydrogen sulfide. Biochem Pharmacol 1980;29:2885–2887.
26. Picton R, Eggo MC, Merrill MJS, et al. Mucosal protection against sulphide: importance of the enzyme rhodanese. Gut 2002;50:201–205.
27. Suarez FL, Furne JK, Springfield J, et al. Production and elimination of sulfur-containing gases in the rat colon. Am J Physiol 1998;274:G727–G733.
28. Teague B, Asiedu S, Moore PK. The smooth muscle relaxant effect of hydrogen sulphide in vitro: evidence for a physiological role to control intestinal contractility. Br J Pharmacol 2002;137:139–145.
29. Sidhu R, Singh M, Samir G, et al. L-Cysteine and sodium hydrosulphide inhibit spontaneous contractility of isolated pregnant rat uterine strips in vitro. Pharmacol Toxicol 2001;88:198–203.
30. Hayden LJ, Franklin KJ, Roth SH, et al. Inhibition of oxytocin-induced but not angiotensin-induced rat uterine contractions following exposure to sodium sulfide. Life Sci 1989;45:2557–2560.
31. Zhao W, Wang R. H_2S-induced vasorelaxation and underlying cellular and molecular mechanisms. Am J Physiol 2002;283:H474–H480.
32. Sun YD, Benishin CG. K^+ channel openers relax longitudinal muscle of guinea pig ileum. Eur J Pharmacol 1994;271:453–459.
33. Kruszyna H, Kruszyna R, Smith RP. Cyanide and sulfide interact with nitrogenous compounds to influence the relaxation of various smooth muscles. Proc Soc Exp Biol Med 1985;179:44–49.
34. Sorensen J, Tiedje JM, Firestone RB. Inhibition by sulphide of nitric oxide reduction by denitrifying Pseudomonas fluorescens. Appl Environ Microbiol 1980;39:105–108.
35. Roediger WE, Babidge W. Nitric oxide effect on colonocyte metabolism: co-action of sulfides and peroxide. Mol Cell Biochem 2000;206:159–167.
36. Johnston M, Jankowski D, Marcotte P, et al. Suicide inactivation of bacterial cystathionine gamma synthase and methionine gamma lyase during processing of L-propargylglycine. Biochemistry 1979;18:4690–4701.
37. Porter DW, Nealley EW, Baskin SL. In vivo detoxification of cyanide by cystathionase gamma-lyase. Biochem Pharmacol 1996;27:941–944.
38. Uren JR, Ragin R, Chaykovsky M. Modulation of cysteine metabolism in mice—effects of propargylglycine and L-cysteine-degrading enzymes. Biochem Pharmacol 1978;27:2807–2814.
39. Reed DJ. Cystathionine. Methods Enzymol 1995;252:92–102.
40. Yu S, Sugahara K, Nakayama K, et al. Accumulation of cystathionine, cystathionine ketime, and perhydro-1,4-thiazepine-3,5 dicarboxylic acid in whole brain and various regions of the brain of D,L-propargylglycine-treated rats. Metabolism 2000;49:1025–1029.
41. Pfeffer M, Ressler C. Beta-cyanoalanine, an inhibitor of rat liver cystathionase. Biochem Pharmacol 1967;242:2299–2308.
42. Braunstein AE, Goryachenkova EV, Tolosa EA, et al. Specificity and some other properties of liver serine sulphhydrase: evidence for its identity with cystathionine-β-synthase. Biochim Biophys Acta 1971;242:247–260.
43. Eto K, Ogasawara M, Imemura K, et al. Hydrogen sulphide is produced in response to neuronal excitation. J Neurosci 2002;22:3386–3391.
44. Abe K, Kimura H. The possible role of hydrogen sulphide as an endogenous neuromodulator. J Neurosci 1996;16:1066–1071.
45. Dayal S, Bottiglieri T, Arning E, et al. Endothelial dysfunction and elevation of S-adenosylhomocysteine in cystathionine beta-synthase-deficient mice. Circ Res 2001;88:1203–1209.
46. Cho ES, Hovanec-Brown J, Tomanek RJ, et al. Propargylglycine infusion effects on tissue glutathione levels, plasma amino acid concentrations and tissue morphology in parenterally-fed growing rats. J Nutr 1991;121:785–794.

21 Interaction of Hydrogen Sulfide and Adenosine Triphosphate-Sensitive Potassium Channels in Vascular Smooth Muscle Cells

Rui Wang

CONTENTS

SUMMARY

K_{ATP} channels link adenosine triphosphate (ATP) production to cellular functions by modulating membrane potentials and transmembrane ion flux. In vascular smooth muscle cells (VSMCs), opening of K_{ATP} channels in plasma membrane leads to membrane hyperpolarization and muscle relaxation. The functionality of K_{ATP} channels in mitochondrial membrane affects the redox status of cells and the outcome of ischemic damage. The modulation of K_{ATP} channels by endogenous substances and pharmacological agents has been known. The interaction of hydrogen sulfide (H_2S) with K_{ATP} channels in VSMCs attracted a great deal of attention. By stimulating K_{ATP} channels, H_2S lowered blood pressure of rats and relaxed vascular smooth muscles. Whole-cell patch-clamp studies revealed that H_2S increased K_{ATP} channel currents and hyperpolarized membrane of single VSMCs. Because the known second-messenger systems are not apparently altered by H_2S, a direct interaction of K_{ATP} channels and H_2S has been assumed. Among mechanisms that underlie the effect of H_2S on K_{ATP} channels are altered ATP metabolism, generation of thiyl free radicals, and sulfuration of K_{ATP} channel proteins. Unmasking the molecular mechanisms for the effect of H_2S on the structure and function of K_{ATP} channels will help researchers to understand the cellular effects of H_2S. This advance in

From: *Signal Transduction and the Gasotransmitters: NO, CO, and H_2S in Biology and Medicine* Edited by: Rui Wang © Humana Press Inc., Totowa, NJ

knowledge will also have the potential to reveal a novel sulfuration mechanism by which many membrane proteins can be directly modified by H_2S.

Key Words: Blood pressure; Hydrogen sulfide; K_{ATP} channel; smooth muscle cells; sulfuration.

1. INTRODUCTION

As discussed in many chapters of this book, hydrogen sulfide (H_2S) is endogenously generated in mammalian cells. Like other gasotransmitters such as nitric oxide (NO) and carbon monoxide (CO), H_2S at physiologically relevant levels affects structures and functions of the human body at molecular, cellular, tissue, and system levels. Unlike NO and CO, the cellular and molecular mechanisms for the physiological effects of H_2S have been largely unknown. Activation of the soluble guanylyl cyclase (sGC)/cyclic guanosine 5'-monophosphat (cGMP) pathway is the major cellular event bestowed with the biological effects of NO and CO. Stimulation of big-conductance K_{Ca} channels independent of the known second messengers, on the other hand, also partially explains the effects of NO and CO. In this case, chemical modifications of K_{Ca} channel proteins by NO or CO, the processes known as *S*-nitrosylation and carboxylation, respectively, change the configuration of the channel complex and lead to increased opening probability. Initial attempts at decoding the cellular mechanisms for the effect of H_2S were also targeted at the known second messengers, especially the sGC/cGMP system. K_{Ca} channel was also a rationalized candidate for H_2S target. Interestingly, none of these attempts were proved correct. Recent studies point to altered membrane excitability, especially the stimulation of K_{ATP} channels, as the key event in H_2S stimulation. Evidence supporting this view is presented in this chapter. Several putative molecular mechanisms underlying the interaction of H_2S and K_{ATP} channel proteins, including altered adenosine triphosphate (ATP) level, generation of thiol free radicals, and structural modifications of sulfhydryl groups of K_{ATP} channel subunits, are also discussed. A universal and novel mechanism for modulating protein functions by H_2S is proposed.

2. K_{ATP} CHANNELS IN VASCULAR SMOOTH MUSCLE CELLS

K_{ATP} channels are inhibited by intracellular ATP and extracellular sulfonylureas but stimulated by K_{ATP} channel openers (KCOs) *(1)*. The activation of K_{ATP} channels leads to membrane hyperpolarization and a relaxation of vascular smooth muscle cells (VSMCs). Originally discovered in cardiac muscle, K_{ATP} channels were later identified in many other tissues, including pancreatic β-cells, skeletal muscle cells, and many types of VSMCs *(2–8)*. Single-channel conductance of K_{ATP} channels in VSMCs is about 20–50 pS with symmetric [K^+] across cell membrane *(8)*. However, large-conductance K_{ATP} channels were also reported *(9,10)*. The controversy over single-channel conductance may result from multiplicity of the isoforms of K_{ATP} channels and the experimental configurations used.

2.1. Function and Modulation of K_{ATP} Channels in VSMCs

Accumulating evidence has shown that K_{ATP} channels contribute to the maintenance of basal vascular tone in some vascular tissues, including mesenteric arteries *(11)* and coronary arteries *(12)*. Under pathophysiological conditions, K_{ATP} channels in these VSMCs can be activated, and the blood and oxygen supplies to the involved tissues are significantly affected. Examples of these pathophysiological conditions are hypoxia, ischemia, acidosis, and septic shock *(13)*.

K_{ATP} channels are regulated by intracellular ATP, adenosine 5-'diphosphate (ADP), or the ATP/ADP ratio, and by some vasoactive substances such as CGRP, adenosine *(14)*, or NO *(1)*. ADP and many other nucleoside diphosphates (NDPs) in the absence of Mg^{2+} also inhibited the activity of K_{ATP} channels. Low sensitivity to ATP and high sensitivity to NDP are characteristics of K_{ATP} channels in VSMCs. Therefore, K_{ATP} channels in VSMCs are often referred to as K_{NDP} channels. When binding to the pore-forming Kir6.x subunit, intracellular ATP inhibits channel opening (ligand action). By contrast, when ATP is associated with the sulfonylurea receptor (SUR) subunit of the K_{ATP} channel complex, it stimulates the channel *(15,16)*.

The most often used blocker for K_{ATP} channels is glibenclamide. This sulfonylurea drug has been widely used to block both plasmalemmal and mitochondrial K_{ATP} channels. To more specifically block mitochondrial K_{ATP} channels, 5-hydroxydecanoic acid is of choice whereas HMR-1098 is more selective in inhibiting plasmalemmal K_{ATP} channels. Among KCOs that stimulate K_{ATP} channels are pinacidil, cromakalim, and nicorandil *(8,9)*. Pinacidil is equally effective in opening plasmalemmal and mitochondrial K_{ATP} channels. Diazoxide offers a relative specific stimulation of mitochondrial K_{ATP} channels. P-1075, on the other hand, selectively opens plasmalemmal K_{ATP} channels. Different pharmacological sensitivities of plasmalemmal and mitochondrial K_{ATP} channels can be explained by the molecular composition of the K_{ATP} channel complex. K_{ATP} channels reconstituted with Kir6.x and SUR1 subunits are far more sensitive to glibenclamide-related inhibition than the reconstituted Kir6.x/SUR2A channels. SUR1 subunit binds glibenclamide with a dissociation constant (K_d) of about 1 nM, whereas SUR2A subunit has a K_d near 1.2 μM *(17)*. On the other hand, Isomoto et al. *(18)* indicated that SUR2B is also a low-affinity SUR. SUR1 proteins are identified in mitochondria of PC12 cells using Western blotting technique *(19)*. Differential responses of SUR subunits to sulfonylureas may lead to different degrees of channel blockade. KCOs appear also to act on SUR subunits (C-terminal end) of the K_{ATP} channel complex. The response of reconstituted K_{ATP} channels to either diazoxide or pinacidil is correlated with the presence of SUR subtypes. SUR1 and SUR2B C-termini, e.g., are homological, and channels formed from these subunits share similar responsiveness to diazoxide. The Kir6.2/SUR2A channel had essentially no response to diazoxide, and the C-terminus of SUR2A is very different from other SUR subunits.

2.2. Molecular Basis of K_{ATP} Channels in VSMCs

K_{ATP} channels are hetero-octamer complexes of four pore-forming subunits and four regulatory sulfonylurea-binding subunits. Kir6.1 and Kir6.2 belong to a class of inwardly rectifying K^+ channels with two membrane-spanning regions. Both the C- and N-termini of Kir6.1 and Kir6.2 are located inside the cell and are important for intracellular ATP binding and interactions with SUR subunits *(20,21)*. As the pore-forming subunit of the K_{ATP} channel complex, Kir6.1 or Kir6.2 dictates the potassium selectivity, inward rectification, and unitary conductance of the K_{ATP} channels. Whether the expression of Kir6.1 or Kir6.2 alone can elicit functional K_{ATP} channel currents has been debated *(22)*.

SURs are members of the ATP-binding cassette protein superfamily *(17,18)*. SURs have 17 putative transmembrane domains, an extracellular N-terminus, and an intracellular C-terminus *(17,18,23)*. K_{ATP} channel complex is assembled with a 1:1 tetrameric stoichiometry of Kir6.x and SUR subunits $(Kir6.x/SUR)_4$. Although Kir6.x and SUR subunits are structurally distinct, they have to physically interact with each other to constitute functional K_{ATP} channels *(24)*. The binding sites for sulfonylureas and KCOs are on SURs *(25)*.

Most of our knowledge about the tissue-type-specific expression of different K_{ATP} channel subunits is derived from detection of the transcripts of these subunits and from the pharmacological sensitivity of native K_{ATP} channels in different tissues. Thus, Kir6.2 and SUR1 are concluded to be mainly expressed in pancreatic β-cells, Kir6.2, and SUR2A in cardiomyocytes *(26)*, whereas Kir6.1 and SUR2B are functionally paired in VSMCs *(18)*. Different combinations of Kir6x and SURs yield tissue-specific K_{ATP} channels with different electrophysiological and pharmacological features. Kir6.2/SUR1 constitutes K_{ATP} channels in pancreatic β-cells and some neurons, as does Kir6.2/SUR2A in cardiac and skeletal muscles. Kir6.2/SUR2B is the K_{ATP} isoform in non-VSMCs and some neurons *(27)*. Functional K_{ATP} channel complex made of Kir6.1/SUR1 has been suggested in glial cells *(28)* and dentate gyrus granule cells *(29)*. Based on the pharmacological sensitivities of different combinations of K_{ATP} channel subunits to diazoxide/P-1075/glibenclamide/5-HD/HMR-1098, it has been concluded that the Kir6.1/SUR1 combination is the molecular makeup of mitochondrial K_{ATP} channels *(30)*. This notion is further supported by the identification of both Kir6.1 and SUR1 proteins in mitochondria of P12 cells *(19)*.

The Kir6.1/SUR2B combination may be specific for VSMCs since Kir6.1 confers the relative ATP insensitivity of native K_{ATP} channels in these cells (not inhibited by ATP at concentrations lower than 1 m*M*). We have detected the transcripts of Kir6.1, Kir6.2, SUR2B, and SUR1 in rat mesenteric artery smooth muscle cells (SMCs) *(31)*. Furthermore, we have cloned four K_{ATP} subunit genes from mesenteric artery SMCs and accordingly referred to them as rvKir6.1, rvKir6.2, rvSUR1, and rvSUR2B. Their GenBank accession nos. are AB043636, AB043638, AB045281, and AB052294, respectively. It is possible that VSMCs possess multiple types of K_{ATP} channels constructed by Kir6.1 with either SUR1 or SUR2B being the regulatory subunit. It is worth noting that a chimeric Kir6.1-Kir6.2 may also occur in native cells because a chimeric Kir6.1-Kir6.2 coexpressed with SUR2 in HEK-293 cells yields functional K_{ATP} channels *(32)*.

3. ROLE OF K_{ATP} CHANNELS IN H_2S-INDUCED VASORELAXATION

Modulation of K_{ATP} channel activities has been shown with many endogenous substances. Endothelin (ET) inhibits a 30-pS single K_{ATP} channel in porcine coronary artery SMCs *(33)*. Single K_{ATP} channel currents recorded from ventricular myocytes were also reversibly inhibited by ET-1. This effect of ET-1 was largely abolished in myocytes preincubated with pertussis toxin but mimicked by muscarinic receptor stimulation *(34)*. An increase in cellular ATP levels subsequent to the inhibition of adenylate cyclase activities through pertussis toxin–sensitive G-proteins coupled to ET-A receptors was speculated to underlie the inhibitory effect of ET-1 on K_{ATP} channels. NO is among the endogenous stimulators of K_{ATP} channels. By activating the cGMP pathway, NO stimulates K_{ATP} channels and hyperpolarizes membrane of SMCs from rabbit mesenteric arteries *(4)*. Another study showed that sodium nitroprusside (SNP), an NO donor, had no effect on K_{ATP} channel currents in porcine coronary artery SMCs *(35)*. The difference between these two studies has not been solved. Calcitonin gene-related peptide and atrial natriuretic factor also activated K_{ATP} channels in VSMCs. The former is mediated by a cAMP pathway *(35–37)* and the latter, particulate guanylyl cyclase *(38)*. Using the perforated patch-clamp technique, Wu et al. *(39)* recorded a whole-cell K_{ATP} channel current in retinal pericytes located on microvessels freshly isolated from adult rats. Dopamine activated this K_{ATP} channel via the cascade of D1 dopamine receptors, adenylate cyclase, and protein kinase A (PKA). It appears that all these reported endogenous modulators of K_{ATP} channels function through cognate membrane receptors to either change ATP metabolism or alter protein phosphorylation. A direct modulation of K_{ATP} channel protein structure and K_{ATP} channel complex configuration by endogenous substances has been much less clear in comparison to the chemical modification of K_{Ca} channels by NO (*see* Chapter 6) or CO (*see* Chapter 13). The recent progress of the direct interaction of K_{ATP} channels with H_2S is summarized in the following sections.

3.1. Mediation of Cardiovascular Effects of H_2S by K_{ATP} Channels

The importance of K_{ATP} channels in maintaining cardiovascular function can be seen from a recent study in which a bolus injection of pinacidil (2.8 μmol/kg) in anesthetized rats decreased mean blood pressure (BP) by 18 mm Hg *(40)*. If this effect of pinacidil represents the relaxation of peripheral resistance arteries by the opening of K_{ATP} channels, then a similar hypotensive effect of H_2S mediated by the stimulation of K_{ATP} channels would be logically deduced. Zhao et al. *(40)* recently reported that an intravenous bolus injection of H_2S at 2.8 and 14 μmol/kg of body wt transiently decreased the mean arterial BP of anesthetized rats by about 13 and 30 mmHg, respectively. This effect of H_2S was significantly reduced by greater than 80% after glibenclamide was injected into the animals 20 min before H_2S injection. In these anesthetized rats, a bolus intravenous or intraperitoneal injection of glibenclamide alone did not alter mean BP. Therefore, the basal activity of K_{ATP} channels appears not to be essential for the regulation of BP in the presence of many other compensatory vasorelaxant mechanisms. Once these channels are stimulated by H_2S or classical KCOs, significant vasorelaxation would occur and BP would be lowered.

Studies on isolated vascular tissues showed that H_2S induced a concentration-dependent relaxation of phenylephrine-precontracted rat aortic tissues (EC_{50} of 125 ± 14 μ*M*) *(40)*. The in vitro vasorelaxant effect of H_2S has been interpreted as the consequence of K_{ATP} channel activation by this gasotransmitter. First, the potency of H_2S to relax vascular tissues was greatly reduced when the tissues were precontracted with a high concentration of KCl. The maximum vascular relaxation induced by H_2S was 90 or 19% when the tissues were precontracted with 20 or 100 m*M* KCl, respectively *(40)*. It is likely that a high concentration of K^+ in extracellular solution alleviates the H_2S-induced vasorelaxation by reducing the driving force for K^+ outflow *(40)*. Teague et al. *(41)* also reported the inability of NaHS, a H_2S donor, to relax the guinea-pig ileum tissues precontracted with 60 m*M* KCl. Although this line of evidence seems to support the importance of K^+ conductance in determining the effect of NaHS, other observations do not favor the activation of K_{ATP} channels by NaHS in ileum tissues. The contractility of the guinea pig ileum tissues induced by 10 m*M* KCl was not changed by NaHS *(41)*. Whether the 10 m*M* KCl-induced contraction of these ileum tissues could be reduced by other known vasorelaxants such as SNP was not further investigated. Because these experiments were carried out by preincubating tissues with NaHS and then administrating the KCl stimulation, the effects of NaHS on the contractility of ileum tissues precontracted with KCl are unknown.

Second, the vasorelaxant effect of H_2S has been shown to be antagonized by a known K_{ATP} channel blocker, glibenclamide. Neither the K_{Ca} channel blockers charybdotoxin and iberiotoxin nor the Kv channel blocker 4-aminopyridine affected the relaxant effect of H_2S on rat aortic tissues *(40)*. Not only was the H_2S-induced relaxation of rat aortic tissues concentration dependently inhibited by glibenclamide, but also pinacidil produced vasorelaxation similar to H_2S. Wang et al. *(42)* recently reported that H_2S also relaxed peripheral resistant arteries of the rat mesenteric artery bed. They found that pinacidil mimicked, but glibenclamide suppressed, the vasorelaxant effect of H_2S on rat mesenteric artery beds. These observations on the vasorelaxant effects of H_2S on conduit and resistant arteries may be specific for vascular tissues. Teague et al. *(41)* observed that glibenclamide failed to alter the relaxant effect of NaHS on ileum tissues. Based on this observation, it is rationalized that the relaxant effect of H_2S on ileum tissues may be mediated by the cellular targets other than K_{ATP} channels. However, direct electrophysiological or molecular biological experiments are needed before this can be concluded because different isoforms of K_{ATP} channel subunits are expressed in vascular and visceral smooth muscle cells and these isoforms have different pharmacological sensitivities.

3.2. Stimulation of K_{ATP} Channels by H_2S in VSMCs

Direct evidence on the simulation of K_{ATP} channels by H_2S was derived from patch-clamp studies on single VSMCs. K_{ATP} channel currents in rat aortic SMCs were significantly and reversibly increased in amplitude by 300 μM H_2S *(40)*. Glibenclamide *per se* did not change the basal K_{ATP} channel current in aortic SMCs. However, the H_2S-stimulated K_{ATP} channel currents were significantly reduced by glibenclamide (5 μM) to the control level *(40)*.

Activation of K_{ATP} channels would lead to membrane hyperpolarization and consequent closure of voltage-dependent calcium channels. The effect of H_2S on resting membrane potential of different types of cells has been reported. In dorsal raphe serotonergic neurons, H_2S induced membrane hyperpolarization *(43)*. After exposing aortic SMCs to H_2S (300 μM), a 17-mV hyperpolarization occurred within 3 min of H_2S application. The hyperpolarizing effect of H_2S was abolished by glibenclamide although this sulfonylurea drug alone did not alter the resting membrane potential *(40)*.

Tang and Wang *(44)* also studied the modulation of K_{ATP} channels by endogenous H_2S using the whole-cell patch-clamp technique. Treatment of SMCs from rat mesenteric artery with D,L-propargylglycine (PPG) significantly decreased K_{ATP} currents by approx 36%. This effect of PPG is time dependent and reversible. Because PPG has been shown to reduce H_2S production by specifically inhibiting dystathionine γ-lyase (CSE), this study indicates that endogenous H_2S provides a basal stimulus for K_{ATP} channels. Another inhibitor of CSE, β-cyano-L-alanine, inhibited K_{ATP} currents by 51%. In line with the blockade of K_{ATP} channels, PPG treatment also resulted in membrane depolarization of the cultured rat mesenteric artery SMCs. In contrast, treatment with aminooxy acetate, a cystathionine β-synthase (CBS) inhibitor, did not alter K_{ATP} currents in rat mesenteric artery SMCs, further confirming the absence of CBS in these cells. The inhibition of H_2S-generating enzymes will lead to reduced production of H_2S as well as ammonium chloride and pyruvic acid. The role of the latter two end products in the effects of PPG and β-cyano-L-alanine were further investigated. Whereas exogenous H_2S significantly increased K_{ATP} currents, directly applying ammonium chloride or pyruvic acid did not affect K_{ATP} channel currents in these VSMCs. This study provides evidence that endogenously generated H_2S contributes significantly to the regulation of K_{ATP} channels in

VSMCs. The abnormal production of endogenous H_2S might be related to the pathogenesis of cardiovascular diseases such as hypertension and diabetes.

4. MECHANISMS FOR EFFECT OF H_2S ON K_{ATP} CHANNELS

4.1. Is the Effect of H_2S on K_{ATP} Channels Because of a Reduced ATP Production in VSCMs?

Suppressed ATP production by acute intoxication of H_2S has been known. Thus, lowered cellular ATP level would release K_{ATP} channel from inhibition. This might constitute one mechanism for the H_2S-induced stimulation of K_{ATP} channels. This hypothesis is not supported by the fast onset of effect of H_2S on vasorelaxation and K_{ATP} channel activation and the quick reversal of effect of H_2S after the removal of the gas unless ATP production is preciously synchronized with H_2S withdrawal. The technical nature of the whole-cell patch-clamp study also dictates the cellular ATP levels because the recording pipet contains a predetermined ATP concentration. It is hard to imagine that the reduced ATP production, if any, in the presence of H_2S would exert visible changes to the ATP level controlled by the dialyzing pipet solution. Zhao et al. *(40)* intentionally changed the ATP concentration of the pipet solution in their whole-cell studies. Although an increase in ATP concentration decreased the basal K_{ATP} currents, the effects of H_2S on K_{ATP} channels were not related to ATP concentrations.

4.2. Is the Effect of H_2S on K_{ATP} Channels Because of a Redox Reaction?

The activation of K_{ATP} channels by oxygen free radicals has been shown in guinea pig ventricular myocytes. Superoxide anions and hydrogen peroxide increased the opening of single K_{ATP} channels. Hydroxyl radicals induced an even greater activation of cardiac K_{ATP} channels *(45)*.

It has been proposed that reactive sulfur species (RSS) can be formed in vivo under conditions of oxidative stress. Among RSS are disulfide-*S*-oxides, sulfenic acids, and thiyl radicals *(46)*. The activities of RSS may lead to the altered redox status of biological thiols and disulfides. H_2S is a reductant *(47)*. It can reduce other substances and can be oxidized by oxygen. Our recent study on the isolated and in vitro perfused rat mesenteric artery bed showed that the vasorelaxant effect of H_2S was not affected by *N*-acetylcysteine, a potent free-radical scavenger *(42)*. Furthermore, superoxide dismutase and catalase did not alter the vasorelaxant effect of H_2S on isolated aortic tissues *(40)*. Whether the H_2S-increased K_{ATP} channel currents can be affected by free-radical scavengers or thiyl antioxidants has not been tested.

4.3. Is the Effect of H_2S on K_{ATP} Channels Because of Stimulation of Cyclic Adenosine Monophosphate Pathway?

Activation of the cyclic adenosine monophosphate (cAMP)/PKA pathway has been acknowledged as one major mechanism for the stimulation of K_{ATP} channels by many endogenous vasodilators. Falling into this category are calcitonin gene-related peptide, vasoactive intestinal polypeptide, prostacylin and adenosine *(13)*, and dopamine *(39)*. The blockade of K_{ATP} channels by ET, on the other hand, is at least partially mediated by inhibition of the cAMP pathway *(34)*.

Stimulation of the cAMP pathway by H_2S has been shown. Kimura *(48)* reported that H_2S enhanced the production of cAMP in primary cultures of brain cells, neuronal and glial cell lines, and *Xenopus* oocytes. Inhibition of adenylyl cyclase reduced the effect of H_2S on cAMP production. Therefore, the role of the cAMP pathway in the effect of H_2S on *N*-methyl-D-aspartate receptors is indicated.

Whether the vascular effects as well as the stimulation of K_{ATP} channels by H_2S are mediated by the cAMP pathway has not been completely established. In one study, the cAMP pathway was blocked with SQ22536 (100 μ*M*) *(49)* in aortic tissues to inhibit adenylyl cyclase. Subsequent application of H_2S still effectively induced vasorelaxation *(40)*. The cAMP production in the presence of H_2S has not been measured in VSMCs or cardiac myocytes.

4.4. Is the Effect of H_2S on K_{ATP} Channels Because of Stimulation of the cGMP Pathway?

K_{ATP} channels can be stimulated by an active cGMP pathway. This mechanism has been ascribed to underlie the effects of NO *(4)* and isosorbide dinitrate *(50)*. Our previous study showed that although 1H-[1,2,4]oxadiazolo[4,3,-a]quinoxalin-1-one (ODQ) specifically blocked the vasorelaxation induced by SNP, ODQ did not affect the relaxation of rat aortic tissues induced by H_2S *(40)*. This study indicates that at least the vasorelaxant effect of H_2S was not mediated by the cGMP pathway

4.5. Is the Effect of H_2S on K_{ATP} Channels Because of a Direct Modification of Sulfhydryl Groups of K_{ATP} Channel Subunits?

The actual mechanisms for the H_2S-enhanced K_{ATP} channel activity are still under investigation. Currently, two working hypotheses are being tested. A sulfuration model is first proposed. This mechanism is reminiscent of the phosphorylation of proteins except that a sulfate group, rather than a phosphate group, donated by H_2S or the consequently generated thiyl free radicals is linked to the free cysteine residues of targeted proteins. The disulfide bonds formed between the sulfate group and cysteine residues would alter the configuration of targeted protein, leading to functional changes. In the case of K_{ATP} channel proteins, sulfuration would result in the opening of channels. The number of free cysteine residues in K_{ATP} channel proteins and their transmembrane location become essential for this sulfuration model.

The second model predicts the reduction of disulfide bonds of K_{ATP} channel protein by H_2S *(51)*. Probing the interaction of disulfide-bonded cysteines with H_2S will be the key experiment for this model.

Both models involve a direct effect of H_2S on cysteine residues. Where are these targeted residues located? Because Kir6.x subunits are responsible for pore forming and SURx for drug binding, structural modulation of cysteines by H_2S would lead to different changes in K_{ATP} channel currents depending on whether these cysteines are located in Kir6.x or SURx.

5. CONCLUSION

The physiological importance of H_2S in regulating vascular tone has been demonstrated. This vasorelaxant effect is mostly mediated by the H_2S-induced activation of K_{ATP} channels in VSMCs *(40)*. A similar activation of K_{ATP} channels may also underlie the neuronal effect of H_2S *(43)*. H_2S appears to directly increase the activity of K_{ATP} channels without the involvement of known second messengers or free radicals. The structural modification of sulfhydryl groups of K_{ATP} channel subunits, Kir6.x or SURx, by H_2S is an appealing mechanism that speculates the formation of H_2S adduct to K_{ATP} channel complex. More direct evidence for this sulfuration mechanism is needed.

Opening of K_{ATP} channels hyperpolarizes membrane and reduces cell excitability. This represents an ubiquitous regulatory mechanism for excitable cells. Logically, our future research scope should be extended beyond VSMCs to include pancreatic β-cells, skeletal muscle cells, cardiomyocytes, and neurons. The production of endogenous H_2S and the stimulation of K_{ATP} channels by H_2S in these tissues would unmask novel mechanisms for the functional regulation of these tissues.

The effect of H_2S on single-channel currents of K_{ATP} channels has not been studied. At the single-channel level in a cell-free patch, one would be able to determine whether H_2S directly stimulates K_{ATP} channel proteins or the activation of another second messenger is mandatory. Moreover, the interaction of H_2S with different K_{ATP} channel subunits and with the specific amino acid residues of a given K_{ATP} channel subunit will also be handily elucidated using the single-channel recording technique. Native K_{ATP} channels in VSMCs or other types of excitable cells are heterogeneously assembled with different K_{ATP} channel subunits. The effect of H_2S on the selectively expressed K_{ATP} channel subunit(s) in the heterologous expression system should be investigated, which will assist in the identification of specific K_{ATP} channel subunit as the target of H_2S.

ACKNOWLEDGMENT

This work was supported by an operating grant from the Natural Sciences and Engineering Research Council of Canada.

REFERENCES

1. Quayle JM, Nelson MT, Standen NB. ATP-sensitive and inwardly rectifying potassium channels in smooth muscle. Physiol Rev 1997;77:1165–1232.
2. Clapp LH, Gurney AM. ATP-sensitive K^+ channels regulate resting membrane potential of pulmonary arterial smooth muscle cells. Am J Physiol 1992;262:H916–H920.
3. Davie CS, Kubo M, Standen NB. Potassium channel activation and relaxation by nicorandil in rat small mesenteric arteries. Br J Pharmacol 1998;125:1715–1725.
4. Murphy ME, Brayden JE. Nitric oxide hyperpolarizes rabbit mesenteric arteries via ATP-sensitive potassium channels. J Physiol 1995;486:47–58.
5. Noack T, Edwards G, Deitmer R, et al. Potassium channel modulation in rat portal verin by ATP depletion: a comparison with the effects of levcromakalim (BRL 38227). Br J Pharmacol 1992;107:945–955.
6. Russ U, Metzger F, Kickenweiz E, et al. Binding and effects of K_{ATP} channel openers in the vascular smooth muscle cell line, A10. Br J Pharmacol 1997;122:1119–1126.
7. Xu X-P, Lee KS. Characterization of the ATP-inhibited K^+ current in canine coronary smooth muscle cells. Pflügers Arch 1994;427:110–120.
8. Zhang H-L, Bolton TB. Activation by intracellular GDP, metabolic inhibition and pinacidil of a glibenclamide-sensitive K-channel in smooth muscle cells of rat mesenteric artery. Br J Pharmacol 1995;114:662–672.
9. Standen NB, Quayle JM, Davies NW, et al. Hyperpolarizing vasodilators activate ATP-sensitive K^+ channels in arterial smooth muscle. Science 1989;245:177–180.
10. Furspan PB, Webb RC. Decreased ATP-sensitivity of a K^+ channel and enhanced vascular smooth muscle relaxation in genetically hypertensive rats. J Hypertens 1993,11:1067–1072.
11. Quayle JM, Bonev AD, Brayden JE, et al. Pharmacology of ATP-sensitive K^+ currents in smooth muscle cells from rabbit mesenteric artery. Am J Physiol 1995;269:C1112–1118.
12. Samaha FF, Heineman FW, Ince C, et al. ATP-sensitive potassium channel is essential to maintain basal coronary vascular tone in vivo. Am J Physiol 1992;262:C1220–C1227.
13. Brayden JE. Functional roles of K_{ATP} channels in vascular smooth muscle. Clin Exp Pharmacol Physiol 2002;29:312–316.

14. Brayden JE. Potassium channels in vascular smooth muscle. Clin Exp Pharmacol Physiol 1996;23: 1069–1076.
15. Babenko AP, Gonzalez G, Bryan J. Pharmaco-topology of sulfonylurea receptors: separate domains of the regulatory subunits of K_{ATP} channel isoforms are required for selective interaction with K^+ channel openers. J Biol Chem 2000;275:717–720.
16. Yokoshiki H, Sunagawa M, Seki T, et al. ATP-sensitive K^+ channels in pancreatic, cardiac, and vascular smooth muscle cells. Am J Physiol 1998 ; 274:C25–37.
17. Inagaki N, Gonoi T, Clement JP 4th, et al. Reconstitution of IK_{ATP}: an inward rectifier subunit plus the sulfonylurea receptor. Science 1995;270:1166–1170.
18. Isomoto S, Kondo C, Yamada M, et al. A novel sulfonylurea receptor forms with BIR (Kir6.2) a smooth muscle type ATP-sensitive K^+ channel. J Biol Chem 1996;271:24,321–24,324.
19. Tai KK, McCrossan ZA, Abbott GW. Activation of mitochondrial ATP-sensitive potassium channels increases cell viability against rotenone-induced cell death. J Neurochem 2003;84:1193–1200.
20. Proks P, Gribble FM, Adhikari R, et al. Involvement of the N-terminus of Kir6.2 in the inhibition of the K_{ATP} channel by ATP. J Physiol 1999;514:19–25.
21. Tucker SJ, Gribble FM, Zhao C, et al. Truncation of Kir6.2 produces ATP-sensitive K^+ channels in the absence of the sulphonylurea receptor. Nature 1997;387:179–181.
22. Gribble FM, Ashfield R, Ammala C, et al. Properties of cloned ATP-sensitive K^+ currents expressed in Xenopus oocytes, J Physiol (Lond) 1997;498:87–98.
23. Raab-Graham KF, Cirilo LJ, Boettcher AA, et al. Membrane topology of the amino-terminal region of the sulfonylurea receptor. J Biol Chem 1999;274:29,122–29,129.
24. Lorenz E, Alekseev AE, Krapivinsky GB, et al. Evidence for direct physical association between a K^+ channel (Kir6.2) and an ATP-binding cassette protein (SUR1) which affects cellular distribution and kinetic behavior of an ATP-sensitive K^+ channel. Mol Cell Biol 1998;18:1652–1659.
25. Ashcroft FM, Gribble FM. New windows on the mechanism of action of K_{ATP} channel openers. Trends in Pharmacol Sci 2000;21:439–445.
26. Inagaki N, Gonoi T, Clement JP, et al. A family of sulfonylurea receptors determines the pharmacological properties of ATP-sensitive K^+ channels. Neuron 1996;16:1011–1017.
27. Liss B, Bruns R, Roeper J. Alternative sulfonylrea receptor expression defines metabolic sensitivity of K_{ATP} channels in dopaminergic midbrain neurons. EMBO J 1999;18:833–846.
28. Skatchkov SN, Rojas L, Eaton MJ, et al. Functional expression of Kir 6.1/SUR1–K_{ATP} channels in frog retinal Muller glial cells. Glia 2002;38:256–267.
29. Pelletier MR, Pahapill PA, Pennefather PS, et al. Analysis of single K_{ATP} channels in mammalian dentate gyrus granule cells. J Neurophysiol 2000;84:2291–2301.
30. Liu Y, Ren G, O'Rourke B, et al. Pharmacological comparison of native mitochondrial K_{ATP} channels with molecularly defined surface K_{ATP} channels. Mol Pharmacol 2001;59:225–230.
31. Cao K, Tang G, Hu DH, et al. The expression and molecular identification of the ATP-sensitive K^+ (K_{ATP}) channel subunits in rat mesenteric artery smooth muscles. Biochem Biophys Res Commun 2002;296:463–469.
32. Cui Y, Giblin JP, Clapp LH, et al. A mechanism for ATP-sensitive potassium channel diversity: functional co-assembly of two pore-forming subunits. Proc Natl Acad Sci USA 2001;98:729–734.
33. Miyoshi Y, Nakaya Y, Wakatsuki T, et al. Endothelin blocks ATP-sensitive K^+-channels and depolarises smooth muscle cells of porcine coronary artery. Circ Res 1992;70:612–616.
34. Watanuki M, Horie M, Tsuchiya K, et al. Endothelin-1 inhibition of cardiac ATP-sensitive K^+ channels via pertussis-toxin-sensitive G-proteins. Cardiovasc Res 1997;33:123–130.
35. Wellman GC, Quayle JM, Standen NB. ATP-sensitive K^+ channel activation by calcitonin gene-related peptide and protein kinase A in pig coronary arterial smooth muscle. J Physiol (Lond) 1998;507: 117–129.
36. Reslerova M, Loutzenhiser R. Renal microvascular actions of calcitonin gene-related peptide. Am J Physiol 1998;274:F1078–F1085.
37. Szilvassy J, Jancso G, Ferdinandy P. Mechanisms of vasodilation by cochlear nerve stimulation: role of calcitonin gene-related peptide. Pharmacol Res 1999;39:217–220.
38. Kubo M, Nakaya Y, Matsuoka S, et al. Atrial natriuretic factor and isosorbide dinitrate modulate the gating of ATP-sensitive K^+ channels in cultured vascular smooth muscle cells. Circ Res 1994;74:471–476.
39. Wu DM, Kawamura H, Li Q, et al. Dopamine activates ATP-sensitive K^+ currents in rat retinal pericytes. Vis Neurosci 2001;18:935–940.
40. Zhao W, Zhang J, Lu Y, et al. H_2S is an endogenous K_{ATP} channel opener in vascular smooth muscle cells. EMBO J 2001;20:6008–6016.

41. Teague B, Asiedu S, Moore PK. The smooth muscle relaxant effect of hydrogen sulphide in vitro: evidence for a physiological role to control intestinal contractility. Br J Pharmacol 2002;137:139–145.
42. Wang R, Cheng Y, Ndisang JF. Potent vasorelaxant effect of hydrogen sulfide on rat mesenteric artery beds. Circulation 2003;108:IV–165 (abstract).
43. Kombian SB, Reiffenstein RJ, Colmers WF. The actions of hydrogen sulphide on dorsal raphe serotonergic neurons in vitro. J Neurophysiol 1993;70:81–96.
44. Tang G, Wang R. Effects of endogenous H_2S on K_{ATP} Channels and membrane potentials in vascular SMCs from rat mesenteric arteries. Can J Cardiol 2002;18(Suppl B):108B (abstract).
45. Tokube K, Kiyosue T, Arita M. Effects of hydroxyl radicals on K_{ATP} channels in guinea-pig ventricular myocytes. Pflügers Arch 1998;437:155–157.
46. Giles GI, Tasker KM, Jacob C. Hypothesis: the role of reactive sulfur species in oxidative stress. Free Radic Biol Med 2001;31:1279–1283.
47. Kim C, Zhou Q, Deng B, et al. Chromium (VI) reduction by hydrogen sulfide in aqueous media: stoichiometry and kinetics. Environ Sci Technol 2001;35:2219–2225.
48. Kimura H. Hydrogen sulfide induces cyclic AMP and modulates the NMDA receptor. Biochem Biophys Res Commun 2000;267:129–133.
49. Talpain E, Armstrong RA, Coleman RA, et al. Characterization of the PGE receptor subtype mediating inhibition of superoxide production in human neutrophils. Br J Pharmacol 1995;114:1459–1465.
50. Kubo M, Nakaya Y, Matsuoka S, et al. Atrial natriuretic factor and isosorbide dinitrate modulate the gating of ATP-sensitive K^+ channels in cultured vascular smooth muscle cells. Circ Res 1994;74: 471–476.
51. Warenycia MW, Steele JA, Karpinski E, et al. Hydrogen sulfide in combination with taurine or cysteic acid reversibly abolishes sodium currents in neuroblastoma cells. Neurotoxicology 1989;10:191–199.

V

Gasotransmitters, Other Gaseous Molecules, and Cell Metabolism

22 Gasotransmitters as a Novel Class of Metabolic Regulators

Nitric Oxide, Carbon Monoxide, and Nitrous Oxide

Misato Kashiba

CONTENTS

SUMMARY

Gasotransmitters constitute a unique class of biomaterials that are indispensable for maintaining homeostasis of biological systems. Their properties for easy penetration through biomembrane as well as through a cavity of macromolecular structure allow them to access the inner space of receptor proteins and modulate their functions. In addition to their well-known roles as signal transducers, gasotransmitters have been reported to inhibit or activate several metabolic enzymes, thereby controlling metabolism. This chapter summarizes recent information about the roles of gasotransmitters, especially nitric oxide and carbon monoxide, in metabolism. The roles of other potential gasotransmitters are also discussed.

Key Words: Metabolism; oxygen, carbon monoxide; nitric oxide; nitrous oxide; glycolysis; urea cycle; citric acid cycle; amino acid metabolism.

From: *Signal Transduction and the Gasotransmitters: NO, CO, and H_2S in Biology and Medicine* Edited by: Rui Wang © Humana Press Inc., Totowa, NJ

1. INTRODUCTION

Metabolism is the overall process through which living systems acquire free energy to carry out various functions. Metabolic pathways are a series of connected enzymatic reactions that produce specific products. There are more than 2000 known metabolic reactions, each catalyzed by distinct enzymes. Gasotransmitters, a novel class of endogenous gaseous molecules *(1)* that have been discussed in great detail in this book, control metabolism through the regulation of metabolic enzymes by transcriptional and posttranslational modifications.

Among the gases used in the body, oxygen (O_2) has been studied extensively regarding mechanisms for its transport and utilization and metabolism in a quantitative manner. In mammals, approx 95% of molecular O_2 consumed in the body is used as a substrate for cytochrome-*c* oxidase. O_2 contributes to generation of the mitochondrial inner membrane potential, and thus to oxidative phosphorylation. Hypoxia leads to metabolic adaptation by inducing a shift from oxidative to glycolytic pathways. The transcriptional regulator hypoxia-inducible factor (HIF)-1 is an essential mediator of O_2 homeostasis. When O_2 concentration is decreased, HIF-1 exhibits nuclear translocation and binds to the DNA sequence 5'-RCGTG-3' and increases the expression of glycolytic genes such as hexokinase and glyceraldehude-3-phosphate dehydrogenase.

Nitric oxide (NO) is a free-radical species synthesized from oxygen and L-arginine by NO synthase (NOS). This enzyme is located in cytoplasm. Because the enzyme converts L-arginine not only to form NO but also to generate citruline, the pathway constitutes a cytoplasmic shunt for the urea cycle. Because NO is highly reactive, it involves diverse biological actions such as vascular relaxation and neurotransmission. In addition, NO has been reported to inactivate aconitases in vivo. Nitrous oxide (N_2O) has been used clinically as a general anesthetic.

Carbon monoxide (CO) is produced by oxidative degradation of protoheme IX through the action of heme oxygenase (HO). N_2O and CO have been reported to inactivate amino acid-metabolizing enzymes, methionine synthase and cystathionine β-synthase, respectively. This chapter provides an overview of oxidative fuel metabolism and discuss the effect of gas molecules or gasotransmitters on metabolism, which is summarized in **Fig. 1**.

2. GLYCOLYSIS: HYPOXIA AND REGULATION OF GLYCOLYTIC GENE EXPRESSION

Every organism has the ability to sense a reduction in O_2 concentrations. Hypoxia causes both acute and chronic responses. Acute responses involve posttranslational modification of proteins through redox and/or phosphorylation-dephosphorylation mechanisms that occur in seconds to minutes, whereas chronic responses involve alterations in gene expression over minutes or hours *(1)*. Several transcriptional factors have been known to participate in chronic adaptation against hypoxia and to stimulate expression of specific gene products for adaptation to hypoxic or oxidative stress. In mammals, the transcriptional regulator HIF-1 is an essential mediator of O_2 homeostasis *(2,3)*. Under normoxic conditions, levels of HIF-1 are regulated by removal of the HIF-1α subunit through ubiquination and proteasomal degradation. Hypoxia suppresses ubiquination of HIF-1α, and prevents its destruction, facilitating its translocation into the nucleus for binding to hypoxia-responsive elements in promoter regions of varied genes such as erythropoietin *(4)*, vascular endothelial growth factor *(5)*, and HO-1 *(6)*.

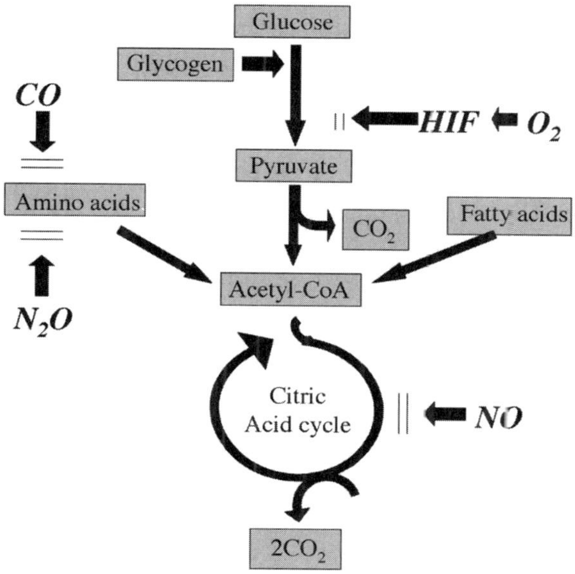

Fig. 1. Key metabolism and the effect of gasotransmitters.

Hypoxia leads to anaerobic adaptation of the cell metabolism by inducing a shift from oxidative to glycolytic pathways. Hypoxia-responsive elements have been identified in a number of genes involved in glycolysis, as shown in **Fig. 2** *(7)*. Expression studies of HIF-1α deficient embryonic stem cells have revealed downregulation of gene expression of glucose transporters and glycolytic enzymes *(8)*.

3. CITRIC ACID CYCLE: INACTIVATION OF ACONITASE BY NO AND ITS DERIVATIVES

Because the free radical gas NO is highly reactive, it reacts with various biomolecules. NO forms dinitrosyl complexes with iron and binds easily with proteins that contain the prosthetic heme group. Soluble guanylate cyclase (sGC) is such an enzyme reacting with NO through its prosthetic group. Binding of NO to sGC activates this enzyme and increases cyclic guanosine monophosphate (cGMP) generation, which largely explains the mechanisms of this mediator's action on vascular smooth muscle. In addition to the prosthetic heme of proteins, NO is known to interact with iron in nonheme protein *(9,10)*. NO reacts with [4Fe-4S] clusters of aconitase and inactivates aconitase. In mammals, the cytoplasmic aconitase serves as an mRNA-binding regulator of iron homeostasis and the mitochondrial aconitase as a catalyst of the energy-yielding reactions of the citric acid cycle *(11,12)*. Mitochondrial aconitase catalyzes the reversible isomerization of citrate and isocitrate, with *cis*-aconitase as an intermediate (**Fig. 3**). Aconitase belongs to the family of iron-sulfur-containing dehydrates whose activities depend on the redox state of the cubane [4Fe-4S] cluster *(13)*. Cysteine residues around the iron cluster are also important for optimal activity *(14)*. NO originally was reported to inactivate aconitase *(15–17)*. However, NO reacts with superoxide anion at a diffusion-limited rate to yield peoxynitrite, which is markedly more potent and efficient than NO in inhibiting aconitase *(18,19)*. Increased synthesis of NO via the upregulation of the inducible NOS is believed

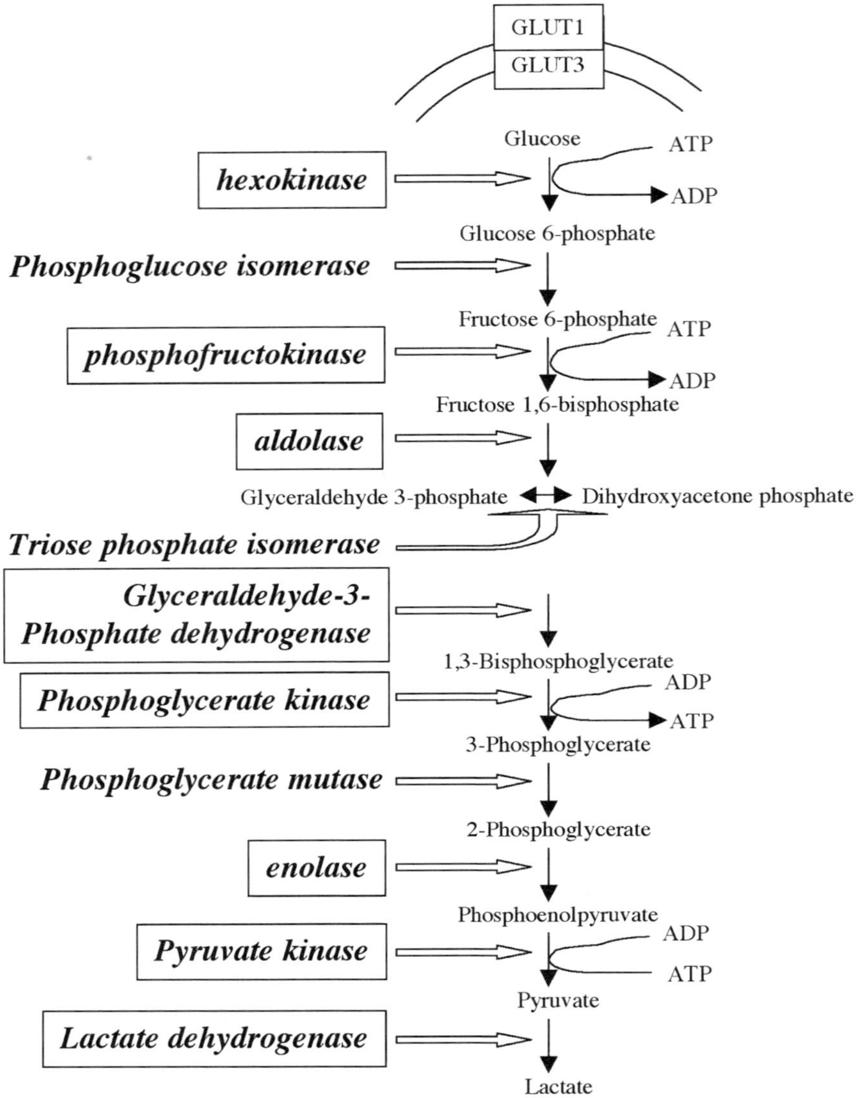

Fig. 2. Transcriptional regulation of glycolysis by HIF-1. The transport of glucose across the membrane and glycolytic enzymes that are transcriptionaly regulated by HIF-1 are show in blue. ATP, adenosine triphosphate; ADP, adenosine 5'-diphosphate.

to play a key role in acute rejection after solid organ transplantation. Pieper et al. *(20)* reported that the activity of aconitase was inhibited by the acute cardiac allograft through the modification of the Fe-S cluster of this protein, as judged by an ESR spectrum. They speculated that inactivation of aconitase by NO may contribute to alloimmune rejection.

4. MODULATION OF SULFUR-CONTAINING AMINO ACID METABOLISM BY CO AND NO

CO is produced by oxidative degradation of protoheme IX through the action of HO (EC 1. 14. 99. 3) *(21)*. The enzyme decomposes protoheme IX by oxidative cleavage of

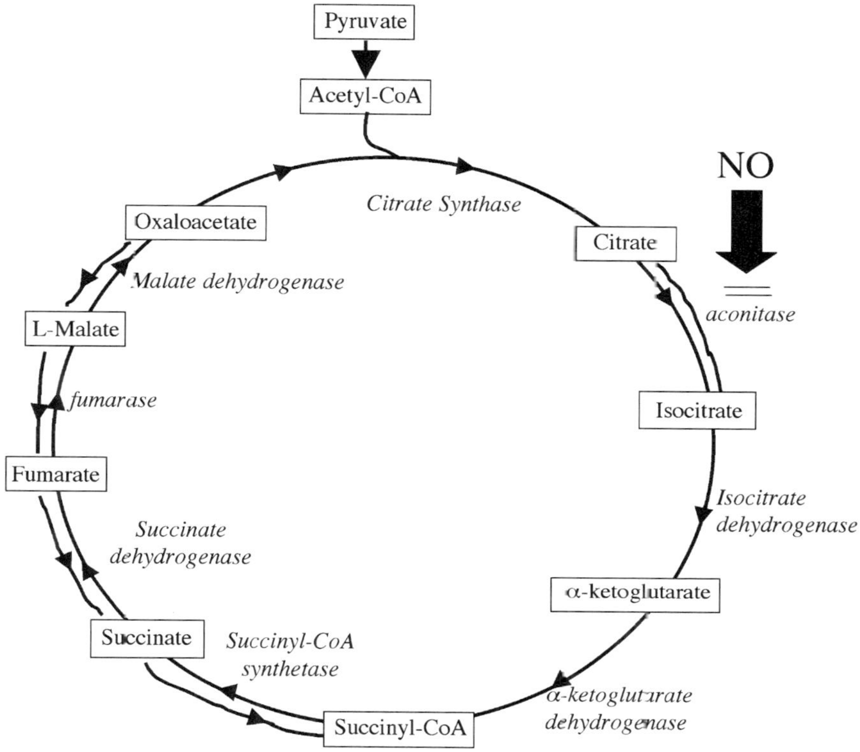

Fig. 3. Reaction of citric acid cycle. NO has been reported to inhibit the activity of aconitase.

its α-methene bridge and generates biliverdin-IXα and divalent iron together with this gas. Biliverdin-IXα is then converted to bilirubin-IXα through the reaction of biliverdin reductase. In mammals, two forms of the HO isoenzymes are responsible for oxidative degradation of heme: HO-1 and HO-2. HO-1 is inducible in response to stressors, while HO-2 is constitutive *(22,23)*. Various stressors such as cytokines, hypoxia, reactive oxygen species, and exposure to heme and heavy metals serve as inducers of HO-1. Detailed mechanisms for transcriptional regulation of HO-1 expression are summarized in a previous review article *(24)*.

Because of its high and reversible properties to ferroheme protein, CO can use sGC as a receptor protein to execute its signaling. sGC is a heme protein and a heterodimeric enzyme converting guanosine 5'-triphosphate to cGMP. It has been demonstrated that iron protoporphyrin is involved in the enzyme activity. There is cogent evidence that NO binding to Fe of the prosthetic heme causes a break in the proximal His-Fe bond, forming a five-coordinated nitrosyl heme complex that is thought to result in conformational changes and a 100-fold increase in cGMP generation. CO also shares a high affinity to the heme iron of the sGC; however, it forms a six-coordinated heme complex with the His-Fe bond remaining intact, presumably inducing smaller conformational changes than observed in the NO-binding to the enzyme. Thus, the potency of CO to activate sGC is far less than that of NO. Based on these data in vitro, Imai et al. *(25)* developed new transgenic mice in which HO-1 gene was preferentially expressed. These animals exhibited systemic hypertension and reduced vasodilatory responses to exogenously applied nitrovasodilators *(25)*. Because these animals maintain functional integrity of sGC, the mechanisms appear to involve a competition between NO and CO at the prosthetic heme; CO serves as an agonist for sGC when local NO is negligible, while acting as an antagonist for sGC when NO is sufficiently generated *in situ*. Kajimura et al. *(26)* reported that under condition in which housekeeping levels of CO were suppressed by the administration of HO inhibitor, retinal cells exhibited enhancement of NO-dependent activation of sGC. These results suggest that different gases can interact with the same receptor protein and modulate its function in vivo.

Recent in vitro experimental evidence shed light on another ferroheme protein, CBS (EC 4.2.1.22), as receptor molecule for CO. CBS is one of two key mammalian enzymes that metabolize cellular homocysteine. Transmethylation, catalyzed by methionine synthase, converts it to methionine, whereas transsulfuration, catalyzed by CBS, yields cystathionine (*see* **Fig. 4**). Loss of CBS activity causes homocystinuria, an autosomal recessive disease characterized by mental retardation, skeletal abnormalities, and vascular disorders with severe thromboembolic complications *(27–29)*. More than 100 mutations have now been described in this gene *(30)*.

CBS contains heme protein. Biophysical approaches including electron paramagnetic resonance spectroscopy *(31–33)*, Raman spectroscopy *(34)*, and extended X-ray absorption fine structure spectroscopy *(33)* revealed that the axial ligand in mammalian CBS is histidine and cysteine. The axial ligands are C52 and H65 in the human sequence *(35)*. The heme in human CBS resembles that found in the bacterial CO-sensing transcriptional activator CooA *(36,37)*. The CO-CooA complex serves as a transcriptional factor that stimulates bacterial replication. Heme in CooA and CBS is six coordinate and low spin in both ferric and ferrous states.

Taoka et al. *(38)* examined the binding of CO to ferrous CBS. The reduction of CBS resulted in a red shift of the Soret band from 428 to 450 nm. The addition of CO to reduced enzyme resulted in conversion of a Soret absorption maximum at 422 nm with an isosbestic point at 434 nm. Binding of CO to ferrous CBS resulted in inhibition of enzyme activity. In a steady-state assay in vitro, complete loss of enzyme activity was observed at a CO concentration of 60 μM, and yielded a K_i of 5.6 μM. NO also binds CBS but is unlikely to serve as an inhibitor: the K_i of the enzyme was estimated to be 320 μM. Whether such a distinct feature of the enzyme activity between the two gases could imply physiological significance of CBS as a CO-specific receptor in vivo remains to be investigated *(39)*.

N_2O has been used clinically as a general anesthetic. The density of the gas is 1.5 times that of air. This gas molecule is stable and rather inert chemically at 37°C. N_2O is formed by both enzymatic and nonenzymatic reduction of NO. In vitro analysis revealed that N_2O is formed by the reaction between NO and thiol *(40)*. Hyun et al. *(41)* reported that NO is reduced to N_2O by the cytosolic fraction of hepatocytes, suggesting the possible formation of this gas in mammalian cells. In bacteria, N_2O is produced during denitrification. NO is reduced to N_2O by NO reductase.

N_2O has been used clinically as a general anesthetic for more than a century. Low potency, low solubility, and rapid induction as well as rapid recovery account for the widespread acceptance of N_2O as one of the safest and least toxic of the inhaled anesthetics *(42,43)*. Jevtovic-Todorovic et al. *(44)* reported that N_2O inhibits both ionic currents and excitotoxic neurodegeneration mediated through the *N*-methyl-D-aspartate receptor. Although N_2O neither serves as a ligand to heme iron nor reacts with thiols, it is detectable

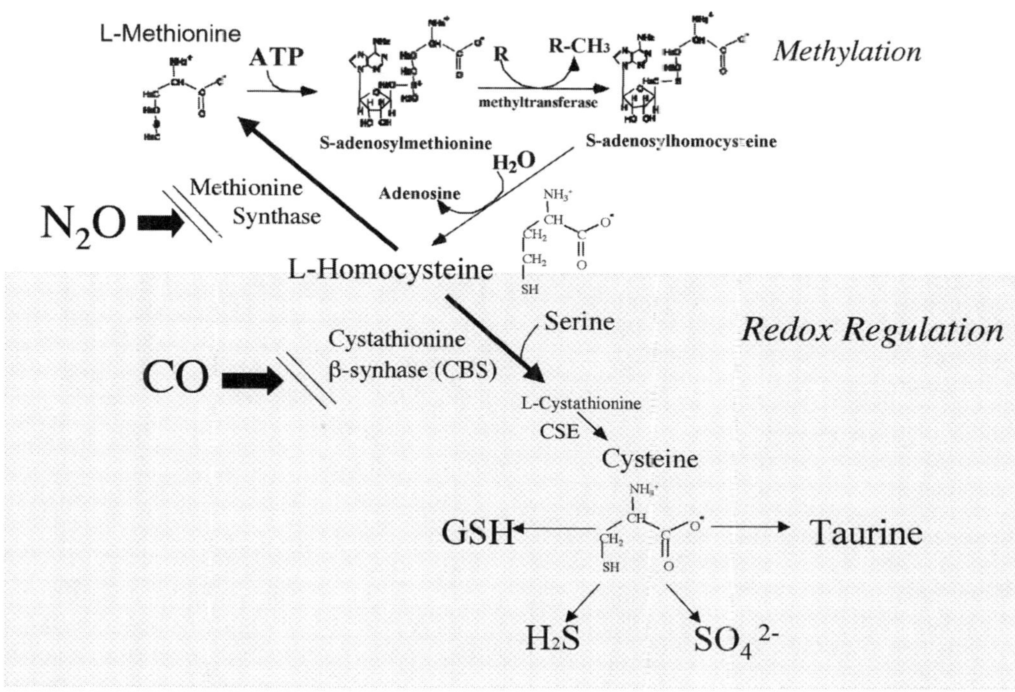

Fig. 4. Sulfur-containing amino acid metabolism. CO has ben reported to inhibit CBS and N_2O to inhibit methionine synthase. CSE, cystathionine γ-lyase.

at the inner structure of heme protein such as hemoglobin, myoglobin, and cytochrome-*c* oxidase. The ability of N_2O to alter the structure and function of the heme proteins was shown by shifts in infrared spectra of cytokine thiols of Hb55 and by partial and reversible inhibition of cytochrome-*c* oxidase. Precise mechanisms for this noncovalent binding between N_2O and the proteins and its link to biological events need further investigation.

N_2O inactivates the cobalamin-dependent enzyme methionine synthase (5-methyltetrahydrofolate-homocysteine methyltransferase, EC 2.1.1.13), causing a block in the remethylation of homocysteine to methionine *(45)*. Methionine synthase catalyzes a folate-dependent reaction, in which 5-methyltetrahydrofolate functions as methyl donor, thereby converting homocysteine to methionine (**Fig. 4**). Cob(I)alamin serves as cofactor in this reaction. N_2O has been reported to inhibit methionine synthase probably through the oxidation of enzyme-bound cob(I)alamin formed during the catalytic cycle *(46)*. Because methionine synthase is one of the two homocysteine-converting enzymes (**Fig. 4**), inhibition of methionine synthase by N_2O causes an increase in the plasma levels of homocysteine. High levels of total homocysteine after exposure to N_2O were described in leukemia patients *(47)*, and in surgical patients undergoing otolaryngology surgery *(48)* or neurosurgery *(49)*. Levels of other sulfur-containing amino acids are also altered by the administration of N_2O *(50,51)*. This effect of N_2O may account for its diverse biological effects, including the megaloblastic changes in human bone marrow *(50,52)* and the antileukemic effect reported in patients *(53,54)* and experimental animals *(55)*.

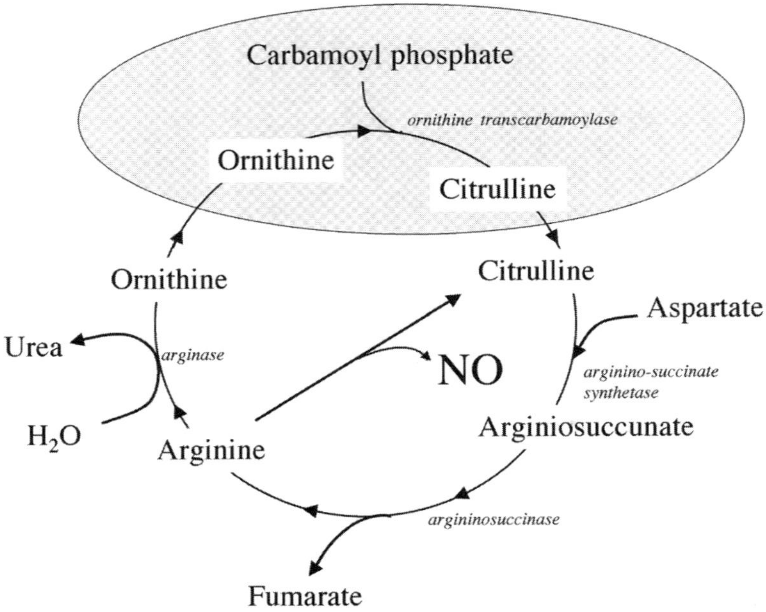

Fig. 5. Urea cycle and its bypass by NOS.

5. UREA CYCLE: EFFECT OF NOS INDUCTION

The free-radical gas NO is synthesized from oxygen and L-arginine by NOS (EC1.14.13.39). In the liver of ureotelic animals, arginine is synthesized from NH_3, CO_2, ornithine, and aspartate by the four enzymes of the urea cycle—carbamylphosphate synthase I, ornithine transcarbamylase, ininosuccinate synthetase, and argininosuccinate lyase—and is hydrolyzed by arginase I to urea and ornithine, forming the cycle (**Fig. 5**). In the presence of various inducers, such as endotoxin shock, inducible NOS (iNOS) is induced and a large amount of NO is produced *(56)*. When NO is synthesized from arginine by the iNOS reaction, citrulline, an intermediate of the urea cycle, is produced. Thus, the urea cycle is bypassed by the NOS reaction. When lipopolysaccharide was administrated to rats and iNOS mRNA was expressed, urea cycle enzymes except for ornithine transcarbamylase were markedly decreased, probably to maintain cellular arginine for increased synthesis of proteins that are critical in endotoxin shock *(57)*.

6. CONCLUSION

In addition to the known gasotransmitters mentioned in this chapter, other gases generated in the body such as CO_2 and SO_2^- could also play physiological roles but with the receptors as the discernible molecular entity remaining unknown. Understanding the whole spectrum of gasotransmitters and their physiological functions is desired. In addition to the six criteria for characterizing gasotransmitters (*see* Chapter 1), additional facts about this class of gaseous molecules merit consideration. First, gases can exert their biological actions through interaction with proteins in multiple ways. These interactions involve covalent binding of gases to prosthetic metal complexes in receptor proteins, their noncovalent binding to the critical region for regulation of the protein function, and their space occupancy by them in and around protein structure that leads to reduced accessibility of other gases to the region. Second, different gases that share a similar chemical structure not only can exert comparable biological actions but often can compete with and are antagonists with each other. The gas-mediated regulatory mechanisms for classic metabolic pathways deserve further studies provided that they shed light on a novel close link among different metabolic pathways, which apparently stand in distal positions to each other in the classic metabolic map.

REFERENCES

1. Wang R. Two's company, three's a crowd—can H_2S be the third endogenous gaseous transmitter? FASEB J 2002;16:1792–1798.
2. Semenza GL. Perspectives on oxygen sensing. Cell 1999;98:281–284.
3. Semenza G. Signal transduction to hypoxia-inducible factor 1. Biochem Pharmacol 2002;64:993–998.
4. Jiang BH, Rue E, Wang GL, Roe R, et al. Dimerization, DNA binding, and transactivation properties of hypoxia-inducible factor 1. J Biol Chem 1996;271:17,771–17,778.
5. Damert A, Ikeda E, Risau W. Activator-protein-1 binding potentiates the hypoxia-inducible factor-1-mediated hypoxia-induced transcriptional activation of vascular-endothelial growth factor expression in C6 glioma cells. Biochem J 1997;327(pt 2):419–423.
6. Lee PJ, Jiang BH, Chin BY, et al. Hypoxia-inducible factor-1 mediates transcriptional activation of the heme oxygenase-1 gene in response to hypoxia. J Biol Chem 1997;272:5375–5381.
7. Semenza GL. Hypoxia, clonal selection, and the role of HIF-1 in tumor progression. Crit Rev Biochem Mol Biol 2000;35:71–103.
8. Iyer NV, Kotch LE, Agani F, et al. Cellular and developmental control of O2 homeostasis by hypoxia-inducible factor 1 alpha. Genes Dev 1998;12:149–162.
9. Drapier JC, Hibbs JB,Jr. Murine cytotoxic activated macrophages inhibit aconitase in tumor cells. Inhibition involves the iron-sulfur prosthetic group and is reversible. J Clin Invest 1986;78:790–797.
10. Lancaster JR Jr., Hibbs JB Jr. EPR demonstration of iron-nitrosyl complex formation by cytotoxic activated macrophages. Proc Natl Acad Sci USA 1990;87:1223–1227.
11. Klausner RD, Rouault TA, Harford JB. Regulating the fate of mRNA: the control of cellular iron metabolism. Cell 1993;72:19–28.
12. Philpott CC, Klausner RD, Rouault TA. The bifunctional iron-responsive element binding protein/cytosolic aconitase: the role of active-site residues in ligand binding and regulation. Proc Natl Acad Sci USA 1994;91:7321–7325.
13. Ramsay RR, Dreyer JL, Schloss JV, et al. Relationship of the oxidation state of the iron-sulfur cluster of aconitase to activity and substrate binding. Biochemistry 1981;20:7476–7482.
14. Hirling H, Henderson BR, Kuhn LC. Mutational analysis of the [4Fe-4S]-cluster converting iron regulatory factor from its RNA-binding form to cytoplasmic aconitase. EMBO J 1994;13:453–561.
15. Stadler J, Curran RD, Ochoa JB, et al. Effect of endogenous nitric oxide on mitochondrial respiration of rat hepatocytes in vitro and in vivo. Arch Surg 1991;126:186–191.
16. Welsh N, Eizirik DL, Bendtzen K, et al. Interleukin-1 beta-induced nitric oxide production in isolated rat pancreatic islets requires gene transcription and may lead to inhibition of the Krebs cycle enzyme aconitase. Endocrinology 1991;129:3167–3173.
17. Drapier JC, Hibbs JB Jr. Aconitases: a class of metalloproteins highly sensitive to nitric oxide synthesis. Methods Enzymol 1996;269:26–36.
18. Hausladen A, Fridovich I. Superoxide and peroxynitrite inactivate aconitases, but nitric oxide does not. J Biol Chem 1994;269:29,405–29,408.
19. Castro L, Rodriguez M, Radi R. Aconitase is readily inactivated by peroxynitrite, but not by its precursor, nitric oxide. J Biol Chem 1994;269:29,409–29,415.
20. Pieper GM, Halligan NL, Hilton G, et al. Non-heme iron protein: a potential target of nitric oxide in acute cardiac allograft rejection. Proc Natl Acad Sci USA 2003;100:3125–3130.
21. Maines MD. Heme oxygenase: function, multiplicity, regulatory mechanisms, and clinical applications. FASEB J 1988;2:2557–2568.
22. Maines MD, Trakshel GM, Kutty RK. Characterization of two constitutive forms of rat liver microsomal heme oxygenase: only one molecular species of the enzyme is inducible. J Biol Chem 1986;261: 411–419.

23. Shibahara S, Muller R, Taguchi H, et al. Cloning and expression of cDNA for rat heme oxygenase. Proc Natl Acad Sci USA 1985;82:7865–7869.
24. Choi AM, Alam J. Heme oxygenase-1:function, regulation, and implication of a novel stress-inducible protein in oxidant-induced lung injury. Am J Respir Cell Mol Biol 1996;15:9–19.
25. Imai T, Morita T, Shindo T, et al. Vascular smooth muscle cell–directed overexpression of heme oxygenase-1 elevates blood pressure through attenuation of nitric oxide-induced vasodilation in mice. Circ Res 2001;89:55–62.
26. Kajimura M, Shimoyama M, Tsuyama S, et al. Visualization of gaseous monoxide reception by soluble guanylate cyclase in the rat retina. FASEB J 2003;17:506–508.
27. Refsum H, Ueland PM, Nygard O, et al. Homocysteine and cardiovascular disease. Annu Rev Med 1998;49:31–62.
28. Mills JL, McPartlin JM, Kirke PN, et al. Homocysteine metabolism in pregnancies complicated by neural-tube defects. Lancet 1995;345:149–151.
29. Clarke R, Smith AD, Jobst KA, et al. Folate, vitamin B12, and serum total homocysteine levels in confirmed Alzheimer disease. Arch Neurol 1998;55:1449–1455.
30. Kraus JP, Janosik M, Kozich V, et al. Cystathionine beta-synthase mutations in homocystinuria. Hum Mutat 1999;13:362–375.
31. Meier M, Janosik M, Kery V, et al. Structure of human cystathionine beta-synthase: a unique pyridoxal 5'-phosphate-dependent heme protein. EMBO J 2001;20:3910–3916.
32. Omura T, Sadano H, Hasegawa T, et al. Hemoprotein H-450 identified as a form of cytochrome P-450 having an endogenous ligand at the 6th coordination position of the heme. J Biochem (Tokyo) 1984;96:1491–1500.
33. Ojha S, Hwang J, Kabil O, et al. Characterization of the heme in human cystathionine beta-synthase by X-ray absorption and electron paramagnetic resonance spectroscopies. Biochemistry 2000;39:10,542–10,547.
34. Green EL, Taoka S, Banerjee R, et al. Resonance Raman characterization of the heme cofactor in cystathionine beta-synthase: identification of the Fe-S(Cys) vibration in the six-coordinate low-spin heme. Biochemistry 2001;40:459–463.
35. Taoka S, Lepore BW, Kabil O, et al. Human cystathionine beta-synthase is a heme sensor protein: evidence that the redox sensor is heme and not the vicinal cysteines in the CXXC motif seen in the crystal structure of the truncated enzyme. Biochemistry 2002;41:10,454–10,461.
36. Shelver D, Thorsteinsson MV, Kerby RL, et al. Identification of two important heme site residues (cytokine 75 and histamine 77) in Cocoa, the CO-sensing transcription factor of Rhodospirillum rubrum. Biochemistry 1999;38:2669–2678.
37. Uchida T, Ishikawa H, Takahashi S, et al. Heme environmental structure of Cocoa is modulated by the target DNA binding: evidence from resonance Raman spectroscopy and CO rebinding kinetics. J Biol Chem 1998;273:19,988–19,992.
38. Taoka S, West M, Banerjee R. Characterization of the heme and pyridoxal phosphate cofactors of human cystathionine beta-synthase reveals nonequivalent active sites. Biochemistry 1999;38:7406.
39. Hori R, Kashiba M, Toma T, et al. Gene transfection of H25A mutant heme oxygenase-1 protects cells against hydroperoxide-induced cytotoxicity. J Biol Chem 2002;277:10,712–10,718.
40. DeMaster EG, Quast BJ, Redfern B, et al. Reaction of nitric oxide with the free sulfhydryl group of human serum albumin yields a sulfenic acid and nitrous oxide. Biochemistry 1995;34:11,494–11,499.
41. Hyun J, Chaudhuri G, Fukuto JM. The reductive metabolism of nitric oxide in hepatocytes: possible interaction with thiols. Drug Metab Dispos 1999;27:1005–1009.
42. Gillman MA, Lichtigfeld FJ. Nitrous oxide acts directly at the mu opioid receptor. Anesthesiology 1985;62:375, 376.
43. Gillman MA, Lichtigfeld FJ. A comparison of the effects of morphine sulphate and nitrous oxide analgesia on chronic pain states in man. J Neurol Sci 1981;49:41–45.
44. Jevtovic-Todorovic V, Todorovic SM, Mennerick S, et al. Nitrous oxide (laughing gas) is an NMDA antagonist, neuroprotectant and neurotoxin. Nat Med 1998;4:460–463.
45. Chanarin I, Deacon R, Lumb M, et al. Cobalamin-folate interrelations: a critical review. Blood 1985;66:479–489.
46. Banerjee RV, Harder SR, Ragsdale SW, et al. Mechanism of reductive activation of cobalamin-dependent methionine synthase: an electron paramagnetic resonance spectroelectrochemical study. Biochemistry 1990;29:1129–1135.
47. Ermens AA, Refsum H, Rupreht J, et al. Monitoring cobalamin inactivation during nitrous oxide anesthesia by determination of homocysteine and folate in plasma and urine. Clin Pharmacol Ther 1991;49:385–393.

48. Christensen B, Guttormsen AB, Schneede J, et al. Preoperative methionine loading enhances restoration of the cobalamin-dependent enzyme methionine synthase after nitrous oxide anesthesia. Anesthesiology 1994;80:1046–1056.
49. Badner NH, Drader K, Freeman D, et al. The use of intraoperative nitrous oxide leads to postoperative increases in plasma homocysteine. Anesth Analg 1998;87:711–713.
50. Nunn JF. Clinical aspects of the interaction between nitrous oxide and vitamin B12. Br J Anaesth 1987;59:3–13.
51. Christensen B, Ueland PM. Methionine synthase inactivation by nitrous oxide during methionine loading of normal human fibroblasts: homocysteine remethylation as determinant of enzyme inactivation and homocysteine export. J Pharmacol Exp Ther 1993;267:1298–1303.
52. Amess JA, Burman JF, Rees GM, et al. Megaloblastic haemopoiesis in patients receiving nitrous oxide. Lancet 1978;2:339–342.
53. Ermens AA, Schoester M, Lindemans J, et al. Effect of nitrous oxide and methotrexate on folate coenzyme pools of blast cells from leukemia patients. Leuk Res 1991;15:165–171.
54. Ikeda K, Aosaki T, Furukawa Y, et al. Antileukemic effect of nitrous oxide in a patient with chronic myelogenous leukemia. Am J Hematol 1989;30:114.
55. Abels J, Kroes AC, Ermens AA, et al. Anti-leukemic potential of methyl-cobalamin inactivation by nitrous oxide. Am J Hematol 1990;34:128–131.
56. Nathan C. Inducible nitric oxide synthase: what difference does it make? J Clin Invest 1997;100: 2417–2423.
57. Tabuchi S, Gotoh T, Miyanaka K, et al. Regulation of genes for inducible nitric oxide synthase and urea cycle enzymes in rat liver in endotoxin shock. Biochem Biophys Res Commun 2000;268:221–224.

INDEX

P

R

S–T

U–X